AF540899

Diseases of Pulse Crops and their Sustainable Management

The Editors

Dr. Samir Kumar Biswas was born on 17th January, 1972 at Nadia district of West Bengal, working as an Associate Professor, Department of Plant Pathology, C.S. Azad University of Agriculture & Technology, Kanpur, Uttar Pradesh. He obtained his B. Sc. (Ag.) degree from BCKV, West Bengal in 1994, M. Sc. (Plant Pathology) in_1997 and Ph.D. (Plant Pathology) in 2001 both from I.A.R.I, New Delhi. He was received merit scholarship in secondary education, JRF (IARI) in M.Sc and SRF (ICAR) in Ph. D. programme. Dr. Biswas is teaching fundamental course of Plant Pathology to under graduate student, basic and advanced course of Plant Pathology to post graduate student. He has published more than 72 research papers in National & International Journal, 02 books (Sustainable disease management of Agricultural Crops & Organic farming and management of biotic stresses), 15 book chapters and 10 articles. He has also guided both M.Sc.(Ag) & Ph. D._students majoring in Plant Pathology and Plant Biotechnology. He has attended several national and international symposia and presented his research paper. He was received the Young Scientist Award-2003 by UPCAR, Lucknow & M. K. Patel Memorial Young Scientist Award by Indian Phytopathological Society, New Delhi in 2011, Excellence in Teaching Award by Astha Foundation and also received Best Poster Award. Dr. Biswas is also associated with various scientific community and organized zonal symposium at Kanpur. He is also a very good sports man and is having 59 certificates from different game & sports. Dr. Biswas is also working as Hostel Warden, Athletes councilor as an additional duty in the University. He also visited the Ohio State University, Columbus, USA under Indo-US Agriculture Knowledge Initiatives work-plan entitled "Teaching and Learning Excellence: A Capacity Building Model".

Dr. Santosh Kumar presently working as Assistant Professor-cum-Junior Scientist in the Department of Plant Pathology, Bihar Agricultural University, Sabour, Bhagalpur, He did his Ph.D. degree in Plant Pathology from G.B. Pant University of Agriculture and Technology, Pantnagar. Dr. Kumar was awarded Senior Research Fellowship, IARI, New Delhi, Rajiiv Gandhi National Fellowship (RGNF) through University Grant Commission (UGC), New Delhi during Ph.D. programme. He has taught many undergraduate (UG) and postgraduate (PG) courses and actively involved in all teaching activity. He was received Young Scientist Award-2015 from Bioved Research Institute, Allahabad. He has published more than 28 research paper, 4 Review papers, 7 book chapters and 32 popular articles in international and national journal of repute. He has also published one book and 2 extension bulletin in mushroom production. He has attended several national and international symposia and presented his research paper and received best poster award in many times. His interest of research includes management of diseases of pulses and rice, biocontrol and mushroom.

Dr. Gireesh Chand Joined the N. D. University of Agriculture and Technology, Faizabad in November 2004 as Assistant Professor and as Associate Professor-cum-Senior Scientist in Bihar Agricultural University, Sabour Bhagalpur (Bihar) in 2011. He did his Ph.D. degree in Plant Pathology from C.S. Azad University of Agriculture and Technology, Kanpur.

Dr. Chand was awarded by P. R. Verma Award-2000, Ph. D. Research Fellowship-2002, Best KVK Award-2005, SPPS Fellow Award-2009, ISHA Best Student-Guide Award-2010, Prof. M. J. Narsingham Award-2011, Young Scientist Award-2013, Best Paper presentation Award-2014, Excellence in Teaching Award-2014 and also visited Beijing, China through the International Financial Support Scheme-2013 by DST, New Delhi and he has research specialization is the study of molecular plant pathology of crop diseases and their management.

He has a decade career dedicated to teaching UG and PG classes, research and extension activities. Guided 8 M.Sc. and 2 Ph.D. Theses and authored / co-authored of 5 Books of Plant Pathology and prepared one Practical Manual also. Published 30 research papers, 12 book chapters, 15 popular articles and presented 35 research papers in National/International Seminars and Symposia and handled more than 10 research projects and he has research specialization is the study of molecular plant pathology of crop diseases and their management.

Diseases of Pulse Crops and their Sustainable Management

— *Editors* —

Dr. Samir Kumar Biswas

Dr. Santosh Kumar

Dr. Gireesh Chand

2016

BIOTECH BOOKS®

ISBN 978-81-7622-357-7

Published by: **BIOTECH BOOKS®**
4762-63/23, Ansari Road, Darya Ganj,
New Delhi - 110 002
Phone: +91-011-23262132
E-mail: biotechbooks@yahoo.co.in

Digitally Printed at **Replika Press Pvt. Ltd.**

PRINTED IN INDIA

Bihar Agricultural University, Sabour, Bhagalpur

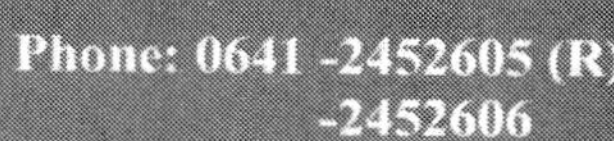

Dr. M.L. Choudhary
Vice-Chancellor

Phone: 0641 -2452605 (R)
-2452606
Fax: 0641 -2452604
Email ID: vcbausabour@gmail.com

Foreword

With the protein rising in food crisis, nutrition for the first time in two decades is high priority on the global agenda. The International Food Policy Research Institute (IFPRI) in its Vision 2020 initiative decided to leverage this moment to inform, influence and catalyze key actors to better use agricultural technology and investment to sustain the pulse production, reduce malnutrition and improve health for the world's vulnerable people. The introduction of modem technology based agricultual systems, in addition to encouraging increased water usage, unbalanced use of chemical fertilizers and huge amount of pesticides are the major problems associated with soil eco-system and environmental degradation. However, high yielding cultiver has lead to mono-cropping of certain pulse reducing farmer's cropping flexibility and decreasing pulse biodiversity. Therefore, it is right time to recognize the necessity of reshaping the pulse production technology by judicial use of natural resources and inputs by conserving resources fer our next generation. Considering the vital role of reshaping of pulse production techologies for increasing the quantity and quality of pulse produce along with management and conservation of resources to promote technically sound, economically viable, eco-friendly, non degradable and socially acceptable productivity.

The book entitled *"Diseases of Pulse Crops and their Sustainable Management"* focuses on abiotic and biotic constraints in pulse production. The chapters emphasizes importance of diseases, symptoms, disease cycle, epidemiology and sustainable management strategies in increasing pulse production for evergreen revolution. I congratulate the editors; Dr. S. K. Biswas, Dr. Santosh Kumar and Dr. Gireesh Chand for their efforts in bringing out this book in such a good shape which is simple and easy to understand. I am confident that this edited book would help the students, teachers, researchers, policy makers and stakeholders to foster agricultural development.

(M. L. Choudhary)

Preface

PULSES are the basic ingredient in the diets of a vast majority of the Indian population, as they provide a perfect mix of vegetarian protein component of high biological value when supplemented with cereals. Pulses are also an excellent feed and fodder for livestock. India is the largest producer, consumer, importer and processor of pulses in the world. The important pulse crops commonly grown in India are chickpea (48 per cent), pigeonpea (15 per cent), mungbean (7 per cent), urdbean (7 per cent), lentil (5 per cent) and fieldpea (5 per cent). For meeting the demand of the growing population, the country is importing pulses to the tune of 2.5–3.5 Mt every year. Strong upward trend in the import of pulses is a cause of concern, since an increase in demand from India has shown to have cascading effect on international prices, thus draining the precious foreign exchange. By 2050, the domestic requirements would be 26.50 Mt, necessitating stepping up production by 81.50 per cent, *i.e.* 11.9 Mt additional produce at 1.86% annual growth rate. This uphill task has to be accomplished under more severe production constraints, especially abiotic stresses, abrupt climatic changes, emergence of new species / strains of insect-pests and diseases, and increasing deficiency of secondary and micronutrients in the soil. The chapters in the book provide a current and detailed account of biotic and abiotic constraints that hamper pulse crops. The book has a comprehensive coverage of occurrence, distribution, and economic importance, symptoms of biotic and abiotic constraints, disease cycle, epidemiology and sustainable management strategies. We hope that this edited book would help the students, teachers, researchers, policy makers and stakeholders to foster agricultural development.

We hope that this compilation would serve as valuable edited book for keen interested in sustainable management of pulse disease. We sincerely thank all the authors for their valuable contributions for providing us valuable published/ unpublished material for shaping the book. All the words in the lexicon will be futile and meaningless if fail to express our reference to family members for their blessing, affection, sacrifice and cheerful cooperation to overcome the hurdle. We thank Biotech Book Publishing House, New Delhi for accepting such valuable publication.

S.K. Biswas

Santosh Kumar

Gireesh Chand

Contents

2016, Diseases of Pulse Crops and their Sustainable Management 1–12
Editors: Samir Kumar Biswas, Santosh Kumar and Gireesh Chand
Published by: BIOTECH BOOKS, NEW DELHI

Chapter 1

Constraints and Strategies in Increasing Pulses Production for Evergreen Revolution

Ch. Srilatha Vani*, P. Ammaji and R. Balazzi Naik

Acharya NG Ranga Agricultural University, ARI, Rajendranagar, Hyderabad, Telangana

Introduction

An adequate, healthy diet must satisfy human needs for energy and all essential nutrients. Furthermore, dietary energy needs and recommendations cannot be considered in isolation of other nutrients in the diet, as the lack of one will influence the others. Thus, requirements for energy will be fulfilled through the consumption of a diet that satisfies all nutrient needs. Protein is one of the most important dietary constituents due to its important role in development of human body. Pulses constitute an important group of food paths in the world. They are the most important source of proteins not only for human population but also for herbivores.

In spite of impressive growth of Indian agriculture, ensuring household food and nutritional security is still a challenge due to imbalanced growth in agriculture based towards wheat and rice. Though production of pulses in the recent decade has increased but is not in pace with the increase in population. Pulses for being a major

* Corresponding Author: E-mail: lathachiluvuri@yahoo.co.in

source of protein in Indian diet and for being resource conserving and environmental friendly, the increase in pulse production will act as a panacea for problems like nutritional security.

Pulses contribute substantially to food production system by enriching the soil through biological nitrogen fixation and improving soil physical conditions. They are rightly called unique jewels of Indian crop husbandry. Their inclusion in the cropping system by Indian farmers is responsible for maintaining soil productivity.

Pulses in India have long been considered as the poor man's only source of protein. Pulses contain more protein than any other plant. They serve as a low-cost protein to meet the needs of the large section of the people. They have, therefore, been justifiably described as 'the poor man's meat'. In general, pulses contain 20 to 28 per cent protein per 100 gm. with the exception of soybean which has as much as 47 per cent. Their carbohydrate content is about 60 per cent per 100 gm. except soybean which has about 30 per cent. Pulses are also fairly good sources of thiamin and niacin and provide calcium, phosphorus and iron. On an average 100 gram of pulses contain energy 345 kcal, protein 24.5 gm., calcium 140 mg. phosphorus 300 mg., iron 8 mg., thiamin 0.5 mg., riboflavin 0.3 mg. and niacin 2 mg.

Table 1.1: List of some Commonly Available Pulses and their Nutritional Value

Pulses and Legumes	*Energy (Kcal)*	*Fiber (g)*	*Carbohydrate (g)*	*Protein (g)*	*Fat (g)*	*Minerals (g)*	*Calcium (mg)*
Gram	372	1.2	59.8	20.8	5.6	2.7	56
Green gram	348	0.8	59.9	24.5	1.2	3.5	75
Green gram whole	334	4.1	56.7	24	1.3	3.5	124
Kidney beans	346	4.8	60.6	22.9	1.3	3.2	260
Moth beans	330	4.5	56.5	23.6	1.1	3.5	202
Peas	315	4.5	56.5	19.7	1.1	2.2	75
Red gram	335	56.5	19.7	1.1	2.2	–	73
Soya bean	432	1.5	57.6	22.3	1.7	4.6	2450
Whole gram	360	3.7	20.9	43.2	19.5	3	202
Black gram	347	0.9	59.6	24	1.4	3.2	154

Source: Directorate of Economics and Statistics, Dept. of Agril. and Coop.

Gross value of output (GVO) of Agricultural products was estimated by the Central Statistics Office, Ministry of Statistics and programme implementation, Government of India during the period from 2004-05 to 2010-11. According to these estimates, the percentage share of pulses to total value of farm crops is 5 per cent in the year 2010-11 as compared to 4 per cent in 2004-05. However, major share is contributed by Cereals, fruits and vegetables.

Item

Among the pulse crops, chickpea has a major share (44 per cent) in value of output of pulse group followed by redgram (18 per cent), greengram (12 per cent), blackgram (11 per cent) and Massor (6 per cent) in 2010-11. A steady increase in the value of output of pulse group was observed from 2004-05 to 2010-11. However, there is a slight decline in value of output of pulses in 2009-10.

Table 1.2: Value of Output of Selected Crops (At constant price) (Rs'00Cr.,)

Agricultural Commodity	*Per cent Share 2004-05*	*2004-05*	*2005-06*	*2006-07*	*2007-08*	*2008-09*	*2009-10*	*2010-11*	*Per cent Share 2010-11*
Pulses	04	202	206	214	228	223	221	263	5
Cereals	30	1398	1451	1527	1607	1639	1525	1660	29
Oilseeds	10	461	528	469	545	492	481	602	10
Sugar	06	284	313	367	367	326	335	368	6
Fruits and vegetables	25	1133	1220	1278	1350	1373	1476	1572	27

Source: Directorate of Economics and Statistics, Dept. of Agril. and Coop.

Table 1.3: Value of Output of Selected Pulse Crops

Agricultural Commodity	*Per cent Share 2004-05*	*2004-05*	*2005-06*	*2006-07*	*2007-08*	*2008-09*	*2009-10*	*2010-11*	*Per cent Share 2010-11*
Chickpea	39	80	83	96	90	107	109	116	44
Redgram	20	40	45	38	51	37	40	48	18
Greengram	9	18	17	19	26	18	11	31	12
Blackgram	10	20	19	23	24	19	20	28	11
Masoor	9	18	17	16	14	17	19	16	6

Source: Directorate of Economics and Statistics, Dept. of Agril. and Coop.

Area and Production of Pulses in different States of India

Pulses are grown in an area of 23.28 million hectares with an annual production of 14.66 million tons. India accounts for 33 per cent of the world area and 22 per cent of the world production of pulses. The major pulse crops grown in India are chickpea, pigeon pea, lentil, moong bean, urdbean and field pea. About 90 per cent of the global pigeon pea, 65 per cent of chickpea and 37 per cent of lentil area falls in India, corresponding to 93 per cent, 68 per cent and 32 per cent of the global production, respectively (FAOSTAT, 2009). Thirty nine per cent of the area in India is under chickpea followed by pigeonpea (21 per cent), mungbean (11 per cent), urad bean (10 per cent) and lentil (7 per cent).

The major pulse crops grown in India include chickpea, pigeonpea, mungbean, urad bean and lentil with respect to area under which they are cultivated. The major pulse growing states in India are Madhya Pradesh, Maharashtra, Rajasthan, Uttar Pradesh, Karnataka and Andhra Pradesh. The area, production and productivity of pulses in different states of India during the year 2010-11 is presented in Table 4. Madhya pradesh contains maximum area (4.94 m/ha) under pulses during the year 2010-11 with a production and productivity of 4.30 million tons and 871 kg/ha, respectively. Maharashtra and Uttar Pradesh stood second and third position in pulse production. However, Bihar and West Bengal stood second and third in pulse productivity. Though the area under pulses is more (3.5 m.ha), the production and productivity are less as compare to other country of the world.

Table 1.4: Area, Production and Productivity of Pulses in India during the Year 2010-11

State	*Area in million ha*	*Rank*	*Production (million tons)*	*Rank*	*Productivity kg/hec*	*Rank*
M.P	4.94	I	4.30	I	871	I
Maharashtra	3.38	II	2.37	II	702	
U.P	2.54	III	1.90	III	748	
A.P	1.93		1.43		740	
Karnataka	2.48		1.12		451	
Rajasthan	3.50		0.71		204	
Gujarat	0.73		0.52		705	
Chhattisgarh	0.81		0.49		604	
Bihar	0.56		0.47		836	II
Orissa	0.87		0.40		461	
Jharkhand	0.32		0.22		709	
Tamilnadu	0.53		0.20		382	
West Bengal	0.18		0.15		826	III
Haryana	0.13		0.10		758	
Others	0.37		0.27		@	
All India	**23.28**		**14.66**		**630**	

@: Since area/production is low, yield rate is not worked out.

Source: Directorate of Economics and Statistics, Dept. of Agril. and Coop.

The contribution to pulse production by different states in India during the year 2010-11 is given in Table 1.5. Madhya Pradesh, Maharashtra, Uttar Pradesh and Karnataka contribute to 73 per cent of pulse production in India. India has the largest cultivated area under pulses but the production and productivity is the least.

With over 25 per cent of global production, India is a dominant player in the global pulses market. It produces 13-17 million tonnes of pulses a year. On the other hand, it consumes around 21 million tonnes a year with the shortfall is being met through imports.

Though India is the world's largest producer of pulses, it imports a large amount of pulses to meet the growing domestic needs. In order to meet the pulse requirement in India, 30.11 lakh tones of pulses were imported during the year 2010-11. The additional requirement of pulses is fulfilled by importing desi chickpea, pigeon pea, moong bean and kidney bean from Burma; kabuli chickpea and peas from Canada, Australia, Mexico, Turkey and Iran and lentil from Nepal and Syria. In the world, India is the largest importer, producer, processor and consumer of pulses. The names of countries and the quantity from where pulses are imported by India during 2010-11 are given in Table 1.6.

Table 1.5: Contribution of Various States to Pulses Production in India during 2010-11

State	*Contribution Percentage*
M.P	29.33
Maharashtra	16.2
U.P	12.96
A.P	9.75
Karnataka	7.64
Rajasthan	4.84
Gujarat	3.54
Chhattisgarh	3.34
Bihar	3.20
Orissa	2.73
Jharkhand	1.50
Tamilnadu	1.36
West Bengal	1.02
Haryana	0.68
Others	1.84

Table 1.6: Quantity of Pulses Imported by India during 2010-11

Sl.No.	*Country*	*Qty (lakh tons)*
1.	Canada	13.91
2.	Myanmar	6.91
3.	USA	2.62
4.	Australia	2.25
5.	Unit.Rep.of Tanzia	1.26
6.	China	0.88
7.	Mozambique	0.64
8.	Malawi	0.39
9.	Kenya	0.31
10.	Namibia	0.22
	Others	0.72
	Total	**30.11**

Source: Department of Agric. and Coop.

Demand and Supply Scenario of Pulse Produce

For the triennium ending 2010-11, the domestic consumption of pulses in India was 186.5 lakh tons. Against this, India produced an average quantity of 158 lakh tonnes. During this period, there was a gap of 28.5 lakh tons of pulses in demand and supply (Table 1.7). This gap was due to higher growth of population as compared to pulse production.

Table 1.7: Production, Demand and Imports of Oulses in India (in million tons)

Year	*Production*	*Imports*	*Demand*
2009-10	14.66	3.51	18.29
2010-11	18.24	2.70	19.08
2011-12	17.09	3.36	19.91
2012-13	18.45	3.84	20.90
2013-14	19.00	1.42	21.77

An analysis by a working group on the 12th Five-Year Plan (2012-13 to 2016-17) shows that despite the rise in production, domestic demand will continue to be more than the supplies in the next few years. It shows that India is expected to produce around 18-21 million tonnes (mt) of pulses till 2016-17, while demand is projected to touch around 22 mt in the next few years. Our rising pulses production will only help in import substitution and from the current 3-3.5 mt of annual imports; it might come down by one-two mt.

Constraints in Pulses Production

Even though India is the largest pulses producer of the world, it imports large amount of pulses from rest of the world. So, it is important to analyze the production and other constraints. Even with the best efforts, pulses production and productivity has been stagnant. Due to stagnant production, the net availability of pulses has come down from 60 gm/day/person in 1951 to 31 gm/day/person (Indian Council of Medical Research recommends 65 gm/day/capita) in 2008.

Due to the low productivity-low input nature, pulses are grown as residual/ alternate crops on marginal lands after taking care of food/income needs from high productivity-high input crops like paddy and wheat by most farmers. Also, they grow as rain fed crops with little or no modern yield enhancing inputs. The low priority accorded to pulse crops may be related to their relatively low status in the cropping system. As a crop of secondary importance, in many of these systems, pulse crops do not attract much of the farmer's crop management attention. In addition to this, these crops are adversely affected by a number of biotic and a biotic stresses, which are responsible for a large extent of the instability and low yields.

The non-availability of seeds of high-yielding varieties in the desired quantities is perhaps one of the major constraints in the expansion of pulses. Although more than 200 improved varieties of pulses have been released since 1970's, its impact hardly gets reflected in the yield. The failure of these varieties to make any real dent in pulse productivity could thus be due to their inherent weakness. In pulses, improved varieties hardly have a yield advantage of 15- 20 percent over the traditional varieties. Even this yield advantage did not attract the farmers to go for these varieties indicating their poor performance at the field level. At present, the very best seed available belong to the same group, which the fanners are already using. Varieties with-better yield advantage and desirable characteristics to suit the varied agro-climatic conditions need to be developed in pulses. The fertilizer use in pulses was very low with chickpea receiving the highest priority and pigeonpea the least.

In pulses there are a number of diseases and insect pests which cause heavy losses resulting in poor production. Though several resistant/tolerant varieties had been developed by research institutions the spread of such varieties in the farmer's fields is very limited. The main reason could be our weak seed production programme. Research on bio-pesticides and incorporation of insect-resistant genes, albeit promising in the field verification trials, is yet to be commercially viable. Chemical pest control seems to be the only option left with the pulse farmers at present. Even though several plant protection chemicals with their method and time of application have been developed the use of pesticides in pulses was very negligible. Either these pesticides may not be giving the desired effect to control the pests or the damage due to these pests may be below the economic threshold level. Hence, a cheaper and effective pesticide to combat the pests in pulses needs to be developed.

Furthermore, there is hardly any visible technological change in pulse farming in the country. This clearly shows that technological stagnation is primarily responsible for the backwardness of pulses in the country as a whole. The important constraints in pulse production are classified into two categories:

1. Production Constraints

Production of major pulses is constrained by both biotic and abiotic stresses. For example, pod borers (*Helicoverpa armigera*), fusarium wilt, root rots, ascochyta blight and botrytis gray mold are some of the major biotic constraints to increase the productivity of chickpea. The major constraints to productivity in pigeonpea are biotic stresses such as pod borer, pod fly, fusarium wilt, and sterility mosaic disease. Similarly, pod borer, aphids, cutworm, powdery mildew, rust and wilt are the major pests and diseases affecting lentil production in India. The richness of legumes in N and P makes them attractive for insect pests and diseases (Sinclair and Vadez, 2012).

Most of the pulses in India are grown in low fertility, problematic soils and unpredictable environmental conditions. More than 87 per cent of the area under pulses is rainfed. Drought and heat stress may reduce seed yields by 50 per cent, especially in arid and semi-arid regions. Another major problem is salinity and alkalinity of soils which is high both in semi-arid tropics and in the Indo-Gangetic plains. With recent changes in the global temperatures, the grain yield is likely to be drastically affected by temperature extremities. Poor drainage/water logging during the rainy season causes heavy losses to pigeonpea on account of low plant stand and increased incidence of Phytophthora blight disease, particularly in the states of UP, Bihar, West Bengal, Chhattisgarh, MP and Jharkhand.

2. Socio-economic Constraints

Even though India is the largest pulses producer in the world, it imports large amount of pulses from rest of the world. Farmers in India treat pulses as secondary crops. Until recently the government also gave less importance to pulses compared to the staple cereals. In general farmer's access to inputs is limited, both because of low purchasing power and accessibility to markets to sell the excess produce of pulses. Because of this situation, the farmers give first priority to staple cereals and cash crops for allocating inputs and the second priority to pulses. As a result, pulses continue to be grown on poor soils with low inputs. In addition, there is lack of policy support and post-harvest innovations related to pulse crops. Availability of quality seed of improved varieties and other inputs is one of the major constraints in increasing the production of grain legumes (David *et al.*, 2002)

It has been estimated that India's population would reach 1.68 billion by 2030 from the present level of 1.21 billion. Accordingly, the projected pulse requirement for the year 2030 is 32 million tons with an anticipated required growth rate of 4.2 per cent (IIPR Vision, 2030). India has to produce not only enough pulses but also remain competitive to protect the indigenous pulse production. In view of this, India has to develop and adopt more efficient crop production technologies along with favorable policies to encourage farmers to bring more area under pulses.

Extension Strategies for Increasing Pulses Production

- ☆ Incorporation of short duration pulse crops like urad bean, mung bean to make the different production system profitable and improve soil health.
- ☆ Intensification of production system through popularizing mung bean cultivation during spring/summer season under irrigated condition.
- ☆ Easy and timely availability of critical input at nearby market.
- ☆ Creation of informal seed village system, where farmer to farmer seed production and distribution chain will ensure easy availability of quality seed.
- ☆ Production of quality seed of improved pulse varieties by private agencies needs to be encouraged to meet the demand of the farmers. Improved seed should be made available to the farmers at his doorstep at reasonable rate.
- ☆ Research efforts to develop drought tolerant high yielding varieties of pulses with pest resistance should be undertaken.
- ☆ It would be profitable to have Dal milling industry instituted between groups of villagers so that proper milling is done soon after harvesting and storage is made of split pulses (Dal) rather than whole seed.
- ☆ Transfer of technology in relation to pulses should be strengthened in farmer participatory mode with active involvement of multidisciplinary team of scientists Necessary efforts should be made to popularize the rural storage of pulses by the farmers under optimum conditions so as to preserve the produce without damage.

- ✩ Private sector may also consider new industries in rural areas like village bakeries, seed agencies, seed village and post harvest technology.
- ✩ Price structure in the market will have to be derived by the Government much in advance to ensure reasonable profits to the pulse growing farmers to encourage them to take up pulses cultivation on a large scale

Strategies Suggested by Directorate of Pulses Development, Technology Mission on Pulses to Overcome Problems in Pulse Production

Availability of Quality Seeds

The Directorate of pulses (DOP) development opined that non-availability of quality seeds in adequate quantity is one of the major constraints in pulse production. The existing seed replacement rate (2 to 5 per cent) has to be increased further to meet the farmer's demand.

Integrated Pest/Disease Management

Among the insect pests affecting pulse crops, on an average, 20-40 per cent crop is annually lost due to damage caused by pod borers in pigeonpea and chickpea. Pod fly is also known to occur losses especially in northern India. Among the diseases inciting pulse crops, wilt and root rots cause heavy loss to pigeonpea and chickpea crops. Effective IPM module is made available by DOP for management of targeted pest and diseases. The IPM module was implemented in pigeonpea and chickpea covering 10 per cent area in the states of M.P., Maharashtra, A.P., Karnataka and Gujarat. Further, the IPM module was extended to pod fly in pigeonpea covering 10 per cent area in the states of U.P., Bihar, Jharkand and West Bengal. For the management of wilt and root rots seed treatment with *Trichoderma* sp. covering 2 lakh ha area of chickpea and 1 lakh ha area of lentil was popularized. However, there is a need to bring more area of pigeonpea and chickpea under IPM umberella and IPM demonstrations against key insect pests and diseases has to be popularized among farmers for technology adoption.

Integrated Nutrient Management

In recent years, wide spread deficiency of sulphur and zinc has been noticed in pulse growing regions, which constraints productivity of pulses. In the major pulse growing areas, 44 districts have shown 40-60 per cent sulphur deficiency and 82 districts with 50-60 per cent zinc deficiency. Very encouraging response to application of S and Zn has been found with cost benefit ratio of 10-21 per cent.

About 40 per cent pulse growing regions have low to medium population of native *Rhizobium*. Seed inoculation with biofertilizer (Rhizobium and PSB) - low cost inputs - can increase pulse productivity by 10-12 per cent. Lack of quality culture in adequate quantity is one of the major constraints in popularization of biofertilizers. Hence, the State departments of Agriculture should ensure timely and adequate supply of Rhizobium and phosphate solubilizing bacterial (PSB) formulations and micronutrients like sulphur and zinc in input deficient districts.

Technology Transfer

Transfer of improved pulse production technologies remains the most neglected component in the past, and consequently the benefit of improved varieties and production technology could not be harnessed. So, there is a need to strengthen the extension strategies for efficient transfer of production technologies through Frontline demonstrations and Block demonstrations involving State Departments of Agriculture (SDA), Krishi Vigyan Kendras (KVK's) for dissemination of knowledge on improved technologies.

Farm Implements

Poor drainage/water stagnation during rainy season causes heavy losses to pigeonpea on account of low plant stand and increased incidence of Phytophthora particularly in the states of U.P., Bihar, W.B., Chattisgarh, M.P. and Jharkhand. Ridge planting has been found very effective in ensuring optimal plant stand and consequently higher yield. A simple ridger already available can effectively be used for this purpose. In problematic areas, simple ridger has to be popularized and made available to the target farmers at subsidized price.

Protective Irrigation

Majority of pulse area (87 per cent) is rainfed and consequently pulses face severe moisture stress with low productivity. Quantum jump in productivity can be achieved by applying come up/life saving irrigation especially in rabi pulses grown on residual moisture. Sprinkler irrigation can be used for most efficient use of scarce irrigation water. Distribution of sprinkler sets for protective irrigation in chickpea in the state of Gujarat, Rajasthan, Haryana and Madhya Pradesh is the need of the hour.

Area Expansion

In Indian agriculture, the major cultivable area in under rice-wheat production system and pulses are considered only as a secondary crops where they are grown after paddy or wheat. The following are some strategies for increasing the pulse area and production in India.

- ☆ Diversification of rice-wheat production system in the states of Punjab, Haryana, Uttar Pradesh and Bihar through chickpea and early pigeon pea
- ☆ Introduction and popularization of short duration varieties of chickpea, lentil, urdbean and mung bean in rice fallows of eastern India (U.P., Jharkhand, Bihar and West Bengal) and coastal peninsula (A.P., Karnataka and Tamil Nadu)
- ☆ Popularization of pulses in intercropping systems
- ☆ Replacing less remunerative cereals by pulses in targeted regions
- ☆ Popularization of urdbean and mungbean during spring/summer as a catch crop in intensive cropping systems of northern India

Post Harvest Management

Pulses suffer heavy losses due to stored grain pests. The quality of seeds stored in the traditional storage structures also deteriorates. Further, there are no small processing units to convert pulse grains into *Dal* and other byproducts. This compels the growers to dispose off their produce immediately after harvest at low price. Low cost IIPR dal mill and metal storage bins are available which need to be popularized.

The major strategies for increasing pulse area and production are presented in the above discussion. In addition to these strategies the following options may also contribute to overcome constraints in pulse production and marketing. These options include:

- ✰ Establishment of marketing intelligence cell for monitoring global prices
- ✰ Creation of special fund to support specific export oriented commodities and its popularization in targeted districts through private partnership.
- ✰ Financial support to NAFED under Price Support Scheme (PSS)
- ✰ Creation of special fund to support strategic research

Conclusion

Importance of pulses in maintaining food security as well as nutritional security has been felt since long. The production of pulses definitely needs to be increased manifold to meet the demand in coming years.

To increase area and production of pulse crops we need crop specific and region specific approaches, which should be adopted in the overall framework of systems approach. The major thrust areas to be addressed are as follows (i) Replacement of cereal crops in the prevailing rice-wheat cropping systems with high yield varieties of pulses. (ii) Inclusion of short duration varieties of pulses as catch crop. (iii) Development of multiple disease and pest resistant varieties. (iv) Reducing storage loses and improving market information and infrastructure. (v) Linking MSP to market prices which can bridge the gap between demand and supply. (vi) Developing high nitrogen fixing varieties, which will play a crucial role in sustainable agriculture, and (vii) Coordination of research, extension and farmers to encourage farmer's participatory research

References

David, S., Mukandala, L. and Mafuru, J. 2002. Seed availability, an ignored factor in crop varietal adoption studies: a case study of beans in Tanzania. *Journal of Sustainable Agriculture* 21: 5-20.

IIPR Vision 2030. 2011. Printed and Published by the Director, Indian Institute of Pulses Research (ICAR), Kanpur-208024.

Parthasarathy, P.R., Birthal, P.S., Bhagavatula S. and Bantilan, M.C.S. 2010. In: Chickpea and Pigeonpea Economies in Asia: Facts, Trends and Outlook, An International crop research institute for Semi-Arid Tropics publication, Patancheru 502 324, Andhra Pradesh, India, p. 1-68.

Shalendra, Gummagolmath, K.C., Sharma Purushottam, Patil, S.M. "Role of pulses in the food and nutritional security in India" http: / / dpd.dacnet.nic.in / strategy / strategy.

Sinclair, T.R. and Vadez, V. 2002. Physiological traits for crop yield improvement in low N and P environments. *Plant and Soil* **245**: 1–15.

Sinclair, T.R. and Vadez, V. 2012. The future of grain legumes in cropping systems. Crop and Pasture Science 26 (3 and 4), http: / / www.diethealthclub.com / calories / calories-in-pulses-and-lentils.html

2016, Diseases of Pulse Crops and their Sustainable Management *13–21*
Editors: **Samir Kumar Biswas, Santosh Kumar and Gireesh Chand**
Published by: **BIOTECH BOOKS, NEW DELHI**

Chapter 2

Climate Change and its Impacts on Management of Pulse Crop Diseases

Amarendra Kumar[1]*, Santosh Kumar[1], Mehi Lal[2] and Sunil Kumar[3]

[1]Department of Plant Pathology,
[3]Department of Agronomy,
Bihar Agricultural University, Sabour, Bhagalpur – 813 210, Bihar
[2]Plant Protection Section, Central Potato Research Institute Campus, Modipuram, Meerut – 250 110, U.P.

Introduction

Various efforts have been made to enhance the productivity of agricultural crops and the improvements in production and productivity tend to be marginal, largely due to the tremendous influence of climate change during the past two decades. The impact of climate change on crop production and productivity in sub-Saharan Africa and South Asia, the two major food-insecure regions, has been extensively reviewed, highlighting important differences in impact between individual crops, continental regions and geographical sub-regions (Knox *et al.*, 2012). Global warming has led to increase in the average global temperatures by 1.2°C over the past century and a further raise of 3°C is estimated to occur by 2100 (Schneider *et al.*, 2007). Pulses

* Corresponding Author: E-mail: kumaramar05@gmail.com

production in the India is expected around 20 million tonnes against 18.34 million tonnes in the previous year with a record growth rate of more than 3 per cent. This production was achieved despite prevailing unusual weather conditions with long cold spell with rains and hailstorms (IIPR, 2013-14). Research on impacts of climate change on plant diseases has been limited, with most work concentrating on the effects of a single atmospheric constituent or meteorological variable on the host, pathogen, or the interaction of the two under controlled conditions. It was reported that climate change could alter stages and rates of development of the pathogen, modify host resistance, and result in changes in the physiology of host-pathogen interactions. The most likely consequences are shifts in the geographical distribution of host and pathogen and altered crop losses, caused in part by changes in the efficacy of control strategies. Recent developments in experimental and modeling techniques offer considerable promise for developing an improved capability for climate change impact assessment and mitigation. Compared with major technological, environmental, and socioeconomic changes affecting agricultural production during the next century, climate change may be less important; it will, however, add another layer of complexity and uncertainty onto a system that is already exceedingly difficult to manage on a sustainable basis. Intensified research on climate change–related issues could result in improved understanding and management of plant diseases in the face of current and future climatic extremes.

Temperature, moisture and greenhouse gases such as CO_2, CH_4, N_2O and O_3 are the major variables of climate change. The average global surface temperature has increased by 0.2°C per decade in the past 30 years (Hansen *et al.*, 2006). The most obvious effect of climate change is on the global mean temperature which is expected to rise between 0.9 and 3.5°C by the year 2100 (IPCC, 2007). The atmospheric concentration of carbon dioxide (CO_2) has reached levels significantly higher than in the last 650 thousand years (Siegenthaler *et al.*, 2005). Similar trends have been observed for methane (CH_4), nitrous oxide (N_2O), and other greenhouse gases (Spahni *et al.*, 2005; IPCC, 2007). Atmospheric CO_2, a major greenhouse gas, has increased by nearly 30 per cent and temperature has risen by 0.3 to 0.6° C. Cold days and nights and frost have become less frequent over most land areas, whereas hot days and nights are becoming more frequent. Variability in rainfall pattern and intensity is expected to be high. Greenhouse gases (CO_2 and O_3) would result in increase in global precipitation of 2 ± 0.5°C per 1°C warming. Variability in Climate is adding a new dimension to manage plant diseases by altering the equilibrium of host-pathogen interactions resulting in either increased epidemic outbreaks or new pathogens surfacing as threats or less known pathogens causing severe yield losses.

Plant-Pathogen Interaction in Climate Change Scenario

The classic disease triangle recognizes the role of climate in plant diseases as no virulent pathogen can induce disease on a highly susceptible host if climatic conditions are not favorable. The three legs of the plant disease triangle namely the host, pathogen, and environment must be present and interact appropriately for plant disease to appear. Climate change implies that the average conditions (mean and/or variability) are changing over time and may never return to those previously experienced. (Coakley, 1998). Climate change could "alter stages and rate of development of the pathogen, modify host resistance, and result in changes in physiology of host pathogen interaction" (Garrett *et al.*, 2006). The climate change and global warming with increase in temperature, moisture and CO_2 levels can impact all three legs of the plant disease triangle in various ways. Climate influences all stages of host and pathogen life cycles as well as development of disease. Climate change and global warming will allow survival of plant and pathogens outside their existing geographical range. The lack of action on climate change not only risks putting prosperity out of reach of millions of people in the developing world, it threatens to roll back decades of sustainable development. Climate change will affect host-pathogen interaction and many new diseases may be emerged. Consequently, the greatest impact of climate change will be on management strategies such as on host resistance, bio control agents, and chemicals etc. Modification in biochemistry, physiology, anatomy, morphology and phenology is well known under elevated CO_2 (Bowes, 1993; Wolfe, 1995). Many of these changes in host physiology can potentially enhance host resistance. Significant increase in rates of net photosynthesis allows increased mobilization of resources into host resistance at elevated CO_2 (Hibberd *et al.*, 1996a). Other changes, including production of papillae and accumulation of silicon at sites of appressorial penetration (Hibberd *et al.*, 1996a); greater accumulation of carbohydrates in leaves; more waxes, extra layers of epidermal cells and increased fibre content (Owensby, 1994); lowered nutrient concentration, leading to partitioning of nitrogen from photosynthetic proteins to metabolism that is limiting to plant growth (Baxter *et al.*, 1994); and greater number of mesophyll cells (Bowes, 1993) can all influence host resistance. Reduced pathogen penetration results from a reduction in stomatal density (Chakraborty *et al.*, 1999) and stomatal conductance at high CO_2 (Hibberd *et al.*, 1996b). Thompson *et al.* (1993) reported a significant reduction in wheat powdery mildew at twice-ambient CO_2, but severity was dependent on nitrogen and water status of plants. Increased lignifications in some forage species at high temperatures (Wilson *et al.*, 1991) can potentially enhance the level of resistance (Strange, 1993).

Effect of Increased CO_2 Concentration on Pathogens

Most of the available data clearly suggests that increased CO_2 would affect the physiology, morphology and biomass of crops (Challinor *et al.*, 2009). Elevated CO_2 and associated climate change have the potential to accelerate plant pathogen evolution, which may affect virulence. Pathogens fecundity increased due to altered canopy environment and was attributed to the enhanced canopy growth that resulted in conducive microclimate for pathogen's multiplication (Pangga *et al.*, 2004). Foliar diseases like Ascochyta blights, Stemphylium blights and Botrytis gray mould can become a serious threat in pulses under the higher canopy density. Increased CO_2 will lead to less decomposition of crop residues and as a result soil borne pathogens would multiply faster on the crop residues. In soybean, elevated concentration of CO_2 and O_3 altered the expression of three soybean diseases, namely downy mildew (*Peronospora manshurica*), brown spots (*Septoria glycines*) and sudden death syndrome (*Fusarium virguliforme*) and plant response to the diseases varied considerably (Eastburn *et al.*, 2010). The concentration of CO_2 and O_3 did not have an effect on sudden death syndrome. Overall, the effects of elevated CO_2 concentration on plant diseases can be positive or negative, although in a majority of the cases disease severity increased (Manning and Tiedemann, 1995).

Effect of Elevated Temperature on the Pathogens

Plant pathogens have varying ranges of temperature requirements that affect the various steps in disease infection cycles such as penetration, pathogen survival, dispersal, epidemic development, and sexual reproduction. Many regions of the India are predicted to increase temperature and more frequent moisture stress (Drought) which is likely to influence plant disease epidemics. For example, dry root rot of chickpea (*Rhizoctonia bataticola*) is becoming more severe in rainfed environments as the host plant is predisposed by moisture stress and higher temperatures during the flowering to pod filling stage (Pande *et al.*, 2010). Data on weather and dry root rot disease collected in India for one decade, showed higher incidence of dry root rot in chickpea varieties that resist Fusarium wilt in years when temperatures exceed 33°C (Pande *et al.*, 2010; Sharma *et al.*, 2010). On the contrary, cooler temperature and wetter conditions are associated with increased incidence of stem rot on soybean (*Sclerotinia sclerotiorum*); blights (*Ascochyta* spp.) in chickpea, lentil and pea; anthracnose in chickpea and lentil (Pande *et al.*, 2010, Panagga *et al.*, 2004). Many models that have been useful for forecasting plant disease epidemics are based on increase in pathogen growth and infection with in specified temperature range. In India, in the last decade the disease scenario of chickpea and pigeon pea has changed drastically; dry root rot (*Rhizoctonia bataticola*) of chickpea and *Phytophthora* blight (*Phytophthora drechsleri* f. sp. *cajani*) of pigeon pea have emerged as a potential threat to the production of these pulses (Pande and Shanna, 2010). Higher risk of dry root rot has been reported in *Fusarium* wilt chickpea-resistant varieties in those years when the temperature increases up to 33° C (Dixon, 2012).

Effect of Moisture Regimes on Pathogens

Moisture can impact both host plants and pathogens in various ways. For example high moisture (rainfall) favours most of the foliar diseases and some soil born pathogens such as *Phytophthora, Pythium, Rhizoctonia solani* and *Sclerotium rolfsii* etc. Forecast models for these pathogens are based on leaf wetness, relative humidity and precipitation. An outbreak of Phytophthora blight of pigeonpea (*Phythophthora drechsleri* f. sp. *cajani*) in Deccan Plateau of India is attributed to erratic and heavy rainfall (>300mm in 6-7 days) leading to temporary flooding (Sharma *et al.*, 2006; Pande *et al.*, 2011). Alternaria blight of pigeon pea is being seen more frequently in recent years in semi-arid tropic regions due to the untimely rainfall.

Impact of Climate Change on Management Strategies

Effectiveness of most crop protection chemicals depends on prevailing climatic conditions. Variation in duration, intensity and frequency of rainfall would impact on the effectiveness of chemicals for managing the diseases. Changes in temperature and precipitation may alter the dynamics of fungicide residues on the crop foliage. Globally, climate change models project an increase in the frequency of intense rainfall events which could result in increased fungicide wash-off and reduced their efficacy managing the diseases. The interactions of precipitation frequency, intensity, and fungicide dynamics are complex, and for certain fungicides precipitation following application may result in enhanced disease control because of a redistribution of the active ingredient on the foliage at two intensities (6 and 30 mm h^{-1}) and found that the higher rate significantly reduced the fungicide residue that could be measured with a chemical assay, but that there was no difference in disease between the two treatments when the leaves were challenged in a bioassay with *Phytophthora infestans*. Data from field experiments and modelling studies suggest precipitation during the post-application period is critical (Wagenet *et al.*, 1994). Precipitation following fungicide application may improve its distribution (Schepers, 1996) but an increase in rainfall intensity can deplete fungicide residue on the foliage (Neuhaus *et al.*, 1974). In addition to the influences of the physical climate on biological control agents, changes in atmospheric composition will modify the community structure of phyllosphere and rhizosphere microflora (Manning and Tiedemann, 1995; Ayres *et al.*, 1996). Temperature and precipitation can alter fungicides residues dynamics in the foliage and degradation of product can be modified. Penetration, translocation and mode of systemic fungicides could affect with CO_2 enriched atmosphere or from different temperature and precipitation conditions (Ghini *et al.*, 2008). Systemic fungicides could have been affected negatively by physiological changes that slow uptake rates, such as smaller stomatal opening or thicker epicuticular waxes in crop plants grown under higher temperature. These fungicides could be affected positively by increasing plant metabolites rates that could be increase fungicides up take (Petzoldt and Seman.www. climate farming). Fungicides Cymoxanil is known to degrade rapidly (Klopping and Delp, 1980). Chen and McCarl (2001) performed a regression analysis between pesticide usage and concluded that per acre pesticides usage average cost for cotton, corn, potato, soybean and wheat were found to increase as precipitation increases, similarly, the pesticides usage average cost for cotton, corn, cotton, soybean and potato also increase as temperature increase, while the pesticides usage cost for wheat decrease. The vulnerability of bio-control agents will surely be higher with climate change, since this is one of the problems when applying antagonists (Garrett *et al.*, 2006). In wheat, increasing temperatures inhibited the expression of resistance genes against *Puccinia graminis* f.sp *tritici* (Harder *et al.*, 1979). Jenns and Leonard (1985) found that resistance to *Bipolaris maydis* Shoemaker in different lines of maize diminished with increase in temperature or decrease in luminance.

Management Strategies to Combat Negative Impact of Climate Change

Climate change would have positive, negative and neutral impact on plant diseases. More research in this field is needed about to identify new opportunities to minimize negative impacts. Positive results will require an improved understanding of the causes, impacts and consequences of climate change from which will decide amelioration and mitigation strategies. The shortage of critical epidemiological data on individual plant diseases needs to be addressed using experimental approaches. In the first instance, studies in a controlled environment may be used to formulate hypotheses and to determine critical relationships to help develop process-based approaches. Field-based research examining the influence of a combination of interacting factors would be needed to provide a more realistic appraisal of impacts (Senft, 1995; Norby *et al.*, 1997). Regional impacts of climate change on plant disease management strategies need a relook using forecasting models and biotechnological approaches in understanding the emerging scenario of host pathogen interactions. Epidemiological knowledge, combined with biophysical and socio-economical understanding are required to deploy resistances and achieve sustainable disease management. There is a need for a greater understanding of the effect of climate, variables on the efficacy of synthetic fungicides, their persistence in the environment, and development of resistance in pathogens populations to the fungicides. Recently, national and international net work is also actively anticipating and responding to biological complexity in the effects of climate change on agriculture and crop diseases (Serge *et al.*, 2011). The primary benefit of such studies to growers will be their ability to control the diseases that become severe as result of climate change, select varieties that are less vulnerable for diseases, and reduce fungicide application. The information will be useful to the crop growers, scientists and extension agencies, NGOs, research planners, and administrators.

Climate Change–The Task Ahead

Impact on an agricultural system must include on- and of farm effects determined at a landscape scale of spatial resolution. In contrast, climate systems operate at a global scale and Global climate models (GCMs) are better at explaining climate at this coarse level of resolution. Historically plant pathology research using site-specific knowledge of individual pathosystems has served well in understanding, predicting and managing diseases. As a discipline, plant pathology is dedicated to ensuring sustainable production in agricultural systems. Environmental variables at the micro-climate level have been utilized at this spatial scale. This difference in the level of understanding between plant pathology and meteorology at the various spatial and temporal scales has hampered inter-disciplinary interaction. The need to bridge this gap in knowledge has been recognized (Kennedy, 1997). In recent years, a number of attempts have been made to downscale GCM outputs to a biologically relevant mesoscale (Martin *et al.*, 1996; Bardossy, 1997; Russo and Zack, 1997). Lack of epidemiologically relevant weather variables has been an impediment to the application of GCMs and other climate models to plant disease modeling. Duration of surface wetness and relative humidity, which critically influence infection and disease development by several plant pathogens, have not been easily obtained from GCM output until recently. Usefulness of remotely gathered site-specific wetness and other data for plant pathology research has been variable provide a more detailed discussion on this topic (Gleason *et al.*, 1997).

Conclusion

Climate change can have positive, negative or neutral impact on individual patho-systems because of the specific nature of the interactions of host and pathogen. As a result it has been difficult to decipher rules of thumb that may be used for specific impact assessment. The uncertainties associated with climate change projections and the difficulties in extracting epidemiologically meaningful environmental variables such as surface wetness for GCMs have contributed to this. When climate change considerations are included deficiencies arise because of lack of detailed knowledge of epidemiology and the relevant meteorological variables needed to predict epidemics at this spatial scale. The benefit of studies on climate change and its impact on pulse disease will be to manage the diseases that become severe as result of climate change, varieties selected that are less vulnerable for diseases, and reduce fungicide application.

References

Annual Report (2013-14). Indian Institute of Pulses Research, Kanpur.

Bardossy, A. 1997. Downscaling from GCMs to local climate through stochastic linkages. *Journal of Environmental Management* 49, 7-17.

Challinor, A.J., Ewert, F., Arnold, S., Simelton, E. and Fraser, E. 2009. Crops and climate change: Progress, trends and changes in simulating impacts of informing adaptation. *J. Exp. Botan.*, 60: 2775-2789.

Chen, C.C., McCarl, B.A. 2001. An investigation of the relationship between pesticide usage and climate change. *Climatic Change*, 50: 475-487.

Coakley, S.M., Scherm, H. and Chakraborty, S. 1999. Climate change and plant disease management. *Annu. Rev. Phytopathol.* 37, 399–426.

Dixon, G.R. 2012. Climate change–impact on crop growth and food production, and plant pathogens. *Can. J. Plant Pathol.*, 34, 362–379.

Eastburn, D.M., Degennaro, M.M., Delucia, E.H., Dermody, O. Mcelrone, and Elevated, A.J. 2010. atmospheric carbon dioxide and ozone alter soybean diseases at SoyFACE. *Global Change Biol.*, 16, 320–330.

Garrett, K.A., Dendy, S.P., Frank, E.E., Rouse, M.N., Travers, S.E. 2006. Climate change effects on plant disease: genomes to ecosystems. *Annual Review of Phytopathology*, 44, 489–509.

Ghini, R., Hamada, E., Bettiol, W. 2008. Climate change and plant diseases. *Sci. Agric.* (Piracicaba, Braz.), 65, 98-107

Gleason, M.L., Parker, S.K., Pitblado, R.E., Latin, R.X., Speranzini, D., Hazzard, R.V., Maletta, M.J., Cowgill, J.R.W.P. and Biederstedt, D.L. 1997. Validation of a commercial system for remote estima- tion of wetness duration. *Plant Disease*, 81, 825-829.

Harder, D.E., Sambomki, D.J.R. Rohringer, Rimmer, S., Kim, W. and Chong, J. 1979. Electron microscopy of susceptible and resistant near-isogenic (sr6/Sr6) lines of wheat are infected by *Puccinia graminist~tici. III* Ultrastucture of incompatible reactions. *Can. J. Bot.*, 57: 2626-2634.

IPCC, 2007: Climate Change: The Physical Science Basis. Contribution of Working Group I to the Fourth Assessment Report of the Intergovernmental Panel on Climate Change [Solomon, S., D. Qin, M. Manning, Z. Chen, M. Marquis, K.B. Averyt, M. Tignor and H.L. Miller (eds.)]. Cambridge University Press, Cambridge, United Kingdom and New York, NY, USA, p. 996.

Jenns, A.E. and Leonard, K.J. 1985. Effects of temperature and illuminance on resistance of inbred lines of corn to isolates of *Bipolaris maydis*. *Phytopathology*, 75: 274-280.

Kennedy, A.D. 1997. Bridging the gap between general circulation model (GCM) output and biological microenvironments. *International Journal of Biometeorology*, 40, 119-122.

Klopping, H.L and Delp, C.J. 1980. 2-Cyano-N-[(ethylamino) carbonyl]-2-(methoxyimino) acetamide, a New fungicide. *J. Agric. Food Chem*. 28: 467-46

Knox, J., Hess, T., Daccache, A. and Wheeler, T. 2012 Climate change impacts on crop productivity in Africa and South Asia. *Environmental Research Letters* **7**, 034032. doi: 10.1088/1748-9326/7/3/034032.

Manning, W.J. and Tiedemann, A.V. 1995. Climate change: potential effects of increased atmospheric carbon dioxide (CO2), ozone (O3), and ultraviolet-B (UVB) radiation on plant diseases. *Environ. Pollut.*, 88, 219–245.

Martin, E., Timbal, B. and Brun, E., 1996. Downscaling of general circulation model outputsÐsimulation of the snow climatology of the French alps and sensitivity to climate change. *Climate Dynamics*, 13, 45-56.

Norby, R.J., Edwards, N.T., Riggs, J.S., Abner, C.H., Wullschleger, S.D. and Gunderson, C.A., 1997. Temperature-controlled open-top chambers for global change research research. *Global Change Biology*. 3, 259-267.

Pande, S. and Sharma, M. 2010. Climate Change: Potential Impact on Chickpea and Pigeonpea Diseases in the Rainfed Semi-Arid Tropics (SAT). In: 5th International Food Legumes Research Conference (IFLRC V) and 7th European Conference on Grain Legumes (AEP VII) April 26-30, 2010- Antalya, Turkey.

Pande, S., Sharma, M., Mangala, U.N., Ghosh, R. and Sundaresan, G. 2011. Phytophthora blight of Pigeonpea [*Cajanus cajan* (L.) Millsp.]: An updating review of biology, pathogenicity and disease management. *Crop Protection*, 30: 951-957.

Pangga, I.B., Chakaraborthy, S. and Yates, D. 2004. Canopy size and induced resistance *Stylosanthes scabra* determine anthracnose severity at high CO2. *Phytopathology*, 94: 221-227.

Russo, J.M. and Zack, J.W. 1997. Downscaling GCM output with a mesoscale model. *Journal of Environmental Management*, 49, 19± 29.

Schneider, S.H., Semenov, S. and Patwardhan, A. 2007. Assessing key vulnerabilities and the risk from climate change. In 'Climate change2007: impacts, adaptation and vulnerability. Contribution of Working Group II to the fourth assessment report of the Intergovernmental Panel on Climate Change'. (Eds ML Parry, OF Canziani, JP Palutikof, PJ van der Linden, CE Hanson) p. 779–810. (Cambridge University Press: Cambridge, UK)

Senft, D. 1995. FACE-ing the future. Agricultural Research 43, 4-6.

Serge, S., Andrew, N., Sparks A.,Willocquet H.L., Duveiller, E., Mahuku, G., Forbes, G., Garrett, K., Hodson, D., Padgham, J., Pande, S., Sharma, M., Yuen, J. and Djurle, A. 2011. International Agricultural Research Tackling the Effects of Global and Climate Changes on Plant Diseases in the Developing World. *Plant Disease.* 59: 1204-1216.

Sharma, M., Mangala, U.N., Krishnamurthy, M., Vadez, V. and Pande, S. 2010. Drought and dry root of chickpea. In: 5th International Food Legumes Research Conference (IFLRC V) and 7th European Conference on Grain Legumes (AEP VII) April 26-30, 2010- Antalya, Turkey.

Sharma, M., Pande, S., Pathak, M., Rao, J.N., Kumar, A., Reddy, M., Benagi, V. I., Mahalinga, D.M., Zhote, K.K., Karanjkar, P.N. and Eksinghe, B.S. 2006. Prevalence of Phytophthora blight of pigeonpea in the Deccan Plateau of India. The *Plant Pathology Journal* 22 (4): 309-313.

Siegenthaler, U., Stocker, T.F., Monnin, E., Luthi, D., Schwander, J., Stauffer, B., Raynaud, D., Barnola, J. M., Fischer, H., Masson-Delmotte, V. and Jouzel, J. 2005. Stable carbon cycle-climate relationship during the late pleistocene. *Science.* 310, p. 1313-1317.

Spahni, R., Chappellaz, J., Stocker, T. J., Loulergue, L., Hausammann, G., Kawamura, K., Fluckiger, J., Schwander, J., Raynaud, D., Masson-Delmotte, V. and Jouzel, J. 2005. Atmospheric methane and nitrous oxide of the late Pleistocene from Antarctic ice cores. *Science.* 310, p1317- 1321.

2016, Diseases of Pulse Crops and their Sustainable Management *23–39*
Editors: **Samir Kumar Biswas, Santosh Kumar and Gireesh Chand**
Published by: **BIOTECH BOOKS, NEW DELHI**

Chapter 3

Abiotic Constraints of Pulse Production in India

Kapil Kumar*, Stuti Solanki, S.N. Singh and M.A. Khan

Department of Crop Physiology,
CSA University of Agriculture and Technology, Kanpur – 208 002, U.P.

Introduction

Mythological, India is the largest producer, consumer and importer of pulses. Even though India is the world's largest pulses producer, but is stagnant a huge shortage of pulses and also prices are not affordable to a large segment of consumers. In instantly need is the development and prevalence of low cast technologies in pulses production, which can be affordable to the common man. Although, from last decade pulses production increased by 3.35 per cent per annum. The cost of cultivation and consequent prices are high to be affordable to common men, to increase production at lower cost is a great challenge (Raddy, 2009). In India, Pulses have long been believed as the poor main's only source of protein. Pulses are major source of proteins among the vegetarians in India and complement the stable cereals in the diets with proteins, essential amino acids, vitamins and minerals. Pulses provide significant nutritional value and health benefits and to reduce several non-communicable disease such as colon cancer and cardiovascular disease (Yude *et al.*, 1993 and Jukanti *et al.*, 2012). Pulses are important source of dietary protein for most of people in developing world including India, because mostly the population is vegetarian. India is the

* Corresponding Author: E-mail: kapilsaini151@gmail.com

largest producer (17.29 million tonnes in 2011-2012) and consumer of pulses in the world but low average productivity (694 kg/ha in 2011-12) for most of the pulses (Anonymous, 2012).

Major pulses grown in India include chickpea or bengal gram (*Cicer arietinum*), pigeon pea or red gram (*Cajanus cajan*), lentil (*Lens culinaris*), urd bean or black gram (*Vigna mungo*), mong bean (*Vigna radiata*), lablab bean (*Lablab purpureus*), horse bean (*Dolichos uniflorus*), pea (*Pisum sativus* var. *arense*), grass bean or khesari (*Lathyrus sativus*) moth bean (*Vigna aconitifolia*), cow pea (*Vigna unguiculata*) and broad bean or faba bean (*Vicia feba*).

Among these chickpea, pigeon pea, mungbean, urdbean and lentil are very popular in our country. In India, pulses are abundantly grown in two seasons: (A) in the warmer / rainy season or kharif (June-October): these are pigeon pea, urdbean, mungbean and cowpea in the cool, dry season or rabi (October-April) are : chickpea, lentil and dry peas. Among various pulse crops chickpea dominates with over 40 per cent share of total pulse production followed by pigeon pea (18-20 per cent), mungbean (11 per cent), urdbean (10-12 per cent), lentil (8-9 per cent) and other legumes (20 per cent) (IIPR Vision, 2030).

Nutritional Value of Pulses

Mostly pulses contain 22-25 per cent protein which is almost twice the protein in wheat and thrice that of rice. Sprouted pulses are especially nutritious. Since sprouted pulses are rich in carbohydrates, protein, fat, fibres and minerals. Its also have good amounts of vitamin B (riboflavin, thiamine and niacin), vitamin C and energy sources. The whole pulses are also very rich in natural fibre that is not really a nutrient, but is helpful in digestion. Since pulses are so nutritious, they are included in a balanced diet. They also are a good source of calories, if you are looking to lose weight, pulses should be consumed in moderation. Both beans and pulses have been observed to be instrumental in reducing blood cholesterol. Increase intake of dried beans can reduced the amounts of lipids stored in the blood serum. Due to the roughage in the whole grams, acids and faecal elimination is aided. Pulses also help to control

Table 3.1: Some of the Commonly Available Pulses, along with their Nutritional Value

Pulses and Legumes	*Energy (Kcal)*	*Fibre (g)*	*Carbohydrate (g)*	*Protein (g)*	*Fat (g)*	*Minerals (g)*	*Calcium (mg)*
Gram	372	1.2	59.8	20.8	5.6	2.7	56
Green gram	348	0.8	59.9	24.5	1.2	3.5	75
Kidney beans	346	4.8	60.6	22.9	1.3	3.2	260
Moth beans	330	4.5	56.5	23.6	1.1	3.5	202
Peas	315	4.5	56.5	19.7	1.1	2.2	75
Red gram	335	56.5	19.7	1.1	2.2		73
Soya bean	432	1.5	57.6	22.3	1.7	4.6	2450
Black gram	347	0.9	59.6	24	1.4	3.2	154

Source: Nutritional Facts Lentils and Lentils Salad for Everyday Meals (2014).

and reduce blood sugar. It is helpful for a diabetes patient to take sprouted pulses or legumes which are not only fibrous, but also have many enzyme inhibitors.

India is world's largest producer of pulses, it imports a large amount of pulses to meet the growing domestic needs. The major pulses imported are desi chickpeas, pigeon pea, urdbean, kidney bean and mungbean which is mostly come from Myanmar. Canada and Australia are major suppliers of dry peas and kabuli chickpea to the Indian market which is about one third of India's pulses import.

Pulse crops reported huge losses due to biotic (these are pests and diseases) and abiotic factors (these are drought, high-low temperature and edaphic factors etc). Some of the studies estimated the losses in the range from 15-20 per cent of normal production (IIPR, 2011).

The production estimates for pulses crops for 2012-2013 are as following:

Tur- 3.07 million tones	Moong- 1.20 million tones
Urd- 1.90 million tones	Gram- 8.88 million tones

Sources: MP:SS:BK:CP: fourth advance estimator (22.7.2013).

Currently major researches and many investigations efforts are addressing a number of biotic stresses, including fungal diseases are the major yield limiting factor in India, while the major abiotic stresses of pulses are those associated with cold, frost, water logging (high-low water conditions), drought, heat, soil pH, salinity, sodicity and boron toxicity.

Phenology

Every pulse crops species have different phenology condition, as like, time of sowing and location. In a particular stress situation, that time at which it occurs in the plant's life cycle will affect plant production through its reduce their morphological characteristics as such plant height, number of leaves and yield components are total biomass, number of pods, number of seeds, test weight, seed weight and harvest index. In generally, life cycle of the pulse crops can be divided into the following five phases, these are,

Phase 1- from planting to seedling emergence.

Phase 2- from seedling emergence to first flowering.

Phase 3 - from first flowering to pod appearance.

Phase 4 - from pod appearance to the beginning of seed formation.

Phase 5 - from seed formation to physiological maturity.

Faba beans and field peas were the earliest to flower and pod compared to chickpeas and lentils. Despite differences in flowering and podding, maturity was similar in the pulse species studied and was largely determined by soil moisture and temperature conditions at the end of the season. The duration of reproductive period (first flower to maturity) was longer in faba beans and field peas.

Drought and Heat Stress

The temperatures for the cool season pulses range between 10°C and 30°C. Temperatures which fall outside the optimum range cause stress. Daily maximum temperature is considered above 25°C as the threshold level for heat stress in cool season pulse crops. The critical temperature for heat tolerance seems to be higher in chickpea than in faba bean, lentil and field pea crops and the reverse are true for cold tolerance. Particularly, pulses are sensitive to heat at the full bloom stage. Even though few days of exposure to high temperatures (30- 35°C), causes heavy yield losses through flower drop and pod abortion. Mostly, observed that reductions in seed yields after brief episodes of high temperatures during seed filling can determine seed set, seed weight and accelerate senescence as well as reduce yield production. Higher soil moisture requirement need in chickpea during germination than lentil, field pea or faba bean. However the critical soil moisture required for seed germination and seedling emergence in chickpea is well below field capacity. Lentil seeds absorb water equal to their weight in less than 36 hours and germinate soon after, but germination is reduced if dehydration occurs thereafter. This makes lentil crop sensitive to early season drought, particularly when planted shallow.

Drought at vegetative stages of growth alone does not appear to cause a significant yield loss in chickpea or field pea. The most sensitive stage to drought is flowering stage which is might be due to the lack of new root growth in pulse crops at this stage of growth. Among the cool season pulse chickpea is considered to be the most drought tolerant because it forms a deep taproot in soil. Chickpea has the biggest ability to tolerate intermittent drought and respond to subsequent rainfall due to its more indeterminate growth habit when compared to other cool season pulse crops. Due to terminal drought, seed yield loss in chickpea can vary between 30 per cent and 60 per cent depending upon the location and climatic conditions during the crop season. Faba bean is very sensitive to drought and losses in potential yield can be up to 70 per cent while lentil is relatively tolerant to drought. These potential losses in yield can be ranged between 6 per cent and 54 per cent. Another pulse crop that is field pea, which also sensitive to drought and yield losses vary from 21 to 54 per cent. But, field pea with the correct phenology has the ability to escape drought and produce respectable yield when compared with other pulses. (IPCC, 2007).

Cold

The major symptoms in winter pulses due to cold effect are reduced leaf production, leaf area expansion and poor dry matter production, chlorosis of leaf and finally necrosis of the older leaves. Daily recommended average temperature between 0°C and 10°C should considered a threshold level for cold or chilling stress in cool season pulses. Chickpea is more sensitive to chilling injury than among pulses like field pea, faba bean or lentil. The reduced ovule fertilization, associated with decline in pollen tube growth and ovule viability, is the major cause for poor seed set in chickpea at low temperatures. Fortunately, recent evaluation of germplasm both in Australia and overseas shows distinct differences in pod and seed set at low temperatures in chickpea. These traits are currently being exploited in the breeding programs.

Frost

The frost injury in winter pulses occurs when daily minimal temperature is below 0°C without snow cover. Frost damage in spring is a major cause of yield loss in crops including pulses in Australia. The most important stress in the freezing process is ice formation and the associated mechanical damage of the tissue. The most sensitive stages are flowering, early pod formation and seed filling of in winter pulses. Frost injury appear to be higher for chickpea than field pea, lentil and faba bean. Among the pulses, field pea seems to be the most susceptible to frost injury than any pulses whereas chickpea is a more indeterminate species

Water-logging

Seed germination is very susceptible to waterlogging. Waterlogging causes common problem that is poor seedling emergence and crop establishment. Among cool season pulses faba bean is relatively tolerant to waterlogging at germination storage compared with lentil, field pea or chickpea. Waterlogging six days after germination of field pea can delay the seedling emergence and finally reduce 80 per cent plant density. Waterlogging has negative effects on root growth more than shoot growth and depresses vegetative growth of plants. Roots of faba bean have greater tolerance to waterlogging than the roots of chickpea, lentil and field pea. After establishment of seedling, the sensitivity of plants to waterlogging stress is generally increases with advancing age. Thus the ability to survive and recover following waterlogging is dependent on the timing of the waterlogging event relative to the stage of growth. Ability to recover generally declines sharply as reproductive growth (flowering and pod filling stage) approaches and seed yield loss resulting from waterlogging can be substantial. Limited data for lentil, field pea and chickpea suggest that reduction in yield can depending upon on species, stage of growth when crops are waterlogged, duration and the extent of root zone affected by waterlogging. Apply some management practices can reduce the effects of waterlogging include the species choice (Faba bean tolerant, lentil highly susceptible), paddock selection, time of sowing, seeding rate and drainage.

Soil pH

Soil is divided in two groups on the basis of soil pH (A) Saline soil: these type soils have less than 15 per cent sodium exchange cations (SEC), pH less than 8.5 but electron conductivity have more than 4 m. mol $cm^{-1}$and, (B) Sodic soils or alkaline soils: these soils have more than 15 per cent sodium exchange cations (SEC), pH more than 8.5 but electron conductivity occurs less than 4 m. mol cm^{-1}. Cool season pulses are sensitive to acid soil conditions and require a neutral to alkaline soil pH for their optimum growth and yield. Among the cool season pulses, lentil is the most sensitive to low pH followed by chickpea, faba bean and field pea. A drop of one pH unit below the threshold value (pH 5.5) can cause more than 86 per cent reduction in seed yield of lentil. Cool season pulses generally appear to be more sensitive to acidity than cereals. Nutrient availability can change dramatically with small changes in soil pH, particularly in mineral soils with low in organic carbon. Toxicity of Al, Fe, and Mn occur when the soil pH falls from 5.5 to 4.5. Root growth of pulses is severely restricted on acid soils.

Deficiencies of some nutrients as like Fe, Zn and Mn occur above pH 8.0. There are very evidence of Fe deficiency in chickpea and lentil while Zn deficiency in field pea, also in chickpea and lentil. On high pH and calcareous soils, Fe deficiency is most commonly occurring nutrient disorder across the cool season pulses. Losses in yield of highly susceptible genotypes of chickpea and lentil range from 22 to 50 per cent. Appearance of Fe - deficient symptoms is highly transient across cool season pulse crops appearing under wet (above field capacity) soil moisture conditions and low temperatures. In cool season pulse crops, screening germplasm grown at high pH (> 8.1) calcareous soils has been very effective method to identify genotypic differences in susceptibility to Fe deficiency.

Salinity and Sodicity

Salinity damage in pulse crops is appeared due to characteristic symptoms of excess ion accumulation. The yellowing of the older leaves and necrosis of the outer margins of leaves are the first symptoms of salinity. The symptoms increase in younger leaves and older leaves die and abscise, as salinity condensable. Salinity quickens anthocyanin pigmentation in leaves and stems in desi chickpea while in kabuli chickpea these tissues become yellow. Salinity symptoms are confused to diagnosis in pulses with symptoms of other nutrient disorders, disease, herbicide damage and drought.

Mostly, it is observed that crop response to salinity changes along with crop age. Lentil, faba bean and field pea are more sensitive at germination than at subsequent growth stages and the converse is true for chickpea. However, it is necessary to establish critical values, relating salt concentrations and their effect on growth and yield reduction to quantify the effect of salinity on plant growth. Lentil and field pea may have largest salinity tolerance than faba bean and chickpea on the suggestion of glasshouse studies and field observations. Salinity tolerance of field pea and lentil are higher comparable to wheat but less than to barley. The flowering is latter stages of chickpea growth, which are more sensitive to salinity than the early vegetative stages. Scope for genetic enhancement of salinity resistance in cool season pulses is limited by the lack of variability in germplasm with the desired levels of resistance. Chickpea and faba bean are sensitive pulse crops which should not be grown on salt affected soils, other tolerant species such as barley should be considered.

Sodicity adversely affects on pulses production by reducing germination and seedling establishment with increasing exchangeable sodium percentage (ESP: 15 – 20 per cent). According to studies in glasshouse and observations of field suggest that chickpea and lentil are more sensitive to sodicity than faba bean and field pea. The level at which 50 per cent reduction of seed yield occur is about 10 ESP for chickpea and 15 ESP for lentil. The seed yield appears more sensitive to sodicity than vegetative growth. Various researchers observed that, the number of pods plant^{-1} and seed weight was severely decreased, whereas the numbers of seeds per pod were little affected in lentil and chickpea. Canola is moderately sodic-tolerant species, can withstand 35 ESP without significant reduction of growth or yield.

Deficiency of Symptoms in Essential Minerals

The nutrient status in plants can be evaluated through visual nutrient deficiency symptoms in plant morphology; it is a very powerful diagnostic tool. One point should keep in mind, that a given individual visual symptom is seldom sufficient to make a definitive diagnosis of a plant's nutrient status. Several of the important classic deficiency symptoms such as leaf scorching, tip burn, chlorosis and necrosis are characteristically associated with more than one mineral nutrient deficiency and also along with other stresses factor that by themselves are not diagnostic for any specific nutrient stress and deficiency symptoms. Albeit, their detection is extremely useful in making an evaluation of nutrient status. In addition, to know the location and timing of these deficiency symptoms is a critical aspect of any nutrient status evaluation for the actual observations of morphological and spectral symptoms.

Sources of Visual Symptoms

Stresses such as biotic, abiotic and edephic *i.e.,* pathogens, salinity and air pollution induce their own characteristic set of visual symptoms. Often, these symptoms closely similar to those of nutrient deficiency. Pathological symptoms can often be separated from nutritional symptoms by their distribution in a population of affected plants. If the plants are under a nutrient stress, all plants of a given type and age in the same environment tend to develop similar symptoms at the same time. However, if the stress is the result of pathology, the development of symptoms will have a tendency to vary between plants until a relatively advanced stage of the pathology is reached.

Five types of deficiency or toxicity symptoms are observed:

1. **Chlorosis** - Yellowing of plant tissue due to limitations on chlorophyll synthesis. This yellowing can be generalized over the entire plant, localized over entire leaves or isolated between some leaf veins (*i.e.* interveinal chlorosis).
2. **Necrosis** - Death of plant tissue sometimes in spots.
3. **Stunting or reduced growth** - New growth continues but it is stunted or reduced compared to normal plants.
4. Accumulation of anthocynanin resulting in a purple or reddish colour.
5. Lack of new growth.

Nutrient deficiencies may not be apparent as striking symptoms such as chlorosis on the plant, especially when mild deficiency is occurring. However, significant reductions in crop yields can occur with such deficiencies. This situation is termed 'hidden hunger' and can only be detected with plant tissue analysis or yield decline. However, experience with growing a specific plant species or variety can greatly help in distinguishing poor crop performance and possible nutrient deficiency symptoms from normal plant growth.

Table 3.2: Functions of Elements in Plant Health

Nutrient		*Function*
Macro nutrients		
Primary nutrients	Nitrogen (N)	Necessary for porphyrins parts of chlorophyll, genetic material (DNA and RNA) formation, stimulates leafy growth and therein greenish and constituent of alkaloids.
	Phosphorous (P)	Necessary for coenzymes (NAD, NADP and ATD), genetic material formation, increases disease resistance, stimulates of root production and early season growth, stimulates of flower and fruit.
	Potassium (K)	Impotent role in stomatal movement and osmotic potential. Older leaves turn yellow initially around margins and die; irregular fruit development.
Secondary nutrients	Calcium (Ca)	Constituents of middle lamella, second messenger of metabolic regulation. Raises soil pH, promotes root hair formation and early growth
	Magnesium (Mg)	Needs in chlorophyll formation and energy metabolism, helps regulate uptake of other elements. It also promotes healthy, disease resistant plants. It is generally available in acid soil.
	Sulphur (S)	Content of sulfhydryl groups, aids in formation of certain oil compounds that give specific odours to some plants and increases oil production in flax and soybeans.
Micro nutrients		
Iron (Fe)		Stimulates the formation of chlorophyll and helps oxidize sugar for energy, also necessary for legume nitrogen fixation. It regulates the respiration in plant's cells.
Zinc (Zn)		Stimulates stem growth and flower bud formation, activation of carbonic anhydrase and alcohol-dehydrogenase and increase auxin synthesis.
Manganese (Mn)		Activator of many respiratory enzymes. Necessary for the formation of chlorophyll.
Molybdenum (Mo)		Needed for nitrogen fixation and nitrogen use in the plant, stimulates plant growth and vigour much like nitrogen. Prosthetic group of nitrate reductase and nitrogenase enzymes.
Boron (B)		Stimulates cell division, flower formation and pollination and facilitate translocation of sugars
Chlorine (Cl)		Needed for photolysis of water and oxygen evolution during photosynthesis, stimulates cell-division in root and leaves and aids water circulation in plants.
Copper (Cu)		Constituent of several oxidising enzymes, stimulates stem development and pigment formation
Cobalt (Co)		Improves growth, water circulation and photosynthesis

Contd...

Table 3.2–*Contd...*

Nutrient	*Function*
Beneficial nutrients	
Sodium (Na)	Increases resistance to drought, increases sugar content in some crops, enhanced cell expansion and transport of pyruvate in C_4- plants.
Silicon (Si)	Increases number of seeds, form polyphenols-complexes and strengthens cell walls of plants.

Sources: All above nutrients and their functions are referred to V.K., Jain (2011) and Taiz and Zeiger (2010).

Table 3.3: Symptoms of Nutrients and their Remedies in Pulses

Primary Elements	*Symptoms*	*Remedies*
Major Nutrients		
Nitrogen (N)	Little new leaf growth, yellow leaves, this being more pronounced in older leaves. Earlier fall leaf drop. New shoots may be red to red-brown (anthocyanin pigment formation). Prolonged dormancy and senescence.	Quick fix: Make weekly foliar applications of fish emulsion or foliar spraying 2 per cent urea.Long term: Apply aged compost, manure, soybean meal or cottonseed meal to the soil once in spring. Seaweed extract will improve the soil environment thus giving nitrogen fixing bacteria a boost.
Phosphorous (P)	Overall dark green with purple, blue or reddish cast to leaves particularly on underside, veins and stems and some plants respond to lack of P with yellowing. Cambial activity is checked. Thickening of tracheidial cells and excess foliage with no flowers can also indicate lack of (P).	Quick fix: Weekly, spray on plant with fish emulsion until symptoms quit. Apply a light soil dressing of wood ashes. Long term: Mix rock phosphate or aged manure into the soil in fall and use of phosphate fertilizers.
Potash (K) Potassium	Shoot may die and sickly looking plants, undersized fruits, leaves showing marginal and interveinal chlorosis. Yellowing starts on older leaves and progresses upwards. Leaves may crinkle, turn brown, and roll upwards. Blossoms may be distorted and small. Plant has little resistance to heat, cold and disease problems.	Quick fix: Spray plant weekly with fish emulsion until symptoms quit. Long term: Use of muriate or sulphate of potash, granite dust or greensand to the soil in fall. Hardwood ashes may be applied to soil anytime to remedy of it.
Secondary Elements	*Symptoms*	*Remedies*
Sulphur (S)	Shoots are stunted. Young leaves develop orange, red or purple colour of pigment. Premature development of lateral buds with dead tips and occur shoot die back.	Long term: Add sulphur or potassium sulphate as necessary. Use precaution of applying sulphur compounds. Sulphur toxicity appears as veinal chlorosis followed by rapid defoliation of the lower leaves.
Magnesium (Mg)	Magnesium lack is characterized almost identically with iron deficiency on the older leaves, generally at the bottom of the plant, show marginal and interveinal reddening or yellowing with leaf base and midrib staying green. Reduce of carotene. Leaves may be brittle and thin with leaf curling and distorted growth.	Use of magnesium sulphate (Epsom salts), We can use it watering with a mix of 1-2 teaspoons or Epsom salts dissolved in 1 gallon of water or using the mix as foliar spray, make three applications for six weeks apart. Other treatments include magnesium sulphate, adding fish meal, basic slag or dolomitic limestone.
Calcium (Ca)	Young leaves are small and distorted with curled back leaf tips. Shoots may be stunted and show some dieback, roots will be stunted.	Long term: Apply of calcium ammonium nitrate or superphosphate.

Contd...

Table 3.3–*Contd...*

Minor Elements	*Symptoms*	*Remedies*
Iron (Fe)	New leaves are the most symptomatic and when condition is most severe they can be all yellow or white but still have green veins as like magnesium. Interveinal white chlorosis and complete bleaching. Overall leaves yellow with green veins leading to marginal scorching or browning of leaf tips. Tips of leaves, especially basal areas of leaflets, intense chlorotic mottling and yellow stem near tip. Iron deficit mostly found at the soil pH is higher than 7.5 means more alkaline.	Test of soil pH to correct 7.0 or lower. In iron-deficient soils, add bone meal or blood meal organic amendments or add iron sulphate or chelated iron liquid or granular inorganic amendments. Foliar spraying ferrous sulphate (0.5 per cent) long with lime (half quantity).
Zinc (Zn)	Plants show resetting and premature shedding. Whitish chlorotic streaks. Mostly zinc deficiencies are occurring on sandy soils which low in organic matter and on organic soils, during cold, wet spring weather. New and intermediate leaves are small, yellow, sometimes with a greyish cast. Small shoots may show rosetting followed by dieback.	Test the soil pH, to sure that the pH is remaining between 5.8 and 6.2. Imbalance pH can inhibit the absorption of zinc and other nutrients. Use of acidic fertilizers. Organic compounds such as zinc chelates (zinc EDTA and zinc NTA). Soil application of zinc sulphate @ 15-30 kg ha^{-1} and foliar spraying of 0.5 per cent $ZnSO_4$ coupled with half content of lime.
Boron (B)	Youngest leaves may be red, bronze or scorched small, thick or brittle and new shoot tips may form as a witches broom. Stems die back, lateral shoots developed and leaves highly tinted purple, brown and yellow. Produce sterile flower, fruit and vegetables may have heart rot. Boron deficiencies are found principally in acid and sandy soils in regions of high rainfall, those with low soil organic matter, more pronounced during drought periods when root activities are restricted.	Apply household borax @ 1 tablespoon borax to 12 quarts of water. Soil application of boron @ 5-10 kg ha^{-1}.
Manganese (Mn)	Similar to N deficiency, leaves display marginal scorching, rolling and reduced width. Dead tissue spots and tissue turn brown are found scattered on the leaves.	Soil application of manganese sulphate @ 15-30 kg ha^{-1} and foliar spraying of 0.5 per cent $MnSO_4$ coupled with half content of lime.

Contd...

Table 3.3–*Contd...*

Minor Elements	*Symptoms*	*Remedies*
Copper (Cu)	Increased phosphorus and iron availability in soils and soil pH decreases copper uptake by plants. Small leaves with necrotic (dead) spots and brown areas near the leaf tips. Reduction of carotene and other pigments.	Soil application of copper sulphate @ 5-10 kg ha^{-1} and 0.1 per cent -0.2 per cent $CuSO_4$ foliar spraying of with half content of lime.
Molybdenum (Mo)	Only a problem with brassicas like broccoli, cauliflower etc in acid soil. Mottling in the older leaves.	Soil application of ammonium molybdenum @ 0.5 to 1 kg ha^{-1} and foliar spraying of 0.01 per cent - 0.02 per cent of ammonium molybdenum.

Source: Bergmann, W. 1993, Blevins, *et al.*, 1998, Cakmac, *et al.*, 1994, Fismes, *et al.*, 2000, Kumar, *et al.*, 2007, V.K., Jain 2011, Taiz and Zeiger, 2010 and Wallace, T. 1961.

Effect of Hormones on Pulses Production

Auxin

Auxin is the first class of major plant hormones discovered to be regulation of plant growth and development at all levels. The most important member of the auxin family is indole-3-acetic acid (IAA), derived from the phenyl-propanoid biosynthesis pathway in plant. The auxin concentration increases in cells undergoing cell elongation, cell differentiation and vascular bundle formation and this site, where found higher concentration of auxin is produced mainly in shoot and move to the root by an active transport process involving auxin efflux protein (IAAH) complexes. This movement is known as basipatel movement. Efflux protein (IAAH) complexes proteins, which may regulate the auxin concentration in the plant. Rhizobium derived from Nod factors and several classes of flavonoids have been reported to have a similar effect on transport.

The higher auxin and low cytokinin ratio in the root is responsible for the initiation of cortical cell divisions and nodule formation. Various plant substances like ethylene, cytokinin and peroxidase could inhibit auxin transport which could lead to local shifts in the plants auxin-cytokinin ratio. During early stage of nodulation increase in early dividing cortical cells and decreased in differentiating nodule primordial and vascular tissue through, inferred auxin level. Similar observation made in *Medicago truncatula AUX1*- like genes (termed as *Mt LAX*). The gene is expressed in nodule primordial at early stage of nodule development. The auxin stimulated lateral root formation while an increase of cytokinin concentration or inhibiting auxin transport induced the development of pseudo-nodules. Auxin maintain dominance of the main root growth over lateral root growth and also the apical dominance of main shoot over the growth of lateral bud and tillers. The herbicide 2, 4-D can also yield auxin-like activity at low concentration (Brewer *et al.*, 1994 and Simon *et al.*, 2010). During the development of leaves and tendrils, which originate from the apical bud, leaflets form in locations with low auxin levels and tendrils form in regions of high auxin, both within the apical bud primordia (DeMason *et al.*, 2004).

Gibberellins

Application of gibberellins promotes cell elongation in pea stem internodes (Brain *et al.*, 1954) and lentil epicotyls (Nitsan *et al.*, 1966). Foliar application of gibberellins markedly increases length and number of internodes formed in semi-dwarf pea plants (Spiker *et al.*, 1976). Finally application of gibberellins inhibitors can consistently reduce length in pea (Ayele *et al.*, 2006b). The plant hormones are tightly promoted the development of leaf and tendrils in legumes. Stem elongation is a shoot morphology similar to that seen when auxins or gibberellins is applied to young seedling. The importance of gibberellins in regulating sunflower shoot morphology is confirmed by the high accumulation of *GA20ox* transcripts during active shoot growth. *GA20ox* is a gene family that encodes an important group of enzyme in Gibberellins synthesis (Carzoli *et al.*, 2009).

Cytokinin

Cytokinins are a group of plant hormones active in cell division and also involved in regulation of many physiological processes during plant development, growth and adaptation to environmental conditions. Cytokinins are implicated as the control of root architecture development including root nodulation. Exogenous application of cytokinins on legume root induced response similar to Nod factors, includes cortical cell divisions, amyloplast deposition and induction of early nodulin gene expression. Exogenous cytokinins application to roots induced formation of pseudo-nodule structures on legumes (Garden Pea and Alfalfa) and even non-legumes (tobacco). Transfer of trans-zeatin secretion gene into Nod bacteria and non-symbiotic bacteria was sufficient to induce nodule-like structure formation at low frequency in alfalfa.

Ethylene

The ability of ethylene to control pea stem elongation is well elucidated (Eisenger 1983). The physical resistance encountered by a growing seedling pushing against the soil surface crust as it attempts to emerge from the soil, promotes ethylene biosynthesis (Goeschl *et al.*, 1966). In pea seedlings, ethylene inhibits cell elongation and promotes cell lateral expansion in the stem tissue, making the stems shorter and thicker, thereby enabling the elongating shoot to more effectively push through the mechanically restrictive soil surface layer (Burg, S.P. 1973 Goeschl *et al.*, 1966 and Morgan *et al.*, 1997).

Ethylene production from the soybean stand gradually increased during the vegetative growth phase and then declines during pod fill. Ethylene can also be involved in abiotic stress response in plants. However, ethylene's role in abiotic stress response varies with the stress and the stress is applied. In soybean plants, gradual development of drought stress did not promote ethylene production or increase levels of ACC, an important precursor of ethylene. In contrast, a rapid development of drought stress increased levels of both compounds (Xu *et al.*, 1993). Dry air that promotes leaf membrane damage can also increase ethylene production and arise levels of ACC. Drought stress is often accompanied by high temperatures, which promote ethylene production, within certain limits (up to 35°C) after which ethylene production is inhibited.

Abscisic Acid

Abscisic acid is a naturally occurring compound in plants, derived from the mevalonic acid biosynthesis pathway that also leads to plant sterols (including brassiosteroids) and gibberellins. In nodulation, it has been reported to play multiple negative roles in nodulation at different stages of development. The application of ABA through root irrigation was first shown to inhibit nodule numbers in *Pisum sativum*. Nodule development is affected by ABA treatment resulting in arrested and small ineffective brownish coloured nodules and early degeneration of bacteriod tissue was noted in developing nodules. The nitrogenase activity of nodules treated with ABA observed lower compared to wild type in *Faba vulgaris,* pea and *Lotus japonica*. The application of ABA stimulated an abrupt stress situation mimicking severe drought which led to leghemoglobin reduction. ABA might create an oxygen barrier in nodules that result in observed decline of nitrogen fixation.

Management of Abiotic Stresses in Pulse Crops

Abiotic stress can be reduced by choosing the best adjusting agronomical practices which include are sowing time, plant density, soil management and appropriate pulse species to ensure sensitive crop stages occur at the most favourable time in the season. In dry land environments with terminal stress, some important points as choosing the optimum sowing time, species and cultivars with appropriate phenology can reduce the effect of frost and drought in pulse crops.

Dealing with stresses caused by extremes in the abiotic environment is to develop cultivars resistant to specific stresses. Breeding for resistance to these stresses is often considered difficult because of the unpredictability of climatic conditions. The effect of the abiotic stress, together with the ability of the plants to tolerate the stress themselves depends on stage of plant growth and the duration of the stress. Drought and heat escape through earliness in flowering and maturity is probably the characteristic most to be widely used by breeders for development of drought resistant pulse variety in low rainfall terminal drought environments. However, germplasm for resistance to various abiotic stresses are not widely available to breeders. But recent studies suggested that wild species of cool season pulses are having desired traits for a number of abiotic stresses. For future studies, it is required to identify desirable genes of concerned germplasm and further transfer to adapted cultivars by conventional and biotechnological approaches to develop abiotic stress resistant cultivars.

References

Anonymous. 2012. Economic survey. Govt. Of India, Ministry of Finance, p 17-19.

Ayele, B.T., Ozga, J.A. and Reinecke, D.M. 2006b. regulation of gibberellins biosynthesis genes during germination and young seedling growth of pea (*Pisum sativum* L.). *J. Plant Growth Regul.* 25: 219-232.

Bergmann, W. 1993. Nutrient deficiencies in crops, Jena: Gustav Fisher Verlag.

Blevins, D.G. and Lukaszewski, K.M. 1998. Boron in plant structure and function *Annu. Rev. Plant Physiol. Mol. Boil.* 49: 481-500.

Brain, P.W. and Henmming, H.G. 1957. A relation between the effects of gibberellic acid and indolyacetic acid on plant cell expansion. *Nature,* 179: 417.

Brewer, G.J., Anderson, M.D. and Urs. N.V.R.R. 1994. Screening sunflower for tolerance to sunflower midge using the synthetic auxin 2,4-dichlorophenoxyacetic acid. *J. Econ. Ent.* 87: 245-251.

Burg, S. P. 1973. Ethylene in plant growth. *Proc. Natl. Acad. Sci. USA* 70: 591-597.

Cakmac, I., Kurz, H. and Marschner, H. 1994. Changes in phloem export of sucrose in leaves in response to phosphorus, potassium and magnesium deficiency in bean plants. *J. Exp. Bot.* 45: 12251-1257.

Carzoli, F.G., Micheloti, V., Fambrini, M., Salvini, M. and Pugliesi, C. 2009. Molecular cloning and organ-specific expansion of two gibberellin *20-Oxidase* genes of *Helianthus annuus, Plant Mol. Biol. Rep.* 27: 144-152.

DeMason, D.A. and Chawla, R. 2004. Roles of auxin during morphogenesis of the compound leaves of pea (*Pisum sativum* L.). *Planta,* 218: 435-448.

Eisenger, W. 1983. Regulation of pea internode expansion by ethylene. AQnn. *Rev. Plant Physiol.* 34: 225-240.

Fismes, J., Vong, P.C., Guckert, A. and Frossard, E. 2000. Influence of sulphur on apparent N-use efficiency, yield and quality of oilseed rape (*Brasssica napus* L.) grown on a calacareous soil. *European J. Agron.* 12: 127-141.

Goeschl, J.D. Rapport, L. and Pratt, H.K. 1966. Ethylene as afactors regulating the growth pea epicotyls subjected to physical stress. *Plant Physiol.* 41: 877-884.

IIPR Vision 2030. 2011. Printed Published by the Director, Indian Institute of Pulses Research (ICAR), Kanpur-208024.

IPCC (Intergovernmental Panel on Climate Change) 2007. Climate Change 2007: Impacts, Adaptation and Vulnerability. In: Contribution of Working Group II to Fourth Assessment Report of the Intergovernmental Panel on Climate Change, M.L. Parry; O.F. canziani; J.P. van der Linden; and C.E. hanson (Eds.), 1000, Cambridge University Press, Cambridge, UK.

Jain, V.K. 2012. Fundamentals of Plant Physiology, fourteenth revised edition, p02-113.

Jukanti, A.K., Gaur, P.M., Gowda, C.L.L. and Chibbar, R.N. 2012. Nutritional quality and health benefits of chickpea (*Cicer arientinum* L.): a review. *British Journal of Nutrition,* 108, S11-S26.

Kumar, A. and Purohit, S.S. 2007. Plant Physiology Fundamental and Application, second enlarged edition. p145-170.

Morgan, P. W. and Drew, M. C. 1997. Ethylene and plant responses to stress. *Physiol. Plant.* 100 620-630.

Nitsan, J. and Lang, A. 1966. DAN synthesis in the elongating nondividing cells of lentil epicotyl and its promotion by gibberellin. *Plant Physiol.* 41: 965-970.

Pandey, S.N. and Sinha, B.K. 2006. Plant Phsiology, fourth edition. p120-139.

Reddy, A. Amarender 2009. Pulses production technology: status and way forward. Economic and Political weekly vol. XLIV No. 52: 73-80.

Siddique, K.H.M., Loss, S.P., Regan, K.L. and Jettner, R.L. 1999. Adaptation and seed yield of cool season grain legumes in Mediterranean environments of south-western Australia. *Australian Journal of Agricultural Research,* 50: 375 – 387.

Simon, S. and Petrasek, J. 2010. Why plants need more than one type auxin. *Plant Sci.* 180: 454-460.

Spiker, S., Mashkas, A. and Yunis, M. 1976. The effect of single and repeated gibberellic acid treatment on internode number and length in dwarf peas. *Physiol. Plant,* 36: 1-3.

Taiz, L. and Zeiger, E. 2010. Plant Physiology, fifth edition. Page: 67-84.Wallace, T. (1961). The diagnosis of mineral deficiencies in plants by visual symptoms. A colour Atlas and Guide. Her Majesty's Stationery Office, London.

Xu, C.C. and Qi, Z. 1993. Effect of drought on hypoxygenase activity, ethylene and ethylene production in leaves of soybean plants. *Acta. Bot. Sin.* 35: 31-37.

Yude, C., Kaiwei, H., Fuji, L., Jie, Y. 1993. The potential and utilization prospects of kinds of wood fodder resources in Yunnan. *Forestry Research,* 6: 346-350.

2016, Diseases of Pulse Crops and their Sustainable Management ***41–63***
Editors: **Samir Kumar Biswas, Santosh Kumar and Gireesh Chand**
Published by: **BIOTECH BOOKS, NEW DELHI**

Chapter 4

Climate-proof Disease Management Strategies in Pulses

Vivek Singh[1]*, Mehi Lal[2] and Mohd. Ali[3]

[1]*Department of Plant Pathology,*
Manyawar Shri Kanshiram Ji University of Agriculture and Technology,
Banda, U.P.
[2]*Plant Protection Section, Central Potato Research Institute Campus,*
Modipuram, Meerut – 250 110, U.P.
[3]*Department of Plant Pathology,*
Sardar Vallabhbhai Patel University of Agriculture and Technology,
Meerut – 250 110, U.P.

Introduction

Greenhouse gas concentrations in the atmosphere after the industrial revolution has increased many fold due to increased use of fossil fuel burning, deforestation and other land use changes and contributed to global climate change. The atmospheric concentration of carbon dioxide (CO_2) has reached levels significantly higher than in the last 650 thousand years (Siegenthaler *et al.*, 2005). Since 2000, the growth rate in CO_2 concentration is increasing more rapidly than in the previous decade. Similar trends have been observed for methane (CH_4), nitrous oxide (N_2O), and other green house gases (Spahni *et al.*, 2005; IPCC, 2007). The average global surface temperature has increased by 0.2°C per decade in the past 30 years. Abnormal changes in air temperature and rainfall, increased intensity and frequency of storms, drought and

* Corresponding Author: E-mail: vsinghiitk@gmail.com

flooding and altered hydrological cycle events are the growing concern to agricultural ecosystems. Increasing climate variability with the change in climate is recognized unequivocally. Global climate change especially increased CO_2 and temperature level (Watson, 2011) is thought to influence or change all the elements of a disease triangle *i.e.*, host, pathogen and environmental factors and their interactions (Legreve and Duveiller, 2010). Changes in the gaseous composition of the air may directly affect the severity of crop disease epidemics through effects on the host, the pathogen or the host–pathogen interaction (Eastburn *et al.*, 2011).

Both climatic seasonality and climate change are relevant drivers of plant disease epidemics and expected to alter the synchrony between crop phenology and disease or pest patterns. This climatic patterns change also do spatial distribution of agro-ecological zones, habitats, distribution patterns of plant diseases and pests which can have significant impacts on food production. A side from climate change, global change also influences change in agriculture includes population growth, labor, migration, energy, policies and trends, as well as shifts in diets (Savary *et al.*, 2011). Climate change (or global change) or shifts in seasonality changes in disease situations already experienced as some minor diseases have now become major diseases in the subcontinent. Current shift in disease scenario in India especially in rice and wheat is a case in point. In rice, bacterial leaf blight (*Xanthomonas oryzae* pv. *oryzae*) has now been a global biotic threat despite of constant efforts to improve resistance through exploitation of host *R*-gene. Sheath blight (*Rhizoctonia solani*), tungro viruses and false smut (*Ustilaginoide virens*) that were of minor importance, has emerged as major problems in most of the rice growing areas. Spot blotch (*Bipolaris sorokiniana*), once unknown or of minor importance, has become a serious problem in wheat (Sinha *et al.*, 2011). Increasing trend in temperature during winter probably makes favourable situation for spot blotch in north western India (Duveiller *et al.*, 2009). Extreme events like cold waves due to climate change are already experienced fact. Prevalence of cold wave conditioning during 2011-12 throughout Northern Western India during peak vegetative stage of wheat growth has become a concern from yellow or stripe rust epidemic as cold weather favours disease occurrence. Yellow rust occurs almost every year but in low magnitude and that too crop is already matured and thus it was not a disease of economic importance. However, due to recurring cold wave yellow rust favourable conditions are now concurring during critical vegetative stages of the crop and thus it has become a threat to wheat production in the country (Sinha *et al.*, 2012). Dry root rot (*Rhizoctonia bataticola*) in chickpea is becoming more severe in rainfed environments due to moisture stress and higher temperatures (Pande *et al.*, 2010). Excess moisture on the other hand favouring some of the dreaded soil-borne diseases caused by *Phytophhora, Pyhthium, Rhizoctonia solani* and *Sclerotium rolfsii* especially in pulses (Sharma *et al.*, 2010). Even with the minor deviations from the normal weather, the efficiency of applied inputs and food production is seriously impaired (Rotter *et al.*, 1999).

According to (Tilman *et al.*, 2002), by 2050, global grain demand will double as a result of population growth. There will be limited arable land available, particularly in areas most affected by population growth, because different sectors are competing for land use. Thus, increasing crop productivity is a major challenge. However, neither planting more productive crops nor cultivating more land area where possible can solely meet the need for more food if yield losses from pests and diseases remain at current levels (Oerke and Dehne, 2004) or even further increase. Therefore, improvements in plant disease and pest control are needed in order to reduce both pre- and postharvest yield losses (Magan *et al.*, 2011). Solutions to pest and disease problems must be specifically tailored to location, crop and the type of pest. They must also consider existing standards for reduced human and or agro- ecosystem exposure to agricultural contaminants such as heavy metals, mycotoxins, plant protection products (PPPs) and harmful microorganisms (Boxall *et al.*, 2009). Hence, the challenge will be to find a balance between public demand for a reduction in PPP use and the need to efficiently control harmful diseases potentially driven by changes in climate (Hannukkala *et al.*, 2007). In view of global population growth, climate change is regarded as a threat to global food production and security. Higher global temperatures, altered precipitation regimes and increases in the frequency of extreme events, particularly in regions expected to suffer greater heat and water stress as a result of projected climate change (Rosenzweig *et al.*, 2001; Ortiz *et al.*, 2008) raise concerns among the scientific community, the public and policy makers.

Systematic quantitative analysis of these effects will be necessary in developing future disease management plans, such as disease resistant breeding, altered planting dates schedules, biological and chemical control, integrated disease management (IDM) and increased monitoring of new disease threats. Plant disease control through host-plant resistance and fungicide treatment makes a major contribution to both climate change mitigation and sustainable crop production systems to ensure global food security (Legreve and Duveiller, 2010). This chapter reveals that certain existing preventive plant protection measures, such as use of a diversity of crop species in a cropping system, adjustment of sowing or planting dates, use of crop cultivars with superior resistance and/or tolerance to disease and abiotic stress, use of effective climate smart bio-pesticides, judicious use of chemical compounds, use of reliable tools to forecast disease epidemics, application of IDM strategies and effective quarantine systems may become particularly important in the future. Effective crop protection technologies are available and will provide appropriate tools to adapt the altered climatic conditions, although the complexity of future risks for plant disease management may be considerable, particularly if new crops are introduced in an new area.

Over time, a large set of powerful tools has evolved to adapt crop production and crop protection to the local climatic situation and changing economic constraints. There have been constant adjustments to conditions differing from region to region, year to year and within season. Ongoing, yet accelerating progress in agricultural technologies, mainly new cultivars, novel plant protection products and agro-technical innovations, has constantly required significant adaptations of the production system by farmers, with increased yield and higher revenues being the main drivers. However, strategies for crop management and protection in terms of climate change scenario are yet to evolve. Based on the response of biological complexity in the effects of climate change on agriculture and crop diseases new practices are to be evolved.

Disease Management Principles under Climate Change Scenario

For yield maximization, plant health management practices are important to avoid ravages particularly from diseases and management must be in principles of 'Prevention is always better than cure'. Efforts must be on prevention of disease onset rather than control the disease after infestation for that plant health monitoring especially for disease and pest is inevitable under changing scenario. Disease monitoring for timely control measures, use of resistant or tolerant genotypes, application of biocontrol agents to reduce soil borne inoculums and need based application of cultural practices become important. Integrated disease management (IDM) approach is essential to minimize economic, health and environmental risks. In this approach, promotion of ecological balance, environmental safety and sustenance of productivity are given due attention. It calls for minimal use of pesticide and only deemed necessary, giving preference to other control methods such as host-plant resistance, cultural practices and biological control. Biological control, agronomic procedure and promotion of cultivation of correct genotype are its main elements. Disease surveillance, quarantine regulatory measures, exclusion and judicious use of chemotherapeutants are all important in harmonizing effective disease management. Crop health surveillance is an important component of IDM system which involves monitoring of disease prevalence and severity, auxiliary crop information and other relevant data for making decision to adopt a particular management strategy. Surveillance primarily focuses on the decision making for proper timing of fungicide usages so as to reduce it to bare minimum. Disease-forecasting models are needed for more pathogens and with further improved quality to be used to guide farmers. Such prediction tools may allow farmers to respond in a timely and efficient manner with PPPs applications. PPPs are used following thorough monitoring and based on established economic damage thresholds. Integrated crop management (ICM) adds to IDM components such as soil tillage, nutrient supply and water management in order to develop sustainable production systems (Juroszek *et al.*, 2008).

Disease Monitoring

Disease forecasting system helps growers in taking decisions regarding disease management with greater precision. Forecasting systems for several diseases are now available to assist the farmers in decision making whether to apply fungicides. Most often forecasting systems are driven by weather factors such as temperature, leaf wetness, rainfall, and relative humidity. Weather data location-specific or regional can be accessed for analysis of disease situations.

Need Based Chemical Control Guided by Disease Forewarning Systems

Plant protection is largely depending on application of chemicals as they are protective as well as curative. Resistance genotypes as are not available against majority of the diseases. Preventive methods like cultural practices and biological agents are partially effective. Fungicides are comparatively available and may have additional beneficial side effect on the physiology of the plant, resulting in enhanced general stress resistance and yield (Wu and Tiedman, 2001). Use of chemicals in disease management is likely to continue to be one of the ways that higher productivity. The availability of a variety of new products, with narrow and broad specificity, offer important disease control options, however, their practical application continues to face the risk of selection resistant pathogen populations. Experience accumulated over the last few decades clearly showed that fungicidal application had a better impact when used within an IDM strategy. However, climate change is expected to affect chemical control. Changes in temperature and precipitation can alter fungicide residue dynamics in the foliage, and the degradation of products can be modified. Alterations in plant morphology or physiology, resulting from growth in a CO_2-enriched atmosphere or from different temperature and precipitation conditions, can affect the penetration, translocation and mode of action of systemic fungicides. Changes in plant growth can alter the period of higher susceptibility to pathogens which can determine a new fungicide application calendar (Coakley *et al.*, 1999; Chakraborty and Pangga, 2004; Pritchard and Amthor, 2005). Therefore, fungicide market is thought to be changed. In USA, per acre pesticide usage average cost for corn, cotton, potatoes, soybeans and wheat were found to increase as precipitation increases (Chen and McCarl, 2001). Similarly, pesticide usage average cost for corn, cotton, soybean and potatoes also increase as temperature increases, while the pesticide usage cost for wheat decreases. In addition public concern has increasingly influenced the fungicide industry in developing effective products with low mammalian toxicity and environmental impact and low residues in food, to meet international health standards and compatibility in integrated disease management programs (Knight *et al.*, 1997).

In India, 52 fungicides are registered for use on agricultural crops. Some of the recently developed fungicides have also been registered for use against different diseases and a good number of the novel action fungicides are currently under evaluation (Thind, 2007). Azoxystrobin and fenamidone have been registerd for control of grape downy mildew and potato blight. Prominent among those being tested against different diseases are mandipropamid, iprovalicarb, benthiovalicarb, fluopicolide, famoxadone, cyazofamid, pyraclostrobin and picoxystrobin. Predominantly, fungicides are applied foliarly or in seed dressing for the control of major and emerging diseases of crop plant (Table 4.1).

Use of Disease Free Materials

Many plant pathogens like bacteria, virus, fungi and nematodes are transmitted by diseased seed or other vegetative propagating parts. Therefore disease free materials should be used for sowing and propagation.

Host Plant Resistance

The use of resistant or tolerant cultivars is easy, cheap, environmentally sound and effective (Dodds and Rathjen, 2010), unless pathogens overcome the resistance. Host plant resistance is an important tool to control diseases of major food crops in developing countries, especially wheat, rice, potato, cassava, chickpea, peanut and

Table 4.1: Fungicides Presently Used for Management of Major and Emerging Disease of Food Crops

Common Name	*Commercial Formulation*	*Dosages*	*Mode of Use*	*Disease Control*
Cabendazim	Bavistin, Derosol	0.15 per cent and 2g/Kg of seed	Foliar spray, seedling dip, seed treatment, soil drench or as post harvest treatment of fungicides.	Powdery mildews, root rots, stem rots, leaf spots.
Carbendazim 12 per cent + mancozeb 63 per cent	Companion	0.15 per cent	Foliar spray, seedling dip, seed treatment, soil drench or as post harvest treatment of fungicides.	Powdery mildews, root rots, stem rots, leaf spots.
Benomyl spray	Benlate, Benomyl	0.1	Seed,soil, Pre and post harvest spray	Powdery mildew, black spot and rots.
Carboxin 37.5 per cent + TMTD 37.5 per cent WS	Vitavax power	2g/kg	Seed treatment	Wilt of chickpea and other seed borne diseases of legume.
Tridemorph	Calixin	0.05 per cent	Foliar spray	Powdery mildews.
Thiophanate	Topsin, Cercobin	0.05 per cent	Foliar spray	Powdery mildew, stem rot and root rot.
Thiophanate methyl rot and root rot.	Topsin M, Cercobin-M	0.05 per cent	Foliar spray	Powdery mildew, cercospora leaf spot, stem
Metalaxyl 35 per cent WS	Apron, Ridomil Ridoxyl	2g/kg	Seed dressing	Downy mildew
Metalaxyl 8 per cent + mancozeb 64 per cent WP	Ridomil MZ 72 WP	2g/kg	Seed dressing	Downy mildew
Propiconazole	Tilt	0.1 per cent	Foliar spray	Leaf spot and powdery mildew.
Hexaconazole	Contaf	0.1 per cent	Foliar spray	Powdery mildew, rust.
Cymoxanil	Curzate		Foliar spray	Downey mildew

Contd...

Table 4.1–*Contd...*

Common Name	*Commercial Formulation*	*Dosages*	*Mode of Use*	*Disease Control*
Captafol	Difolatan		Foliar spray, soil drench	Collar rot and anthracnose.
Captan	Captan 50w Captan 75w Esso fungicide406	0.2 per cent	Seed dressing	Scab, wilt, downy mildew, black rot, damping off, rots, spots.
Zineb	Dithane Z-78	0.2 per cent	Foliar spray	Early blight of potato, rust and spots.
Mancozeb 75 per cent WP	Indofil M-45, Maneb	0.2 per cent	Foliar spray	Rusts of pulses.
Thiram	Thiram, TMTD	2g/Kg of seed	Seed treatment	Rusts and soil borne pathogen like *Pythium, Fusarium* and *Rhizoctonia.*
Copper Oxychloride	Blitox, Fytolan	2.5 per cent	Foliar spray	It is protective fungicide, controls Potato late blight, several leaf spot and leaf blight pathogen in field and horticultural crops. It is also effective against seedling disease of horticultural crops.

cowpea. The use of resistant varieties is very much welcomed by resource poor farmers because it does not require additional cost and it is environment-friendly. Many varieties resistant to Urdbean: anthracnose, cercospora leaf spot, powdry mildew and yellow mosaic virus web blight (Basandari *et al.* (1999), mungbean yellow mosaic virus (Kaushal and Singh, 1989; Mishra and Hota, 1990; Gupta, 2003), cercospora leaf spot and bacterial leaf spot (Singh, 1995), web blight (Kaushal and Singh, 1989), Powdry mildew (Vishwa Dhar *et al.*, 2004); Mungbean: mungbean yellow mosaic virus (Yadav and Bar, 2010), root rot (Chaudhry *et al.*, 2010); lentil: powdery mildew (Tikoo *et al.*, 2005); chickpea: wilt (Dubey and Singh, 2008; Dubey and Singh, 2010), Botrytis grey mould (Haware and Nene, 1982; Pandey *et al.*, 1982; Chaube *et al.*, 1983; Singh *et al.*, 1982); Pigeonpea: *Fusarium* wilt and sterility mosaic (Mandhare *et al.*, 2005; Gwata *et al.*, 2006); common bean: anthracnose (Sharma *et al.*, 2012). Expression of plant resistance is determined by the interaction between genetic factors in the pathogen and the plant, and the environmental impact represented by cultural practices and climate. Future plant disease management strategies should include resistance breeding approaches for broad adaptation to multiple environments. Breeding for resistance against several pathogens should be combined with breeding for tolerance to abiotic stresses such as drought and heat (Legreve and Duveiller, 2010). However, when several plant diseases occur at the same time, multiple host-plant resistance may not be sufficient and needs to be supported by integration of other plant protection methods, such as crop rotation and the use of spatial and temporal crop diversity (Tilman *et al.*, 2002).

Conservation Agriculture

Conservation agriculture is an obvious strategy in the context of climate change is, because it benefits crop resilience in stressed environments (Ortiz *et al.*, 2008). Crop residue management for disease control has gained importance with the expansion of conservation agriculture. More powerful methods should be developed to impede saprotrophic colonization of crop residues by pathogens in order to decrease the carry-over of inoculums between cropping seasons (Melloy *et al.*, 2010). Under worst-case conditions, farmers may consider ploughing in order to turn the soil and bury diseased residues, although the benefits from conservation tillage related to soil structure and water retention are lost. Thus, there may be trade-offs between strategies to cope with climate change and possible penalties resulting from diseases. To run conservation agriculture successfully in the long term, cultivars are needed with better water use efficiency, improved root health and durable resistance to the economically important pathogens that emerge from crop residues or result from adoption of conservation practices (Ortiz *et al.*, 2008).

Biological Control

The most obvious and apparent environment friendly alternative to pesticides is to use naturally occurring biological approaches, to manage agriculturally important pest and diseases. Biological control may be effective either upon introduction by application or through strengthening their natural occurrence. In general, their effectiveness requires specific conducive environmental conditions. The few results obtained focus on climate change impacts on the composition and dynamics of the microbial community of the phyllosphere and the soil, which can be very important for plant health. Grüter *et al.* (2006) concluded that exposition to an environment with a 600 ppm CO_2 concentration did not quantitatively alter the soil bacterial community. However, these same authors concluded that one of the potential effects of climate change is on plant diversity which results in changes in the soil bacterial composition (types of bacteria and frequency of occurrence). Using a FACE experiment to evaluate the effects on saprophyte fungi, Rezacova *et al.* (2005) observed that *Chlonostachys rosea*, an important biological control agent of *Botrytis* and other pathogens, and *Metarrhizium anisopliae*, one of the most important entomopathogens for insect pest control, were strongly associated with the cover crop in a high CO_2 concentration environment. The authors suggest the abundance of these fungi species can indicate an increase in the soil suppressiveness to phytopathogenic fungi and other pests. Ghini and Bettiol (2008) argued that in general the climate change will be beneficial for biological control, both natural and introduced, since the awareness of the society towards environmental problems will demand measures that minimize pollutant emissions. Therefore, the biological equilibrium of agricultural systems will be benefited, leading to an increase in the complexity of the system, and consequently, to biological control. To achieve this, specialists from different agriculture-related areas need to go beyond disciplinary boundaries and position the global climate change impacts in a broader context, including the whole agro-ecosystem.

Fungal Biocontrol Agents

Fungal antagonists of phytopathogenic fungi have been used to control plant disease and 90 per cent of such applications have been carried out with different strains of the fungus. They were found to be effective against soil and air borne plant pathogens (Monte, 2001). Formulations (Table 4.2) of *Trichoderma* as biological control agents are now commercially available (Hjeljord and Tronsmo, 1998).

Bacterial Biocontrol Agents

A number of bacterial species have so far been tried as biocontrol agents; these are *Agrobacterium, Actinoplanes, Alcaligens, Amorphosprangium, Arthobacter, Azotobcater, Bacillus, Cellulomons, Enterobacter, Erwinia, Pseudomonas, Streptomyces* and *Xanthomonas* (Weller, 1988). The use of bacteria like *Pseudomonas* sp., *Bacillus* sp., have been investigated because their properties to reduce antifungal metabolites and protect plants from fungal infection (Moita *et al.*, 2005; Siddiqui *et al.*, 2005; Nourozian *et al.*, 2006). The materials based on microorganisms have following properties: high specificity against target plant pathogens; easy degradability; and low mass production cost. A number of commercial preparations of bacterial biocontrol agents (Table 4.2) are available in market (Weller, 1988; Becker and Schwinn, 1993; Rodger, 1993; Ravindran and Vidhyasekaran, 1996).

Cultural Practices

Cultural measures are mainly preventive which involve agricultural cropping, harvesting and storage, tillage, crop rotation, soil management, growing of resistance varieties planning of land use and other related practices. By proper adjustment of

Table 4.1: Commercial Formulation of Fungal and Bacterial Biocontrol Agents

Commercial Product	*Source*	*Target Pathogen/Disease*	*Crops*	*Mode of Application*
Root Shield	*Trichoderma harzianum* Rifai strain-KRL AG2(T-22)	*Pythium, Rhizoctonia, Fusarium*	Forest trees, Ornamental and food crops	Spray and cut stumps
BINAB T	*Trichoderma harziumum/ Trichoderma polysporium*	Wood decay fungi	Trees	Spray, wound
Promote	*Trichoderma harzianum* and *Trichoderma viride*	*Pythium, Rhizoctonia Fusarium* (Transplanted trees)	Tree, shrub, ornamental	Soil application
Trichodex	*Trichoderma harzianum'*	*Plasmopara, Colletotrichum, Monilin,* (Various plant)	Various	Soil application, spray
Soil gard (Glio gard)	*Gliocladium* (*Trichoderma virens* GL-21)	*Rhizoctonia solani, Pythium* (Ornamental and food crops)	Ornamental and food crops	Sulurry, seed treatment
Monitor SD	*Trichoderma* spp.	Soil borne plant pathogens	Food crops	Seed treatment
Monitor WP	*Trichoderma* spp.	Soil borne plant pathogens	Food crops	Spray
Trichoderma 2000	*Trichoderma* spp.	*Rhizoctonia solani, Sclerotium rolfsii, Pythium* spp., *Fusarium* spp.	Ornamental and food crops	Seed treatment, spray
Kodiak	*B. subtilis* GB03	*Rhizoctonia solani, Fusarium Alternaria*	Cotton and legumes	Slurry to seed
Companion	*Bacillus subtilis* strain GB03	*Pythium, Phytophthora Fusarium Rhizoctonia*	Many in green house and nursery	Drench at planting time
Dagger G.	*P. fluorescence*	*Rhizoctonia, Pythium*	Field crops vegetables	Seed treatment
Blight ban A506	*P. fluorescence* A506	Frost damage, *Erwinia amylovora*	Stone fruit potatoes, tomatoes	Spray
Histick N/T	*B. subtilis* Str. MB 1600	*Fusarium, Rhizoctonia* and *Aspergillus*	Legumes	Slurry to seed
Galltrol	*Agrobacterium radiobacter* strain 84	*A. tumefaciens* (Crown gall)	Fruit and ornamental	Slurry to seeds,seedling, drench

cultural practices, crop diseases can be avoided or at least adverse effect be minimized. However, successful use of cultural practices for disease control can be made only when a complete knowledge of the nature of pathogen and its behaviour in different conditions of the environment is known. Suitable adjustment in cultural practices can modify the environment in such a manner that it becomes unfavorable for the pathogen and disease development (Khoury and Makkouk, 2010). The impact of such cultural practices on diseases is known to be more influential within a shorter time than the long –term climate changes, which render them efficient tools to counterbalance the potential increases in disease risks. Cultural methods not only serve in promoting the healthy growth of the crop, but are also effective in directly reducing inoculums potential (pruning, rouging, crop rotation, ploughing, etc.) and in enhancing the biological activities of antagonists in the soil (crop rotation and mulching). Agronomic and cultural management of BGM has been demonstrated in India, Bangladesh and Nepal (Pande *et al.*, 2002). All cultural practices are not possible to be adapted in climatic change. Therefore, some effective cultural practices which are more adaptable and effective given below:

Proper Selection of Geographical Area

Selection of optimal site for crop species and introduction of new crop species would benefit from climate change and elevated level of atmospheric CO_2 (Wolf *et al.*, 2008). Areas where pathogen is absent or geographical area in which pathogen affected adversely may be selected. Many fungal and bacterial diseases are more severe in wet areas such as pythium rot, root rot, web blight, bacterial rot than dry areas. Crops which are susceptible to these diseases, if grown in wet areas, are likely to be affected by plant pathogens. *Rhizoctonia solani*, the pathogen responsible for the web blight of urdbean is more severe and reduced huge yield loss in wet areas in region where rain occurs for long durations during the pod filling stage. So select an area with less rainfall during flowering for the cultivation of Pearl millet.

Selection of Field

The selection of suitable area or field particularly for soil -borne pathogens, if the same field is chosen for cultivating a specific crop, there is every likelihood of the disease appearing in severe form due to the buildup of the inoculums potential of the pathogen. Hence, it is advisable to not to grow the crop in the same field, where in previous year, it was infected with a pathogen capable of surviving in the soil. The drainage conditions of the field are also important. Low lying, waterlogged fields favor such diseases such as downy mildew of pearl millet.

Crop Rotation

Crop rotation is one of the recommended agricultural managements for increasing crop production as well as disease control, especially those caused by soil inhibiting plants pathogens (Hwang *et al.*, 2009). Crop rotation helps to control many soil borne diseases like mosaics, root rots, nematodes and wilts. Enhanced diversity in cropping system may reduced the risk of diseases that otherwise, in monoculture, would become more severe as a result of climate change. Therefore, crop rotation, including the integration of cover crops and intercropping must be practiced in reducing specific disease risks associated with expected climate change. Many residue-borne plant diseases caused by necrotrophic pathogens can be managed through crop rotation and other agronomic practices designed to reduce inoculums levels (Melloy *et al.*, 2010).

Mixed Cropping and Intercropping

Mixed cropping and inter cropping might be another way to reduce disease risks, for example, by slowing down epidemic rates, thereby facilitating disease control with fungicide, such as shown for late blight of potato (Garret *et al.*, 2001). Particularly under high disease pressure, these methods were not sufficiently effective and had to be combined with other integrated disease management (IDM) measures, such as genetic resistance and fungicide application (Wolf, 1985; Garrett *et al.*, 2001). Mixed cropping and inter cropping, in field experiments frequently have only small effects on disease severity. However, their cumulative effects over many growing season could be important (Garrett *et al.*, 2001), such as maintaining durability of resistance genes. More research is needed to evaluate the usefulness of cultivar mixed and inter cropping to cope with potential disease risks associated with changing climate. Mixed cropping reduces the economic losses from diseases. It also results in the efficient use of nutrients, soil moisture and solar radiation. Devi and Chhatery (2013) showed that pigeonpea intercropped with sorghum is effective in suppressing *Fusarium udum* population and reduction of wilt disease. It had been also seen that lowest wilt incidence in chickpea obtained with intercropping and mixed cropping with linseed. The reduction in disease incidence in a mixed crop can be attributed to following causes.

- ☆ In mixed cropping, the number of host plants is reduced, with the result that the disease cannot spread due to mixture of plants.
- ☆ Roots of resistant crops acts as barrier to pathogens.
- ☆ Release of some toxic substance from secondary crops suppress the growth of pathogens. For example Sorghum plant roots exudate Hydrogen cyanide (HCN) which is toxic to *Fusarium udam* that causes arhar wilt.
- ☆ Due to less number of hosts, the air-borne and foliar pathogens are reduced.
- ☆ By proper selection of crops for the mixed cropping, soil environment can be made unfavourable for the pathogen.

Time of Sowing and Harvesting

Manipulation of planting date is prime importance as a pest management tactic, to render the crop less vulnerable to the pest. Pathogens are able to infect susceptible plants by only under certain environmental conditions. A slight delay in sowing is helpful in this case. Changing planting and D or harvest dates of annual crops and short-lived perennial can be an effective, low cost option to render the crop less vulnerable to pests and diseases or adverse abiotic conditions (Srivastava *et al.*, 2010). However, the practicability of changing planting and harvesting dates is dependent on region, cultivar, magnitude of climate change, and market situation (Wolfe *et al.*, 2008). Under certain circumstances delayed planting as avoidance strategy against pathogens may become less reliable under climate change (Garrett *et al.*, 2006), for example, if in temperate climates mild conditions prevail in autumn and early winter. Srivastava *et al.* (2010) concluded that more low-cost adaptation strategies, such as changing sowing date and cultivar, should be explored to reduce the vulnerability of crop production to climate change, although these can have trade-offs. Delayed harvesting of crops in temperate climate is more prone to contamination by pathogens. Haware and McDonald (1992) reported that delayed sowings reduced blue green mould (BGM) incidence even in susceptible cultivars, but significantly reduced the grain yields. Singh (1997) also observed that the late sown crop (around 20 Nov.) showed significantly low incidence of BGM.

Mulching and Soil Amendments

Mulch or crop residues can also influence disease incidence by altering the immediate environment. Water retention, nutrient enrichment of the soil, decrease in soil temperature, weed inhibition and seedling protection are all effects of mulching that can influence disease development. Spread of soil-borne diseases that rely on splash-dispersal can be reduced, but other diseases, for which the altered environment is favourable, can be enhanced. The addition of organic amendments to the soil can also reduce disease influence by increasing the activity of competing or predatory micro-organisms in the soil, although it can also increase the incidence or severity of diseases caused by pathogens that thrive on the amendment. Mayur *et al.* (2003) reported that chickpea wilt incidence was significantly reduced by amending the soil with deoiled mustard cake, groundnut cake and farm yard manure.

Soil Solarization

A clear plastic sheet preferably white spread over the soil trap solar heat, can reduce soilborne diseases, insect, nematode, and many weed seeds. Treatment should occur during summer's high air temperature and intense solar radiation. Keep the soil damp during the solarization process, and keep the plastic for several weeks. The longer treatment period, the higher pathogen-killing rates and the deeper effectiveness of soil solarization (Garibaldi, 1987). Solarization controls important soilborne pathogens such as *Verticillium dahliae, Fusarium* spp., *Phytophtora cinnamomi, Sclerotium rolfsii, Rhizoctonia solani, Agrobacterium tumefaciens,* and *Streptomyces scabies*; nematodes such as several root-knot nematode species (*Meloidogyne* spp.) and lesion nematodes (*Pratylenchus* spp.), and numerous weeds *e.g.*, bermudagrass, barnyardgrass, nightshades, and some pigweed species (Katan, 1981; Strand, 2000). Climate warming may facilitate the use of soil solarization, both in the greenhouse and open field, because it could be used successfully in more pathogen–plant systems and more regions, with the heat penetrating to deeper soil layers and the mulching periods (usually 4 weeks or longer) becoming shorter. However, in regions where global warming is associated with reduced rainfall, effectiveness of soil solarization might be lower, as dry soils provide reduced efficacy, unless irrigation will be applied. Thus, global warming may improve the efficacy and handling of certain disease control methods, particularly those requiring high temperature, such as soil solarization. The effectiveness of soil solarization in controlling wilt disease in pigeon pea over control was due to increase in soil temperature (45°C) below the polythene sheet. This increase in soil temperature with solarization has been considered as the major driving force for the various biological and physiological changes in the soil that affect plant growth. The major legumes, such as bean, pea and lentil also suffer from heavy yield losses due to *Fusarium* diseases including root rot. Soil solarization greatly reduced the population densities of *Fusarium* species as the major soil borne fungal pathogens.

Quarantine Regulations

Quarantine systems are the basis for trans-boundary pest and disease prevention through control of the sources of pest introductions and incursion. Quarantine can be defined as a legal restriction on the movement of agriculture commodities for the purpose of exclusion, prevention or delay in the spread of plants pests and disease in uninfected areas. Pest and disease surveys as well as pest risk analysis (PRA) studies are critical for preparing and updating quarantine pest lists. Some pathogens may be more harmful in a new environment (Garrett *et al.*, 2010). There will be new opportunities for crops and cultivars to be introduced in regions and locations where they have not been grown before, but effective system, such as diagnostic tools, must be in place to detect and follow invading pathogens and monitor their behavior under such altered conditions (Boland *et al.*, 2004). Use of climate matching tools and geographical information systems may assist quarantine agencies in determining the threat posed by a given pathogen in different regions under current and future climate shifts (Coakley *et al.*, 1999; Legreve and Duveiller, 2010). In some cases, emerging plant pathogens can be excluded from a country by regulatory actions at the nation's borders. Use of Geographical Information System (GIS) and climate matching tools may assist quarantine agencies in determining the threat posed by a given pathogen under current and future climates. This approach was used by Sansford and Baker (1998) to assess the risk of establishment of Karnal bunt in the cereal-growing region of the European Union. In India, domestic quarantine measures exist for three diseases, wart (*Synchytrium endobioticum*) of potato, bunchy top (virus) and mosaic of banana with a view to preventing the spread of these diseases.

Limitations of Present Available Disease Management System

Delayed/Adjusting Planting Dates Less Effective

Under certain circumstances, delayed planting as an avoidance strategy against pathogen may become less reliable under climate change (Garrett *et al.*, 2006) for example, if in temperate climate mild conditions prevail in autumn and early winter.

Increased Vulnerability of Biocontrol Agents

Garrett *et al.* (2006) assumed that vulnerability of BCAs will be higher under climate change, because if climate variability becomes greater it would impose difficulties on the survival and activity of applied antagonists. If appropriate temperature and moisture are not consistently available, BCA populations may fail to reduce disease incidence and severity, and may not recover as rapidly as pathogen populations when conducive conditions recur.

Reduced Efficacy of Chemical Control

Changes in temperature, rainfall, wind speed, soil D air moisture and light conditions can influence the effectiveness of PPP applications (Bedos *et al.*, 2002; Ziska and Runion, 2007). These environmental factors may alter fungicide dynamics in the soil (Monkiedje *et al.*, 2007) and in the foliage, including uptake, degradation and volatilization (Bedos *et al.*, 2002). Precipitation during the post-application period is particularly critical. It may either improve fungicide distribution or deplete the fungicide layer on the foliage, altering the fungicide's efficacy (Chakraborty *et al.*, 2000). According to Wolfe *et al.* (2008), more frequent rainfall events projected for the northern latitudes during winter and spring may trigger more frequent fungicide applications, because of the difficulty of keeping contact fungicides on the plant canopy. Uptake, translocation and mode of action of systemic fungicides could be negatively affected by plant morphological responses, such as reduced stomatal density and smaller stomatal openings or thicker epicuticular waxes on the leaves, in response to elevated CO_2 increased air temperature. This may reduce or slow down the uptake rates of systemic fungicides, although increased plant metabolic rates at warmer temperatures could increase fungicide uptake (Coakley *et al.*, 1999).

Development of Resistance to Fungicide

Frequent application and higher doses of fungicides are more experienced under climate change situation and make the conditions favourable for development of new strain of pathogen. After continuous and widespread use of these chemicals, several plant pathogens have developed strains that are resistant to certain fungicides. Several of the important fungal pathogens, *e.g.*, *Alternaria, Botrytis, Cercospora, Colletotrichum, Fusarium, Verticillium, Sphaerotheca, Phytophthora, Pythium,* and *Ustilago*, are known to have produced strains resistant to one or more of the systemic fungicides.

IDM is Location and Crop Specific

Future Outlook

Over time, a large set of powerful tools has evolved to adapt crop production and crop protection to the local climatic situation and changing economic constraints. There have been constant adjustments to conditions differing from region to region, year to year and within season. Ongoing, yet accelerating progress in agricultural technologies, mainly new cultivars, novel plant protection products and agro-technical innovations, has constantly required significant adaptations of the production system by farmers, with increased yield and higher revenues being the main drivers. However, strategies for crop management and protection in terms of climate change scenario are yet to evolve. Based on the response of biological complexity in the effects of climate change on agriculture and crop diseases new practices are to be evolved.

Since, regional impact of climate on plant disease will be more, disease management strategies will require adjustment. It is probable that climate change will significantly influence plant disease management (Ghini *et al.*, 2008; Juroszek and Tiedemann, 2011), degradation and uptake of PPPs (Bloomfield *et al.*, 2006), average external costs of PPPs use (Koleva and Schneider, 2009), and their environmental distribution and toxicity (Miraglia *et al.*, 2009; Noyes *et al.*, 2009). Significant contributions can derive from better field monitoring of diseases and pests, improved timing of PPPs application and better systems for delivering PPPs to their targets (Strand, 2000). This includes in situ and ex situ conservation of genetic diversity in crop species, including their wild relatives, in order to increase the availability of genes to breed for resistance to existing and new pathogens and to abiotic stresses (Garrett, 2008). The requirement to breed new cultivars with more durable resistances within shorter time periods will possibly give biotechnological approaches a key role in crop adaptation to changing climate. Accumulation of minor genes and race-non-specific resistance to pathogens by conventional breeding programmes will remain equally important to achieve durable resistance in diverse environments. Preventive plant protection measures may become particularly important under climate change. These may include greater heterogeneity in cropping systems (Christen, 2008) that reduce disease risks, use of superior cultivars resistant and D or tolerant to abiotic and biotic stress (Ordon, 2008), and reliable tools for forecasting pathogen occurrence in order to respond in a timely manner to plant pathogens (Tiedemann and Ulber, 2008). Disease-forecasting models based on weather data can help to identify the meteorological factors (and the time period) which are significantly correlated with disease (Coakley *et al.*, 1988). These types of disease-forecasting models can be combined with general circulation models in order to simulate future scenarios of disease epidemics, although most general circulation models operate on larger scales of resolution. Down-scaling climate models can contribute to bridge this gap (Soussana *et al.*, 2010). Based on future scenarios of disease epidemics, disease management practices can be suggested and D or improved (Salinari *et al.*, 2006).

From a food-security perspective, emphasis must now shift from impact assessment to developing adaptation and mitigation strategies and options. Two broad areas of empirical investigation will be essential; firstly, to evaluate under climate change the efficacy of current cultural, biological and chemical control tactics including disease resistant secondly, to include future climate scenarios in all research aimed at developing new tools and tactics (Chakraborty and Newton, 2011).Transgenic solutions must receive serious consideration in integrated disease management strategies food security. Breeding or selection plant types as long term strategy and available practices like use of resistance sources as well as varieties that are less vulnerable for diseases, possible practices of reducing inoculums, use of biocontrol agents, escaping diseases and fungicides as short term strategy are the possible ways to combat losses from diseases under climate change scenario.

References

Backer, J.O. and Schwinn, F.J. 1993. Control of soil-borne pathogens with living bacteria and fungi. Status and Outlook. *Pesticide Sciences*, 37: 355-363.

Basandrai, A. K., Gartaw, S. L., Basandrap, D. and Kalin, V. 1999. Blackgram (*Phaseolus mungo*) germplasms evaluation against different diseases. *Indian Journal of Agricultural Sciences* **69** (7): 506-508.

Bedos, C., Cellier, P., Calvet, R., Barriuso, E. and Gabrielle, B. 2002. Mass transfer of pesticide into the atmosphere by volatilization from soils and plants: overview. *Agronomie*, 22: 21–33.

Bloomfield, J.P., Williams, R.J., Gooddy, D.C., Cape, J.N. and Guha, P. 2006. Impacts of climate change on the fate and behaviour of pesticides in surface and groundwater – a UK perspective. *Science of the Total Environment*, 369: 163–77.

Boland, G.J., Melzer, M.S., Hopkin, A., Higgins, V. and Nassuth, A. 2004. Climate change and plant diseases in Ontario *Canadian Journal of Plant Pathology*, 26: 335–50.

Boxall, A.B.A., Hardy, A. and Beulke, S. 2009. Impacts of climate change on indirect human exposure to pathogens and chemicals from agriculture. Environmental Health Perspectives 117, Human Health 2, 90–104.

Chakraborty, S. and Newton, A.C. 2011. Climate change, plant diseases and food security: an overview. *Plant Pathology*, 60: 2-14.

Chakraborty, S. and Pangga, I.B. 2004. Plant disease and climate change.In: Gillings M, Holmes A, eds. *Plant Microbiology*. London, UK: BIOS Scientific, 163–80.

Chakraborty, S., Tiedemann, A.V. and Teng, P.S. 2000. Climate change: potential impact on plant diseases. *Environmental Pollution*, 08: 317–26.

Chaubey, H. S., Beniwal, S. P. S., Tripathi, H. S. and Nene, Y. L. 1983. Field screening of chickpea for resistance to Botrytis gray mold. *InternationalChickpeaNewsletter*.8, 20-21.

Chen, C.C. and McCarl, B.A. 2001. An investigation of the relationship between pesticide usage and climate change. *Climatic Change*, 50: 475–87.

Choudhary, S., Pareek, S. and Saxena, J. 2010. Management of dry root rot of greengram (*Vigna radiata*) caused by *Macrophomina phaseolina*. *Indian Journal of Agricultural Sciences*, 80 (11): 988–92, November 2010.

Christen, O. 2008. Langfristige Trends and Anpassung der Anbausysteme an den Klimawandel. In: Tiedemann AV, Heitefuss R, Feldmann F, eds. Pflanzenproduktion imWandel – Wandel im Pflanzenschutz. Braunschweig, Germany: DPG Selbstverlag, 57–64.

Coakley, S.M. 1988. Variation in climate and prediction of disease in plants. *Annual Review of Phytopathology*, 26: 163–81.

Coakley, S.M., McDaniel, L.R. and Line, R.F. 1988. Quantifying how climatic factors affect variation in plant disease severity: a general method using a new way to analyze meteorological data.*Climatic Change*, 12: 157–75.

Coakley, S.M., Scherm, H. and Chakraborty, S. 1999. Climate change and plant disease management. *Annual Review of Phytopathology*, 37: 399–426.

Cook, R.J. and Baker, K.F. 1983. The nature and practice of biological control of plant pathogens. St. Paul: American Phytophathological Society, p. 539.

Devi, T. R. and Chhetry, G. K. N. 2013. Effect of certain agronomic practices for the management of wilt of pigeonpea in Manipur agro-climatic conditions. *IOSR Journal of Engineering*. 3(4): 21-23.

Dodds, P.N. and Rathjen, J.P. 2010. Plant immunity: towards an integrated view of plant–pathogen interactions. *Nature Reviews Genetics*, 11: 539–48.

Dubey, S.C. and Singh, S.R. 2008. Virulence analysis and oligonucleotide fingerprinting to detect diversity among Indian isolates of *Fusarium oxysporum* f. sp. *ciceris* causing chickpea wilt. *Mycopathologia* 165: 389-406.

Dubey, S.C., Singh, S.R. and Singh, B. 2010. Morphological and pathogenic variability of Indian isolates of *Fusarium oxysporum* f. sp. *ciceris* causing chickpea wilt. *Archives of Phytopathology and Plant Protection* 43(2): 174-190.

Duveiller, E. and Sharma, R.C. 2009. Genetic improvement and crop management strategies to minimize yield losses in warm non-traditional wheat growing areas due to spot blotch pathogen *Cochliobolus sativus*. *J. Phytopathol*. 157: 521-534.

Eastburn, D.M., McElrone, A.J. and Bilgin, D.D. 2011. Influence of atmospheric and climatic change on plant-pathogen interactions. *Plant Pathology*, 60, 54–69.

Garibaldi, A. 1987. Soil solarization: possibilities and limitations in Mediterranean countries. In: Cavalloro R, ed. Integrated and Biological Control in Protected Crops. Proceedings of a meeting of the EC Experts' Group, Heraklion, 24–26 April 1985.Rotterdam, the Netherlands: Balkema, 24–26.

Garrett, K.A. 2008. Climate change and plant disease risk. In: Relman DA, HamburgMA, Choffnes ER,Mack A, eds. Global Climate Change and ExtremeWeather Events: Understanding the Contributions to Infectious Disease Emergence.Washington, USA: National Academies Press, 143–55.

Garrett, K.A. Jumpponen, A. and Gomez Montano, L. 2010. Emerging plant diseases: .what are our best strategies for management?. In: Kleinman DL, Delborne J, Cloud-Hansen KA, Handelsman J,eds. Controversies in Science and Technology Vol. 3: From Evolution to Energy. New Rochelle, NY, USA: Mary Ann Liebert, Inc, 152–60.

Garrett, K.A., Dendy, S.P., Frank, E.E., Rouse, M.N. and Travers, S.E. 2006. Climate.change effects on plant disease: genomes to ecosystems. *Annual Review of Phytopathology*, 44: 489–509.

Garrett, K.A., Nelson, R.J., Mundt, C.C., Chacon, G., Jaramillo, R.E. and Forbes, G.A. 2001. The effects of host diversity and other management components on epidemics of potato late blight in the humid highland tropics. *Phytopathology*, 91: 993–1000.

Ghini, R., Hamada, E. and Bettiol, W. 2008. Climate change and plant diseases. *Scientia Agricola*, 65: 98–107.

Gupta, O.M. 2003. Resistance to *Mungbean yellow mosaic virus*, phenotypic characters and yield components in Urdbean. *Indian Pnytopeth*. 56 (1): 110-111.

Gwata, E. T., Silim, S. N. and Mgonja, M. 2006. Impact of a new source of resistance to fusarium wilt in pigeonpea. *J. Phytopath.*, **154:** 62.

Hannukkala, A.O., Kaukoranta, T., Lehtinen, A. and Rahkonen, A. 2007. Late-blight epidemics on potato in Finland, 1933–2002; increased and earlier occurrence of epidemics associated with climate change and lack of rotation. *Plant Pathology,* 56: 167–76.

Haware, M.P. and McDonald, D. 1992. Integrated management of botrytis gray mould of chickpea. p. 3-6. In: *Botrytis gray mould of chickpea, (Eds. Haware, M.P., Faris, D.G. and Gowda, C.L.L.).* ICRISAT, Patancheru, A.P., India).

Haware, M.P. and Nene, Y. L. 1982. Screening chickpea for resistance to botrytis grey mould. *International Chickpea Newsletter.* 6: 17-18.

Hjeljord, L. and Tronsmo, A. 1998. *Trichoderma* and *Gliocladium* in Biocontrol: an review, In: C.P. Kubicek and G.E. Harman (ed.) *Trichoderma and Gliocladium.* Taylor and Franas Ltd. London, U.K. p. 135-151.

Hwang, S. F., Ahmed, H. U., Gossen, B. D., Kutcher, H. R., Brandt, S. A., Strelkov, S. E., Chang, K. F., Turnbull, G. D. 2009. Effect of crop rotation on the soil pathogen population dynamics and canola seedling establishment. *Plant Pathology Journal.* **8**: 106-112.

IPCC, 2007. Summary for Policymakers. Pages 7-22 in: Climate Change 2007: Impacts, Adaptation, and Vulnerability. Contribution of Working Group II to the Fourth Assessment Report of the Intergovernmental Panel on Climate Change. M. L. Parry, O. F. Canziani, J. P. Palutikof, P. J. van der Linden, and C. E. Hanson, eds. Cambridge University Press, Cambridge, UK.

Juroszek, P., Lumpkin, T.A. and Palada, M.C. 2008. Sustainable vegetable production systems. *Acta Horticulturae,* 767: 133–49.

Katan, J. 1981. Solar heating (solarization) control of soilborne pests. *Annual Review of Phytopathology,* 19: 211–236.

Kaushal, R.P. and Singh, B.M. 1989. Evaluation of blackgram (*Phaseolu3 mungo*) gennplasm for multiple disease resistance. *Indian Journal of Agricultlll'al Sciences* 59 (10): 726-727.

Khoury, W.E.I. and Makkouk, K. 2010. Integrated Plant disease management in mdeveloping countries. *Journal of Plant Pathology,* 92(S4): 35- 42.

Knight, S.C., Anthony, V.M., Brady, A.M., Greenland, A.J., Heaney, S.P., Murray, D.C., Powell, K.A., Schulz, M.A., Spinks, C.A., Worthington, P.A. and Youle, D. 1998. Rationale and prospective on the development of fungicides. *Annual Review Phytopathology,* 35: 349-372.

Koleva, N.G. and Schneider, U.A. 2009. The impact of climate change on the external cost of pesticide applications in the US agriculture. *International Journal of Agricultural ustainability,* 7: 203–16.

Legreve, A. and Duveiller, E. 2010. Preventing potential diseases and pest epidemics under a changing climate. In: Reynolds MP, ed. Climate Change and Crop Production. Wallingford, UK: CABI Publishing. p. 50–70.

Magan, N., Medina, A. and Aldred, 2011. Possible climate-change effects on mycotoxins contamination of food crops pre-and post harvest. *Plant Pathology,* 60: 150-163.

Mandhare, V. K., Suryawanshi, A. V., Tagad, L. N. and Jamadgani, B. M. 2005. Field reaction of pigeonpea cultivars to fusarium wilt and sterility mosaic disease in Maharastra, India. *JNKVV Research Journal* **38:** 90-91.

Mayur *et al.*, 2003. *Research on Crops* 4(1): 141-143.

Melloy, P., Hollaway, G., Luck, J., Norton, R., Aitken, E. and Chakraborty, S. 2010. Production and fitness of *Fusarium pseudograminearum* inoculum at elevated carbon dioxide in FACE. *Global Change Biology,* 16: 3363–73.

Miraglia, M., Marvin, H.J.P. and Kleter, G.A. 2009. Climate change and food safety: an emerging issue with special focus on Europe. *Food and Chemical Toxicology,* 47: 1009–21.

Mishra, D. and Hota, A. K. 1990. Reaction of mung and urdbeall cultivars to yellow mosaic vims under natural conditions. *Legume Research* 13: 47-49.

Monkiedje, A., Zuehlke, S., Maniepi, S.J.N. and Spiteller, M. 2007. Elimination of racemic and enantioenriched metalaxyl based fungicides under tropical conditions in the field. *Chemosphere,* 69: 655–63.

Monte, E. 2001. Understanding *Trichoderma* between biotechnology and microbial ecology. *Int. Microbiology.* 4: 1-4.

Nourozian, J., Etebarian, H.R. and Khodakaramian, G. 2006. Biological control of *Fusarium graminearum* on wheat by antagonistic bacteria Songklanakarin J. Sci. Technol., 2006, 28 (S1): 29-38.

Noyes, P.D., McElwee, M.K. and Miller, H.D. 2009. The toxicology of climate change: environmental contaminants in a warming world. *Environment International,* 35: 971–86.

Oerke, E.C. and Dehne, H.W. 2004. Safeguarding production – pests, losses and crop protection in major crops. *Crop Protection,* 23: 275–285.

Ortiz, R., Sayre, K.D. and Govaerts, B. 2008. Climate change: can wheat beat the heat? *Agriculture, Ecosystems and Environment,* 126: 46–58.

Pande, S. Desai, S. and Sharma, M. 2010. Impacts of climate change on rainfed crop diseases: current status and future research needs. In: National Symposium on Climate Change and Rainfed Agriculture, India. Hyderabad, 18–20 February, 2010. CRIDA, 55–59.

Pandey, M. P., Beniwal, S. P. S. and Arora, P. P. 1982. Field reaction of chickpea varieties to chickpea grey mould. *International Chickpea Newsletter.* 7: 13.

Pritchard, S.G. 2011. Soil organisms and global climate change. *Plant Pathology,* 60: 82–99.

Ravindran, R. and Vidhyasekaran, P. 1996. Development of a formulation of *Pseudomonas fluorescens* PfAIR2 for management of rice sheath blight. *Crop Protection,* 15: 715-721.

Rezacova, V., Blum, H., Hrselova, H., Gamper, H., Gryndler, M. 2005. Saprobic microfungi under *Lolium perenne* and *Trifolium repens* at different fertilization intensities and elevated atmospheric CO2 concentration. *Global Change Biology,* 11: 224-230.

Rosenzweig, C., Iglesias, A., Yang, X.B., Epstein, P.R. and Chivian, E. 2001. Climate change and extreme weather events: implications for food production, plant disease, and pests. Global Change and Rosenzweig C, ParryML, 1994. Potential impact of climate change on world food supply. *Nature,* 367: 133–8.

Rotter, R. and Van de Geijn, S.C. 1999. Climate change effects on plant growth, crop yield and livestock. *Climate Change,* 43: 651-681.

Salinari, F., Giosue, S. and Tubiello, F.N. 2006. Downy mildew (*Plasmopara viticola*) epidemics on grapevine under climate change. *Global Change Biology* 12, 1299–307.

Savary, S., Nelson, A., Sparks, A.H., Willocquet, L., Duveiller, E., Mahuku, G., Forbes, G., Garrett, K.A., Hodson, D., Padgham, J., Pande, S., Sharma, M., Yuen, J. and Djurle A, 2011.International Agriculture Research Tackling the Effects of Global and Climate Changes on Plant Diseases in the Developing World. *Plant Diseases,* 94: 1204-1216.

Sharma, M., Mangala, U.N., Krishnamurthy, M., Vadez, V. and Pande, S. 2010. Drought and dry root of chickpea. In: 5th International Food Legumes Research Conference (IFLRC V) and 7th European Conference on Grain Legumes (AEP VII) April 26-30, 2010, Antalya, Turkey.

Sharma, P.N., Kajal banyal, Rana, J.C. Ruby Nag, Sharma, S.K. and Anju Pathania 2012. Screening of common bean germplasm against *Colletotrichum lindemuthianum* causing bean anthracnose. *Indian Phytopath.* 65 (1): 99-101.

Siegenthaler, U., Stocker, T.F., Monnin, E., Lüthi, D., Schwander, J., Stauffer, B., Raynaud, D., Barnola, J.M., Fischer, H., Masson-Delmotte, V. and Jouzel, J. 2005. Stable carbon cycle-climate relationship during the late Pleistocene. *Science,* 310: 1313-1317.

Singh, D. P. 1995. Breeding for resistance to diseases in pulse crops. (*in*) *Genetic Research and Education: Current Trends and the Fifty Years,* pp 339-420. Sharma B (Ed). Indian Society of Genetics and Plant Breeding, New Delhi.

Singh, G. 1997. Epidemiology of Botrytis grey mould of chickpea. p. 47-50. *In: Recent advances in research on Botrytis grey mould of chickpea (Eds. Haware, M. P., Lenne, J. M., and Gowda, C. L. L.).* ICRISAT, Patancheru, Andhra Pradesh, India.

Singh, G.; Kapoor, S. and Singh, K. 1982. Screening of chickpea for grey mould resistance. *International Chickpea Newsletter.* 7: 13-14.

Sinha, P., Jain, R.K. and Sharma, P. 2011. Plant disease monitoring for adaptation strategies under climate change scenario: In Plant Pathology in India: Vision 2030. Indian phytopathological Society, Division of Plant Pathology, IARI, New Delhi. p. 121-127.

Soussana, J.F., Graux, A.I. and Tubiello, F.N. 2010. Improving the use of modelling for projections of climate change impacts on crops and pastures. *Journal of Experimental Botany* 61, 2217–28.

Spahni, R., Chappellaz, J., Stocker, T., Loulergue, L., Hausammann, G., Kawamura, K., Fluckiger, J., Schwander, J., Raynaud, D., Masson-Delmotte, V. and Jouzel, J. 2005. Atmospheric Methane and Nitrous Oxide of the Late Pleistocene from Antarctic Ice Cores, Science, 310, 1317–1321.

Srivastava, A. Kumar, S.N.and Aggarwal, P.K. 2010. Assessment of vulnerability of sorghum to climate change in India. *Agriculture, Ecosystems and Environment* 138: 160–169.

Strand, J.F. 2000. Some agrometeorological aspects of pest and disease management for the 21st century. *Agricultural and Forest Meteorology,* 103: 73–82.

Thind, T.S. 2007. Changing cover of fungicide umbrella in crop protection. *Indian Phytopathology*. 60: 421-433.

Tiedemann, A.V. and Ulber, U. 2008. Vera¨ndertes Auftreten vonKrankheiten und Scha¨dlingen durch Klimaschwankungen. In: Tiedemann AV, Heitefuss R, Feldmann F, eds. Pflanzenproduktion imWandel – Wandel im Pflanzenschutz. Braunschweig, Germany: DPG Selbstverlag, 79–89.

Tikoo, J.L., Sharma, B., Mishra, S.K. and Dikshit, H.K. 2005. Lentil (*Lens culinaris*) in India: present status and future perspectives. *Indian Journal of Agricultural Sciences* 75: 539-562.

Tilman, D., Cassman, K.G., Matson, P.A., Naylor, R. and Polasky, S. 2002. Agricultural sustainability and intensive production practices. *Nature,* 418: 671–677.

Vishwa, D., Singh, R.A. and Gurha, S.N. 2004. Integrated disease management in pulse crops. Pages 324- 344. In: Pulses in New perspective (eds. Masood Ali., Singh B.B., Shiv Kumar and Vishwa Dhar). Indian Society of Pulses research and development, IIPR, Kanpur.

Watson, P.J. 2011. Is there evidence yet of acceleration in mean sea level rise around mainland Australia? *Journal of Coastal Research,* 27: 368–377.

Weller, J.M. 1988. Biological control of soil borne plant pathogens in the rhizosphere with bacteria. *Annual Review of Phytopathology,* 26: 379-407.

Wolfe, D.W., Ziska, L., Petzoldt, C., Seaman, A., Chase, L. and Hayhoe, K. 2008. Projected change in climate thresholds in the Northeastern U.S.: implications for crops, pests, livestock, and farmers. *Mitigation and Adaptation Strategies for Global Change,* 13: 555–575.

Wolfe, M.S. 1985. The current status and prospects of multiline cultivars and variety mixtures for disease management. *Annual Review of Phytopathology,* 23: 251–273.

Wu, Y.X. and Tiedemann, A.V. 2001. Physiological effects of azoxystrobin and epoxiconazole on senescence and the oxidative status of wheat. *Pesticide Biochemistry and Physiology,* 71: 1–10.

Yadav, M.S. and Brar, K.S. 2010. Assessment of yield losses due to Mungbean yellow mosaic India virus and evaluation of mungbean genotypes for resistance in South-West Punjab. *Indian Phytopath,* 63 (3): 318-320 (2010).

Ziska, L.H. and Runion, G.B. 2007. Future weed, pest, and disease problems for plants. In: Newton PCD, Carran RA, Edwards GR, Niklaus PA, eds. Agroecosystems in a Changing Climate. Boca Raton, FL, USA: CRC Press, 261–287.

2016, Diseases of Pulse Crops and their Sustainable Management 65–86
Editors: Samir Kumar Biswas, Santosh Kumar and Gireesh Chand
Published by: BIOTECH BOOKS, NEW DELHI

Chapter 5

Bio-fertilizers: A Tools for Enhancement of Pulse Production

S.K. Singh[1], Ashok Kumar Singh[1], J.N. Srivastava[2]* and V.K. Razdan[1]

[1]Department of Plant Pathology, Sher-e-Kashmir University of Agricultural Sciences and Technology, Jammu, J&K

[2]Department of Plant Pathology, Bihar Agricultural University, Sabour, Bhagalpur, Bihar

Introduction

Bio-fertilizers are preparations containing efficient strain of nitrogen fixing microorganisms used for seed or soil application by increasing the number of microorganisms in rhizospheric soil and consequently improve the nitrogen fixing ability. Bio-fertilizers are different from chemical fertilizers *i.e.* Bio-fertilizers on application remain in soils; they multiply and keep benefiting the growing crops. They do not get leached, volatilized or lost by denitrification as in case of chemical fertilizers.

1. They supplement fertilizer supplies for meeting the nutrients needs of crops.
2. They can add 20-200 kg/ha (by fixation) under optimum conditions and solubilize/mobilize 30-50 kg/ha.

* Corresponding Author: E-mail: j.n.srivastava1971@gmail.com

3. They liberate growth promoting substances and vitamins and help to maintain soil fertility.
4. They suppress the incidence of pathogens and control diseases.
5. They increase the crop yield by 10-50 per cent; fixers reduce depletion of soil nutrients and provide sustainability to the farming system.
6. They are cheaper, pollution free and based on renewable energy sources.
7. They improve soil physical properties, tilth and soil health in general.

Application Technology of Bio-fertilizer

Bio-fertilizers application is being popularized as safe alternative to conventional chemical fertilizers. Air has 78 per cent Nitrogen and 88,000 t/ha of N in the world. Earth can port billions of bacteria that can trap the atmospheric nitrogen from the air and put into absorvable form of the plant. *Azospirillum* forms an associative symbiosis with many plants and can be applied as seed treatment and soil application in crops like rice, sugarcane, pulses, soybean and vegetables. Legume inoculants and Arbuscular mycorrhizae (AM) are also effective in crop yield improvement. *Bacillus* sp. and *Pseudomonas* sp. synthesizes the insoluble form of phosphorus. The combined application of PSB, *Rhizobium* and PGPR greatly enhanced the biomass production, uptake of nutrients and yield.

- Bio-fertilizer has to be properly applied to the seed or in the soil. The following guidelines are important for effective use:
- To purchase the right type of bio-fertilizers of good quality from a reputed source and use it before the expiry date.
- The bio-fertilizers should be host specific and location wise adoptive.
- They act better in the soil pH 5.5-6.5. Alkaline soils are not suitable.
- They should be stored/placed in shaded, well furnished and ventilated place.
- Before preparing or mixing in the seed/slurry, it should not be dried in the sun.

On a worldwide basis it is estimated that about 175 million tons of nitrogen per year is added to soil through biological nitrogen fixation (BNF). The term bio means living; so biofertilizers refer to living, microbial inoculants that are added to the soil. These biofertilizers are products consisting of selected and beneficial microrganisims, which are known to improve plant growth through supply of plant nutrients. Biofertilizers include environment-friendly fertilizers with organisms such as: (a) Rhizobium strains for legumes, (b) Azotobacter strains for nonlegume crops, (c) Azospirillum, (d) Blue Green Algae, (e) Azolla. (f) VAM strains for use in agriculture, horticulture, and plantation crops, and (g) phosphorus solubilizing bacteria (PSB) - phosphorus dissolving bacteria strains. Most common among these are symbiotic mycorrhiza and Rhizobium, which deliver plant nutrition, disease resistance, and tolerance.

Nitrogen Fixing Microorganism

Rhizobium

Beijerinck in Holland was first to isolate and cultivate a microorganisms from nodules of legumes in 1888 and named at *Bacillus radicicola* which is now placed in Bergeys Manual of Systematic Bacteriology under the genus *Rhizobium*. Rhizobia are gram negative, symbiotic, N_2 fixing bacteria which fix nitrogen in association with roots of leguminous plants. Cells are rod shaped, motile with one polar, sub polar or 2-4 peritrichous, non-spore forming, non-acid fast and aerobic. They are 0.5-0.9 micron x 1.2-3.0 micron in size and can be cultured on Yeast mannitol agar media. An effective nodule seems larger and pinkish due to presence of red colored pigment leghaemoglobin which is found between bacteriods and membrane envelopes. Family Rhizobiaceae contains five genera *i.e. Rhizobium, Bradyrhizobium, Azorhizobium, Agrobacterium* and *Phyllobacterium*. Mostly they are found in the nodules of leguminous plants, someone as saprophytes in the soil. These bacteria convert atmospheric nitrogen into nitrogenous compounds (nitrates) which are easily available to the plants. Fixed nitrogen is not only utilized by the bacteria but also by the host plants. In return a bacterium obtains carbohydrates from the leguminous plants. They are beneficial to each other and this process is known as symbiosis. A lakhs of bacteria's are found in each associative nodule of the plant. In Nitrogen cycle, decomposition of nitrogenous compounds occurs through few nitrifying bacteria, which oxidises nitrite to nitrate as well as nitrobacter which oxidized nitrate into nitrate. Oxidation of ammonia takes a major place, which is utilized by the higher plants and that is recyclable by synthesis of organic nitrogenous compounds. Further, denitrifying bacteria oxidizes into N_2 and N_2 gas thus returning these back to the atmosphere. Hence, these bacteria increase the fertility of soil by fixing nitrogen. Nif genes are responsible (coded) for regulation and synthesis of nitrogenase enzyme. There are many possibilities to transfer nif genes in non leguminous crops like preparation of such types of mutants that can form nodules in non-leguminous crops. There is also a great possibility to transfer lectin and leghaemoglobin of leguminous crops. By an estimate, annual demand of nitrogen will be 200×10^6 ton. By this demand 40×10^6 ton from inorganic fertilizers and some will be fixed by *Rhizobium* in leguminous crops. So far biotechniques will be suitable medium to meet this demand. The most widely used bio-fertilizer is *Rhizobium*, which colonize the roots of leguminous plants by forming nodules. In the absence of leguminous crops, population of *rhizobium* in the soil diminishes. Most strains of native rhizobia are more effective in comparison to other non-native strains in nitrogen-fixation. Their ability of Nitrogen-fixation is 100-300 kg/ha in one crop season and restore nitrogen for the following crop. Artificial seed inoculation is often needed to restore population of effective and specific strain of *Rhizobium* near the root zone.

An increase of 20-32 per cent in yield can be obtained by inoculation of specific strain of *Rhizobium* in pulse crop. The symbiotic association can generally meet 80 per cent of nitrogen requirement of pulse crop.

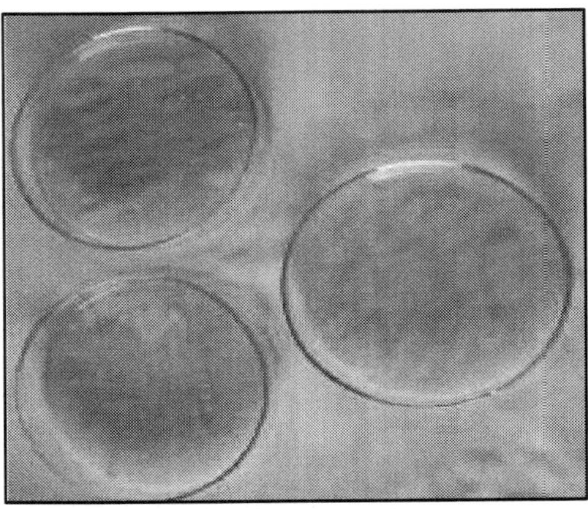
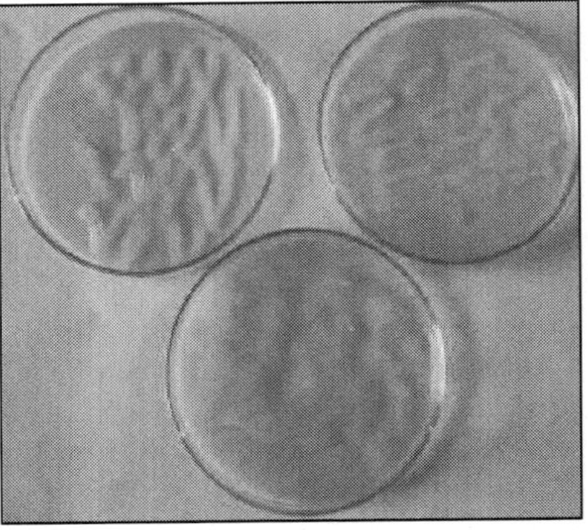

Figure 5.1: Different *Rhizobium* Isolates on Chickpea Isolated from P.R.S.S., Samba.

Table 5.1: Cross Inoculation Group of *Rhizobium*

Sl.No.	*Rhizobium Species*	*Cross Inoculation Group*
1.	*R. leguminosarum bv. viciae*	Peas
2.	*R. phaseoli*	Beans
3.	*R. trifolii*	Clover
4.	*R. meliloti*	Alfalfa
5.	*R. lupini*	Lupinii
6.	*R. japonicum*	SoybeanCowpea

Method for Isolation and Seed Treatment with *Rhizobium*

Pink and healthy nodules of pulse crops are taken for this purpose and washed thoroughly under tap water to remove gross surface contamination. The nodules should be exposed momentarily to 95 per cent ethanol and immersed in 0.1 per cent acidified $HgCl_2$ for 1-3 minutes. It has to be washed thoroughly in at least six changes of sterile distilled water. Surface sterilized nodules are crushed aseptically on a sterile glass slide and the exudates streaked from the nodule on to the surface of Yeast Extract Mannitol Agar plates. The petri plates are incubated at 28±2°C for 5-7 days. Isolated colonies of *Rhizobium* are picked up and transferred directly to the slopes of Yeast Extract Mannitol Agar or streaked again to get isolated colony (Vincent, 1970). The cultures are purified by streaking method and isolated colonies are transferred to the YMA slants in culture tubes. Characteristics of colonies are recorded and cultures are identified by using Gram's staining technique. Shape and size of rhizobial cells are measured under oil immersion with light microscope. They are Gram (-), rods, 0.5-0.9 x 1.2-3.0 micron meter in size, motile colonies circular, opaque, white convex and 4-5 mm in diameter in 6-7 days.

For inoculation of legume seeds with rhizobia, actively growing cultures (72 hours old) should be mixed in peat and rhizobial populations in the concentration of about 10^7-10^8 cells / g of peat should be prepared. Pea seeds are coated with the help of 10 per cent sucrose solution and gum arabic. After few hours of coating seeds are then sown in the field / pots.

Azospirillum

Beijerinck in 1925 reported a nitrogen fixing bacteria under the name *Spirillum lipoferum*. Later on, Tarrand *et al.*, realized its potentiality to fix nitrogen and named it as *Azospirillum* having close associative symbiosis with roots of higher plants. The bacteria grow better under deficient oxygen levels and fix Nitrogen 10-40 kg / ha.

It is a free living nitrogen fixing bacteria having higher N-fixing potentiality. The bacterium is gram (-) ve, motile, generally vibriod in shape and contains poly-Beta-hydroxy butyrate granules. When grown in N-free medium, it behaves as microaerophilic and when supplied with fixed nitrogen it behaves as aerobes. In addition to nitrogen it also produces hormones, antibacterial compounds and biosynthetic compound siderophore. It promotes early tillering, reproductive growth and increases the filling rate of grains and the grain weight per plant at harvest. Inoculative response of *Azospirillum* found beneficial with different crops such as maize, sorghum, sunflower, bhindi, cotton and jowar. An increase in grain yield of 5-22 per cent in cereals, 30 per cent in millets and 50 per cent in fodder were recorded with application of *Azospirillum*. 500 gms of *Azospirillum* is mixed with 500ml of rice ganzi or gum arabica to make a slurry and sown after shade drying. *Azospirillum* could be isolated from the roots of tropical grass (*Digitaria decumbens*) in a semi-solid N2-free medium. Surface sterilization of roots by 70 per cent alcohol should be done and then creation of micro-aerophilic (low oxygen requirements) in the medium.

Azotobacter

The bacterium, *Azotobacter* are rod shaped, relatively large organisms, measuring 2.0-7.0 x 1.0-2.5 μ and pertrichous flagella. The species can be isolated in a Jensen's medium (N-free agar medium) by dilution plating method. Ten gms of soil samples are transferred to 100 ml sterile distilled water and mixed thoroughly by shaking the flasks for 5 minutes. Serial dilutions are made by using sterile distilled water. Jensen's medium is prepared and poured into sterile Petri-plates. One ml sample from the appropriate dilutions are spread evenly over cooled agar medium in the plates. The plates are incubated at 30°C in an incubator for 3-4 days. *Azotobacter* colonies appear as flat, soft, milky and mucoid on agar plates.

Blue-Green Algae

The blue-green algae (BGA) and Azolla (a delicate water fern) are important groups of micro organisms capable of fixing atmospheric nitrogen. They act as bio-fertilizers, grow most abundantly in tropical and subtropical regions and are common in flooded rice paddy soils. Their possible role in the nitrogen accumulation in soil and providing the growing rice plants with considerable amounts of fixed-N has been reported in several parts of the world. The blue-green algae represent a self-supporting system of carrying out both photosynthesis and biological nitrogen fixation. They provide all that nitrogen which is given by chemical fertilizers, besides providing numerous other benefits like the increase uptake of water, reduce uptake of heavy metals etc.

The amount of nitrogen fixed by blue-green algae is dependent on the algal species as well as physiological and environmental conditions. The amount of nitrogen fixed by BGA under laboratory condition varies markedly according to the algal species, the type of medium, incubation period, and the growth condition. This amount ranges between 22-270mg/100ml medium. The amount of nitrogen fixed by the BGA under field conditions ranges between 15-80kg/ha dependence on algal species, fertilization, soil and environmental conditions. The blue-green algae liberate large quantities of ammonia as well as variety of organic compounds and some growth promoting substances.

Phosphate Solubilizing Micro-organisms

Next to nitrogen, phosphorus is the important nutrient which is needed for growth of plants and microorganisms. Most of the soil phosphorus is in unavailable form and hardly about 1-2 per cent of it is incorporated into the above ground parts of the plants. Generally phosphorus is supplied to the soil in the form of chemical fertilizers as S.S.P./D.A.P./murate of potash, out which 10-20 per cent is only utilized by the plants, rest remains in the soil as insoluble phosphate in the form of rock phosphate and tri-calcium phosphate. Plants take up phosphorus in the ionic form such as HPO_4^-, $H_2PO_4^-$ and PO_4^{--}. Phosphorus is also supplied to the plants in the form of organic sources such as decomposed plant and animal materials. Phosphorus solublizing microorganisms includes various bacterial, fungal and actinomycetes forms which help convert insoluble inorganic phosphates into simple and soluble forms. Members of *Pseudomonas, Micrococcus, Bacillus, Flavibacterium, Penicillium. Fusarium, Sclerotium* and *Aspergillus* are some of the phosphate solubilizing micro-organisms. They normally grow in a medium containing insoluble tri-calcium phosphate, apatite, rock phosphate, $FePO_4$ and $AlPO_4$ as sole source of phosphate. The initial isolation of phosphate solublizers is made by using a medium suspended with insoluble phosphates such as tri-calcium phosphates. The production of clearing zones around the colonies of the organisms is an indication of the presence of phosphates solubilizing organisms. Such cultures are isolated, identified and the extent of the solubilization determined quantitatively. Several rock phosphate dissolving bacteria, fungi, yeasts and actinomycetes were isolated from soil samples collected from rock phosphate deposits and rhizospheric soils of different leguminous crops. Many isolates solublized a very high quantity of rock phosphates *e.g. Pseudomonas rathonis* and *Bacillus polymyxa* and fungal isolates *Aspergillus awamori, Penicillium digitatum, Aspergillus niger* and *Schwanniomyces occidentalis*. The efficient microorganisms have shown consistently their capability to solublize chemically fixed soil phosphorus and rock phosphate from different places like Mussoorie, Udaipur, Matoon, Singhbhum, Morocco, Gafsa and Jordan. In India field trials were conducted in different soil types to evaluate the efficacy of seed inoculation with Phosphobacterin (*Bacillus megeterium* var. *phosphaticum*). By developing improved techniques for isolation of rock phosphate solublizing micro-organisms and by systemic investigations new efficient bacteria such as *Pseudomonas striata, Bacillus polymyxa* and fungi like *Aspergillus awamori* have been selected for preparation of carrier based inoculant in addition to several basic studies as for example IARI' Microphos culture. Bacteria can be used in neutral to alkaline soils and fungi can function better in acidic soils. The efficient cultures have shown capacity to solublize insoluble inorganic phosphate such as rock phosphate, tri-calcium phosphate, and iron and aluminum phosphates by production of organic acids. They can also mineralize organic phosphate compounds present in organic manures and soils. The above cultures were tested in multi-location field trials in wheat, paddy, gram, lentil and potato.

Mycorrhiza **(Myco=fungus+rhiza=rhizoids)**

This is an association of a fungus with roots of higher plants. In nature most of the plants have been found associated with mycorrhizal fungi. Nearly a Century back (1885) a German botanist A.B. Frank found that the roots of most of the plants are colonized by the fungi and transformed into the fungus root organ and called it as Mycorrhiza. Mycorrhiza results from a mutuality symbiosis between roots and certain fungi. The fungus gets sugars (monosaccharide) which are converted into trehalose, mannitol and glycogen; B-vitamins which stimulate spore germination and in return supplies the plant with nutrients (P, N, K, Cu Mg, Zn, and N), hormones etc. and protect it from root pathogens. It is estimated that trees invest 10 per cent of photosynthates to their mycobionts. 90 per cent of vascular plants are normally associated with AM fungi. The four major types of mycorrhiza are i) Ectomycorrhiza ii) Vesicular arbuscular mycorrhiza iii) Ericoid mycorrhiza and iv) Orchidaceous mycorrhiza.

Ectomycorrhizas (ECM) are associations where basidiomycetes and other fungi form short swollen lateral roots covered by mantle hyphae. These roots have Hartig net hyphae around cells in the epidermis or cortex. Arbuscular Mycorrhizae (AM) is found to occur on roots of most of the food and horticultural crops and tropical trees. They infect more plants than any other fungi and are necessary for growth and development of many plants. The mycorrhizal associations are formed by the fungi belonging to the genera *Glomus, Gigispora, Acaulospora, Sclerocystis, Entophospora* and *Scutellospora*. The genera are separated at the first level on the manner of spore formation, as inferred from the morphology of spore and spore bearing structures within the chlamydosporic taxa. Further generic separation is based on sporocarp morphology. Genus *Gigaspora* forms no vesicle, hence we have reserved the vesicle and in spite of VAM, presently they are considered as AM fungi. The spores of *Glaziella, Glomus* and *Sclerocystis* are regarded as chlamydospores *i.e.* specialized asexual resting cells. Each genus has been described in detail by Geredemann and Trappe (1974). In genus *Glomus*, chlamydospores are formed at the hyphal tip, usually one per tip, but in *G. fuegianum* several spores emerge from a swollen hyphal tip. At the time of maturity, the spore contents are separated from the attached hypha by a septum or by occlusion with deposit of wall material. Two or more hyphae may be attached to a single spore in some species. Spores of most *Glomus* species are borne singly in soil but in some species they may form in the cortex of the roots or in sporocarps. Most sporocarps of *Glomus* species are non-organized conglomerations of spores. According to Hacskaylo (1972), mycorrhiza corrects not only phosphorus deficiency but also any nutrients deficiency that are needed for growth and development of the plants. Mycorrhizal plants develop extensive root system as compared to non-mycorrhizal plants, which ensures the plant with increased availability of water and nutrients, thereby helping the plant for better growth and development. VAM fungi enhance the root growth, expand the absorptive capacity of the root system for nutrients and water and alter the cellular processes in host root system. Mycorrhizal plants usually have more vascular bundles, hence lignifications in the xylem vessels is greater. Mycorrhiza strengthens the cell walls of increasing lignifications and production of other polysaccharides. A stronger vascular system will increase the flow of nutrients, imparts greater mechanical strength and diminish effects from vascular pathogens. They impart to their a host a variety of benefits which includes increased growth and yield due to enhanced nutrients acquisition, water relation, pH tolerance, disease and pest resistance. The most common effect of Mycorrhiza is increased uptake of immobile nutrients, notably phosphorus from soil. The extrametrical mycelium of mycorrhizal fungi acts in effect as an extension of the root system, more thoroughly exploring the soil volume. The Phosphorus depletion zone around a non-mycorrhizal root extends to only 1-2 mm, approximately the length of a root hair whereas extra radical hyphae of mycorrhizal fungi extends 8 cm or more beyond the root making the phosphorus in this greater volume of soil available to the host. Soil inoculums are produced using, a traditional sterile soil-sand (1:1) techniques containing highly infective AM propagules which are multiplied simultaneously in barley and maize crop. In furrow application, inoculums should be placed 3 cm below the seed @ 1.9 tones/ha. It can be used in field crops like wheat, barley, maize, chilli, jowar, bajra and tomato; horticultural crops like litchi, mango, citrus and marigold and forest tree species like leucaena, turmeric and acacia.

Orchid mycorrhizas consist of coils hypae within roots or stems of plants in the family Orchidaceae. Young orchid seedlings and some adult plants which lack chlorophyll are entirely dependant on mycorrhizal fungi for their survival. Ericoid mycorrhizae have hyphal coils in outer cells of the narrow "hair roots" of plants in the plant order Ericales. These associations also occur in thicker roots of Australian members of the Epacridaceae. There are a number of nonmycorrhizal genera which are important in agriculture and horticulture including members of the families-Chenopodiaceaea, Amaranthaceae, Caryophyllaceae, Polygonaceae, Brassicaceae, Scrophulariaceae, Commelinaceae, Juncaceae, Cyperaceaea.

Obligatory Mycorrhizal Plants

Plants which do not survive to reproductive maturity without being associated with mycorrhizal fungi in the soils (or at the fertility levels) of their natural habitats. These species consistently support mycorrhizal colonization throughout most of their young roots (if inoculum is sufficient).

Facultative Mycorrhizal Plants

Plants that benefit from mycorrhizal associations only in some of the least fertile soils in which they naturally occur. In ecosystem surveys, inconsistent mycorrhizas or low levels of mycorrhizal colonization (less than 25 per cent) have been used to designate facultatively mycorrhizal species when soil fertility levels could not be manipulated.

Non-mycorrhizal Plants

Plants with roots that consistently resist colonization by mycorrhizal fungi, at least when they are young and healthy. Nutrient levels and other soil properties and mycorrhizal propagule dynamics can also reduce mycorrhizal formation, but usually do not prevent it completely.

Mycorrhizai Effect on Crop Growth

Being an obligate parasite, mycorrhizae can not be cultured in artificial media. In every soil mycorrhizae are present; sporocarps can be collected by Sieving and Decanting method (Gredmann and Trappe, 1974) and then picking up through needle under binocular microscope. It can be multiplied in soil with alternate host crops *viz.*, jowar, bajra, maize, barley, wheat and other grasses. Singh and Srivastava (2007) studied interaction of different varieties with three mycorrhizal species (Tables 5.2 and 5.3). Among those varieties, Trapper showed maximum colonization (61.52 per cent) with *Glomus mosseae* followed by *G. monosporum*. *G. monosporum* showed maximum colonization with cultivar 5064 followed by PG3, HUDP15 and Trapper. This indicates that cultivars Trapper, 5064, PG3, Rachna and HUDP15 respond efficiently with mycorrhizal fungi. Variety HFP-4 showed poor performance with *G. mosseae*, HUP16 with *G. monosporum* and HUP2 with *G. constrictum*. This variability might be due to genetic make up of the host plant. Active mineralization in rhizoshere by other PGPR may also influence the mycorrhizal effect (Barea and Jaffries, 1995).

Table 5.2: Effect of Mycorrhizal Inoculation on Shoot Weight (g) of Pea Cultivars

Cultivars	*G. mosseae*	*G. monosporum*	*G. constrictum*	*Average*	*Control*
HFP-4	1.93 (28.67)	1.13	1.57 (4.44)	1.54	1.50
KPSD-1	1.93 (34.65)	1.33	1.70 (18.60)	1.66	1.43
PG-3	1.37	2.43 (73.81)	1.53 (9.52)	1.78	1.40
HUDP-15	1.27 (2.70)	1.60 (29.73)	1.30 (5.41)	1.39	1.23
HUDP-16	3.43 (71.67)	1.67	1.50	2.20	2.00
Rachna	2.47 (39.62)	1.57	1.77	1.94	1.77
HUP-2	2.07 (21.57)	1.10	1.73 (1.96)	1.63	1.70
HUP-16	1.77 (8.16)	0.78	1.40	1.32	1.63
5064	1.73 (18.18)	1.08	1.27	1.36	1.47
Trapper	2.63 (64.58)	0.67	1.13	1.48	1.60

Source: Singh and Srivastava (2007).

Figures in parentheses shows per cent increase in dry shoot weight.

Table 5.3: Per cent Root Colonization of Pea Cultivars with different *Glomus* spp.

Cultivars	*G. mosseae*	*G. monosporum*	*G. constrictum*	*Average*
HFP-4	25.0	25.3	26.7	28.7
KPSD-1	29.6	26.1	28.5	28.0
PG-3	31.7	43.7	26.7	34.0
HUDP-15	27.3	31.0	35.0	31.1
HUDP-16	28.3	30.4	28.3	29.0
Rachna	40.3	27.7	36.6	38.2
HUP-2	26.0	26.7	21.6	24.8
HUP-16	31.5	22.1	26.7	26.7
5064	35.7	58.6	31.7	42.0
Trapper	61.5	29.3	32.6	41.1
Average	33.7	33.1	27.4	32.1

Source: Singh and Srivastava (2007).

When *Rhizobium* and Mycorrhizae were inoculated along with *Pseudomonas*, this combination not only affected the nodulation but there was a significant increase in grain yield as compared to control or *Rhizobium-Glomus* individually or in combination (Table 5.4). Ross and Harper (1970) recorded 29 per cent increase in yield in soybean grown in fumigated field plots, due to double inoculation with endomycorrhizae + *Rhizobium* over single inoculation with only *Rhizobium*. While the principle effect of mycorrhizae on nodulation is undoubtedly phosphate mediated, but the fungus may have some secondary effects, possibly of hormone nature. Hence, there is synergistic effect of symbionts on the host plant and this effect is more pronounced when PGPR are dominating in the rhizosphere. Singh and Srivastava (2010) showed in his study that there is significant varietal interaction with *Pseudomonas* strains and inoculation of *Pseudomonas* may be able to improve pea nitrogen fixation and yield (Table 5.4).

Singh *et al.* (2004) found that colonization of *Glomus mosseae* and *Pseudomonas fluorescens* (mutant Pf2) increased the growth of three pea cultivars; Rachna, HUDP16 and HUP2. Plants of all the cultivars showed greater accumulation of phenolic acids in treatments than the control. Mycorrhizal colonization and total phenolics accumulation were closely correlated with disease intensity on pea plants (Table 5.5).

Benefits of Mycorrhizal Associations to Plants

1. Increased plant nutrient supply by extending the volume of soil accessible to plants.
2. Increased plant nutrient supply by acquiring nutrient in a form that would not normally be available to plants.
3. Some ECM and ericoid fungi have the capacity to breakdown phenolic compounds in soils which can interfere with nutrient uptake.

Table 5.4: Effect of *Rhizobium*, *G. mosseae* and *Pseudomonas* on Nodulation and Grain Yield of Pea var. Rachna

Treatment	*No.of Nodules/ Plant*	*Percent of Nodules Showing str-Resistance and Agglutination (+) Reaction*	*Per cent Increase in Nodules over Control*	*Dry Weight of Nodules/ Plant (g)*	*Per cent Increase in Dry Weight of Nodules*	*Yield per Plot (g)*	*Per cent Increase in Yield*
G. mosseae	29.0	–	14.5	0.04	103.5	378.97	23.9
Rhizobium mutant	39.0	40.0	53.9	0.09	350.9	362.07	18.3
Pf-2 mutant	38.7	–	52.6	0.09	366.7	356.03	16.4
G. mosseae + Rhizobium mutant	53.3	30.0	110.5	0.12	508.8	399.03	30.4
G. mosseae + Pf-2 mutant	52.0	–	105.3	0.12	549.1	394.93	29.1
Rhizobium mutant + Pf-2 mutant	72.3	50.0	185.5	0.25	1235.1	407.70	33.3
G. mosseae + Rhizobium mutant + Pf-2 mutant	70.0	80.0	176.3	0.21	1015.8	430.47	40.7
Control	25.3	10.0	–	0.02	–	305.95	–

Source: Singh and Srivastava (2009).

Table 5.5: Effect of Mycorrhizal Colonization on Total Phenolics and Disease Intensity of *Erysiphe pisi*, 45 Days after Inoculation (Singh *et al.*, 2004)

Treatment	*Mycorrhizal Colonization Per cent*		*Total Phenolics (µg of Tannic Acid Equivalents/g Fresh Leaf wt.*	*Disease Intensity (per cent)*
	(-) E. pisi	*(+) E. pisi*		
cv. 'Rachna'				
G. mosseae	78.49±2.3	77.32±3.7	87.6±2.8	36.7±2.6
Control	–	–	57.9±3.8	55.2±3.6
cv. 'HUDP16'				
G. mosseae	69.36±1.3	77.16±5.6	77.09±1.9	28.7±3.6
Control	–	–	48.17±3.6	47.2±4.3
cv. 'HUP2'				
G.mosseae	73.08±4.6	71.31±3.1	94.00±1.7	31.6±4.7
Control	–	–	69.29±4.3	56.5±2.9

(-) *E. pisi*: without infection; (+) *E. pisi*: with infection.

4. Root colonization by ECM and VAM fungi can provide protection from parasitic fungi and nematodes.
5. Non-nutritional benefits to plants due to changes in water relations, phytohormone levels, carbon assimilation, etc. have been reported, but are difficult to interpret.
6. Mycorrhizal benefits can include greater yield, nutrient accumulation and / or reproduction success.
7. Mycorrhizae can increase growth, changes to root architecture, vascular tissue, etc.
8. Suppression of competing non-host plants, by mycorrhizal fungi has been observed.
9. Networks of hypae supported by dominant trees may help seedlings become estalished or contribute to the growth of shaded understorey plants.
10. Significant amounts of carbon transfer through ECF fungus mycelia connecting different plant species has been measured.This could reduce competition between plants and contribute to the stability and diversity of ecosystems.
11. Nutrient transfer from dead to living plants can occur.

Vermi Compost

It is 100 per cent pure eco-friendly organic fertilizer. This organic fertilizer has nitrogen, phosphorus, potassium, organic carbon, sulphur, hormones, vitamins, enzymes and antibiotics which help to improve the quality and quantity of yield. It is observed that due to continuous use of chemical fertilizer soil losses its fertility and gets salty day by day. To overcome such problems natural farming is the only remedy and Vermi compost is the best solution.

Biocompost

It is eco-friendly organic fertilizer which is prepared from the sugar industry waste material which is decomposed and enriched of with various plants and human friendly bacteria and fungi. Biocompost consists of nitrogen, phosphate solubilizing bacteria and various useful fungi like decomposing fungi, *Trichoderma viridea* which protects the plants from various soil borne disease and also help to increase soil fertility which results to a good quality product to the farmers.

Benefits of Biofertilizers

- Aid in replenishing and maintaining long-term soil fertility by providing optimal conditions for soil biological activity.
- Suppress pathogenic soil organisms.
- Degrade toxic organic chemicals.
- Stimulate microbial activity around the root system significantly increasing the root mass and improving plant health.
- Increase the available nitrogen for plants far in excess of their own content by stimulating the growth of natural soil microorganisms. These soil microorganisms metabolize nitrogen from the air to multiply. When they die (some microorganisms have a life-span of less than 1 hour) the nitrogen is then released to the soil in a form that is readily available to the plants, animals and its adoptions plants.
- Interact with other soil organisms and biodegradable components in the soil to supply essential nutrients such as nitrogen, phosphorus, calcium, copper, molybdenum, iron, zinc, magnesium and moisture to the plants.
- Aid in solubilizing manganese. Manganese is thought to play a significant role in both disease resistance and plant growth.
- Increase crop yields by both enhanced growth and by protection because enhanced plant growth is accompanied by reduced stress and improved disease resistance.
- Initiate and accelerate the natural decomposition of crop residue turning it into humus.
- Effectively control incidents of fungal disease including pathogens on fruits and vegetables.
- Provide protection against disease associated with numerous fungi. In some environments, they produce peptides which inhibit the growth of fungi. In others, through a process know as mycoparasitism, they grow toward the hyphae of fungi, coil around them and degrade the cell walls.
- Significantly increase yield and reduce incidents of disease in fruit, vegetables, root crops, flowers, trees, shrubs, turf, grain ornamental crops and more.

- Provide protection (directly or indirectly) against collar rots, silver leaf, european canker, damping off, root infecting fungus, die back, dead arm disease, etc.
- Improve soil porosity, drainage and aeration, reduce compaction and improve the water holding capacity of the soil thereby helping plants resist drought and produce better crops in reduced moisture conditions. One estimate indicates that a 5 per cent increase in organic matter quadruples the soils ability to hold and store water.
- Promote the break up unproductive soil, turning it into a productive growing medium.
- Stimulate seed germination and root formation and growth.
- Promote improved drainage.
- Improve soil aeration.
- Increase the protein and mineral content of most crops.
- Produce thicker, greener and healthier crops.
- Produce plants with increased sugar flavor and nutrient content.
- Improve seed germination.
- Reduce input costs.
- Aid in the development of root systems that produce stronger healthier plants more able to resist pests and drought conditions.
- Increase a soil microorganism population which in turn increases the uptake of nutrients from soil to plants.
- Improve oxygen assimilation in plants.
- Aid in rebuilding depleted soil.
- Aid in balancing soil pH.
- Aid in reducing soil erosion.

Mesorhizobium

Several species were moved from *Rhizobium* to this genus. It currently consists of 19 species.

Genus	Reference
Mesorhizobium albiziae	Wang *et al.*, 2007
Mesorhizobium alhagi	**New** (Chen *et al.*, 2009)
Mesorhizobium amorphae	—
Mesorhizobium australicum	**New** (Nandasena *et al.*, 2009)
Mesorhizobium caraganae	Wang *et al.*, 2007
Mesorhizobium chacoense	—

Contd...

Contd...

Genus	Reference
Mesorhizobium ciceri	formerly *Rhizobium ciceri*
Mesorhizobium gobiense	—
Mesorhizobium huakuii	formerly *Rhizobium huakuii*
Mesorhizobium loti	formerly *Rhizobium loti*, Type species
Mesorhizobium mediterraneum	formerly *Rhizobium mediterraneum*
Mesorhizobium metallidurans	Vidal *et al.* (2009)
Mesorhizobium opportunistum	**New** (Nandasena *et al.*, 2009)
Mesorhizobium plurifarium	—
Mesorhizobium shangrilense	**New** (Lu *et al.*, 2009)
Mesorhizobium septentrionale	—
Mesorhizobium tarimense	—
Mesorhizobium temperatum	—
Mesorhizobium tianshanense	*formerly Rhizobium tianshanense*

Ensifer (formerly *Sinorhizobium*)

The *Sinorhizobium* genus was described by Chen *et al.*, in 1988. However some recent studies show that *Sinorhizobium* and the genus *Ensifer* belong to a single taxon. *Ensifer* is the earlier heterotypic synonym (it was named first) and thus takes priority. This means that all *Sinorhizobium* spp. must be renamed as *Ensifer* spp. according to the Bacteriological code. The taxonomy of this genus was verified in 2007 by Martens *et al.*, The genus currently consists of 15 species.

Ensifer abri	**D**
Ensifer americanum	—
Ensifer arboris	—
Ensifer fredii	formerly *Rhizobium fredii*, Type species
Ensifer indiaense	**D**
Ensifer kostiense	—
Ensifer kummerowiae	—
Ensifer medicae	—
Ensifer meliloti	formerly *Rhizobium meliloti*
Ensifer mexicanus	New (Lloret *et al.*, 2007)
'Sinorhizobium morelense'	**E**
Ensifer adhaerens	**A**
Ensifer saheli	**B**

Ensifer terangae	**C**
Ensifer xinjiangense	

A: This name is a earlier heterotypic synonym of *Sinorhizobium morelense*

B: This species is also known as *Sinorhizobium sahelense*

C: This species is incorrectly known as *Sinorhizobium teranga*

D: Although these species have been published (Ogasawara, *et al.*, 2003), they have not yet been included in the "Validation List" of the International Journal of Systematic and Evolutionary Microbiology.

E: This species is distinct from *Ensifer adhaerens* but cannot yet be named *Ensifer*. See Martens *et al.*, 2007 for details.

Bradyrhizobium

The *Bradyrhizobium* genus was described by Jordan in 1982. It currently consists of 8 species.

Bradyrhizobium canariense	—
Bradyrhizobium elkanii	—
Bradyrhizobium iriomotense	**New** (Islam *et al.*, 2008)
Bradyrhizobium japonicum	formerly *Rhizobium japonicum*, Type species
Bradyrhizobium jicamae	**New** (Ramírez-Bahena *et al.*, 2009)
Bradyrhizobium liaoningense	—
Bradyrhizobium pachyrhizi	**New** (Ramírez-Bahena *et al.*, 2009)
Bradyrhizobium yuanmingense	—

Azorhizobium

The *Azorhizobium* genus was described by Dreyfus *et al.*, in 1988. It currently consists of 2 species.

Azorhizobium caulinodans	Type species
Azorhizobium doebereinerae	formerly *Azorhizobium johannae*

Methylobacterium

The *Methylobacterium* genus currently contains only one rhizobial species.

Methylobacterium nodulans

Burkholderia

The *Burkholderia* genus currently contains seven named rhizobial members and others as *Burkholderia* sp.

Burkholderia caribensis	—
Burkholderia cepacia	—
Burkholderia mimosarum	—
Burkholderia nodosa	**New** (Chen *et al.*, 2007)
Burkholderia phymatum	—
Burkholderia sabiae	**New** (Chen *et al.*, 2008)
Burkholderia tuberum	—

Vandamme *et al.*, 2007 notes that recent experiments revealed that *nodA* and *nif* genes are not longer detected on these strains and he suggests that the strains lost the characteristic, or the original observation was based on experimental error. Thank you Paulina Estrada de los Santos for this observation.

Cupriavidus

Cupriavidus formerly *Wautersia,* formerly *Ralstonia,* has recently undergone several taxonomic revisions. This genus currently contains a single rhizobial species.

Cupriavidus taiwanensis

Devosia

The *Devosia* genus currently contains only a single rhizobial species.

Devosia neptuniae

Herbaspirillum

The *Herbaspirillum* genus currently contains a single rhizobial species.

Herbaspirillum lusitanum

Ochrobactrum

The *Ochrobactrum* genus currently contains two rhizobial species.

Ochrobactrum cytisi
Ochrobactrum lupini

Phyllobacterium

The *Phyllobacterium* genus currently contains three rhizobial species.

Phyllobacterium trifolii	—
Phyllobacterium ifriqiyense	—
Phyllobacterium leguminum	—

F Note: isolated from nodules, but not proven to nodulate

Shinella

The *Shinella* genus currently contains a single rhizobial species.

Shinella kummerowiae	**New** (Lin *et al.*, 2008)

Bio-Control Agents

Biological control is the mechanisms or practices induced by other organisms to restrict the activities of pests under natural conditions. It is not based on the principle of only eradication of the pathogen population but also minimize the losses at optimum level or Economic Threshold Level. A number of bio-control agents are evolved having different mode of action as antibiosis, competition and hyperparasites. Growing of Resistant varieties and cross protection is also a major aspect in this regard. They are safer to environment or not creating any kind of disturbance in the ecological balance of nature. They have no broad spectrum and specific in their action on target organisms. Losses in biodiversity, environmental degradation and resistant problems on target organisms eliminated or nullified due to behavior and nature of bioagents.

Some Registered Microorganisms for Bio-control

In U. S. A., U. S. Environmental Protection Agency (E.P.A.) has registered at least 6 microorganisms as "Microbial Pesticide" [Microorganisms that colonize plants or plant parts] which is used in more than 10acre (3.9ha.) of wheat-.

1. *Agrobacterium radiobacter* K-84- developed by Kerr Associates (1980) in Australia against crown gall disease.
2. *Pseudomonas flurescens* (Trade name Dagger G) used against *Rhizoctonia* and *Pythium* (Damping off of cotton).
3. *Gliocladium virens* (Trade name Glio Gard) used against seedling diseases of ornamentals.
4. *Trichoderma harzianum* (Trade name F-Stop).
5. *Trichoderma harzianum/polysporum* (Trade name BINAB-T) used against wood decay.
6. *Bacillus subtilis* (Trade name Kodiak) used as seed treatment to control soil borne diseases.

Biological Control Organization

Commonwealth Institute of Biological Control (CIBC) is a unit of CAB with its headquarters at Trinidad (West Indies). CAB has its unit's throught world with one at Bangalore (India).

Trichoderma

Trichoderma a bio-pesticide which controls soil borne pathogens like *Fusarium, Pythium* and other diseases. It belongs to Sub-division Deuteromycotina which is placed in fungi imperfectii. It is mostly photo-sensitive and sporulates in concentric patterns of alternating rings in response to diurnal alternation of light and darkness. It secretes metabolites Trichodermin-1 and Trichodermin- 4 which degrades the cell wall of the pathogen and acts as antibiosis. They also produce various enzymes eg. exo- and endo-gluconases,chitinase,cellobiose that are helpful to degrade organic matter in soil results in release of nutrients.

Trichoderma Based Formulation Products

1. BINAB-T pellets (T. *harzianum/T. polsporum*) manufactured by bioinnovation AB (BINAB) in sweads and used as wood protectant in sweadery and against Dutch elm disease in U.K (Belgium).

BINAB-T SEPPIC- used for control of *Verticillium malthousei* in mushroom cultivation.

T. viride = *Hypocrea rufa* Webster, 1969

T. virens = *H. gelatinosa* Webster, 1969.

Rifai, M.A.-divided *Trichoderma* in a species.

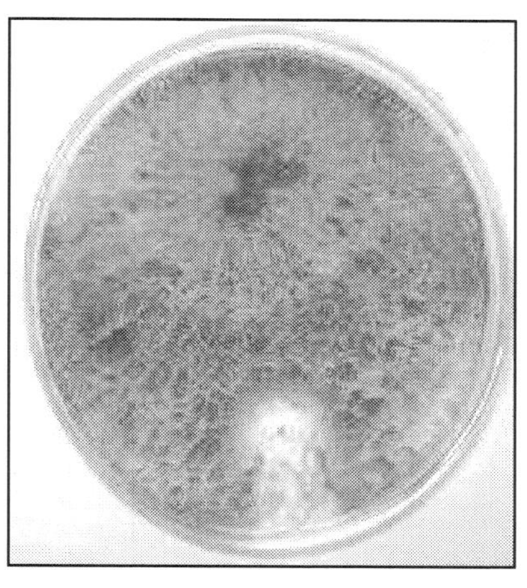

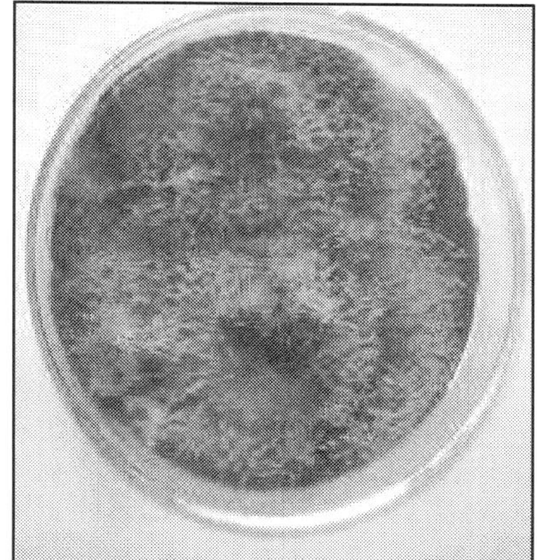

(a) Early stage

(b) Later stage.

Figure 5.2: A Native *Trichoderma* Suppressing the Growth of *Fusarium oxysporum f.sp.ciceri.*

Verticillium

Verticillium lecanii is a carrier based myco-pesticides which contains conidia of *Verticillium lecanii*. It is an effective bio-control agent for the most of sucking pest like mites, whiteflies, thrips, aphids, nematodes and *Verticillium lecanii* eat mosquito bug. The wettable powder formulation of this product is sprayed to control crop pests. It is a carrier based product containing the colony forming units (2x 10^7spores/gram) of *Verticillium lecanii*.

Mode of Action

When the conidia of *Verticillium lecanii* come in contact with insect-pests, the spore germinates and penetrates through the cuticle and multiplies in the haemocoel of the insect. This inhibits the metabolism and ultimately leads to total paralysis of all physiological functions. This ultimately results into death of pest

References

Brundrett, M. 1991. Mycorrhizae in natural ecosystems. *Adv.Eco.Res.* **21**: 171-212.

Chun Tao Gu, En Tao Wang, Chang Fu Tian, Tian Xu Han, Wen Feng Chen, Xin Hua Sui and Wen Xin Chen 2008. *Rhizobium miluonense* sp. nov., a symbiotic bacterium isolated from Lespedeza root nodules. *Int J Syst Evol Microbiol.***58**: 1364-1368.

Chen, Weimin, Zhu, Wenfei, Bontemps, Cyril, Young, J. Peter W., Wei, Gehong 2009. *Mesorhizobium alhagi* sp. nov., isolated from root nodules of the wild legume *Alhagi sparsifolia. Int J Syst Evol Microbiol.* 0: ijs.0.014043-0

Chen, Wen-Ming, James, Euan K., Coenye, Tom, Chou, Jui-Hsing, Barrios, Edmundo, de Faria, Sergio M., Elliott, Geoffrey N., Sheu, Shih-Yi, Sprent, Janet I., Vandamme, Peter. 2006. *Burkholderia mimosarum* sp. nov., isolated from root nodules of *Mimosa* spp. from Taiwan and South America. *Int J Syst Evol Microbiol.* **56**: 1847-1851.

Chen, W.M., de Faria, S.M., Chou, J.H., James, E.K., Elliott, G.N., Sprent, J.I., Bontemps, C., Young, J.P., Vandamme, P., 2008. *Burkholderia sabiae* sp. nov., isolated from root nodules of *Mimosa caesalpiniifolia. Int J Syst Evol Microbiol.* **58(9):** 2174-9.

Frank, A.B., 1885. Ueber die auf Murrelsumbioge beruhende Ernahrung gewisser Baume durrch unterirdische Pilze, Ber, *Deut.Botan.Gen.*3: 128-145.

Garcia-Fraile, Paula, Rivas, Raul, Willems, Anne, Peix, Alvaro, Martens, Miet, Martinez-Molina, Eustoquio, Mateos, Pedro F., Velazquez, Encarna. 2007. *Rhizobium cellulosilyticum* sp. nov., isolated from sawdust of *Populus alba Int J Syst Evol Microbiol.* **57**: 844-848.

Geredemann, J.W. and Trappe, J.M., 1974. The Endogonaceae in the Pacific Northwest. *Mycologia Mem.*, 5: 1-76.

Guan, Su Hua, Chen, Wen Feng, Wang, En Tao, Lu, Yang Li, Yan, Xue Rui, Zhang, Xiao Xia, Chen, Wen Xin. 2008. *Mesorhizobium caraganae* sp. nov., a novel rhizobial species nodulated with *Caragana* spp. in China. *Int J Syst Evol Microbiol.*2008 **58**: 2646-2653.

Han, Tian Xu, Wang, En Tao, Wu, Li Juan, Chen, Wen Feng, Gu, Jin Gang, Gu, Chun Tao, Tian, Chang Fu, Chen, Wen Xin. 2007. *Rhizobium multihospitium* sp. nov., isolated from multiple legume species native of Xinjiang, China. *Int J Syst Evol Microbiol.* **58**: 1693-1699.

Hacskaylo, E. 1972. Mycorrhizae: The ultimate in reciprocal parasitism. *Bioscience* **22**: 577-582.

Islam, M.S., Kawasaki, H., Muramatsu, Y., Nakagawa, Y., and Seki, T. 2008. "*Bradyrhizobium iriomotense* sp. nov., isolated from a tumor-like root of the legume *Entada koshunensis* from Iriomote Island in Japan." *Biosci. Biotechnol. Biochem.* **72**: 1416-1429.

Lioret L, Ormeno-Orrillo E, Rincon R, Martinez-Romero J, Rogel-Hernandez MA, Martinez-Romero E. 2007. *Ensifer mexicanus* sp. nov. a new species nodulating *Acacia angustissima* (Mill.) Kuntze in *Mexico. Syst Appl Microbiol.*

Lin, Dong Xu, Wang, En Tao, Tang, Hui, Han, Tian Xu, He, Yu Rong, Guan, Su Hua, Chen, Wen Xin 2008. *Shinella kummerowiae* sp. nov., a symbiotic bacterium isolated from root nodules of the herbal legume *Kummerowia stipulacea. Int J Syst Evol Microbiol* **58**: 1409-1413

Lu, Yang Li, Chen, Wen Feng, Han, Li Li, Wang, En Tao, Zhang, Xiao Xia, Chen, Wen Xin, Han, Su Zhen, 2009. *Mesorhizobium shangrilense* sp. nov., isolated from root nodules of *Caragana* spp. *Int J Syst Evol Microbiol* 0: **ijs.0.007393-0**

Martens, Miet, Delaere, Manuel, Coopman, Renata, De Vos, Paul, Gillis, Monique, Willems, Anne. 2007. Multilocus sequence analysis of *Ensifer* and related taxa *Int J Syst Evol Microbiol.*, **57**: 489-503.

Nandasena, Kemanthi G., O'Hara, Graham W., Tiwari, Ravi P., Willems, Anne, Howieson, John G. 2009. *Mesorhizobium australicum* sp. nov. and *Mesorhizobium opportunistum* sp. nov., isolated from *Biserrula pelecinus* L. in Australia *Int J Syst Evol Microbiol.*, **59**: 2140-2147.

Peng, Guixiang, Yuan, Qinghua, Li, Huaxing, Zhang, Wu, Tan, Zhiyuan 2008 *Rhizobium oryzae* sp. nov., isolated from the wild rice *Oryza alta. Int J Syst Evol Microbiol*, **58**: 2158-2163

Ramirez-Bahena, Martha Helena, Peix, Alvaro, Rivas, Raul, Camacho, Maria, Rodriguez-Navarro, Dulce N., Mateos, Pedro F., Martinez-Molina, Eustoquio, Willems, Anne, Velazquez, Encarna *Bradyrhizobium pachyrhizi* sp. nov. and *Bradyrhizobium jicamae* sp. nov., isolated from effective nodules of Pachyrhizus erosus *Int J Syst Evol Microbiol.*, **59**: 1929-1934.

Ramirez-Bahena, Martha Helena, Garcia-Fraile, Paula, Peix, Alvaro, Valverde, Angel, Rivas, Raul, Igual, Jose M., Mateos, Pedro F., Martinez-Molina, Eustoquio, Velazquez, Encarna 2008. Revision of the taxonomic status of the species *Rhizobium leguminosarum* (Frank 1879) Frank 1889AL, Rhizobium phaseoli Dangeard 1926AL and *Rhizobium trifolii* Dangeard 1926AL. *R. trifolii* is a later synonym of *R. leguminosarum*. Reclassification of the strain *R. leguminosarum* DSM 30132 (=NCIMB 11478) as *Rhizobium pisi* sp. nov. *Int J Syst Evol Microbiol.*, **58**: 2484-2490.

Ross, J.P. and Harper, J.A., 1970. Effect of *Endogone* mycorrhizae on soybean yields.*Phytopathology*, **60**: 1552-1556.

Singh, D.P., Srivastava, J.S., Bahadur Amar, Singh, U.P and Singh, S.K. 2004. Arbuscular mycorrhizal fungi induced biochemical changes in pea (*Pisum sativum*) and their effect on powdery mildew (*Erysiphe pisi*).*Journal of Plant diseases and Protection*, **2:** 111.

Singh, U.P. and Ray, A.B. 1999. Antifungal activity of %3-Alstovenine,a plant alkaloid isolated from Alstonia venenata.*Folia Microbiol.* **44(5),** 510-512. Singh, S. K and Srivastava, J.S.(2007) Response of pea cultivars to rhizobia and mycorrhizal fungi.*Journal of Food legumes.* **20(1):** 87-89.

Singh,S. K and Srivastava, J. S.2009. Response of pea cultivars to rhizobia and mycorrhizal fungi. *Journal of Food legumes.* **22** (4): 283-287.

Sophie Mantelin, Marion Fischer-Le Saux, Frederic Zakhia, Gilles Bena, Sophie Bonneau, Habib Jeder, Philippe de Lajudie and Jean-Claude Cleyet-Marel 2006. Emended description of the genus *Phyllobacterium* and description of four novel species associated with plant roots: *Phyllobacterium bourgognense* sp. nov., *Phyllobacterium ifriqiyense* sp. nov., *Phyllobacterium leguminous* sp. nov. and *Phyllobacterium brassicacearum* sp. nov. *Int J Syst Evol Microbiol.*, **56 (4)**: 827.

Tian Xu Han, Li Li Han, Li Juan Wu, Wen Feng Chen, Xin Hua Sui, Jin Gang Gu, En Tao Wang and Wen Xin Chen 2008. *Mesorhizobium gobiense* sp. nov. and *Mesorhizobium tarimense* sp. nov., isolated from wild legumes growing in desert soils of Xinjiang, China. *Int J Syst Evol Microbiol.* **58(11):** 2610-2618.

Valverde, Angel, Igual, Jose M., Peix, Alvaro, Cervantes, Emilio, Velazquez, Encarna 2006. *Rhizobium lusitanum* sp. nov. a bacterium that nodulates *Phaseolus vulgaris. Int J Syst Evol Microbiol.* **56**: 2631-2637.

Vidal, Celine, Chantreuil, Clemence, Berge, Odile, Maure, Lucette, Escarre, Jose, Bena, Gilles, Brunel, Brigitte, Cleyet-Marel, Jean-Claude. 2009. *Mesorhizobium metallidurans* sp. nov., a metal-resistant symbiont of *Anthyllis vulneraria* growing on metallicolous soil in Languedoc, France. *Int J Syst Evol Microbiol.*, **59**: 850-855.

Vincent, J.M. 1970. International Biological Programme Handbook No.15.A manual for the practical study of root nodule bacteria. *Oxford Blackwell Scientific Publication.*

Wang, Feng Qin, WChen, Wen-Ming, de Faria, Sergio M., James, Euan K., Elliott, Geoffrey N., Lin, Kuan-Yin, Chou, Jui-Hsing, Sheu, Shih-Yi, Cnockaert, M., Sprent, Janet I., Vandamme, Peter 2007. Burkholderia nodosa sp. nov., isolated from root nodules of the woody Brazilian legumes *Mimosa bimucronata* and *Mimosa scabrella. Int J Syst Evol Microbiol.*, **57**: 1055 - 1059.

Wang, Feng Qin, Wang, En Tao, Liu, Jie, Chen, Qiang, Sui, Xin Hua, Chen, Wen Feng, Chen, Wen Xin. 2007. *Mesorhizobium albiziae* sp. nov., a novel bacterium that nodulates *Albizia kalkora* in a subtropical region of China. *Int J Syst Evol Microbiol.*, **57**: 1192-1199

Zurdo-Pineiro, Jose Luis, Rivas, Raul, Trujillo, Martha E., Vizcaino, Nieves, Carrasco, Jose Antonio, Chamber, Manuel, Palomares, Antonio, Mateos, Pedro F., Martinez-Molina, Eustoquio, Velazquez, Encarna. 2007. *Ochrobactrum cytisi* sp. nov., isolated from nodules of Cytisus scoparius in Spain. *Int J Syst Evol Microbiol* **57**: 784-788.

2016, Diseases of Pulse Crops and their Sustainable Management 87–101
Editors: Samir Kumar Biswas, Santosh Kumar and Gireesh Chand
Published by: BIOTECH BOOKS, NEW DELHI

Chapter 6

Agronomic Management Practices to Control Pulse Diseases

Ratneswar Poddar[1]* *and Partha Deb Roy*[2]

[1]*Department of Agronomy, BCKV, Mohanpur, West Bengal*
[2]*Jute Research Station, Katihar, BAU, Sabour, Bihar*

After Independence, India won the backdrop of the famine occurred in Bengal during 1942-43 and became the nation which occupies the first or second position in terms of production and area in several major crops. The green revolution entailed the use of improved technologies, HYV, irrigation facilities, mechanization, use of chemical fertilizer and pesticides. But despite the productivity of grains, poverty and hunger persist while land degradation and environmental damage are prevalent and unbeaten. FAO estimated that near about 900 millions of people suffered from chronic hunger worldwide, of which 98 per cent belongs to the under developed and developing countries (FAO, 2008). Our first Prime Minister, Jawaharlal Nehru said early in 1948, "*Everything else can wait but not agriculture*". So, it requires increased food production which can be achieved by enhancing inputs, knowledge and skills of farmers'. However, the majority of chronically hunger is seen to the developing countries that use the marginal land for crop production, deficient in nutrient. Besides, lack of access to inputs and other product markets, as well as financial set up to procure costly chemicals fertilizer and other agro chemicals that might enhance the productivity of the land, make the situation more difficult.

Sustainable agriculture thus emerged as alternative agricultural systems to combat with the existing management practices for the resource poor farmers and at

* Corresponding Author: E-mail: ratneswar89@gmail.com

the same time it ensures the environmental security. This system involves a combination of inter related soil, plant and livestock production practices in combination with the discontinuation or reduced the use of external inputs that are potentially harm full to the environment as well as health of the farmers. Instead of, it emphasized the use of the technique that integrate and adapt to the local natural process such as nutrient recycling, biological nitrogen fixation, soil regeneration and natural enemies of pest, into food production process (Pretty *et al.*, 2003). To achieve the sustainability in crop production, pulse crops are the prime one, which not only increased the soil fertility but minimize our malnutrition by providing protein and essential amino acids. Pulse crops belong to the family Leguminaceae. The commonly grown pulse crops in India are given below:

Table 6.1: Pulse Crop Grown in India

Common Name	*Botanical Name*	*Common Name in India*
Chickpea or Bengal gram	*Cicer arietinum*	Chana
Pigeon pea or Red gram	*Cajanus cajan*	Arhar or tur
Green gram or mung bean	*Vigna radiata*	Mung
Black gram or urd bean	*Vigna mungo*	Urd or mash kolai
Lentil	*Lens esculenta*	Masoor
Cow pea	*Vigna sinensis*	Lobia
Field pea or pea	*Pisum sativum*	Mutter
Moth bean	*Vigna acontifolia*	Moth
Horse gram	*Dolichos biflorus*	Kulthi
Lathyrus or grass pea	*Lathyrus sativa*	Khessary
Beans	*Phaseolus vulgaris*	Rajmash
Soybean	*Glycine max*	Soybean

Specific Features of Pulse

Pulse can be grown widely from tropics to temperate regions. They posses some specific feature, which makes them very important in cropping system.

- ☆ Pulse can be utilized both as food and fodder in subsistence cropping system.
- ☆ They are grown as a sole crop, intercrop, catch crop, relay crop, cover crop and green manure crop *etc.*, under sequential/mono-cropping in different agro-ecological regions.
- ☆ They fix atmospheric N_2 in their root nodules and improve the soil fertility.
- ☆ In the cropping systems of dry areas, pulses are predominant due to their low input requirements and capacity to withstand drought and consequently perform relatively better than other crops in the fragile and harsh climate prevailing of the regions.

- ☆ They add considerable amount of organic matter through root biomass and leaf fall, helps in mobilization of nutrients, protect the soil against erosion and improve the microbial biomass and keep the soil productive and alive by bringing qualitative changes in physical, chemical and biological properties.
- ☆ Pulse seeds are important source of protein in vegetarian diet.

Constraints in Pulse Production

The problems that is associated with the low production of pulses are discussed below:

1. **Unfavourable weather condition:** Pulses are generally very much susceptible to water logging condition. Continuous and heavy rains cause great damage in rainy season pulses. Rain just after the sowing adversely affects the germination. Rains at the time of ploughing are harmful for most of the pulses.
2. **Abnormal soil condition:** Abnormal soil such as saline-alkaline soil, acidic soils, and water logged soils are not at all suitable for the cultivation of these crops.
3. **Varietal constraints:** The main varietal constraints associated with:
   - (*a*) Lack of high yielding varieties
   - (*b*) Dropping of the flower
   - (*c*) Unavailability of short duration variety
   - (*d*) High susceptibility to insect pest and other diseases.
   - (*e*) Poor response to input
4. **Seed constraints:** Seed constraints involve the following:
   - (*a*) Poor quality seed
   - (*b*) Inadequate availability
   - (*c*) High price
   - (*d*) Not supply at proper time.
5. **Agronomic constraints:** The main constraints in pulse production due to agronomic deficiency are:
   - (*a*) Improper sowing of time
   - (*b*) Low seed rate
   - (*c*) Defective method of sowing
   - (*d*) Inadequate inter-culture
   - (*e*) Insufficient irrigation
   - (*f*) Sowing under utera condition
6. **Neglected use of *Rhizobium* culture**
   - (*a*) Non-availability of efficient strain
   - (*b*) Ignorance about advantages of culture
   - (*c*) Improper storage facility

7. Non adoption of plant protection measure by poor farmers.
8. Inadequate extension facilities.
9. The poor farmers are not able to use sufficient inputs for the production of pulses

So, among the different constraints which play important role in the low productivity of pulses, insect pest and diseases are the prime one. Pulses are subjected to attack by numerous pathogens both in the roots and the aerial parts of the plant. Among biotic stresses, diseases are often the major factor limiting the yield of pulses and they are economically important wherever these crops are grown. The importance of single biotic and abiotic stresses in cool-season pulse crop in the most important eco-geographic world regions has been reviewed by Saxena (1993). But the pattern of incidence of this pest is changing day by day due to the variability in changing climate.

Climate Change and Changing Scenario of Insect Pests and Diseases of Pulse Crops

Analysis of weather and climatologically data of the last four decades indicate there was highly significant variation in some parameters for particular weekly averages across decades and over years for all the nine locations namaely Pantnagar, Kanpur, Faizabad, Samastipur/Dholi, Rahuri, Parbhani, Patancheru, Bangalore, Coimbatore, which indicated that at those locations there has been significant change in climate for those parameters. There was significant rise in morning relative humidity and maximum temperature at Kanpur, Rahuri and Bangalore which might have impacted the rise in *Helicoverpa* infestation on pigeonpea and chickpea at those locations. The rise in temperature at Kanpur may have been beyond the tolerance limit of *Aceria cajani*, the mite vectoring Sterility Mosaic Virus of pigeonpea, which could have influenced decline in the disease there. On the other hand, the weather factors might have shifted in favour of the vector at Bangalore that may have resulted in rise of the disease on the crop there. Climate change may have also influenced Phytophthora blight incidence at Kanpur and Pantnagar in mutually opposite directions. While Alternaria blight on pigeonpea in Andhra Pradesh, Cercospora leaf spot on mungbean and urdbean in Karnataka, Stemphylium blight on lentil and chickpea in some parts of India are on rise, which could be due to the effects of climate change (*Annual Report 2011-12*, IIPR, Kanpur).

Important Diseases in Pulse

The grain legumes that are grown worldwide are prone to attack by a large number of plant pathogens like fungi, bacteria, phyto-plasmas and viruses which result in severe economic losses globally. Among these, fungi and viruses are the largest and perhaps the most important groups affecting all parts of the plant at all stages of growth. The important diseases that cause harmful effect to the *rabi* and *kharif* pulses are presented in the Table 6.2.

Table 6.2: List of the Major Diseases in Pulses

Name of the Legume	*Name of the Disease*	*Causal Organism*
Cold season Pulses		
Chickpea	Wilts	*Fusarium oxysporum* f.sp. *ciceris*
	Ascochyta blight	*Ascochyta rabiei*
	Grey mold	*Botrytis cinerea*
	Sclerotinia blight	*Sclerotina sclerotiorum*
	Stunt	Bean leaf roll luteovirus (BLRV)
Lentil	Rust	*Uromyces viciae-fabae*
	Ascochyta blight	*Ascochyta lentis*
	Wilt	*Fusarium oxysporum* f.sp. *lentil*
	Anthracnose	*Colletotrichum truncatum*
Field pea	Powdery mildew	*Erysiphe polygoni*
	Downy mildew	*Peronospora viciae*
	Wilt	*Fusarium oxysporum* f.sp. *pisi*
	Rust	*Uromyces viciae-fabae*
Faba bean	Ascochyta blight	*Ascochyta fabae*
	Chocolate leaf spot	*Botrytis cinerea/Botrytis fabae*
	Rust	*Uromyces viciae-fabae*
Warm season pulses		
Pigeonpea	Sterility mosaic	Pigeonpea sterility mosaic virus
	Wilt	*Fusarium oxysporum* f. sp. *udum*
Green gram and Black gram	Yellow vein mosaic	yellow mosaic virus
	Cercospora leaf spot	*Cercospora cruenta, C. canescens*
	Powdery mildew	*Erysiphe polygoni*
	Web blight	*Rhizoctonia solani*
Cowpea	Cowpea golden mosaic	Cowpea golden mosaic virus
	Cercospora leaf spot	*Cercospora canescens* and *Pseudocercospora cruenta*
	Cowpea aphid-borne mosaic	Cowpea aphid-borne mosaic virus

Management of Diseases of Pulse

The main aim for a researcher is to combat diseases as it reduces the crop yield near about 50 per cent or above in most of the pulses. For this the chemical control of pest management where ever applicable, get popularity and are quite common in farmers' and practice in the management of grain legume diseases has been developed and greater emphasis was on identifying, evaluating, and integrating location specific components of integrated disease management (IDM). In general, IDM has followed the principles of IPM (Jeger, 2000). Besides this some agronomic manipulation of the crop cultivation can also minimize or even control the diseases that are difficult to manage. Thus, it has no monetary investment but by this practice we can able to harvest higher amount of economic product. Some important agronomic practices which helps to manage pulse disease to some extent or fully are discussed below.

1. Selection of Suitable Variety

For the successful harvest of a crop, the choice of suitable variety for sowing is very important. The following factors should be account in the choice of a suitable variety:

1. The variety should posses higher potential yield
2. It should tolerate the aberrant weather conditions (Drought, Flooding)
3. It should have high germination percentage
4. Should match with the growing season
5. It should be pest and diseases resistant

Table 6.3: Some Resistance Cultivar of Pigeon-pea for Important Diseases

Disease	*Resistant Cultivar*
Wilt	YCP 4769', 'ICP 71 18'. 'ICP 7182', 'ICP 8862', 'ICP 9120', 'ICP 9168', 'ICP 9174', 'ICP 9177', 'ICP 10269', 'ICP 10858', 'ICP 11299', 'ICP 12731 ', 'ICP 2745', 'ICP 12758', 'ICPL 3', 'ICPL 8343', 'ICPL 84008', 'ICPL 88047', 'ICPL 9048', 'BWR 254', 'BWR 270', 'G33', 'G 36', 'G 52', 'DPPA 85-2', 'DPPA 85-3', 'DPPA 85-13', 'DPPA 85-14'.
Sterility mosaic virus	'ICPL 6997', 'ICPL 7035', 'ICPL 7197', 'ICPL 7264', 'ICPL 7867', 'ICPL 7998', 'ICPL 8094', 'ICPL 8362', 'ICPL 8862', 'ICPL 10976', 'ICPL 10977', 'ICPL 0996', 'ICPL 11049', 'ICPL 11204', 'ICPL 11206', 'ICPL I1207', 'ICPL 1123 I', 'ICPL 11297', 'BSMR I', 'BSMR 2', 'Bahar', 'DA 1I', 'Pusa 14', 'Pusa 15', 'Pusa 17', Pusa 18', 'ICPL 335', 'ICPL 342', 'ICPL 366', 'ICPL 88034', 'DPPA 85-2', 'DPPA 85- 3', D?PA 85-14', 'DPPA 85-15', 'BWR 159', 'MA 3', 'MA 6'
Phytophthora blight	'WBR 80-20-1', 'KPBR 80-2', 'WBR 8-2-2', 'ICP 8103', 'ICP 9252', 'ICP 10958'.
Wilt + sterility mosaic virus	'ICP 4769', 'ICP 7035', 'ICP 7867', 'ICP 8860', 'ICP 8862', 'ICP 9174', 'ICP 11297', 'DPPA 83-1 I', 'DPPA 85-12', 'DPPA 85-13', 'DPPA 85-14', 'BWR 159', 'PR 5149
Sterility mosaic virus + *Phytophthora* blight	'ICP 8103', 'KPBR 8-2-2', 'KPBR 80-2'.
Wilt + sterility mosaic virus + *Phytophthora* blight	'ICPL 43', 'ICPL 44', 'ICPL 45'

Source: Dhar and Chaudhary, 2001.

So, selection of variety, suitable for the growing period is very important. It has been seen that in long duration pigeon pea, all the existing varieties are susceptible to at least one of the major diseases *viz., Fusarium* wilt, Phytophthora stem blight and sterility mosaic disease (SMD). Secondly, the variety should posses the characteristics of diseases resistance. For chickpea cultivation the main disease that reduces the yield is wilt. If we select the wilt disease resistant variety of chickpea like JG 315, Avrodhi, DCP 92-3, JG 74, BG 372, KWR 108 automatically yield will be higher.

Similarly, now a day's scientists have developed diseases resistant variety for most of the pulses crop. Efforts were made to develop high yielding varieties with multiple disease resistance. Thus by selecting the proper variety disease of the pulses can be reduced and it has no extra monetary input.

2. Choice of Suitable Date for Sowing

It is also a non monetary input but by slight alteration of the sowing time the severity of diseases in pulses can be reduced markedly. Generally pulses that are sown early often perform best in drier years, particularly when sown into stubble, and in wider rows but do not sow too early. Early sowing results in excessive vegetative growth due to more residual moisture and low to moderate temperatures causing more disease incidence, whereas late sowing reduces vegetative growth and hence lower disease incidence.

Studies at IIPR, Kanpur during 2002-03 revealed that sowing of lentil at the end of October results in least disease intensity by 13.45 per cent and higher grain yield by 1322 kg ha^{-1} in rust susceptible variety, Lens 830 (Table 6.4). Therefore, sowing of crop at the end of October or in first week of November is best option for managing rust disease in lentil.

Table 6.4: Effect of Sowing Dates on Development of Rust in Lentil (2002-03)

Sowing Dates	*Susceptible Variety, Lens 830*				*Yield Loss (per cent)*
	Unprotected		*Protected*		
	Per cent Rust	*Yield (kg ha^{-1})*	*Per cent Rust*	*Yield (kg ha^{-1})*	
29/31 October	13.45	1322	7.80	1468	9.94
5/7 November	18.85	1138	10.00	1475	22.89
12/14 November	38.05	1151	22.25	1391	17.04
19/21 November	60.30	872	38.55	1171	28.40
26/28 November	69.60	831	38.95	1103	28.03
3/5 December	78.15	724	40.35	937	28.76

Early planted chickpea crops usually face more disease attack. Several studies have suggested that higher disease control and yield are obtained when the planting is delayed until the last week of October (Chandra *et al.*, 1974). Sangar and Singh (1994) studied the effect of sowing dates and pea varieties on the severity of rust, powdery mildew and yield on peas cv. Khaparkheda, Rachna and JP 789 sown on 5 or 25 Oct., 14 Nov. or 4 Dec. Delaying sowing after 5 October increased the incidence of *Uromyces fabae* and *Erysiphe polygoni* infestations and decreased yield. The variety Khaparkheda gave the highest seed yield (1.54 t/ha) and had the lowest incidence of these two diseases.

The lower disease incidence in late-sown crop was considered to be due to low temperature prevailing during the period of late-sown crop. Whereas, early sowing during the first fort night of November is the best time of sowing to reduce the rust as well as powdery mildew disease severity in pea crop.

Table 6.5: Effect of Sowing Date, on Rust and Powdery Mildew in Pea (2002-04)

Treatment	*Rust Severity (per cent)*	*Powdery Mildew Severity (per cent)*	*Yield (kg/ha)*
1st November	3.5	6.1	2677
16th November	6.0	7.7	2542
1st December	10.4	11.2	2100
CD at 5 per cent	4.0	3.0	–

3. Land Preparation

For successful establishment of the crop, the land where the crop is to be grown should be well prepared. Generally small seeded crop require fine seed bed where as bold seeded crops are grown well in rough seed bed condition. The land preparation before sowing of the crop is very important to control the diseases in pulses. Generally deep summer ploughing is very important to control the wilt disease in chickpea. During land preparation the method of tillage also take important role for the reduction of diseases. It has been reported that Charcoal rot caused by *Macrophomina phaseolina* was detected in soybean under conventional tillage with mould board plough, with an incidence of 54 per cent, whereas in the, remaining tillage treatments like direct seeding (DS), direct seeding + scarifier (DS+S), minimum tillage with disc harrow, the disease was not present (Perez-Brandan *et al.*, 2012)

4. Seed Treatment

Seed treatments are one of the important strategies to reduce the yield losses from disease in pulse crops. Even though disease is reduced on resistant varieties, seed treatments can enhance their performance. When the pulse seeds are treated with a fungicide, it reduces the establishment of seed-borne diseases in crops and also protects seed from infection by soil-borne fungi. Seed dressings with fungicide also protect seedlings from external airborne infection for the first few weeks. It is very important in reducing the subsequent establishment and spread of disease within crops. A maximum of four to six weeks after sowing through seed treatments, the plant prevent the incidence of diseases but do not provide absolute control.

Table 6.6: Effect of Seed Treatment on Wilt and Grain Yield

Seed Treatment	*Per cent Wilt Reduction*	*Per cent Yield Increase*
T. viride + Vitavax (4g + 2g kg^{-1})	32.6	59.47
Carbendazim (Bavistin - 2g kg^{-1})	47.3	115.13
Thiram (2g kg^{-1})	32.8	96.90
Thiram + Carbendazim (Bavistin) (2g+1g)	53.6	156.34

From the experiment it has been observed that in pea and lentil seedling blight, early root rots, *and* Ascochyta blight can be effectively controlled through seed treatment. Treating the chickpea seed with registered seed dressings controls seedling disease, improves crop establishment and increases yield. Laboratory tests confirmed the seedling disease to be caused by *Botrytis cinerea* is readily controlled with seed treatments provided they are applied correctly. Various fungicides and bioagents were tried as seed treatment to control the disease. Study at IIPR, Kanpur showed that Carbendazim + thiram and bio-agent, *Trichoderma viride* in combination with Vitavax were the best treatment for reducing wilt incidence and increasing the grain yield.

5. Depth of Sowing

Depth of sowing of pulses generally varies according to the type of crop, type of soil, the diseases likely to be present in the soil. There is a maximum depth at which the pulse crop can be grown safely just to avoid poor establishment and lower seedling vigour. Sowing of the seed above or below the suggested range will delay emergence and make attack to the soil borne pathogen more vigorously. Generally, the optimum seeding depths are ¼ to ½ inch on clay and loam soils and ½ to 1 inch on sandy soils. Planting of chickpea seeds at proper depth (10-12 cm) was helpful in reducing the incidence of wilt (Singh and Sandhu, 1973), while shallow sown crop seemed to attract more disease (Sugha *et al.*, 1994)

6. Canopy Modification

Crop canopies modification can also suppress plant diseases in some pulse crops. This modification can be achieved by changing the crop geometry which can be possible by alteration of crop spacing. Evidence shown that plants spaced at 15-20 cm had much higher disease incidence of wilt than those spaced at 7.5 cm; this was attributed to the shallower root system in widely spaced plants which were susceptible to wilt when subjected to moisture stress condition. (Bahl, 1976). Modifications in canopy can reduce foliar diseases in field pea and chickpea. There is a trend towards wider row spacings for pulses for better foliar aeration and early disease control. Andrabi *et al.*, 2011 reported that maximum disease incidence at row to row spacing of 20 cm and this could be attributed to the higher crop density and excessive crop canopy which helped in conserving moisture and humidity that fa-voured the infection

Table 6.7: Effect of Row to Row Spacing on Incidence of Chickpea Wilt Complex Under Field

Row to Row Spacing	*Per cent Disease Incidence (PDI)*	*Yield [q ha^{-1}]*
20 cm	29.17 (29.86)	8.10
30 cm	23.08 (28.46)	11.52
40 cm	20.49 (26.58)	6.04
50 cm	17.35 (24.56)	4.96
CD (p = 0.05)	3.45	0.37

Source: ndrabi *et al.*, 2011.

* Figures in parentheses are arc sine transformed values.

and further disease development and secondary spread of disease. It has been reported that larger crop canopy restrict the movement of air which prevents the quick evaporation of moisture conserved by the plants. It also indirectly helps in germination of inoculum, infection of the host plant and further development of the disease.

Snehlata *et al.* (2010) reported that rust severity in pea gradually decreased with gradual increase in row spacing (40x10cm^2) with maximum rust severity (52.13 per cent) was recorded at 30 x7.5 cm^2 while minimum of 48.5 per cent at 40 cm spacing. severity of powdery mildew, gradually increased with gradual increase in row spacing. Maximum powdery mildew severity in pea (59.35 per cent) was recorded at 40 x10cm^2 mildew.

Table 6.8: Effect of Spacing on Severity of Wilt, Rust and Powdery Mildew of Field Pea

Treatments	*Wilt (Per cent)*			*Rust (Per cent)*			*Powdery Mildew (Per cent)*		
	Y1	*Y2*	*Mean*	*Y1*	*Y2*	*Mean*	*Y1*	*Y2*	*Mean*
Spacings (cm)									
30 × 7.5	12.20	11.81	12.00	53.00	51.27	52.13	59.14	45.55	52.34
30 × 10	13.30	11.68	12.55	52.24	49.91	51.07	62.53	51.06	56.79
40 × 10	14.75	12.77	13.76	50.64	46.36	48.50	63.33	55.38	59.35
CD(P=0.05)	NS	NS	NS	1.28	1.13	1.19	2.32	2.11	2.25

7. Irrigation

Pulses are generally grown on limited soil moisture condition. The two important think that should be kept in mind during the water management are- i) Method of irrigation and ii) Time of irrigation. Irrigation requirements are primarily determined by crop water requirements, but also depend on the characteristics of the irrigation system, management practices and the soil characteristics in the irrigated area. In flooding method, all the surface soil remain wet where as the improved method of irrigation like sprinkler or drip method not only reduce the water requirement but also control from excessive wetness. It has been reported that high soil moisture and humidity in crop canopy influences development of powdery mildew and rust in pea. In chickpea it was recorded that one irrigation before flowering decreases wilt disease incidence and increases yield. However, when the number of irrigations increased the incidence of root rot or wilt disease increased by 2 to 3 fold. Study at IIPR, Kanpur revealed that irrigation after 90 days of sowing significantly increased powdery mildew and rust in pea crop. Irrigation in December-February increased downy mildew. Irrigation after 110 days of sowing further increased powdery mildew and rust diseases. For obtaining maximum yield, irrigation at 90 DAS is most important (Table 6.9). Avoiding waterlogged soils can reduce Phytophthora blight in pigeon pea also.

Table 6.9: Influence of Irrigation Schedule on Diseases and Yield (2003-04) in Pea

Irrigation Schedule	*Powdery Mildew (Per cent)*	*Rust (Per cent)*	*Downy Mildew (Per cent)*	*Alternaria Leaf Spot (Per cent)*	*Yield (kg ha^{-1})*
Pre-sowing	2.0	3.1	2.0	0.6	2807
Pre-sowing + 2nd Feb	3.6	6.9	4.5	1.8	2892
Pre-sowing + 20th Feb	5.7	10.4	6.2	2.5	3148
Pre-sowing + 20th Jan and 20th Feb	6.1	13.1	5.5	2.9	3145
Pre-sowing + 20th Dec, 2nd Feb and 10th Mar	8.8	14.5	5.8	1.9	2823
CD (P = 0.05)	3.3	3.4	2.0	0.9	300

8. Fertilizer Management

Fertilizer management is one of the important agronomic practices for the higher productivity of crop. The productivity of crop depend on the method of application, time of application and form of fertilizer to be applied, the crop productivity is determined. It has been seen that careful fertilizer management can also reduce the disease incidence in pulse crop too. At present integrated nutrient management come into attention as in this practice organic form of fertilizer is also mixed with inorganic form and it was reported that the utilization of organic form of fertilizer can reduce the soil borne diseases. The incidence and severity of wilt disease in pigeon pea was reduced to some extent in the organic manure amended soil as compared to non amended soils (Tables 6.10). Some workers also reported the effectiveness of neem leaves amended soil in reducing *Fusarium udum* population. It is well known fact that the decomposition of the organic matter helped in alternation of the physical, chemical and biotic conditions of the soil and thereby reduce the pathogen number. Micro nutrient like molybdenum and cobalt take part in reducing the diseases like root rot and wilt in lentil as reported by El-Hersh *et al.*, 2011. Mayur *et al.* (2003) reported that chickpea wilt incidence was significantly reduced by amending the soil with deoiled mustard cake, groundnut cake and farm yard manure.

Table 6.10: Effect of Organic Amendments on Wilt Disease of Early Maturing Variety of Pigeon Pea

Treatments	*Disease Incidence (per cent)*		*Mean*	*Disease Severity (per cent)*		*Mean*	*Per cent Disease Control*
	2009	*2010*		*2009*	*2010*		
FYM	68.33	80.07	74.2	64.5	78.58	71.54	22.59
PM	70.75	81.59	76.17	66.86	79.18	73.02	20.53
FYM+ PM	74.97	84.2	79.58	72.46	80.52	76.49	16.97
Control	94.70	97.2	95.85	92.84	96.68	94.91	–
CD at 5 per cent	15.17	–	–	15.74	–	–	–

Source: Devi *et al.*, 2013.

9. Soil Solarization

Soil Solarization is an effective technique used primarily for the disinfestations of soil. The technique was developed and described over three decades ago by a group of Israeli scientist. It is a simple, safe, and effective alternative to the toxic, costly soil pesticides and to control many damaging soil pests. Radiant heat from the sun is the lethal agent involved in soil solarization. A clear polyethylene mulch or tarp is used to trap solar heat in the soil. Over a period of several weeks to a few months, soil temperatures become high enough to kill many of the damaging soil pests. White or black plastic usually does not transmit enough solar radiation to raise soil temperatures to lethal levels for many soil pests. Thinner sheets (0.5 to 1 milimeter) are less costly, but they tear or puncture more easily. It is to be ensured that the moisture levels are adequate for working the soil before laying the plastic sheet. If the soil is dry, water the areas to be solarized before laying sheet because most soil pests are more sensitive to high temperatures in wet soil than in dry soil. This technique is very helpful in controlling pathogen problems such as root knot nematode, sheath blight and other soil borne diseases. Soil solarization has been used worldwide as an IPM technique for fruit trees, vegetables and grain legumes. The effectiveness of soil solarization in controlling wilt disease in pigeon pea over control was due to increase in soil temperature (45°C) below the polythene sheet. The increase in soil temperature with solarization has been considered as the major driving force for the various biological and physiological changes in the soil that affect plant growth. The major legumes, such as bean, pea and lentil also suffer from heavy yield losses due to Fusarium diseases including root rot. Soil solarization greatly reduced the population densities of *Fusarium* species as the major soil borne fungal pathogens.

Table 6.11: Effect of Soil Solarization on Wilt Disease of Pigeon Pea

Treatments	*Disease Incidence (per cent)*		*Mean*	*Disease Severity (per cent)*		*Mean*	*Per cent Disease Control*
	2009	*2010*		*2009*	*2010*		
Soil Solarization	60.15	60.22	52.0	55.98	56.08	56.03	37.21
Control	94.7	97.2	93.76	92.84	96.98	94.91	
t value	*6.43	–	–	*6.23	–	–	–

* Significant at <5 per cent.

Source: Devi *et al.*, 2013.

10. Plant Growth Promoting Rhizobacteria

Plant growth promoting rhizobacteria (PGPR) are defined as root- colonizing bacteria *i.e. Bacillus subtilis* and *Pseudomonas fluorescens* that cause either plant growth promotion or biological control of plant diseases. Plant pathogens such as fungi, bacteria, viruses, nematodes etc., which cause various diseases in crop plants are controlled by PGPR (Raupach *et al.*, 1996; Vidhyasekaran *et al.*, 2001; Viswanathan and Samiyappan, 2002). Mechanisms of bio-control may be competition or antagonisms; however, the most studied phenomenon is the induction of systemic resistance by these rhizobacteria in the host plant (Ramamoorthy *et al.*, 2001). PGPR control the damage to plants from pathogens by a number of mechanisms including, competing the pathogen by physical displacement, secretion of siderophores to prevent pathogens in the immediate vicinity from proliferating, synthesis of antibiotics and variety of small molecules that inhibit pathogen growth, production of enzymes that inhibit the pathogen and stimulation of the systemic resistance in the plants. PGPR may also stimulate the production of biochemical compounds associated with host defense (Niranjan Raj *et al.*, 2005). Biological control using PGPR is a desirable alternative to chemical control of vascular wilt.

11. Crop Rotation/Intercropping Practice

Crop rotation is one of the recommended agricultural managements for increasing crop production as well as disease control, especially those caused by soil inhibiting plants pathogens (Hwang *et al.*, 2009). Several investigators have study the effect of crop rotation of different crops and root exudates for control of plant diseases. A crop grown as part of a rotation is less vulnerable to pests, compared to crops in monocultures. Growing the same crop, or those susceptible to the same disease, in the same place year after year, results in the build-up of disease organisms in the soil. Intercropping can reduce pest problems by making it more difficult for pests to find crops, and by providing habitat for beneficial organisms. Study at IIPR, Kanpur revealed that intercropping pea with faba bean reduced rust and powdery mildew, possibly due to better aeration and less humidity inside the crop canopy (Table 6.12). On the other hand, intercropping pea with lentil increased lentil rust as a result of higher humidity due to the shade effects of pea.

Table 6.12: Influence of Legume Hosts on Pea Diseases (2003-04)

Cropping System	*Rust (Per cent)*	*Powdery Mildew (Per cent)*
Field pea sole at 40 cm. row spacing	9.8	12.6
Pea + Lathyrus (1:1) at 30 cm. row spacing	6.7	11.7
Pea + Lentil (1:1) at 25 cm. row spacing	3.9	8.4
Pea + Fababean (1:1) at 25 cm. row spacing	3.8	8.9
Pea+Lentil + Lathyrus (1:1:1) at 25 cm row spacing	4.0	9.6
Pea + Lentil + Fababean + Lathyrus (1:1:1:1) at 25cm row spacing	9.7	6.7
CD (P = 0.05)	2.45	4.05

Devi *et al.*, 2013 showed that pigeonpea intercropped with sorghum is effective in suppressing *Fusarium udum* population and reduction of wilt disease. It had been also seen that lowest wilt incidence in chickpea obtained with intercropping and mixed cropping with linseed.

12. Conservation Agriculture Practice

Conservation agriculture is becoming popular due to its potential for enhanced productivity and cost savings among small scale farmers in developing countries. It includes zero tillage, raise bed planting, minimum tillage etc. Through this conservation tillage practice some soil borne pathogen can be reduced that cause serious diseases in pulse crop. *Fusarium* or other soil born pathogen due to less disturbances in soil does not get expose and their activity also become low. Evidence showed that zero tillage practice invites less disease incidence than the conventional tillage practiced crop.

So, from the above different methods of cultivation practices involved in legume production, it can be concluded that with the help of this agronomic practices along with Integrated Pest Management (IPM), the diseases occurred in legume can be cured up and make the crop production more sustainable.

References

Andrabi, M., Vaid, A., Razdan, V. K. 2011. Evaluation of different measures to control wilt causing pathogens in chickpea. *Journal of Plant Protection Research*. **51**(1): 55-59.

Annual report 2011-12. Indian Institute of Pulses Research, Kanpur, p. 22.

Bahl, P. N. 1976. *Indian Journal of Genetics*. **36**: 351-352.

Chandra, S. *et al*., 1974. *Indian Journal of Genetics*. **34**: 257-262

Devi, T. R., Chhetry, G. K. N. 2013 Effect of certain agronomic practices for the management of wilt of pigeonpea in Manipur agro-climatic conditions. *IOSR Journal of Engineering*. **3**(4): 21-23.

Dhar Vishwa and Chaudhary R G. 2001. Disease resistance in pulse crops - Current status and future approaches. *I: Role of Resistance in Intensive Agriculture*, pp 144-57. Nagarajan, S and Singh, D P (Eds). Kalyani Publishers, Punjab, Ludhiana, India.

El-Hersh, M. S., Abd El- Hai, K. M. and Ghanem, K. M. 2011. Efficiency of Molybdenum and Cobalt Elements on the Lentil Pathogens and Nitrogen Fixation. *Asian Journal of Plant Pathology*. 5(3): 102-114.

FAO, 2008. *The state of food insecurity in world*: high food price and food security threats and opportunities, Rome, FAO.

Gohel V, Maisuria V, Chhatpar H.S. 2006. Utilization of various chitinous sources for production of mycolytic enzymes by *Pantoea dispersa* in bench-top fermentor. *Enzyme Microb. Technol*. **40**: 1608-1614.

Hwang, S. F., Ahmed, H. U., Gossen, B. D., Kutcher, H. R., Brandt, S. A., Strelkov, S. E., Chang, K. F., Turnbull, G. D. 2009. Effect of crop rotation on the soil pathogen population dynamics and canola seedling establishment. *Plant Pathology Journal*. **8**: 106-112.

Jeger, M. J. 2000. Bottlenecks in IPM. *Crop Protection* **19:** 787-792.

Kumari, S., Kumar, S., Tripathi, H.S. and Singh, V., 2014. Effect of dates of sowing and spacing on wilt rust and powdery mildew disease and yield of field pea. *Ann. Pl. Protec. Sci.*, **22(1):** 190-239.

Mayur *et al.*, 2003. *Research on Crops* **4(1)**: 141-143.

Niranjan Raj, S., Shetty, H.S., and Reddy, M.S. 2005. Plant growth-promoting Rhizobacteria: Potential green alternative for Plant productivity., Springer, Dordrecht, The Netherlands *Z.A.* Siddiqui (ed.), *PGPR: Biocontrol and Biofertilization*. 197-216

Perez-Brandan, C., Arzeno, Jose, L., Huidobro, J., Grumberg, B., Conforto, C., Hilton, S., Bending, G. D., Meriles, J. M., Vargas-Gil, S. 2012. Long-term effect of tillage systems on soil microbiological, chemical and physical parameters and the incidence of charcoal rot by *Macrophomina phaseolina* (Tassi) Goid in soybean. *Crop Protection*. **40**: 73-82

Pretty, J. N., Morison, J. I. L. and Hine, R. E., 2003. Reducing food poverty by increasing agricultural sustainability in developing countries. *Agriculture, Ecosystem and Environment* **95**: 217-234.

Ramamoorthy, V., and Samiyappan, R., 2001. Induction of defense related genes in *Pseudomonas fluorescens* treated chili plants in response to infection by *Colletotrichum capsici. Journal of Mycology and Plant Pathology* **31**: 146-155.

Raupach, G. S. and Kloepper, J. W., 1998. Mixtures of plant growth-promoting rhizobacteria enhance biological control of multiple cucumber pathogens. *Phytopathology*, **88**: 1158-1164.

Sangar, R.B.S. and Singh, V.K.1994. Effect of sowing dates and pea varieties on the severity of rust, powdery mildew and yield. *Indian J. Pul. Res.* **7**: 88-89.

Saxena, M. C., 1993. The challenge of developing biotic and abiotic stress resistance in cool-season food legumes eds. K.B. Singh and M.C. Saxena In: Breeding for Stress Tolerance in Cool-Season Food Legumes. ICARDA-Wiley-Sayce, Chichester, p. 3-14.

Singh, K. B. and Sandhu, T. S. 1973. Cultivation of Gram in Punjab. Punjab Agriculture University, Ludhiana, India.

Singh, K. K., Ali, M. and Venkatesh, M.S., 2009. Pulses in Cropping Systems. Technical Bulletin, IIPR, Kanpur, p. 5.

Sugha, S.K. *et al.*, 1994. *Indian Journal of Mycology and Plant Pathology*. **24**: 97-102

Vidhyasekaran, P., Kamala, N., Ramanathan, A., Rajappan, K., Paranidharan, V., and Velazhahan, R., 2001 Induction of systemic resistance by *Pseudomonas fluorescens* Pf1 against *Xanthomonas oryzae* pv. *oryzae* in rice leaves. *Phytoparasitica*, **29**: 155–166.

Viswanathan, R and Samiyappan, R., 2002. Induced systemic resistance by fluorescent pseudomonads against red rot disease of sugarcane caused by *Colletotrichum falcatum*. *Crop Protection*. **21**: 1-10.

Windling R, Fawcett H.S. 1936. Effect of patial sterilization by stem or formalin on damping off sikka spruce. *Trans. Br. Mycol. Soc.* **35**: 258-262.

2016, Diseases of Pulse Crops and their Sustainable Management 103–115
Editors: **Samir Kumar Biswas, Santosh Kumar and Gireesh Chand**
Published by: **BIOTECH BOOKS, NEW DELHI**

Chapter 7

Nutrient Management in Pulse Crops

***Krishnendu Ray*[1]* *and Jajati Mandal*[2]**

[1]*Department of Agronomy, Bidhan Chandra Krishi Viswavidyalaya, Mohanpur, Nadia-741252, West Bengal*

[2]*Department of Soil Science and Agricultural Chemistry, Bihar Agricultural University, Sabour, Bhagalpur, Bihar – 813 210*

Introduction

About 65 per cent of the people in India are dependent on agriculture. The population pressure of the country is increasing day-by-day without any possibility of increase in the existing farm area. Hence to feed the enormous population of the country from the limited cultivable area, proper crop management practices are of prime importance. With the introduction of modern high yielding varieties (HYVs) and improved irrigation facilities, consumption of chemical fertilizers has been increased (Pathak *et al.*, 2010). However, the nutrient use by different field crops is still not up to the mark. So to ensure the best management practices (BMPs) for the crops and to facilitate better nutrient use efficiency (NUE), nutrient management is very much important. Among the different field crops cultivated in India, pulses held significant position. The production of pulses in India is about 18.34 million tonnes from an area of about 23.26 million hectares with average productivity 789 kg/ha (Economic survey, 2013). The most important thing about pulses is that they take so little and gives so much to the soils. They are significant in restoration and maintenance

* Corresponding Author: E-mail: krishnenduray.bckv@gmail.com

of soil fertility. Pulses add about 0.5-1.5 tonnes of organic matter to the soil in the form of their roots left after harvest of the crop (Table 7.1).

Table 7.1: Amount of Organic Matter Added to the Soils by some Pulses

Sl.No.	*Crop*	*Green Matter (t/ha)*	*N per cent*	*Total Added Nitrogen*
1	Blackgram	12.0	0.41	38.3
2	Greengram	8.0	0.53	34.5
3	Lentil	5.6	0.70	32.8
4	Pea	20.1	0.36	59.4
5	Cowpea	15.2	0.51	52.4

Source: Singh, 2009, pp. 154.

Besides, the net build-up of the soil organic carbon (SOC) stock depends largely on existing cropping systems and management practices which ensure the amount of crop residue returned to soil. The optimum levels of SOC can be managed through appropriate crop rotation (Wright and Hons, 2005), nutrient management (Schuman *et al.*, 2002) and tillage methods (Bhattacharyya *et al.*, 2008). Because of their ability for atmospheric nitrogen fixation, leaf shedding ability and greater below-ground biomass, pulses contribute a significant amount of organic carbon to soil. About 10.7 per cent increase in SOC content in intensive pulse-based cropping systems have been reported (Newaj and Yadav, 1994). In another study, Singh and Sandhu (1980) observed 6.3 per cent increment of SOC due to inclusion of greengram and cowpea in a maize-wheat system.

Pulses play an important role in Indian agriculture as they restore soil fertility by fixing atmospheric nitrogen through their nodules. Nitrogen fixation is an important process by which nitrogen from the sterile inorganic molecular form in the atmosphere is fixed and converted to an organic form (Mengel and Kirkby, 1979, pp. 331). Biological nitrogen fixation (BNF) is brought about by free living microorganisms (*Azotobacter, Clostridium, Beijerinckia, Achromobacter* and *Pseudomonas* etc.) and by microorganisms (*Rhizobium* sp.) which live in symbiosis with higher plants mainly leguminous pulses. It has a tremendous importance in agriculture. It contributes a significant portion of indigenous nitrogen supply (INS) from soil. Apart from BNF, pulses also add N to the soil through green manuring in both two ways *viz.* green manuring in-situ and green leaf manuring.

However, the production of pulse crops in our country is not sufficient enough to meet the domestic demand of the population. Hence, there is an ample scope for enhancement of the production and productivity of these crops by proper agronomic practices. Several strategies have been initiated to boost the productivity of pulses. One such is application of nutrients from organic and inorganic sources for exploiting genetic potential of the crop. It also necessitates nutrient application at critical stages. Different essential macro and micronutrients play significant role in increasing the yield and quality in pulses (Chandrasekhar and Bangarusamy, 2003). Management of these nutrient elements in different types of major and minor pulses has been jot down below with an intension to make the nutrient management aspect in pulse crops more convenient and efficient.

Major Pulses

Chickpea (*Cicer arietinum* L.)

Chickpea is the most important among pulse crops occupying largest area (6.4 million hectare) and production (5.10 million tonnes) in India (Pramanik *et al.*, 2009). It is one of the ICRISAT's five crops (sorghum, pigeonpea, pearl millet, groundnut, and chickpea). Chickpea fulfills its nitrogen requirement from symbiotic nitrogen fixation. Thus it does not respond well to higher dose of nitrogen. A starter dose of 20-25 kg N/ha is usually recommended for its cultivation. Phosphorus plays a very vital role in root development, nodulation, growth and yield of the crop. Basal dose of 50-60 kg P_2O_5 should be applied to secure good yield of the crop. The effect of phosphorus is more pronounced when applied in conjunction with starter nitrogen, *rhizobium* and irrigation. It is considered to be less responsive to phosphorus fertilizer application than cereals. Under field conditions, the effectiveness of fertilizer phosphorus is expected to be strongly influenced by the method of application because of the limited mobility of phosphorus and its fixation by soil colloids. For Alfisol and Vertisols, it is usually considered that deep placement of phosphorus at 12-15 cm in band would be more effective for higher yield as it places phosphorus in a zone where root proliferate and thus minimizes phosphorus fixation. Response of the crop to applied potassium is inconsistent. Potassium application upto 20 kg/ha has been found to improve the yield. Total uptake of N, P and K by chickpea generally varies from 20-40, 4-8 and 20-35 kg/ha, respectively. Application of integrated source of nutrient has been observed to significantly increase crop yield. About 68 per cent more seed yield (over control) of succeeding chick pea grown after sorghum was obtained due to the residual effect of 25 per cent N through farmyard manure (FYM) + 25 per cent N through vermicompost + 50 per cent N through recommended dose of fertilizer (RDF) (100:50:50 :: N: P_2O_5: K_2O) applied to preceding forage sorghum in *kharif* season (Nawale *et al.*, 2009). Gawai and Pawar (2007) also reported the increase in seed yield of chickpea due to application of organic and inorganic fertilizer in combination. Sulphur nutrition has favorable effect on growth parameters which leads to greater nutrient uptake, efficient partitioning of metabolites and adequate translocation and accumulation of photosynthates. Lal *et al.* (2014) reported increased growth attributes, nodulation, leghaemoglobin content, nitrogenase activity, yield components and seed yield of the crop with sulphur fertilization.

Among the micronutrients, zinc and iron play an important role in chickpea production. Application of 25 kg $ZnSO_4$/ha at the time of sowing along with the macronutrients is recommended. The effect of Fe is more pronounced in the presence of Zn. One foliar spray of 2 per cent solution at 30 days after sowing (DAS) is recommended for improving productivity of the crop. Singh (2001) reported the yield response of chickpea by 350, 330 and 360 kg/ha to boron, iron and zinc, respectively.

Lentil (*Lens culinaris* L.)

Low productivity of lentil may be ascribed to many reasons, but inadequate and imbalanced fertilization is the major factor. In most of the lentil-growing areas, poor soil fertility is one of the severe constraints limiting expression of its full production potential. To augment the productivity of this crop, use of balanced fertilization by application of chemical and organic sources along with biofertilizers, *viz. Rhizobium* and *phosphate solubilising bacteria* (PSB) is of great importance at this stage. The crop responds well to application of starter nitrogen of 20-25 kg/ha and moderate level of phosphorus (50-60 kg/ha). Application of potassium is totally based on soil test value. However, in an experiment conducted by Sahay *et al.* (2013), yield and protein content of lentil was found to increase significantly even upto 90 kg/ha. Mandal *et al.* (2009) observed maximum yield of the crop with soil test-based fertilizer recommendation of 20 kg N, 50 kg P, 50 kg K, 40 kg S, 1.5 kg B/ha and Zn as Zn-EDTA (0.05 per cent) at 6 weeks of the crop. In another study conducted by Singh *et al.* (2013), lentil was grown as succeeding crop after rice in Indo-Gangetic plain. Four different levels (0, 20, 30 and 40 kg/ha) of sulphur and four different levels (0, 4, 5, 6 kg/ha) of Zn was applied in preceding rice crop and effect of these nutrients in the successive lentil was studied. They observed 23.23 per cent higher (over control) lentil yield when the crop was given 30 kg S/ha in combination with 6 kg Zn/ha. Combined application of inorganic N, P, K and S (20, 17, 16, 20 kg/ha) + vermicompost 2 t/ha has been reported to significantly increase the growth, yield attributes, seed yield and net returns of the crop over absolute control (Ram *et al.*, 2013). Integrated use of recommended dose of inorganic fertilizer, FYM or vermicompost and biofertilizers tend to increase the seed yield further over their sole applications, which could be due to the combined and synergistic effect. Furthermore, manures contain high amounts of organic matter which increases the moisture retention of the soil and improves dissolution of nutrients particularly phosphorus. In another study, application of RDF (20 kg N + 40 kg P_2O_5/ha) + seed inoculation with *Rhizobium leguminosarum* and PSB (*Bacillus* sp.) each @ 500 g/ha was observed to give higher number of nodules/plant, nodule dry weight, shoot dry weight, plant height, lentil yield and benefit: cost (B: C) ratio (Singh *et al.*, 2010). Among the beneficial nutrients, cobalt has been found to have significant effect on lentil. It is essential for nodulating bacteria for fixing atmospheric nitrogen. Growth and yield attributes, yield (seed and straw) and uptake of N and P of lentil has been found to increase up to application of 4 kg Co/ha, but S uptake decreased with gradual increase in Co application (Sahay *et al.*, 2013). Recent studies have suggested the use of bio-fortification of lentil seed to address the micronutrient deficiencies. They observed that agronomic bio-fortification approaches can be used to complement genetic techniques to improve lentil yield. These strategies include the application of inorganic fertilizers containing Fe-chelates, and the improvement of solubilization and mobilization of Fe in the soil through intercropping and crop rotations, as well as soil microorganisms (DellaValle *et al.*, 2013).

Pea

Pea (*Pisum sativum* L.) is one of the most important pulse crops and also the cheapest source of dietary protein. It is an efficient N-fixer and thus improves soil fertility and can serve as a useful component in any viable cropping system, and for the same reason it does not require higher doses of nitrogen. Application of 20-30 kg N/ha as starter dose can meet plant requirement before formation of nodules.

Application of 60-70 kg P_2O_5/ha and 30-40 kg K_2O/ha can bring about better growth and yield of the crop. Combined use of both organic and inorganic sources of nutrients through integrated nutrient management (INM) has proved to augment the yield and quality of the crop. Response of gardenpea towards nitrogen fertilization is more in comparison with fieldpea and thus the former can be supplied with about 50-60 kg N/ha. Sepehya *et al.* (2012) conducted experiment on INM for gardenpea where they found significantly higher yield of both green pod and stover with application of FYM or vermicompost @ 10 t/ha in combination with 75 per cent recommended dose of NPK (recommended dose was 50, 60 and 60 kg of N, P_2O_5 and K_2O/ha) produced higher yield over that in NPK or FYM alone. Application of 15 kg S/ha was reported to have better effect on the crop Singh (1999). It may be increased upto the extent of 30 kg S/ha as the element is of much importance in terms of better root growth, dry matter production and glutathione content. A good response of pea has been observed to the application of 0.5 kg Mo/ha (Tandon, 2002, pp. 47). In a pot experiment conducted by Singh *et al.* (2012), the effects of cobalt, boron, molybdenum on yield of and nutrient uptake by pea under graded levels of soils fertility was studied. Some astonishing results were obtained from the experiment. Micronutrient application augmented the growth, yield and nutrient uptake of the crop at every successive level. Sole application of Co boosted up the crop performance more than that of B and Mo. Combined application of these micronutrients proved superior to their sole applications.

Pigeonpea [*Cajanus cajan* (L.) Millsp.]

Pigeonpea is the most important rainy season grain legume in India. It occupies 16-17 per cent of total pulse area but accounts for only 17 per cent of pulse production in the country (Kumar *et al.*, 2013). One of the important reasons for this low yield of the crop is negligence of proper fertilizer application. On soils with low nitrogen status, application of 20-30 kg N/ha as starter dose have been proved to be fruitful. However, the crop can respond upto 50 kg N/ha. Significant increase in seed and stalk yields of pigeonpea was recorded after application of 50, 75 and 100 kg NPK/ha along with 12.5 kg Zn/ha and 10 kg S/ha (Umesh *et al.*, 2013). Despite of the higher seed yield obtained with 80 kg P_2O_5/ha, it was statistically on par with the combined application of 40 kg P_2O_5/ha + PSB + VAM (Kumar *et al.*, 2013). The later treatment recorded higher moisture extraction pattern, moisture indices and B: C ratio. Response of pigeonpea to phosphorus is generally low. A possible reason why pigeonpea responds less to phosphorus application is its dependence on vesicular arbuscular mycorrhizal (VAM) associations that permit better utilization of available soil phosphorus especially at low levels of available phosphorus. However, survival of the crop on Alfisol without addition of VAM or phosphorus may be attributed to its special ability to utilize sparingly soluble available phosphorus in Alfisol. Alfisols have most of their phosphorus in an iron-bound form whereas in Vertisol, there is a greater proportion of calcium-bound phosphorus. Field experiments have proved that pigeonpea can grow much better than other crops with Fe-P which mirrored the better efficiency of the crop to utilize Fe-P. Umesh and Shankar (2013) reported an increase in the uptake of macronutrients (N, P and K) when the crop was fertilized with 50, 75 and 100 kg N, P and K/ha, respectively along with 12.5 kg Zn/ha and 10 kg S/ha. Deshbhratar *et al.* (2010) indicated a significant increase in seed yield of pigeonpea after 20 kg S/ha and 50 kg P_2O_5/ha with common dose of nitrogen @ 30 kg/ha. In that study, the increase in seed yield was 102.77 per cent as compared to control.

There is a lack of information on the nutrient management of *rabi* pigeonpea. Singh (2009) has suggested to use about 30, 50, 20 and 10 kg of N, P_2O_5, $ZnSO_4$ and S, respectively. Among the 30 kg N/ha, 20 kg should be supplied at basal and the rest 10 kg N/ha at 30 days after sowing.

Greengram [*Vigna radiata* (L.) Wilczek]

Greengram is of greater important not only as a low input requiring pulse crop but also as an valuable crop in most of the major cropping systems in India. The crop yields better with 15-20 kg N and 40-50 kg P_2O_5/ha. However, both the autumn and summer greengram performed best when N, P and K @ 20: 40: 40 kg/ha (with sulphur and phosphorus from single super phosphate) was applied (Banik and Sengupta, 2013). In another experiment conducted in Gwalior, M.P., application of 37.5 N: 90 P_2O_5: 5.0 Zn Kg/ha along with 30 kg S/ha through ammonium sulphate have resulted better seed yield, protein content, uptake of N, P, K, Zn and S and highest B: C ratio (Sasode, 2008). When grown after wet seeded rice applied with both organic and inorganic source of nutrients, greengram does not require to fertilized additionally, and at this condition the crop can yield upto 596 kg/ha (Vennila and Jayanthi, 2007). The application of organic manure to preceding rice crop has residual effect by continuous mineralization and supply of N to succeeding greengram.

Summer Greengram is new introduction mainly to utilize the land and water resources which otherwise remains unexploited during summer season. In an experiment conducted by Mandal *et al.* (2005), 30, 60 and 40 kg of N, P_2O_5 and K_2O, respectively recorded the best result in terms of yield, NPK uptake, gross and net return. Different growth and yield attributing characters were highest and the flower drop/plant was lowest in summer greengram applied with foliar spray of 1.5 per cent urea and 1.5 per cent di-ammonium phosphate (DAP) before flowering (Deshmukh *et al.*, 2013). In an experiment, conducted by Meena *et al.* (2012), summer greengram was grown after baby corn-potato without any nutrient application. Baby corn fertilized with $N_{60}P_{13}K_{17}$ + FYM (equivalent to 60 kg N/ha, containing 0.48 per cent N, 0.22 per cent P_2O_5 and 0.51 per cent K_2O; applied at 10 days prior to sowing) and potato with $N_{60}P_{17}K_{42}$ + FYM (equivalent to 60 kg N/ha) recorded highest seed yield of succeeding Greengram. The result was at par when baby corn was given with $N_{60}P_{13}K_{17}$ + bio-compost (equivalent to 60 kg N/ha, containing 1.61 per cent N, 1.45 per cent P_2O_5 and 2.70 per cent K_2O; applied at 10 days prior to sowing) and potato with $N_{60}P_{17}K_{42}$ + FYM (equivalent to 60 kg N/ha).

Blackgram [*Vigna mungo* (L.) Hepper]

Blackgram is an important short duration pulse grown throughout India. In different predominant intercropping systems, blackgram has been proved to enhance total productivity and land equivalent ratio (LER) besides the benefits of suppressing weeds and spreading the risk involved. The production of blackgram is not sufficient enough to meet the domestic demand of the population. There is an ample scope for augmentation of the production and productivity of blackgram by proper agronomic practices. Nutrient management is one of them. Proper fertilization is essential to improve the productivity of blackgram. Application of 15-20 kg N/ha as starter dose is sufficient for a better yield. Crop perform better when is applied with 50-60 kg P_2O_5 and 30-40 kg K_2O/ha. Application of 60 kg P_2O_5 and 40 kg S/ha yielded highest plant height, number of nodules and number of leaves per plant, yield, nutrient and protein content (Mir *et al.*, 2013). In addition to the RDF, foliar spray of 40 ppm naphthalene acetic acid (NAA) + 0.5 per cent chelated micronutrient + 2 per cent DAP at 35 and 50 DAS recorded significantly higher growth components like plant height, number of branches/plant, leaf area index, total dry matter production, grain yield, nutrient uptake, net returns and B:C ratio (Shashikumar *et al.*, 2013).

Soybean (*Glycine max* L.)

Soybean is an energy-rich grain legume containing about 40 per cent protein and 19 per cent oil in the seeds. However, presence of lipoxidase (off flavour producing enzyme) in soybean has limited its use as a pulse crop in India. Application of 30-40 kg N, 70-80 kg P_2O_5 and 40-50 kg K_2O/ha is recommended as basal under irrigated conditions. Low rate of nitrogen application is advised due to the capacity of the crop to meet a part of its nitrogen requirement through symbiotic nitrogen fixation with *Rhizobium japonicum*. However, some researchers have also proved that the nitrogen fixed by soybean is not as much as to promote growth and development, thus necessitates an application of about 60 kg N/ha in many cases (Singh, 2009, pp. 214). Under rainfed conditions, application of 20 kg N, 40-50 kg P_2O_5 and 20-30 kg K_2O/ha as basal is optimum. The crop responds to varying rate of applied sulphur from 10 kg to about 40 kg S/ha. In Zn deficient soils, application of 20 kg $ZnSO_4$/ha is recommended. For Mn deficient soils, foliar application of 1 per cent $MnSO_4$ twice (one at 20 DAS and another at 40 DAS) has been reported to correct Mn deficiency.

Minor Pulses

Cowpea [*Vigna unguiculata* (L.) Walp.]

Cowpea, an important grain legume in rainfed regions and marginal areas of the tropics and subtropics, is particularly important in India for its use as pulse, vegetable and fodder. Short growing cycle of the crop makes it important in different predominant cropping system. Cowpea reduces the soil erosion, enriches the soil by atmospheric nitrogen fixation and reduces the weed growth by smothering effect, and therefore, it is considered as important legume cum vegetable crop. The crop can fix more than 50 per cent of its N from N-fixation (Khan *et al.*, 2002). It generally requires about 20-25, 50, 30-40 and 20 kg of N, P_2O_5, K_2O and S/ha for better productivity. The recommended dose of fertilizers (25:60:40 kg NPK/ha) recorded highest yield in Arunachal Pradesh (Choudhary and Kumar, 2013). Result of the experiment conducted by Pravakar *et al.* (2013) revealed that application of phosphorus @ 60 kg/ha and manganese @ 10 kg/ha recorded significantly higher plant height, green foliage, dry matter and concentration of N, P, K and Mn. Application of 60 kg K_2O and 40 kg Zn/ha have also been recorded to increase yield, nutrient uptake and protein content of cowpea (Chavan *et al.*, 2012). Parasuraman (2001) conducted experiment to study the influence of seed pelleting with diammonium phosphate (DAP) and potassium dihydrogen phosphate (KH_2PO_4) in comparison with DAP spray in rainfed cowpea in North-Western Tamil Nadu. Soil application of recommended inorganic fertilizers (12.5: 25: 0 :: N: P: K kg/ha) + 2 per cent DAP spray twice (first at flowering and second at 15 days after first spray) showed much better growth attributes, seed yield, net income and B: C ratio over seed pelleting with DAP and KH_2PO_4. Inclusion of Fe @ 4kg/ha in the recommended fertilizer dosage improved the yield of cowpea at Jobner (Mahriya and Meena, 1999).

French Bean (*Phaseolus vulgaris*)

French bean or rajmash is an exceptional leguminous crop which does not fix nitrogen biologically with native rhizobia, and thus the fertilizer N requirement of this crop is very high. However, Martinez-Romero *et al.* (1991) have identified *Rhizobium tropici* as novel rhizobia to nodulate french bean root. The BR-885 strain of *Rhizobium tropici* was found more efficient over strain BR-865 and fixed 7.85 per cent higher nitrogen in unfertilized french bean (Jha and Singh, 2013). The former one also recorded higher leaf area index, grain yield, and N and P uptake. In the same experiment, application of 40 Kg N/ha was found the best with biological nitrogen fixation (BNF) point of view. But when such nodulating rhizobia is not available, the recommended dose of 120: 60: 40 :: N, P_2O_5, K_2O kg/ha gives higher yield of the crop. Combined application of 100 per cent RDF + 50 per cent N through vermicompost recorded higher yield of the crop (Kumar *et al.*, 2009). In the same experiment, combined effect of biofertilizers (*Rhizobium* + *Bacillus* + *Pseudomonas*) and micronutrients (Zn @ 5 kg/ha and Fe @ 1 kg/ha) recorded significantly better growth characters, yield attributes, yield, harvest index, nutrient uptake and B: C ratio. An investigation was carried out at Karnataka to study the residual effect of zinc and boron applied for transplanted rice on french bean. Application of RDF (120: 80: 60 :: N: P_2O_5: K_2O kg/ha) + Zinc (Zinc sulphate) @ 18 Kg/ha + boron (boric acid) @ 4 Kg/ha brought about significantly the highest residual impact on growth and yield of the crop (Hamsa and Puttaiah, 2012). Recently studies have been made on the ameliorating effect of Cd toxicity of the crop through foliar spray of Fe (Chatterjee, 2013).

Lathyrus (*Lathyrus sativus* L.)

Lathyrus, commonly known as khesari or teora or grass pea or chickling pea, is generally grown as minor pulse crop on residual fertility and soil moisture. It is mainly grown as paira crop and sown in the rice field about 15 days before harvest of the crop. The crop may be fertilized with 20, 50, 20 and 30 kg of N, P_2O_5, K_2O and S/ha.

Horsegram (*Macrotyloma uniflorum* L. Verde)

It is an important minor pulse crop and being used for human consumption and cattle feed owing to its higher protein content. The crop requires about 10-20 kg N, 30 kg P_2O_5 and 30 kg K_2O/ha. Application of 12.5 kg N and 25 kg P_2O_5/ha before sowing is also recommended (Chandrasekaran *et al.*, 2010, pp.579). Response of horsegram genotypes to plant densities and phosphorus levels was studied at Raichur, Karnataka

(Keshava *et al.*, 2007). Among phosphorus levels tried, 60 kg P_2O_5/ha, being on par with 30 kg P_2O_5/ha, recorded higher seed yield. The same experiment has also documented the higher total nitrogen and phosphorus uptake with application of 30 and 60 kg P_2O_5/ha, respectively. Lenka *et al.* (1989) recorded that the application of 10 kg N and 20 kg P_2O_5/ha under rainfed and 20 kg N and 60 kg P_2O_5/ha under irrigated condition is optimum for horsegram.

Mothbean

It is an important arid legume and can tolerate both drought and heat. In arid region, under harsh environment, it is the ultimate choice for realization of sustained production. The crop occupies about 54 per cent of the cultivated area under pulses in Rajasthan with a very low productivity of 250 kg/ha (Jat *et al.*, 2011). Main reason behind this is the low adoption of the improved technologies. Optimum nutrient requirement for the crop is 15-20 kg N and 30 kg P_2O_5/ha. Potassium application should be based on soil test value. Application of 30 kg P_2O_5/ha, being on par with 60 kg P_2O_5/ha, significantly increased the growth attributes, effective nodules and dry weight of nodules, seed yield, net returns and nutrient (N, P, K and Zn) concentration in seed (Yadav *et al.*, 2012). The same experiment also recorded application of 4 kg Zn/ha (on par with 6 kg Zn/ha) to be optimum for growth, yield, economics and nutrient uptake of the crop.

References

Banik, N.C. and Sengupta, K. 2013. Increased production of mungbean (*Vigna radiata*) through nutrient management and utilization of residual soil productivity for raising succeeding crop. *Indian Journal of Agronomy* 58(4): 560-563.

Bhattacharyya, R., Kundu, S., Ved Prakash, and Gupta, H.S. 2008. Sustainability under combined application of mineral and organic fertilizers in a rainfed soybean wheat system of the Indian Himalayas. *European Journal of Agronomy* 28: 33-46.

Chandrasekaran, B., Annadurai, K. and Somasundaram, E. 2010. A Textbook of Agronomy. New Age International Publishers, New Delhi, India.

Chandrasekhar, C.N. and Bangarusamy, U. 2003. Maximizing the yield of mung bean by foliar application of growth regulating chemicals and nutrients. *Madras Agricultural Journal* 90(1-3): 142-145.

Chatterjee, J. 2013. Amelioration of cadmium effects on fruit quality of french bean with iron. *Indian Journal of Agricultural Biochemistry* 26(2): 155-159.

Chavan, A.S., Khafi, M.R., Raj, A.D. and Parmar, R.M. 2012. Effect of potassium and zinc on yield, protein content and uptake of micronutrients on cowpea [*Vigna unguiculata* (L.) Walp.]. *Agricultural Science Digest* 32(2): 175-177.

Choudhary, V.K. and Kumar, P.S. 2013. Nutrient budgeting, economics and energetics of cowpea under potassium and phosphorus management. *SAARC Journal of Agriculture* 11(2): 129-140.

DellaValle, D.M., Thavarajah, D., Thavarajah, P., Vandenberg, A., Glahn,R.P. 2013. Lentil (*Lens culinaris* L.) as a candidate crop for iron biofortification: Is there genetic potential for iron bioavailability? *Field Crops Research* 144: 119-125.

Deshbhratar, P.B., Singh, P.K., Jambhulkar, A.P. and Ramteke, D.S. 2010. Effect of sulphur and phosphorus on yield, quality and nutrient status of pigeonpea (*Cajanus cajan*). *Journal of Environmental Biology* 31(6): 933-937.

Deshmukh, S.B., Raundal, P.U., Kunjir, N.T. 2013. Effect of foliar spray of nutrients on growth, yield and quality of summer Greengram. *Bioinfolet* 10(3B): 1060-1064.

Economic Survey, 2013. Statistical Appendix. Ministry of Home Affairs. Government of India [Available at: http: / /indianbudget.nic.in].

Gawai, P.P. and Pawar, V.S. 2007. Nutrient balance under INMS in sorghum–chickpea cropping sequence. *Indian Journal of Agricultural Research* 41: 137-141.

Hamsa, A. and Puttaiah, E.T. 2012. Residual effect of zinc and boron on growth and yield of french bean (*Phaseolus vulgaris* L.)-rice(*Oryza sativa* L.) cropping system. *International Journal of Environmental Sciences* 3(1): 167-171.

Jat, H.L., Singh, P. and Lakhera, J.P. 2011. Constraints in adoption of moth bean production technology in arid zone of Rajasthan. *Rajasthan Journal of Extension Education* 19: 185-189.

Jha, A.K. and Singh, R. 2013. The influence of interaction between *Rhizobium tropici* and fertilizer N on nutrient uptake, growth and yield of French bean (*Phaseolus vulgaris* L.) under salt stress. *Progressive Horticulture* 45(1): 222-228.

Keshava, B.S., Halepyati, A.S., Pujari, B.T. and Desai, B.K. 2007. Yield and nutrient uptake of horsegram (*Macrotyloma uniflorum Lam*. Verde) as influenced by genotypes, plant densities and phosphorus levels. *Karnataka Journal of Agricultural* Sciences 20(1): 4-5.

Khan, D.F., Peoples, M.B., Chalk, P.M. and Herridge, D.F. 2002. Quantifying below ground nitrogen of legumes. 2. A comparison of 15N and non-isotopic methods. *Plant and Soil* 239: 277-289.

Kumar, P., Rana, K.S., Ansari, M.A. and Hariom. 2013. Effect of planting system and phosphorous on productivity, moisture use efficiency and economics of sole and intercropped pigeonpea (*Cajanus cajan*) under rainfed conditions of northern India. *Indian Journal of Agricultural Sciences* 83(5): 549-554.

Kumar, R.P., Singh, O.N., Singh, Y., Dwivedi, S. and Singh, J.P. 2009. Effect of integrated nutrient management on growth, yield, nutrient uptake and economics of french bean (*Phaseolus vulgaris*). *Indian Journal of Agricultural Sciences* 79(2): 122-128.

Lal, B., Rana, K.S., Rana, D.S., Priyanka Gautam, Shivay, Y.S. and Ansari, M.A. Meena, B.P. and Kuldeep Kumar. 2014. Influence of intercropping, moisture conservation practice and P and S levels on growth, nodulation and yield of chickpea (*Cicer arietinum* L.) under rainfed condition. *Legume Research* 37(3): 300-305.

Lenka, D., Behura, A.K. and Nayak, R.K. 1989. Effect of irrigation and fertilization on yield of horsegram and uptake of nitrogen and phosphorus in acid laterite soils. *Orissa Journal of Agricultural Research* 2(1): 10-15.

Mahriya, A.K. and Meena, N.L. 1999. Response of phosphorus and iron on growth and quality of cowpea [*Vigna unguiculata* (L.) Walp.]. *Annals of Agro-Bio Research* 4(2): 203-205.

Mandal, M.K., Pati, R., Mukhopadhyay, D. and Majumdar, K. 2009. Maximization of lentil (*Lens culinaris*) yield through management of nutrients. *Indian Journal of Agricultural Sciences* 79(8): 645-647.

Mandal, S., Biswas, K.C. and Jana, P.K. 2005. Yield, economics, nutrient uptake and consumptive use of water by summer greengram (*Vigna radiata* L.) as influenced by irrigation and phosphorus application. *Legume Research* 28(2): 131-133.

Martinez-Romero, E., Segovia, L., Mercante, F.M., Franco, A.A., Graham, P. and Pardo, M.A. 1991. *Rhizobium tropici*, a novel species nodulating *Phaseolus vulgaris* L. beans and Leucaena sp. Trees. *International Journal of Systematic Bacteriology* 41: 417-426.

Meena, S.R., Kumar, A., Jat, N.K., Meena, B.P., Rana, D.S. and Idnani, L.K. 2012. Influence of nutrient sources on growth, productivity and economics of baby corn (*Zea mays*)-potato (*Solanum tuberosum*)-mungbean (*Vigna radiata*) cropping system. *Indian Journal of Agronomy* 57(3): 217-221.

Mengel K. and Kirkby E.A. 1979. *Principles of plant nutrition*, 2nd edition, International potash institute, Bern, Switzerland.

Mir, A.H., Lal, S.B., Salmani, M., Abid, M. and Khan, I. 2013. Growth, yield and nutrient content of blackgram (*Vigna mungo*) as influenced by levels of phosphorus, sulphur and phosphorus solubilizing bacteria. *SAARC Journal of Agriculture* 11(1): 1-6.

Newaj, R. and Yadav, D.S. 1994. Changes in physico-chemical properties of soil under intensive cropping systems. *Indian Journal of Agronomy* 39: 373-378.

Parasuraman, P. 2001. Effect of seed pelleting with diammonium phosphate and potassium dihydrogen phosphate and foliar spray with diammonium phosphate on growth and yield of rainfed cowpea (*Vigna unguiculata*). *Indian Journal of Agronomy* 46(1): 131-134.

Pathak, H., Mohanty, S., Jain, N. and Bhatia, A. 2010. Nitrogen, phosphorus and potassium budgets in Indian agriculture. *Nutrient Cycling in Agroecosystems* 86: 287-299.

Prabhakar, S., LAL, M., Kherawat, B.S., Ravichandra, K., Kumar, A., Agrawal, M.C. and Singh, A.P. 2013. Effect of phosphorus, manganese and growth regulator (NAA) on yield and uptake of nutrients by cowpea (*Vigna unguiculata*). *Indian Journal of Agricultural Sciences* 83(6): 656-661.

Pramanik, S.C., Singh, N.B. and Singh, K.K. 2009. Yield, economics and water use efficiency of chickpea (*Cicer arietinum*) under various irrigation regimes on raised bed planting system. *Indian Journal of Agronomy* 54(3): 315-318.

Ram, B., Punia, S.S., Tetarwal, J.P., Meena, D.S. and Tomar, S.S. 2013. Effect of organic and inorganic sources of nutrients on productivity, profitability and soil health in lentil (*Lens culinaris*) under Vertisols. *Indian Journal of Agronomy* 58(4): 564-569.

S.S. Nawale, A.D. Pawar, B.M. Lambade and N.S. Ugale. 2009. Yield maximization of chick pea through INM applied to sorghum-chickpea cropping sequence under irrigated condition. *Legume Research* 32(4): 282-285.

Sahay, N., Singh, S.P. and Sharma, V.K. 2013. Effect of cobalt and potassium application on growth, yield and nutrient uptake in lentil (*Lens culinaris* L.). *Legume Research* 36(3): 259-262.

Sasode, D.S. 2008. Response of greengram [Vigna radiata (L.) WILCZEK] to fertility levels and sulphur sources application. *Agricultural Science Digest* 28(1): 18-21.

Schuman, G.E., Janzen, H.H. and Herrick., J.E. 2002. Soil carbon dynamics and potential carbon sequestration by range lands. *Environmental Pollution* 116: 391-396.

Sepehya, S., Bhardwaj, S.K., Dixit, S.P. and Dhiman, S. 2012. Effect of integrated nutrient management on yield attributes, yield and NPK uptake in garden pea (*Pisum sativum* L.) in acid Alfisol. *Journal of Food Legumes* 25(3): 247-249.

Shashikumar, Basavarajappa, R., Salakinkop, S.R., Manjunatha Hebbar, Basavarajappa, M.P. and Patil, H.Y. 2013. Influence of foliar nutrition on performance of blackgram (*Vigna mungo* L.) nutrient uptake and economics under dry land ecosystems. *Legume Research* 36(5): 422-428.

Singh, A.K., Meena, M.K., Bharati, R.C. and Gade, R.M. 2013. Effect of sulphur and zinc management on yield, nutrient uptake, changes in soil fertility and economics in rice (*Oryza sativa*)-lentil (*Lens culinaris*) cropping system. *Indian Journal of Agricultural Sciences* 83(3): 344-348.

Singh, D.K., Singh, A.K., Singh, M., Bordoloi, L.J. and Srivastava, O.P. 2012. Production potential and nutrient uptake efficiency of pea (*Pisum sativum* L.) as influenced by different fertility levels and micronutrients. *Journal of the Indian Society of Soil Science* 60(2): 150-155.

Singh, G. and Sandhu, H.S. 1980. Studies on multiple cropping II. Effect of crop rotation on physical and chemical properties of soils. *Indian Journal of Agronomy* 25: 57-67.

Singh, G., Aggarwal, N. and Khanna, V. 2010. Integrated nutrient management in lentil with organic manures, chemical fertilizers and biofertilizers. *Journal of Food Legumes* 23(2): 149-151.

Singh, M.V. 1999. Sulphur management for oilseed and pulse crops. Bulletin No. 3. Indian Institute of Soil Science, Bhopal, p. 54.

Singh, M.V. 2001. Micronutrient status of Indian soils and crop responses to their application. Paper in National Seminar on Biofertilizers and Micronutrients, New Delhi, India.

Singh, S.S. 2009. Crop management (under irrigated and rainfed conditions). Kalyani publishers, New Delhi, India.

Tandon, H.L.S. 2002. Nutrient management recommendations for pulses and oilseeds. Fertilizer Development and Consultation Organization, New Delhi, India.

Umesh, M.R. and Shankar, M.A. 2013. Yield performance and profitability of pigeonpea (*Cajanus cajana* L.) varieties under different nutrient supply levels in dryland *Alfisols* of Karnataka. *Indian Journal of Dryland Agricultural Research and Development* 28(1): 63-69.

Umesh, M.R., Shankar, M.A. and Ananda, N. 2013. Yield, nutrient uptake and economics of pigeonpea (*Cajanus cajan* L.) genotypes under nutrient supply levels in dryland Alfisols of Karnataka. *Indian Journal of Agronomy* 58(4): 554-559.

Vennila, C. and Jayanthi, C. 2007. Residual effect of integrated nitrogen management of wet seeded rice + dhaincha dual cropping system on yield and nutrient uptake of Greengram. *Agricultural Science Digest* 27(4): 288-290.

Wright, A.L. and Hons, F.M. 2005. Tillage impacts on soil aggregation and carbon and nitrogen sequestration under wheat cropping sequences. *Soil and Tillage Research* 84: 67-75.

Yadav, L.R., Choudhary, P., Santosh, Sharma, O.P. and Choudhary, M. 2012. Effect of phosphorus and zinc on yield and economics of mothbean under semiarid conditions. *Journal of Food Legumes* 25(4): 361-363.

2016, Diseases of Pulse Crops and their Sustainable Management 117–130
Editors: Samir Kumar Biswas, Santosh Kumar and Gireesh Chand
Published by: BIOTECH BOOKS, NEW DELHI

Chapter 8

Diseases of Chickpea and their Management

***Prabhat Kumar*[1]*, *Santosh Kumar*[2], *Mehi Lal*[3] *and J.N. Srivastava*[4]**

[1]*Betelvine Research Centre, Islampur, Nalanda – 801303 (Bihar Agricultural University, Sabour, Bihar)*
[2]*Jute Research Station, Katihar, Bihar Agricultural University, Sabour – 854 105, Bhagalpur, Bihar*
[3]*Plant Protection Section, Central Potato Research Institute Campus, Modipuram, Meerut – 250 110, U.P.*
[4]*Department of Plant Pathology, Bihar Agricultural University, Sabour, Bhagalpur, Bihar*

Introduction

Chickpea (*Cicer arietinum* L.) is one of the most important grain legumes of India, commonly known as *Gram or Bengal gram*. It is cultivated in an area of about 8.32 million hector with the production of 7.70 million tonnes and productivity of 925 kg ha^{-1} (FAO, 2012). It is the third most important food legume in the world after dry bean (*Phaseolus vulgaris* L.) and field pea (*Pisum sativum* L.) and is cultivated over 45 countries belonging to Asia, Africa, Americas and Oceania (Vishwadhar and Gurha, 1998). Chickpea is used as an important source of protein in human nutrition and cattle feed. It is also considered to have medicinal effects and is used for blood

* Corresponding Author: E-mail: prabhathau@gmail.com

purification. It contains 21 per cent protein, 61.5 per cent carbohydrate and 4.5 per cent fat. It is also rich in calcium, iron and niacin. Malic and oxalic acid collected from green leaves are prescribed for intestinal disorders. It improves soil fertility by biological nitrogen fixation.

Chickpea is a crop of both tropical and temperate regions. *Kabuli* type is grown in temperate regions while the *Desi* type chickpea is grown in the semi-arid tropics. The chickpea crop is attacked by nearly 172 pathogens including 67 fungi, 22 viruses, 3 bacteria, 80 nematodes and mycoplasma in all over the world (Nene *et al.*, 1996), but only few of them have the potential to devastate the crop. Some important diseases which cause severe losses in chickpea are aschochyta blight, fusarium wilt, botrytis grey mold, dry root rot and rust. The crop's yield losses by individual insects and diseases were estimated about 5 to 10 per cent in temperate regions and 50 to 100 per cent in tropical regions (Van Emden *et al.*, 1988).

Ascochyta Blight

Ascochyta blight, caused by *Ascochyta rabiei* (Pass.) is the most important disease of chickpea in almost all the countries. In India, the disease is appeared in severe in North-western part of Uttar Pradesh, Punjab and Haryana.

Symptomatology

The fungus attacks in above-ground parts of the plant including leaves, stems and pods at any stage of plant growth, but plants are most susceptible to disease during flowering. Brown, circular spots with brownish red margin appear on leaves and pods of affected plants (Figure 8.1a). On petioles and stem, the spots are elongated in shape (Figure 8.1b). The spots on leaves coalesce turning the leaf completely brown. On green pods, the circular lesions have dark margin where black dot like bodies appear known as pycnidia (Figure 8.1c). The pycnidia are arranged in concentric circles. The elongated lesions on stem and petioles also bear black dots and may girdle the stem. The parts above the lesions droop and wilt. If the stem is girdled at the base, the whole plant will show wilting. During wet weather, the disease spread very fast and may cover the whole field.

Causal Organism

Ascochyta rabiei (Pass.) Labrousse is the causal pathogen of Ascochyta bligh. The perfect stage of the fungus is *Didymella rabiei* (Kov.) Von Arx. The mycelium of the pathogen is septate. The pycnidia develop on stem, leaves and seed pods are dark brown, globose and measure 140-200μ in length. Conidia are formed within pycnidium and remain viable for long period of time. The pycnidia absorb water, swell and release conidia. Several strains of the pathogen are known.

Epidemiology

The pathogen overwinters on plant debris left in the field and also on seeds which serve the source of primary inoculums. Ascospores were also found to play a role in the initiation of the disease. Secondary spread of the pathogen is through conidia which are disseminated by rain splash, insects and contact of healthy and diseased leaves and by the movement of man and animals. Other hosts such as

(a) Symptom on leaves (b) Symptom on stems

(c) Symptom on pods

Figure 8.1: Symptoms of Aschochyta Blight of Chickpea.

cowpea and bean are also infected by the pathogen and serve as a source of inoculum. The disease development is favoured at 9-24°C with 10 h or more wetness. Wet, windy weather favour rapid spread of disease spread. *A. rabiei* showed variation in morphological, physiological and pathological characters. The disease builds up and spreads fast when night temperatures are around 10°C, day temperatures around 20°C, and rains are accompanied by cloudy days. Excessive canopy development also favours blight development.

Disease Management

Disease free seed is a pre requisite for effective disease management. Seed treatment with combination of bavistin + thiram (1:2 ratios) @ 2.5g/kg of seed for eradication of internally and externally seed pathogen. The biocontrol potential of fungal antagonists,

Chaetomium globosum, *T. viride*, and *Acremonium implicatum* were explored under *in vitro* and *in vivo* conditions. *C. globosum* caused 48.6 per cent reduction in colony diameter and 70.9 per cent reduction in pycnidiospores germination under *in vitro* conditions, whereas, its post inoculation spray reduced 73.1 per cent disease. Foliar applications of zineb, maneb, captan and daconil also reduced the severity of disease. Desi accessions H00-108, GL 92024 ICC 4475, ICC 6328 and ICC 12004; and Kabuli ILC 3864, ILC 3870 and ILC 4421ILC 200 and ILC 6482 showed resistance to blight. The cultivars- Gaurav, GNG 146, GNG 469, PBG 1 and L 551 were found resistant to the disease. Resistant cultivars secreted lesser amount of malic acid and posses more glandular hairs as compared to susceptible ones.

Fusarium Wilt

Chickpea production is severely affected by Fusarium wilt. It is caused by *Fusarium oxysporum* f. sp. *ciceris* in most of the chickpea growing areas of the world which causes annually, 10 to 15 per cent yield losses of chickpea (Jalali and Chand, 1992). In India, wilt alone causes of average 10 per cent loss annually and it is prevalent in all chickpea growing states (Singh and Dahiya, 1973). However, it was observed that early wilting causes, 77-94 per cent losses while late wilting causes, 24-65 per cent loss (Haware and Nene, 1980). The incidence of wilt was reported, 14.1 to 32.0 per cent from the states of Delhi, Punjab, Haryana, Rajasthan and Jharkhand (Dubey *et al.*, 2008). It causes complete loss in grain yield, if the disease occurs in the vegetative and reproductive stages of the crop (Haware and Nene, 1980; Haware *et al.*, 1990; Halila and Strange, 1996; Navas-Cortes *et al.*, 2000).

Symptomatology

The disease manifests as mortality of young seedlings (within 25 to 30 days after sowing) to wilting or death of adult plants. Seedlings that die due to wilt disease can be confused with other diseases of wilt complex, if not examined carefully. Fusarium wilt infected seedlings collapse and lie flat on the ground retaining their dull green colour. Adult plants show typical wilt symptoms of drooping of petioles, rachis and leaflets. All the leaves turn yellow and then light brown (Figure 8.2a). The roots of the wilting plants do not show any external rotting but when split open vertically, dark brown discoloration of internal xylem is seen (Figure 8.2b). Vascular discolouration is observed on longitudinal splitting of stem. Seeds harvested from wilted plants were lighter and duller than those from healthy plants (Haware and Nene, 1980).

Causal Organism

The fungus produces white to light orange aerial mycelium and sporodochia on incubated seed. The mycelium is profusely branched, covers the entire seed and is white to light pink in colour. Sporodochia are rarely produced, but if present, the aerial mycelium completely covers it. Micro conidia are abundant, and are produced on short, unbranched monophialides (micro conidiophores) in small, dry, false heads. Micro conidia are hyaline, single celled, over to cylindrical straight to slightly curved, and measuring 2.5 to 3.5 x 5 to11 μ in size. Macro conidia are fewer then micro conidia and produced on branched macro conidiophores. They are fusoid with pointed ends, hyaline septate 3 to 5 septa and measure 3.5 to 4.5 x 25 to 65 μ.

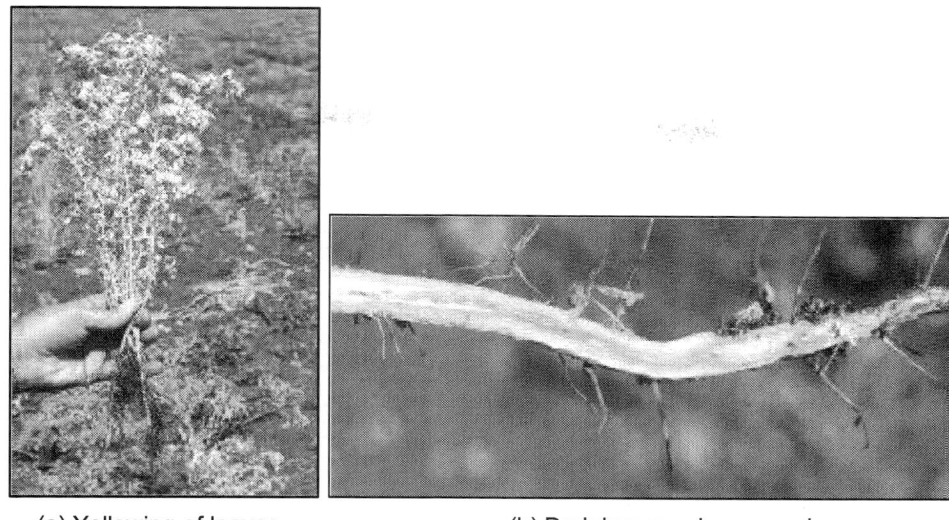

(a) Yellowing of leaves (b) Dark brown xylem vessel

Figure 8.2: Fusarium Wilt Symptoms of Chickpea.

Chlamydospores are usually intercalary and produce singly in pairs or in chains. They are globose to sub-globose, thick walled and smooth surfaced. Chlamydospores like swellings are often seen also on the hyphae.

Epidemiology

The Fusarium wilt is soil and internally seed borne disease. It can survive in soil more than 6 years as facultative saprophyte in the absence of susceptible host (Haware *et al.*, 1986). Haware *et al.* (1982) showed the fungus to be in the helium of the seed in the form of chlamydospore like structures. The primary infection is through chlamydospores or mycelia. The conidia of the fungus are short lived; however, the chlamydospores can remain viable up to next crop season. Plant species other than chickpea may serve as symptomless carriers of the disease. Gupta (1991) reported *Vigna radiata, V. mungo, Cajanus cajan, Pisum sativum* and *Lens culinaris* as symptomless carriers of the disease. The pathogen may also parasitize several weeds such as *Cyperus rotendus, Tribulus terrestris, Convolvulus arvensis* and *Cardiospermum halicacabum* (Nene *et al.*, 1980). The soil type, reaction, moisture and temperature are known to influence disease development. Rachana *et al.* (2002) reported that black soil support highest wilt incidence (75.5 per cent). Wilt incidence in sandy-loam, red and clay soil was found to be 64.4, 59.9 and 46.6 per cent, respectively. According to Sugha *et al.* (1994), soil temperature in the range of 24.8- 28.5°C and soil moisture above 25 per cent within the water holding capacity of soil were most conducive for chickpea wilt. Below 17 °C, infection remains restricted in the root without any wilt symptoms. The importance of the soil temperature has also been substantiated by the observation that late sowing of the crop reduces the incidence of the disease.

Disease Management

Deep ploughing during summer and removal of host debris from the field reduces inoculums levels. The soil inoculums can be reduced by addition of 15-20 tonnes of farmyard manure with *Trichoderma* sp. @ 4-5 kg / ha before sowing (Singh and Dubey, 2007). The species of *Trichoderma* have been evaluated against the wilt pathogen and have exhibited greater potential in managing chickpea wilt under glasshouse and field conditions (Kaur and Mukhopadhayay, 1992). Earlier, Padwick (1941) also observed that a species of *Trichoderma* was highly antagonistic to wilt pathogen of chickpea. Mane (1995) isolated and evaluated 88 different fungal and bacterial agents against *F. oxysporum* f. sp. *ciceris* and four isolates of *Trichoderma* spp., three isolates of *Pseudomonas fluorescence,* one isolate of *Acrophilophora* sp. and three isolates of *Gliocladium* spp. showed antagonistic activity *in vitro.* The filamentous fungi, *Trichoderma* have attracted the attention because of their multiprong action against various plant pathogens (Harman *et al.,* 2004). Seed dressing with Benlate T (Benomyl + thiram) eradicated seed borne inoculums (Haware *et al.,* 1978). The disease can be managed by seed treatment with various seed dressing fungicides. The seed treatment with Bavistin + thiram (1:1) @ 2.5 g / kg seed before sowing decreased seedling mortality 7 per cent and increased seed germination 11.2 per cent and grain yield 2.8 Q / ha (Pal and Singh, 1993). Under field conditions, maximum wilt reduction (28.3 per cent) in cultivar Pusa 256 was observed when *T. viride* was applied as seed coating with talc as carrier with gum. Nikam *et al.* (2007) tested four oilseed cakes and observed that groundnut cake followed by neem seed and castor cake were found to be most effective in checking percent wilt incidence by 61.91, 52.39 and 47.62 per cent, respectively as against control. Haware *et al.* (1978) reported seed-borne inoculums can be eradicated by seed dressing with Benlate (benomyl 30 per cent + thiram 30 per cent) at 0.25 per cent rate. Nikam *et al.* (2007) revealed that foliar sprays of thiram followed by carbendazim and captan proved to be effective in checking the wilt incidence by 42.46, 38.10 and 33.34 per cent, respectively as against control (100 per cent wilting). Kolte *et al.* (1998) effectively controlled chickpea wilt with seed treatment by *Rhizobium, T. viride, T. harzianum* and *Azotobacter sp*. The identification and use of host plant resistance has the great potential in the long-term management of wilt. The genotypes H 99-9, Pusa 212, JG 315, JG 322, PCS 1(Sel.ICCV-11), PCS 2 (Sel.KPG 142-1), PCS 5 (Sel.BGD-112), PCS 6 (Sel. Pusa- 1073), H01-36 and PCS 8 (Sel.H82-2) showed resistant (<10 per cent wilt incidence) reaction (Dubey and Singh, 2008; Dubey and Singh, 2010). In addition to these a large number of wilt resistant cultivars namely Avrodhi, Haryana Channa 1, BGD 72, BGM 547, GNG 469 (Smrat), GNG 663 (Vardan), RSG 693 (Aadhar), KPG-59 (Uday), K 3256 (Pragati), Phule G-87207 (Vishal), Phule G 9425-9, JG 322, GPF 2, PBG 1, Pusa 372 and Pusa 1053 (Chamtkar) were also identified

Botrytis Grey Mould (BGM)

Botrytis grey mould (BGM) is the most important disease of chickpea. The occurrence of Botrytis grey mould on chickpea was first reported by Shaw and Ajrekar in 1915. The epidemic of the disease have been reported during 1967-68, from chickpea growing parts of West Bengal, Bihar, Uttar Pradesh, Rajasthan, Haryana, Punjab and Himachal Pradesh (Singh, 1997). The disease was responsible for heavy losses in the Indo- Gangetic plains of India during 1979-1982 (Grewal and Laha, 1982) and caused 70-100 per cent losses in yield at Central state farm, Hisar and several parts of Punjab.

Symptomatology

All the aerial parts of chickpea are susceptible to the disease with growing tips and flowers being the most vulnerable (Bakr *et al.*, 1997; Haware, 1998). Initial symptoms appear on stem, leaves, inflorescence and pods as grey or dark brown lesions covered with erect hairy sporophores (Figure 8.3). Stem lesions are 10-30 mm long which later girdles the stem completely. Tender branches break off at the points where grey mould causes rotting. Affected leaves and flowers turn into a rotting mass. In the field, the disease first appears in isolated patches when the crop has achieved maximum canopy and the morning, relative humidity is very high with low temperature. As the disease advances, patches of disease plant become more prominent, spreading slowly in the entire field. According to Laha and Grewal (1983) symptoms appeared on leaflets, petioles and growing tips as water soaked lesions. The lesions are brown and limited in size. However, under conditions of high humidity leaflets got blighted and bear abundant fungal fructifications.

Figure 8.3: Botrytis Grey Mold Symptoms on Chickpea .

Causal Organism

BGM of chickpea is caused by *Botrytis cinerea* Pers. Ex. Fr. It is a necrotrophic fungus; belonging to order flyphales and Family Moniliaceae *B. cinerea* is dominant on chickpea crops. *B. cinerea* grown on potato dextrose agar (PDA) has a white, cottony appearance, which turns light grey with age. The mycelium is septate, brown and 8-10μm wide. Young hyphae are thin and hyaline. Conidia and conidiophores are not in pycnidia or acervuli. Conidiophores lighter brown than hyphae, with hyaline tip, septate, 8-24μm wide. Tips of conidiophores or their branches are slightly enlarged and bear small pointed sterigmata. Conidia are hyaline, one- celled oval or globose or short cylindrical and borne in clusters at the tips of conidiophores branches.

Epidemiology

Reports on epiphytotics of botrytis grey mould from different parts of the world indicated the existence of definite and efficient mechanisms of survival of the pathogen from one season to another. The information regarding the survival and epidemiology is scanty so far as the botrytis grey mould of chickpea is concerned. BGM can devastate chickpea, resulting in complete yield loss in years of extensive winter rains and high humidity (Pande *et al.*, 1982; Reddy *et al.*, 1993).

Disease Management

Haware and McDonald (1992) reported that delayed sowings reduced BGM incidence even in susceptible cultivars, but significantly reduced the grain yields. Singh (1997) also observed that the late sown crop (around 20 Nov.) showed significantly low incidence of BGM. Combination of wider row spacing, intercropping with linseed and two spray application of carbendazim @ 0.2 per cent significantly reduced BGM severity and increased grain yield of chickpea and linseed. Bakr *et al.* (1993) reported that seed treatment with bavistin + thiram (1:1), indofil M-45, thiabendazole, ronilan, rovral, bavistin @ 0.3 per cent controls seed borne inoculum of *B. cinerea.* Foliar spray with ronilan, bavistin + thiram combination @ 0.1 per cent or bavistin alone @ 0.2 per cent provided complete protection to chickpea plants against aerial infection by *B. cinerea* (Grewal and Laha, 1982). Haware *et al.* (1997) reported that one spray with vinclozolin (0.2 per cent) at the time of flowering in the integrated management system reduced BGM incidence. Haware *et al.* (1999) reported biocontrol potential of *T. viride* isolate T-15 (isolated from chickpea rhizosphere) on *B. cinerea* in chickpea under controlled environmental conditions.

Dry Root Rot

Dry root rot disease of chickpea is caused by *Rhizoctonia bataticola* (Taub) Butler or *Macrophomina phaseolina* (Maubl.) Ashby. It has been reported from Australia, Ethopia, California, India, Iran, Lebanon, Mexico Pakistan, Syria and Turkey (Nene, 1979; Allen, 1983). Wilt and dry root rot diseases alone affecting the productivity of chickpea, 10-35 per cent loss in yields (Mahendra, 1998). *Rhizoctonia bataticola* is a polyphagous soil borne pathogen infecting over 500 plant species worldwide causing huge losses.

Symptomatology

The first symptom of the disease is yellowing of the leaves within a day or two; such leaves droop and in the course of the next two or three days, plant showed completely dried. The tissues were weakened and break off easily. The affected roots are dark brown to black and usually dry, unless the soil is wet. The tap root is quite brittle, show shredding of the bark and can be broken easily (Figure 8.4). If the plant were pulled out from the soil and examined the basal stem and main root system showed extensive rooting with most of the lateral roots destroyed. In advance cases minute dark black sclerotial bodies can be seen on the surface of the root, as well as in the pith. If the plants are pulled from the soil and examined, the basal stem and the main roots may show dry rot symptoms.

Figure 8.4: Dry Root Symptom on Chickpea.

Causal Organism

The fungus lacks fruiting bodies and spores. The mycelium is light-brown, thick in which black sclerotia are formed. Sclerotia are variable in form, small and loosely connected by mycelial threads.

Epidemiology

The fungus is seed and soil-borne, facultative parasite and may survive in the soil in the form of sclerotia for long time (Dingra and Sinclair, 1994). Soil borne inoculum is more important in causing infection and disease development. It produces pycnidia when the atmospheric temperature is above 30° C and the pycniospores remain viable for over a year. The fungus is mainly a soil-dweller and spreads from plant to plant through irrigation water, tools and implements and cultural operations. The sclerotia and pycniospores may also become air borne and cause further spread of the pathogen. The disease appears suddenly when ambient temperatures are between 25 and 30 °C.

Disease Management

Drenching the affected plants and the infested soil with Bordeaux mixture or other effective fungicide may help in reducing the inoculums potential. Field sanitation measures, including cutting down the diseased plants and burning them and deep ploughing in summer during the following season help to reduce the diseases intensity. Drought should be avoided. Sowing should always be done on the recommended time. Seed treatment with a mixture of carbendazim 1.5 g and thiram 1.5 g per kg of seed and with *Trichoderma viride* formulation + 3 g thiram per kg seed can reduce the disease incidence. Nagamani *et al.* (2011) observed that seed treatment with carbendazim @ 2 g/kg of seed+ seed treatment with *T. viride* @ 4 g/kg of seed +soil application of FYM was found most effective to control root rot. Hence integrated management of the disease using bio-control agents and chemicals is the best alternative (Ramarethinum *et al.*, 2001).

Chickpea Rust

Chickpea rust is caused by, *Uromyces ciceris-arietin*. Rust is not known to cause as widespread damage on chickpea as other diseases of chickpea. it can occasionally be serious when conditions during crop growing are favourable. In a particularly favourable year, an epidemic of rust on Bengal gram in Karnataka, India, caused incidences of up to 90-100 per cent.

Symptomatology

The symptom appears initially on the leaves as small, round or ellipsoidal, cinnamon-brown, powdery pustules. These pustules tend to coalesce. Sometimes a ring of small pustules can be seen around larger pustules which occur on both leaf surfaces but more frequently on the lower one (Figure 8.5). Occasionally pustules can be seen on stems and pods especially when infection is severe. Severe infection results in premature defoliation and possible death of the entire plant.

Figure 8.5: Rust on Lower Leaflet of Chickpea.

Causal Organism

The fungus produces the uredial and telial stages on chickpea. There is no any alternate host of the pathogen. The uredospores are globose to sub-globose, brownish yellow in colour, with minute spines on the walls. They measure 20-28 μ in diameter with 3-4 μ thick cell wall and contain 4-8 germ-pores. The teliospores are also similar to uredospores, but round to ovate, rough and thick walled, with thickened apex. While the uredospore germinates readily, the teliospores are not known to germinate. Its pycnidial and aecial stage are not known.

Epidemiology

Cool and moist weather conditions favours rust build up although rain is not essential for its development. The symptoms usually become conspicuous later in the growing season although epiphytotics may occur earlier in the season when conditions are favourable.

Disease Management

Cultural practices such as field sanitation, seed selection, crop rotation and early sown crops help to escape infection. Foliar spray of fungicides such as Mancozeb (0.2 per cent a.i.), Bayleton (0.05 per cent a.i) and Calixin (0.2 per cent a.i.) are found effective against the pathogen.

References

Allen, D.J. 1983. *The pathology of tropical food legumes.* John Wiley and Sons, New York, 413 p.

Bakr, M.A., Hossain, M.S., and Ahmed, A.U. 1997. Research on Botrytis grey mould of chickpea in Bangladesh. p. 15-18. In: *Recent advances in research on Botrytis grey mould of chickpea (Eds. Haware, M.P., Lenne, J.M., and Gowda, C.L.L).* ICRISAT, Patancheru, Andhra Pradesh, India.

Bakr, M.A., Rahman, M.M., Ahmed, F. and Kumar, J. 1993. Progress in the management of Botrytis gray mold of chickpea in Bangladesh. p. 17-18. In: *Recent advances in research on Botrytis gray mold of chickpea: summary proceedings of the second Working Group Meeting to discuss collaborative research on Botrytis Gray Mold of Chickpea* (Eds. Haware, M.P.; *Gowda, C.L.L. and McDonald, D.*). ICRISAT, Patancheru, A.P. (India).

Dhingra, O.D. and Sinclair, J.B. 1994. *Basic plant Pathology* Methods. CRS press, London, p. 443.

Dubey, S.C. and Singh, S.R. 2008. Virulence analysis and oligonucleotide fingerprinting to detect diversity among Indian isolates of *Fusarium oxysporum* f. sp. *ciceris* causing chickpea wilt. *Mycopathologia,* 165: 389-406.

Dubey, S.C., Singh, S.R. and Singh, B. 2010. Morphological and pathogenic variability of Indian isolates of *Fusarium oxysporum* f. sp. *ciceris* causing chickpea wilt. *Archives of Phytopathology and Plant Protection,* 43(2): 174-190.

Grewal, J.S. and Laha, S.K. 1982. Chemical control of botrytis grey mould of chickpea. *Indian Phytopath.,* 36: 516-520.

Gupta, O. 1991. Symptomless carriers of chickpea vascular wilt pathogen (*Fusarium oxysporum* f. sp. *ciceris*). *Legume Res.* 14: 193-194.

Halila, M.H. and Strange, R.N. 1996. Identification of the causal agent of wilt of chickpea in Tunisia *Fusarium oxysporum* f. sp. *ciceris* race 0. *Phytopath. Medit.,* 35: 67-74.

Harman, G.E., Howell, C.R., Viterbo, A., Chet, I. and Lorito, M. 2004. *Trichoderma* species-opportunistic, avirulent plant symbionts. *Nature Reviews* 2: 43-56.

Haware, M.P. 1998. Diseases of chickpea. p. 473-516. In: *The pathology of food and pasture legumes* (*Eds. Allen, D.J. and Lenne, J.M.*). ICARDA, CAB International, Wallingford, UK.

Haware, M.P. and McDonald, D. 1992. Integrated management of botrytis gray mould of chickpea. p. 3-6. In: *Botrytis gray mould of chickpea,* (*Eds. Haware, M.P., Faris, D.G. and Gowda, C.L.L.*). ICRISAT, Patancheru, A.P., India).

Haware, M.P. and Nene, Y.L. 1980. Influence of wilt at different stages on the yield loss in chickpea. *Trop. Grain Legume Bull.,* 19: 38-40.

Haware, M.P. and Nene, Y.L. 1982. Symptomless carriers of the chickpea wilt Fusarium. *Plant Dis.* 66: 250-251.

Haware, M.P., Jimenez-Diaz, R.M., Amin, K.S., Phillips, J.C. and Halila, H.1990. Integrated management of wilt and root rots of chickpea. p. 129-137. In: *Chickpea in the Nineties: Proceedings of the second international work shop on chickpea improvement*. ICRISAT, Patancheru, India.

Haware, M.P., Mukherjee, P.K., Lenne, J.M., Jayanthi, S., Tripathi, H.S. and Rathi, Y.P.S. 1999. Integrated biological-chemical control of Botrytis gray mould of chickpea. *India Phytopath.,* 52: 174-176.

Haware, M.P., Nene, Y.L. and Natarajan, M. 1986. Survival of *Fusariunz oxysporum* f. sp. *ciceri* in soil in the absence of chickpea. *National seminar on management of soil-borne diseases of crop plants.* TNAU, Coimbatore, 8-10 Jan. 1986. Pl (Abstr.).

Haware, M.P., Nene, Y.L. and Rajeshwari, R. 1978. Eradication of *Fusariutn oxysporum* f. sp. *ciceri* transmitted in chickpea seed. *Phytopathology* 68: 1364-1367.

Haware, M.P., Tripathi, H.S., Rathi, Y.P.S., Lenne, J.M. and Jayanthi, S. 1997. Integrated management of botrytis gray mould of chickpea: cultural, chemical, biological, and resistance options. p. 9-12. In: *Recent advances in research on Botrytis grey mould of chickpea* (*eds. Haware, M.P., Lenne, J.M. and Gowda, C.L.L.*). ICRISAT, Patancheru, Andhra Pradesh, India.

Jalali, B.L. and Chand, H. 1992. Chickpea wilt. p. 429-444. In: *Plant Diseases of International Importance. Vol. 1. Diseases of Cereals and Pulses.* (*Eds. Singh, U.S. Mukhopadhayay, A.N., Kumar, J. and Chaube, H.S.*). Prentice Hall, Englewood Cliffs, NJ.

Kaur, N.P. and Mukhopadhayay, A.N. 1992. Integrated control of chickpea wilt complex by *Trichoderma* spp. and chemical methods in India. *Trop. Pest Management,* 38: 372-375.

Kolte, S.O., Thakre, K.G., Gupta, M., Lokhande, V.V. 1998. Biocontrol of *Fusarium* wilt of chickpea (*Cicer arietinum*) under wilt sick field condition. National Symposium on management of soil and soil borne diseases 9-10th Feb., 1998. ISOPP, P. 22.

Laha, S.K. and Grewal, J.S. 1983. Botrytis blight of chickpea and its perpetuation through seed. *Indian Phytopatholgoy* 36: 630-634.

Mahendra, P. 1998. Diseases of pulse crops, their relative importance and management. *J. Mycol. Plant. Pathol.,* 28(2): 114-122.

Mane, S.S. 1995. Studies on *Fusarium oxysporum f. sp. ciceri* causing chickpea wilt with special reference to its management by bio-agents. Ph.D. thesis, IARI, New Delhi. p. 72.

Nagamani, P., Viswanath, K. and Kiran babu, T. 2011. Management of dry root rot caused by Rhizoctonia bataticola (Taub.) Butler in Chickpea. *Current Biotica* 5(3): 364-369.

Navas-Cortes, J.A., Hau, B. and Jimenez-Diaz, R.M. 2000. Yield loss in chickpea in relation to development to Fusarium wilt epidemics. *Phytopathology* 90: 1269-1278.

Nene, Y.L. 1979. Diseases of chickpea. In: *Proc. International seminar on chickpea improvement*, ICRISAT, Hyderabad, India, p. 171-187.

Nene, Y.L. and Sheila, V.K. 1996. ICRISAT News Letter, Proceeding on the international workshop on chickpea improvement, Feb-March, 1996. p. 172-180.

Nene, Y.L., Kannaiayan, J., Haware, M.P. and Reddy, M.V. 1980. Review of the work done at ICRISAT on soil-borne diseses of piegeonpea and chickpea. In: *Proceedings of the consultants group discussion on the resistance to soilborne diseases of Legumes*, 8-11 January 1979, ICRISAT Center, Patancheru, A.P., India: ICRISAT. p. 3-39

Nikam P.S., Jagtap, G.P. and Sontakke P.L. 2007. Management of chickpea wilt caused by*Fusarium oxysporium* f. sp. *Ciceri*. *African Journal of Agricultural Research* 2(12): 692-697.

Padwick, G.W. 1941. Report of the Imperial Mycologist. Scient. Rept. Agric. Res. Inst., New Delhi, 1937-38. p. 94-101.

Pal, M. and Singh, B. 1993. Channe koo Uktha Rog Se Bachayen. *Kheti* 47(6): 24-25.

Pandey, M.P., Beniwal, S.P.S. and Arora, P.P. 1982. Field reaction of chickpea varieties to chickpea grey mould. *International Chickpea Newsletter* 7: 13.

Ramarethinam, S., Morugesan, N. and Marimuthu, S. 2001. Compatibility Studies of fungicides with *Trichoderma viride* used in the Commercial formulation- Bio-CURF-F. *Pestology* 25: 2-6.

Reddy, M.V., Ghanekar, A.M., Nene, Y.L., Haware, M.P., Tripathi, H.S., and Rathi, Y.P.S. 1993. Effect of vinclozolin spray, plant growth habit and inter-row spacing on botrytis gray mold and yield of chickpea. *Indian Journal of Plant Protection* 21: 112-113.

Sharma, K.D., Chen, W.D. and Muehlbauer, F.J. 2005. Genetics of chickpea resistance to five races of Fusarium wilt and a concise set of race differentials for *Fusarium oxysporum* f. sp. *ciceris*. *Pl. Dis.* 89: 385-390.

Singh, B. and Dubey, S.C. 2007. Channe Kaa Mallyni Rog- Bachaw Ke Uppaya. *Kheti* 60(8): 18-20.

Singh, G. 1997. Epidemiology of Botrytis grey mould of chickpea. *In: Recent advances in research on Botrytis grey mould of chickpea* (*Eds. Haware, M. P., Lenne, J. M., and Gowda, C. L. L.*). ICRISAT, Patancheru, Andhra Pradesh, India, p. 47-50.

Singh, K.B. and Dahiya, B.S. 1973. Breeding for wilt resistance in Chickpea. Symposium on problem and breeding for wilt resistance in Bengal gram. Sept. 1973 at IARI, New Delhi, p. 13-14.

Sugha, S. K., Kapoor, S. K. and Singh, B. M. 1994. Factors influencing Fusarium wilt of chickpea (*Cicer arietinum* L.). *Indian J. Mycol. Plant Pathol.* 24: 97-102.

Van Emden, H.F., Ball, S.L. and Rao, R. 1988. Pest diseases and weed problems in pea lentil and faba bean and chickpea. p. 519-534. In: *World Crops: Cool season Food Legumes (ed. R.J. Summerfield)*. Kluwer Academic Publishers. Dordrecht, Netherland.

Vishwadhar and Gurha, S.N. 1998. Integrated Management of Chickipea Diseases. P. 249. In: *Integrated Pest and disease Management (Eds. K. Rajeev, K.G. Upadhyay, B.P. Mukerji, Chamola and O.P. Dubey)*. APH Publishing Co., New Delhi. (India).

2016, Diseases of Pulse Crops and their Sustainable Management 131–148
Editors: Samir Kumar Biswas, Santosh Kumar and Gireesh Chand
Published by: BIOTECH BOOKS, NEW DELHI

Chapter 9

Diseases of Mung Bean and Urd Bean and their Management

Sangita Sahni[1]*, Bishun D. Prasad[2] and Sunita Kumari[3]

[1]*Department of Plant Pathology, T.C.A., Dholi, Muzaffarpur, Bihar*
[2]*Department of Plant Breeding and Genetics, B.A.C, Sabour, Bihar*
[3]*Krishi Vigyan Kendra, Kishanganj, BAU, Sabour, Bihar*

Introduction

Mungbean [*Vigna radiate* (L.) Wilczek] and Urdbean [*V. mungo* (L.) Hepper] are the important pulse crops in India after chickpea and pigeonpea. These are also widely cultivated throughout Southern Asia like India, Pakistan, Sri Lanka, Bangladesh, Thailand, Laos, Vietnam, Indonasia, China and Taiwan. In India these crops are cultivated in three different seasons, *viz., kharif, rabi* and summer. It is grown as sole relay crop in rice fallows during *rabi* season in Andhra Pradesh, Tamilnadu, Karnataka and Orissa and sole catch crop during spring/summer season in Uttar Pradesh, Bihar, West Bengal, Jharkhand, Punjab, Haryana and Rajasthan. However, maximum area is under *kharif* cultivation as intercropping with sorghum, pearl-millet, maize, cotton, castor, pigeonpea etc., Short maturity duration (<60 days) make the crop ideal for catch cropping, intercropping and relay cropping. These crops are grown principally for its high protein seeds that are used as human food.That can be prepared by cooking, fermenting, milling or sprouting, they are utilized in making soups, curries, bread, sweets, noodles, salads, boiled dahl, sprouts, bean

* Corresponding Author: E-mail: sangitampp@gmail.com

cake, confectionery, to fortify wheat flour in making vermicelli and many other culinary products like sabut dhal, dhal, papad, namkeen, halwah, and vari *etc.* (Singh *et al.*, 1988). The protein is comparatively rich in lysine, an amino acid that is deficient in cereal grains. They complement each other and hence enhance the food quality. Besides being a rich source of protein, these are also important for sustainable agriculture and enriching soil organic matter through biological nitrogen fixation. India is the largest producer of mungbean and account 54 per cent of the world production and covers 65 per cent of the world acreage. Mungbean is grown on about 3.43 million hectares with annual production of 1.71 million tons. Similarly, Urdbean is grown on about 3.30 million hectares with annual production of 1.83 million tons (AICRP, 2012-13). The average yield fluctuates between 300 to 500 kg/ha for a decade in India. The yield losses (5-100 per cent) reported due to various biotic stresses which is responsible for the fluctuation in the average yield. The biotic stresses like diseases, incited by fungi, bacteria, viruses, and nematodes are the major limiting factors for higher yield. Therefore, there is a need to correct identification, diagnosis and adaptation of suitable management strategies against different diseases of these crops. A brief account of the most important diseases of these crops in India, including the causes, symptoms, management of these destructive diseases, are discussed here. These diseases are responsible for reducing overall production as well as quality of the crop produce.

Mungbean Yellow mosaic

Causal Organism

Yellow mosaic disease is caused by Mungbean yellow mosaic virus (MYMV), a member of Geminivirus group transmitted by whitefly (*Bemisia tabaci* Gen.), is a most destructive disease of mungbean and urdbean in India as well as in other countries in Asia. MYMV incidence was first reported from the fields of IARI, New Delhi by Nariani (1960). Nene (1968) named it mungbean yellow mosaic virus. The paired particles of the causal virus measure 30 x 15 nm having ssDNA (Honda *et al.*, 1981). The MYMV is not transmitted through sap (Nariani, 1960), seed or soil (Nair, 1971). The whitefly is a very efficient vector as it can acquire and inoculate the virus in certain hosts within 10 to 15 minutes. For 100 percent transmission, 10 viruliferous whiteflies per plant are required (Nair, 1971; Nene, 1973). The disease found on several alternate and collateral host which act as primary sources of inoculums. Rathi and Nene (1974) found the host range of MYMV to be restricted to species belonging to the families *Leguminosae, Compositae* and *Gramineae*. In India, this virus cause more severe yellow mosaic disease in urdbean than mungbean (Williams *et al.*, 1968).

Symptoms

The first visible sign of the disease is the appearance of yellow spots scattered on young leaves, which increase with time leading to complete yellowing. The next trifoliate leaf emerging from the growing apex showed irregular alternating yellow and green patches which also turn yellow. These colour changes in affected plants are so conspicuous that the disease can be detected in the field from a distance. The

Symptoms

The disease affects both the vegetative growth and yield components of these plants (Beniwal and Chaubey, 1979; Kadian, 1982; Kolte and Nene, 1973 and Ilyas *et al.*, 1992). The disease is characterized by the appearance of extreme crinkling, curling, puckering and rugosity of leaves, stunting of plants and malformation of floral organs (Kolte and Nene, 1973). The crinkling is observed on some branches while others remain apparently healthy (Brar and Ratual, 1986). Pollen production, fertility and subsequent pod formation is severely reduced with affect on seed weight and size of seeds in infected plants leading to decrease in yield (Nene, 1972). The virus has been reported to decrease grain yield from 35 to 81 per cent depending upon genotype and time of infection (Bashir *et al.*, 1991).

Management

1. Seeds from diseased crops should not be used.
2. Treat the seeds with imidacloprid 70 WS@ 5ml/kg or give solar Seed treatment by soaking seed in water for 3-4 hours and then exposure to solar heat from 12 to 4 p.m. in May and June.
3. Cultivation of resistant varieties:

   Mungbean: D-3-9, K 12, ML 26, RI 59 and T44 RII

   Urdbean: HUP 27, 102, 164 and HUP 315.
4. Rogue out the infected plants to avoid contact between healthy and diseased plants during intercultural operations.
5. Application of one foliar spray of insecticide (dimethoate 30 EC @ 1.7ml/ha) on 30 days after sowing.

Mosaic Mottle

This disease is caused by Bean Common Mosaic Virus which belongs to potyvirus group. The mosaic mottle of urdbean and mungbean is common in India as well as South-east Asian countries (Tsuchizaki *et al.*, 1986). It can be transmitted by sap, mechanically and by seed (Shahare and Raychaudhary, 1963; Nene, 1972). Singh and Nene (1978) also reported its transmission by aphids, *Aphis craccivora* and *A. gossypii*. Host range of the virus is confined to the family *Leguminosae* (Srivastava *et al.*, 1969). However, urdbean is more susceptible than mungbean.

Symptoms

The disease is characterized by a mosaic pattern of irregular broad patches of light and dark green areas and blistering and puckering of leaf blade. The size of the leaf gets reduced and margins show upward rolling. The leaves become rough and brittle. Affected plants show reduction in overall growth and often display excessive branching. In cases of severe infection the whole inflorescence is changed into leaf like structures, thereby causing 100 percent loss in seed yield (Nene, 1972).

Management

1. Use diseased free seeds

Figure 9.3: Symptoms of Bean Common Mosaic Potyvirus in Mungbean.

A) Infected plants showing green mosaic areas and downward cupping along the main vein of each leaflet. B) Advance stage of Bean Common Mosaic potyvirus infection. Infected leaves showing green veinbanding, blistering and malformation.

http://vegetablemdonline.ppath.cornell.edu/PhotoPages/Bean/Viruses/BeanVirus1.htm

2. Rogue out the infected plants to avoid contact between healthy and diseased plants during intercultural operations.
3. Foliar spray of insecticide (dimethoate 30 EC @ 1.7ml/ha) on 30 days after sowing.

Leaf Curl

Causal Organism

This disease is caused by Tomato Spotted Wilt Virus, is an important potential killer of mungbean and urdbean plants (Nene, 1972). The virus is transmitted by sap, grafting and the thrip, *Frankliniella occidentalis* (Amin *et al.*, 1985).

Symptoms

Nene described the symptoms of this disease for the first time in 1968. Chlorosis will develop around some lateral veins and their branches near the margin of the youngest trifoliate leaf. The leaves show downward curling of margin, sometimes rolling and twisting of young leaves can also be observed. If plants infected early after sowing, they remain stunted and majority of these die due to top necrosis within two weeks, however, plants infected in late stages of growth do not show severe curling and twisting of the leaves but show conspicuous veinal chlorosis. The infected plants produce few pods which contain small seeds.

Management

1. Foliar spray of metasystox @ 0.1 per cent on first appearance of the disease and subsequently at 10 days interval.
2. Grow resistant/tolerant varieties

   Urdbean: N 212 and Khargone 3

   Mungbean: Pant mung 3

Cercospora Leaf Spot

Causal Organism

Cercospora leaf spot (CLS) is caused by several species dominated by *Cercospora canesens* and *Cercospora cruenta* which may cause severe losses of yield under warm and humid weather conditions. Mungbean is more susceptible to this disease than urdbean. The fungus survives on the infected seeds and crop debris (Grewal, 1978). Rath and Grewal (1973) observed heavy sporulation at 27 ° C temperature and 96 percent relative humidity.

Symptoms

Leaf spots develop with a somewhat circular to irregular shape. The central area will turn tan or gray with reddish brown or brown to dark brown margin. Lesions vary in size depending on the isolate and the host (Ilag, 1978). The petioles, stems and pods also get affected by the pathogen. During favourable condition the spots increase in size during flowering and increase is most rapid at the pod-filling stage

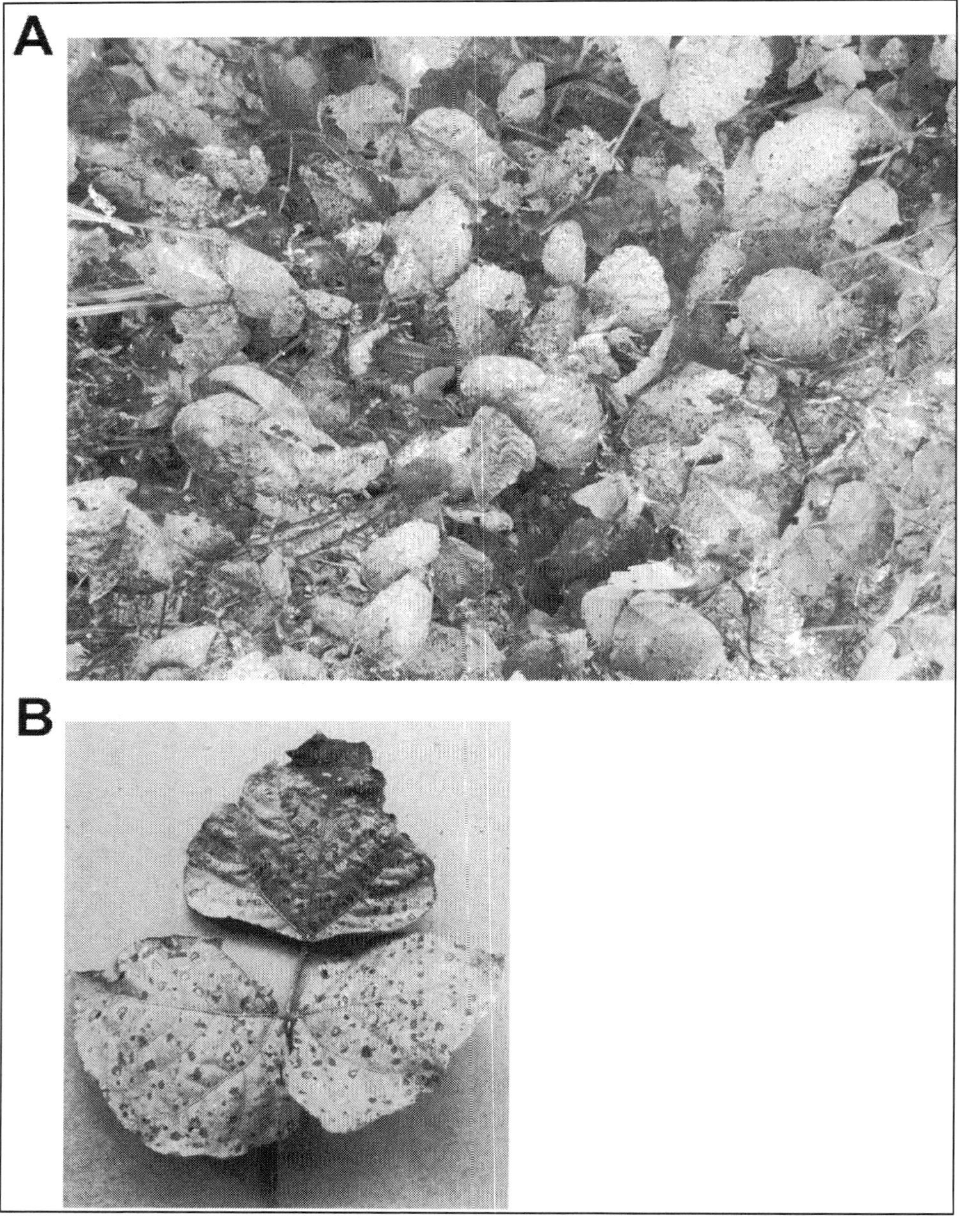

Figure 9.4: Symptomatology of Cercospora Leaf Spot Disease in Mungbean.
A) Field view showing severity of disease. B) Cercospora leaf spot on upper leaf surface of mungbean.

lead to defoliation. The size of pods and seeds is reduced resulted reduced the grain yield (Grewal, 1978). Singh *et al.* (2000) reported yield losses extent 50 per cent in severely diseased field.

Management

1. Destruction of infected crop debris and avoiding the collateral hosts in the vicinity of the crop would greatly help in reducing the incidence of the disease.
2. Crop rotation with non leguminous crops.
3. Treat the seeds with thiram or captan @ 2.5g/kg of seed
4. Cultivation of resistant varieties:

   Mungbean: LM 113, LM 168, LM 170 and JM 171

   Urdbean: Naveen, Jawahar, Urd-3, Gujarat Urd-1 and Barkha
5. On appearance of the symptoms spray with carbendazim 50 WP @1.0 g/l or mancozeb 45 WP @ 2.0 g/l or copper oxychloride @ 3 to 4 g/l. Subsequent spray should be done after 10 to 15 days, if required.

Powdery Mildew

Causal Organism

Powdery mildew (PM), caused by the pathogen *Erysiphe polygoni* DC, is one the most destructive and wide spread disease of mungbean and urdbean in India and south east Asian countries. The disease serious problem in all the areas of the country having rice based cropping systems (Abbaiah, 1993). It occurs almost every year causing considerable yield loss in India (Legapsi *et al.*, 1978). The fungus is obligate, parasite, and spreading on the surface of the host and sending haustoria into the epidermal cells to obtain nutrients. Host range of this fungus restricted to species belonging to the family *Leguminosae.* The fungus survives in its conidial form or cleistothecia on the infected host tissues which act as a source of primary inoculum. The secondary spread of the pathogen is occured through air borne conidia. Severe infection by the fungus is occured in the cool, dry months where the yield losses owing to PM have been estimated to be around 20-40 per cent (Reddy *et al.*, 1994). In India, the losses due to powdery mildew of urdbean and mungbean is more in winter sown crop as compared to rainy season crop.

Symptoms

The disease appears on all the part of plants above soil surface. Disease initiates as faint dark spots, which develop into small white powdery spots, coalescing to form white powdery coating on leaves, stems and pods. At the advance stages, the color of the powdery mass turns dirty white. In case of severe infection, defoliation takes place. The disease induces forced maturity of the infected plant causing heavy yield losses.

Management

1. The diseased plants should be detected and destroyed.
2. Delayed sowing of mungbean and urdbean with wider spacings considerably reduce the disease severity.
3. Cultivation of resistant varieties:

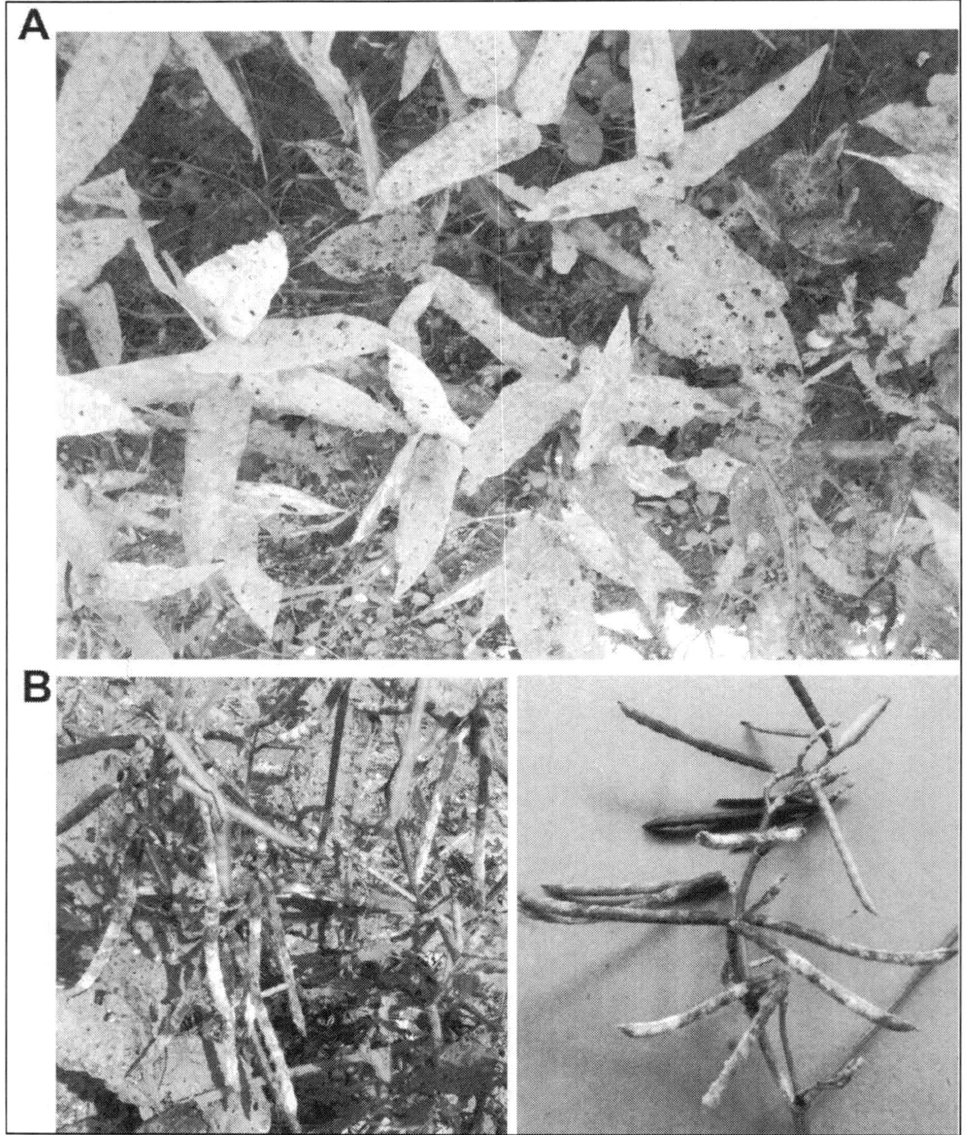

Figure 9.5: Symptom of Powdery Mildew of Urdbean.
A) Infected leaf of urdbean; B) Infected pods in urdbean.

Mungbean: LM 223, LM 24, P115, ML 131, MI 322, ML 337, ML 395 SS1, JRUM 1, TARM 1 and AVRDC 1381

Urdbean: COBG10, LBG 648, 17, Prabha, IPU 02-43, AKU 15 and UG 301

4. Spray with NSKE @ 50 g/l or neem oil 3000 ppm @ 20 ml/l twice at 10 days interval from first appearance of disease or spray with eucalyptus leaf

extract 10 per cent at initiation of the disease and 10 days later also if necessary or Spray with water soluble sulphur 80 wp @ 4 kg/l or carbendazin 50 WP @ 1 g/l (0.05 per cent), benlate (0.05 per cent) and topsin-M (0.15 per cent) as solutionwise to minimize the disease.

Anthracnose

Causal Organism

Five species of *Colletotrichum* are known to attack mungbean and urdbean but *C. lindemuthianum* and *C. capsici* are wide spread and cause severe infections under favourable environment conditions. The pathogen survives from one crop season to the next on infected seeds and crop residue (Singh *et al.*, 1981). Intermittent rains at frequent intervals favour the epidemic development of the disease. The optimum temperature and relative humidity for disease development are 17-24°C and 100 per cent, respectively.

Symptoms

The disease appear on the above ground parts of plant *i.e.*, foliage, stems and pods. The characteristic symptoms of this disease are circular, brown, sunken spots with dark centers and bright red orange margins on leaves and pods. In severe infection, affected part withers off. Infection just after germination causes seedling blight, resulted poor population of the plant on the field.

Management

1. Seed treatment with the hot water at 58° C for 15 minutes has been found effective in checking the seed borne infection and increasing proportion of seed germination.
2. Seed treatment with thiram 80 per cent WP @ 2 g/l or captan 75 WP @ 2.5 g/l helps in eliminating the seed borne infection.
3. Spray the crop with 0.2 per cent zineb 80 per cent WP @ 2 g/l or ziram 80 per cent WP @ 2 g/l at first appearance of symptoms on the crop and repeat after 15 days of first spraying.

Macrophomina Blight

Causal Organism

Microphomina blight is caused by *Macrophomina phaseolina.* The pathogen is also causes root rot, collar rot, seedling blight, stem rot, leaf blight, pod and seed infection. In pre-emergence stage, the fungus causes seed rot and mortality of germinating seedlings. In post-emergence stage, seedling blight disease appears due to soil, water or seed-borne infection. Fungus produces numerous jet black colour sclerotia that survive in soil and host residue for long time and become source of primary infection. The pathogen is also carried through the infected seeds. The fungus is having wide host range. It perpetuates freely and become virulent when optimum pre-disposing conditions in the host exist. Dark brown to black pycnidia are formed on the diseased spots and pycniospores coming out of pycnidia may contribute to aerial spread of the disease. The pathogen is most favoured at a temperature of 30°C and 15 per cent moisture.

Symptoms

The disease is difficult to identify in initial stages. However, dark lesions are formed on the main stalk near soil level, forming localized dark green patches. The tissues of the affected portions become weak and shredded easily. Decay of secondary roots and shredding of the cortex region of the tap root are prominent symptoms. If the plants pull out, the basal stem and root may show dry rot symptoms. Black dot like sclerotia are formed on the surface and below the epidermis on the outer tissue of the stem and root. The disease develops rapidly and causes severe infestation under high temperature and water stress conditions.

Management

1. Basal application of zinc sulphate @ 25kg/ha or neem cake @ 150 kg/ha or soil application *Pseudomonas fluorescens* (1×10^{10}cfu/g) or *Trichoderma viride* (1×10^{8} cfu/g) @ 2.5 kg/ha + 50 kg of well decomposed FYM at the time of sowing helps in prevention of the disease.
2. Seeds treated with *Trichoderma* (1×10^{8}cfu/g) 5 – 10 g/kg of seed or captan 75 WP @ 2.5 g/l and thiram 80 per cent WP @ 2 g/l before sowing provides significant protection of the plant.
3. The diseased plants should be uprooted and destroyed so that the sclerotia do not form or survive.
4. Spray with carbendazim 50 WP @ 1.0 g/l at an interval of 15 days with the appearance of the symptoms.

Web Blight

Causal Organism

Web blight caused by *Rhizoctonia solani Kuhn.*(Teleomorph: *Thanatephorous cucumeris*) is one of the most important disease of mung bean. It causes considerable damage by reducing seed quality and yield.The disease is eidely distributed in Punjab, Haryana, Bihar, Rajasthan, Uttaranchal, Uttar Pradesh, West Bengal, Himanchal Pradesh and Jammu and Kashmir in India (Saksena and Dwivedi, 1973). The intensive crop cultivation and modified agro-practices have increased the populations of *R. solani* in soil and gradually built up new disease problems. The pathogen is a soil and seed borne pathogen and has wide hosts range. It forms sclerotia in/on soil and survives for a long period in the absence of a host either as sclerotia or thick walled brown hyphae in plants debris. The pathogen causes considerable yield loss in mungbean and urdbean in India (Dubey, 2003). Yield loss up to 57 per cent in mungbean was reported from Iran (Kaiser, 1970).

Symptoms

The symptoms of web blight occur on roots, stems, petioles and pods, but the disease is the most destructive on foliage. It causes seed decay, pre and post emergence

Figure 9.6a: Leaves of Mungbean Showing Typical Web Blight.

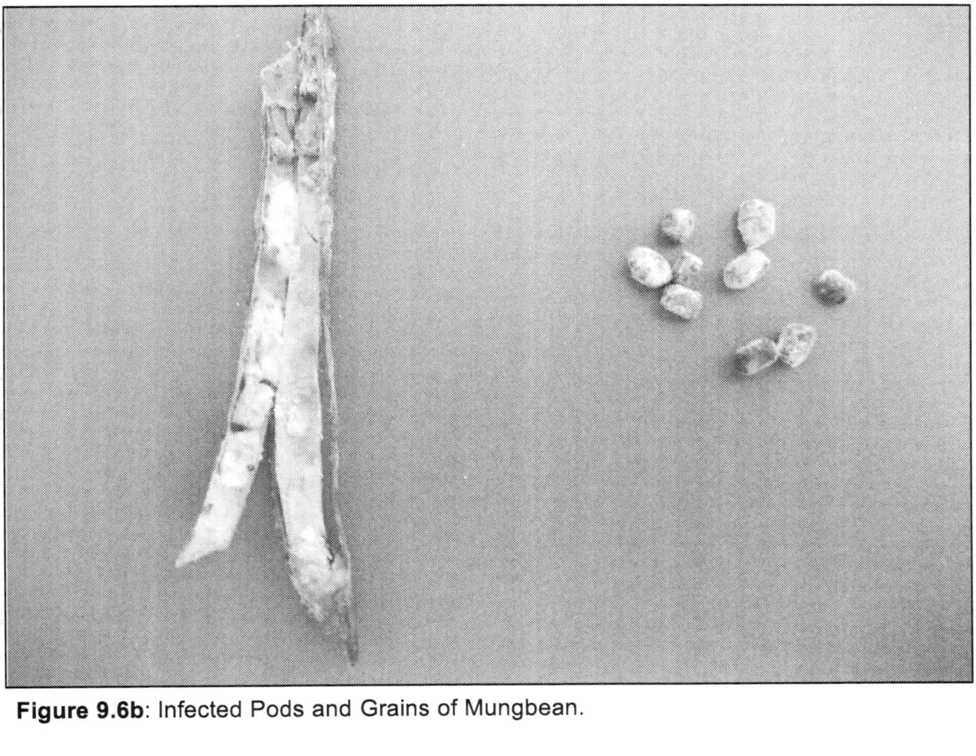

Figure 9.6b: Infected Pods and Grains of Mungbean.

mortality. Normaly, occurs during second and third week of plants growth. The first symptoms appear as small circular brown spots on the primary young leaves. These spots enlarge, often show concentric banding and surrounded by irregular conspicuous water soaked areas. The lesion expands and coalesces and white mycelial fungal growth can be seen under surface of infected leaves and young branches. The mycelium on infected leaves appears as spider web thus suggested the name web blight disease (Figure 9.6a). The pods are also vulnerable to infection at all stages of their development. In initial stage, light irregular specks appeared on the young pods, the spots were dark brown, more or less circular, srunken with darker margin in mature pods. Gradually the spots increased in number and size to cover the entire pods. The grains in the infected pods become small, shrived and light in colour (Figure 9.6b).

Epidemiology of Disease

Dubey (1997) reported that 26-28° C temperature and 90-100 per cent relative humidity (RH) favoured the development of disease. Moisture is alsoan important factor which effect on increasing inoculum potential of *R. solani*. The development of disease incited by *R. solani* on aerial parts of host plant is dependent upon the availability of free moisture or conditions near 100 per cent relative humidity (Braker and Martison, 1970). Basidiospores are an efficient source of inoculums as they are produced in large number and subsequently disseminate by air.

Survival and Disease Cycle

R. solani is primarily soil-borne in nature and initial inoculums come mainly from soil splashes on to the leaves during heavy rains. The pathogen can survives for a long period in the absence of a host either as sclerotia or thick walled brown hyphae in plants debris (Boosalis and Scnaren, 1959). The fungus is attracted to the plant by chemical stimuli released by a growing plant and/or decomposing plant residue. The process of penetration of a host can be accomplished in a number of ways. Entry can occur through direct penetration of the plant cuticle/epidermis or by means of natural openings in the plant. Hyphae will come in contact with the plant and attach to the plant by which they begin to produce an appressorium which penetrates the plant cell and allows for the pathogen to obtain nutrients from the plant cell. The pathogen can also release enzymes that break down plant cell walls, and continue to colonize and grow inside dead tissue. This breakdown of the cell walls and colonization of the pathogen within the host is what forms the sclerotia. New inoculum is produced on or within the host tissue, and a new cycle is repeated when new plants become available.

Management

1. Proper sanitation and burying the infected leaves immediately after harvest will reduce the primary inoculums.
2. Crop rotation, which help in controlling the disease to a greater extent.
3. Time of sowing of crop should be adjusted in such a way that susceptible stage of crop to be avoid rainy season.

4. Avoid the thick canopy of the crop by using proper seed rate
5. Cultivation with resistant varieties
6. Seed treatment with Carbendazim or thiophanate methyl were found best in controlling seedling mortality of mung bean caused by *R. solani*
7. Foliar spray of bavistin 0.05 per cent along with seed treatment with bavistin (0.2 per cent) is highly effective in reducing web blight.

Bacterial Leaf Spot

Causal Organism

This disease is caused by *Xanthomonas phaseoli* (Smith) Dowson, is a gram negative, rod shape bacteria. The bacteria survive in the seeds, plant debris and on the other host plants during off season. Warm and humid weather is favourable for disease development. The optimum temperature for the growth of the bacterium is 30-33 ° C.

Symptoms

The disease is characterized by small, brown and dry raised spots develop on leaves and stem. Leaf spots first appear as superficial eruption and gradually invade the tissues, giving corky or rough appearance. Leaves become yellow with advancement of disease causes premature defoliation occurs. The stem and pods also get infected.

Management

1. Seed treatment with streptomycin sulphate @ 500 ppm or captan @ 0.3 per cent or bleaching power @0.025 per cent.
2. Three protective spray of streptocycline @ 100 ppm or zineb @ 0.3 per cent or benomyl @ 0.2 per cent or three spray of streptomycin @ 0.025 per cent + 0.1 per cent carbendazim is effective in managing the disease.

References

Abbaiah, K. 1993. Development of powdery mildew epidemics in urd bean in relation to weather factors. *Indian Journal of Pulses Res.*, **6**: 186-188.

Ahmad, Z., M. Bashir and T. Mtsueda. 1997. Evaluation of legume germplasm for seed-borne viruses. *In:* Harmonizing Agricultural Productivity and Conservation of Biodiversity: Breeding and Ecology. *Proc. 8th SABRAO J. Cong. Annu. Meeting Korean Breeding Soc., Seoul, Korea.*, p. 117-120.

AICRP 2012-13. All India Coordinated Research Project on MULLaRP, Project Coordinator's Report in Annual Group Meet at TNAU, Coimbatore., 2012-13.

Amin, P.W. 1985. Apparent resistance of groundnut cultivar Robut 33-1 to bud necrosis disease. Pant Dis. 69: 718-719.

Baker, R. and Martison, C.A. 1970. In: *Rhizoctonia solani, biology and pathology*. (Ed. J.R. Parmeter, Jr), University of California Press, Berkeley, 172-188 p.

Bashir, M., S. M. Mughal and B.A. Malik. 1991. Assessment of yield losses due to leaf crinkle virus in urdbean (*Vigna mungo* (L) Hepper). *Pak. J. Bot.*, **23**: 140-142.

Beniwal, S. P. S. and N. Bharathan. 1980. Beetle transmission of urdbean leaf crinkle virus. *Indian Phytopathol.*, **33**: 600-601.

Beniwal, S.P.S., Chaubey, S.N. 1979. Urdbean leaf crinkle virus: Effect on yield contributing factors, total yield and seed charcters of urdbean (*Vigna mungo*). *Seed Res.* **7**: 125-181.

Boosalis, M. G. and Scharen, A. L. 1959. *Phytopathology*. 49: 192 –198.

Brar, J.S. and Rataul, H.S. 1987. Evidence against the transmission of urdbean leaf crinkle virus (ULCV) in mash, Vigna mungo (L.) through insects- a field approach. *Indian J. Ent.*, **49**: 57-63.

Dhingra, K.L. 1975. Transmission of urdbean leaf crinkle virus by two aphid species. *Indian Phytopathol.* **28**: 80–82

Dhingra, K.L. and Chenulu, V.V. (1985) Effect of yellow mosaic on yield and nodulation of soybean. *Indian Phytopathology*, **38**: 248-251.

Dhingra, K.L., Chenula, V.V. 1981. Studies on the transmission of urdbean leaf crinkle and chickpea leaf reduction viruses by *Aphis crassivora* Koch. *Indian Phytopath.*, **34**: 38-42.

Dubey, S. C. 2003. Integrated management of web blight of urd / mung bean by bio-seed treatment. *Indian Phytopath.*, **56**: 34-38.

Dubey, S.C. 1997. Influence of age of plants, temperature and humidity on web blight development in groundnut. *Indian Phytopath.*, 50(1): 119-120.

Grewal, J.S. 1978. Diseases of mungbean in India. First Intl. Mungbean Symp. Proc., Univ. Philippines, 1977, p. 165–168

Honda, Y.M., Twaki, Y., Saito, P. Thangmeearkom, P. Kittisak, K. Deema, N. (1981) Mechanical transmission, purification and some properties of whitefly transmitted mungbean yellow mosaic virus in Thailand. *Plant Disease.*, **65**: 801-04.

Ilag, L.L. 1978 Fungal diseases of mungbean in the Philippines. First Intl. Mungbean Symp. Proc., Univ. Philippines, 1977, p. 154–156

Ilyas, M.B., Haq, M.A., Iftikhar, K. 1992. Studies on the responses of growth components of urdbean against leaf crinkle virus. *Pakphyton.*, **4**: 51-56.

Kadian, O.P. 1980. *Studies on leaf crinkle disease of urdbean* (*Vigna mungo* (*L.*) *Hepper*), *mung bean* (*V. radiata* (*L.*) *Wilczek*) *and its control*. Ph.D Thesis, Dept. Plant Pathology, Haryana Agric. Univ., Hisar. 177 p.

Kadian, O.P. 1982. Yield loss in mungobean and urdbean due to leaf crinkle disease. *Indian Phytopath.*, **35**: 642-644.

Kaiser, N. J. 1970. Rhizoctonia stem canker disease of mungbean in Iran. *Plant Dis Rep.*, **54**, 240-50.

Kolte, S.J. and Nene, Y.L. 1973. Studies on the symptoms and mode of transmission of leaf crinkle virus of urdbean (*Phaseolus mungo*) *Indian Phytopath.*, **25**: 401-404.

Lagapsi, B.M., Capiton, E.M. and Hubbell, J.N. 1978. AVRDC., Phillipine, programme studies. First international symposium on mungbean.

Nair, N.G. 1971. *Studies on the yellow mosaic of urdbean caused by mungbean yellow mosaic virus.* Ph.D. thesis, U.P. Agric. Univ., Pantnagar, India

Narayansamy, P. and T. Jaganthan. 1973. Vector transmission of black gram leaf crinkle virus. *Madras Agric. J.*, **60**: 651-652.

Nariani, T.K. 1960. Yellow mosaic of mung (*Phaseolus aureus*). *Indian Phytopathology*, **13**: 24-29.

Nene, Y.L. 1968. Annual report (No. 1) project, FG-In-358, U.P. Agric. Univ., India.

Nene, Y.L. 1972. A survey of viral diseases of pulse crops in Uttar Pradesh, G.B. Pant Univ. Agric. Tech., Pantnagar, *Res. Bull.* **4.**

Nene, Y.L. 1973. Viral diseases of some warm weather pulse crops in India. *Plant Disease Reporter*, **57**: 463-467.

Nene, Y.L. 1973. Viral diseases of some warm weather pulse crops in India. *Plant Dis. Rep.* **5**: 463–467.

Nene, Y.L., Srivastava, S.K. and Naresh, J.S. 1972. Evaluation of urdbean (*Phaseolus mungo* L.) and mungbean (*Phaseolus aureus* L.) germplasms and cultivars against yellow mosaic virus. *Indian J. Agric. Sci.* **42**: 251–254.

Rath, G.C. and Grewal, J.S. 1973. A note on *Cercospora* leaf spot of *Phaseolus aureus*. *Indian J. Mycol. Plant Pathol.* **3**: 204–207.

Rathi YPS. 2002. Epidemiology, yield losses and management of major diseases of Kharif pulses in India. In: Plant Pathology and Asian Congress of Mycology and Plant Pathology, Oct.-1-4, 2002. University of Mysore, Mysore, India.

Rathi, Y.P.S. and Nene, Y.L. 1974. The additional hosts of mungbean yellow mosaic virus. *Indian Phytopathol.* **27**: 429–430.

Reddy, K.S., S. E. Pawar, and C. R. Bhatia. 1994. Inheritance of powdery mildew (*Erysiphe polygoni* DC) resistance in mungbean (*Vigna radiate* L.Wilczek). *Theoretical and Applied Genetics.* **88**: 945-948.

Rishi, N. 1990. Seed and crop improvement of northern Indian pulses (*Pisum* and *Vigna*) through control of seed- borne mosaic viruses *Final Technical Report, (US-India Fund) Dept. Plant Pathology, CCS Haryana Agric. Univ. Hisar, India*, p. 122.

Saksena, H. K. and Dwivedi, R. P. 1973. *Indian J. Farms Sci.* **1**: 58 - 61.

Shahare, K.C.and Raychaudhary, S.P. 1963. Mosaic disease of urd (*Phaseolus mungo* L.) *Indian Phytopathol.* **16**: 316–318.

Singh, A., A. Sirohi, and K. S. Panwar. 1998. Inheritance of mungbean yellow mosaic virus resistance in urdbean (*Vigna mungo*). *Indian J. Virol.* **14**: 89-90.

Singh, A.K and Srivastava,S.K.1985. Nodular physiology of urdbean as affected by urdbean mosaic virus.V. Effect on some enzymematic activity-*Phyton* (Austria) **25**: 213-217.

Singh, D.P. 1980. Inheritance of resistance to yellow mosaic virus in blackgram (*Vigna mungo* (L.) Hepper). *Theor. Appl. Genet.* **57**: 233–235

Singh, D.P. 1981. Breeding for resistance to diseases in greengram and blackgram. *Theor. Appl. Genet.* **59**: 1-10.

Singh, G., Kapoor, S. and Singh, K. 1988. Multiple disease resistance in mungbean with special emphasis on mungbean yellow mosaic virus. In: Mungbean, Proceedings of the second International Symposium on Mungbean, Shanhua, Asian Vegetable Research and Development. Shanmuga Sundaram, S. (Ed.)

Singh, J.P., 1980. Effect of virus diseases on growth component and yield of mungbean and Urdbean. *Indian Phytopatholo.* **8**: 405-08.

Singh, R. N. and Nene, Y.L. 1978. Further studies on the mosaicmottle disease of urdbean, *Indian Phytopath.* **31**: 159-162.

Singh, R.A., De, R.K., Gurha, S.N. and Ghosh. A. (2000). Yellow mosaic disease of mungbean and urdbean. In: *Advances in Plant Disease Management,* (Eds.) U. Narain, K. Kumar and M. Srivastave. Advance Publishing Concept, New Delhi, India, p. 337-348.

Srivastava, K.M., Verma, G.S. and Verma, H.N. 1969. A mosaic disease of blackgram (*Phaseolus mungo*). *Sci. Cult.* **35**: 475-476.

Tsuchaizaki, T., Iwaki, M., Thongameearkom, P., Sarindu, N. and Deema, N. 1996. Bean common mosaic virus isolated from mungbean (*Vigna radiata*) in Thailand. *Technical Bulletin of the Tropical Agriculture Research Centre,* **21**: 184-188.

Williams, F.J., Grewal, J.S. and Amin, K.S. 1968. Serious and new diseases of pulse crops in India in 1966. *Plant Dis. Rep.* **52**: 300–304.

2016, Diseases of Pulse Crops and their Sustainable Management 149–158
Editors: Samir Kumar Biswas, Santosh Kumar and Gireesh Chand
Published by: BIOTECH BOOKS, NEW DELHI

Chapter 10

Diseases of Pigeonpea (Arhar) (*Cajanus cajan* L. Millsp.) and their Management

Ajay Kumar[1], *Mahesh Singh*[1], *Gireesh Chand*[2]* *and Ramesh Nath Gupta*[2]

[1]*Department of Plant Pathology,*
Narendra Deva University of Agriculture and Technology,
Kumarganj, Faizabad – 224 229, U.P.
[2]*Department of Plant Pathology,*
Bihar Agricultural University, Sabour, Bhagalpur – 813 210, Bihar

Introduction

Pigeonpea [*Cajanus cajan* (L.) Millsp.] commonly known as red gram or arhar is one of the major food legume crops of tropics and sub-tropics. Pigeonpea is the fifth prominent pulse crop in the world and second most important pulse crop after chickpea in India. India has largest area under pigeonpea 3.90 m ha with a total production and productivity of 2.89 mt and 741kg/ha, respectively (DAC, 2011). Pigeonpea is an often cross pollinated crop and having chromosome number 2n =22. The crop can be considered as unique because it is a legume and a woody shrub. It is a rich source of protein and supplies a major share of protein requirement to vegetarian population of the country. It has an inherent capacity to withstand environmental

* Corresponding Author: E-mail: gireesh_76@rediffmail.com

stresses (especially drought) making it one of the most sought after crops in plant introduction trials aimed at bringing new areas under cultivation. It contributes C, N and P economy of the soil (Fujita *et al.*, 2004; Kumar Rao *et al.*, 1987; Rego and Nageswara Rao, 2000) which enhancing their performance even under marginal input. Pigeonpea is tolerant to low P supply, acid soils and having high capacity for incorporation of external P into organic P (Fujita *et al.*, 2004). Its critical requirement of P concentration for dry matter production is low compared to other major proteinous crops like soybean [*Glycine max* (L.)] (Adu-Gyamfi *et al.*, 1990).

Pigeonpea crop suffers from over 210 pathogens (83 fungi, 4 bacteria, 19 viruses and mycoplasma and 104 nematodes) reported from 58 countries (Reddy *et al.*, 1990; Nene *et al.*, 1996). The major diseases that assume significant importance include wilt (*Fusarium udum* Butler), sterility mosaic (Pigeonpea sterility mosaic virus) and phytophthora blight (*Phytophthora drechsleri* Tucker f. sp. *cajani* Kannaiyan *et al.*). Among these, wilt is the most serious disease causing irreversible losses and lethal damage to crop. At present, farmers mainly grow landraces of pigionpea and it is possible that they have some degree of tolerance to most of the pathogens. This situation could change once the diverse landraces are replaced by a few improved cultivars.

Fusarium Wilt

In India, the infestation occurs in almost every state in which pigeonpea is cultivated, especially in Rajasthan, Maharashtra, Madhya Pradesh, Uttar Pradesh and Bihar. The wilt is also problematic in many countries like Kenya, Malawi and Tanzania. The continuous cultivation of pigeonpea in the same field every year results upto 50 per cent plant mortality due to the disease. In Bihar and U.P., 5-10 per cent loss of standing crop is common feature every year (Kannaiyan *et al.*, 1984). The disease incidence is more pronounced during flowering and pod formation stages.

Symptoms

Symptoms are more pronounced and more damaging is revealed when plants have grown up after rainy season. The susceptible cultivars are attacked in early stage of about 5-6 weeks old. The symptoms are variable. The first typical symptom is premature yellowing of leaves and next is wilting or withering of leaves of diseased plants. Wilting is characterized by gradual, sometimes sudden, yellowing, withering and drying of leaves followed by drying of the entire plant or some of its branches. Patches of diseased plants are scattered throughout the field indicating locations of pathogens where infection is started. Examination by split open of main root and the base of stem and infected plant after pulling it out from the soil, the black or brown longitudinal streaks are present. These black or brown streaks represent the vascular bundles of the infected plants plugged with mycelia and fructifications of the fungi. The intensity of browning or blackening decreases from the base to the tip of the plant. Partial wilting of the plant is a clear indication of fusarium wilt, and distinguishes this disease from drought, termite damage and phytophthora blight that kill the whole plant. Partial wilting is mainly concerned with lateral root infection, while total wilt is due to tap root infection

Causal Organism

The wilt of pigeon pea is caused by *Fusarium udum* Butler. It is facultative parasite in nature and can survive in soil for many years without host through chlamydospores. The mycelium is restricted to the vascular tissues which is inter and intracellular, hyaline, branched, geniculate and septate. The profuse growth of mycelium within the xylem vessels completely plugged with the lumen of the vessels and thus, the free flow of water is checked resulting in disease development. Mycelium produces three types of spores within host tissues *viz.* macro conidia, micro conidia and chlamydospores. The micro conidia are small, elliptical or curved, thin walled, one or two septate and size 5-15 x 2-4 μm. The macro conidia are linear, sickle shaped, pointed at both ends, thin walled, 3-4 septate and measures 15-50 x 3-5 μm. The chlamydospores are usually formed in chains within the tissues of the host plant. They are round or oval, thick walled and can remain viable for longer period.

The perfect stage *Gibberella indica,* described by Rai and Upadhyay (1982) is usually found on exposed roots and collar region of the stem upto height of 35 cm above ground level. The mature perithecia are dark violet, superficial, aggregate, sub-globose to sub-cylindrical, measures 60-80 x 6-10 μm, broader in middle with short stalk, narrow apex and a central apical pore. The ascospores are elliptical to ovate size 10-17 x 5-7 μm, hyaline, generally 2-celled, rarely 3-4 celled and constricted at the septa.

Disease Cycle

Fusarium udum is facultative parasite and lives saprophytically in the soil without host for a longer time. The fungal pathogen attacks host roots by germtubes arising from asexual spores and reached in vascular tissues to establish infection. The fungus can be grow and multiply in the vascular tissues and causing partial or complete wilting of the host plant. The host plants die from wilt within few days or weeks. When the crop is harvested, plants are cut at the soil level leaving the infected root system and stubble to infest the soil. Being saprophytic, the fungus continues to grow and multiply, remains in soil until the next crop season. The fungus can survive at pH 4.0-9.0 and soil temperature upto 35°C. If the crop grows continuously on the same field, the incidence of disease increases every year.

Disease Management

The Fusarium wilt disease is mainly soil borne in nature and it is difficult to control wilt of pigeonpea like other soil borne fungal diseases. There are several methods to manage the severity of the disease. The crop rotation or mixed cropping with sorghum or tobacco for long duration for 3-5 years help in reducing the virulence of the pathogen even it present saprophytically in the soil. Seed treatment with carbendazim and *Trichoderma* spp.(1:4) may provide protection in the early stage of plant growth. The most effective aspect of disease management is cultivation of disease resistant or tolerant varieties of pigeon pea like Asha, Narendra arhar1,Malviya arhar13, NP 15, ICPL 96058, ICP 4769, ICP 7118, ICP 7035, ICP 7182, ICP 8863, ICP 9168, ICP 10958 and ICP 11299. At ICRISAT, three genotypes ICPL 96047, ICPL 96061 and ICPL 96046 were found resistant to fusarium wilt, powdery mildew and phyllody. Amendment of soil with roots of certain leguminous crops (sweet clover), molasses and oil cakes (groundnut cake) have significantly increased the antibiotic (bulbiformin) production by *Bacillus subtilis* and a reduction of 88 per cent wilt incidence. Seedlings obtained resistance against fusarium infection when seeds were inoculated with *Bacillus subtilis* before sowing. Mori (2003) observed that mustard cake was most effective in inhibiting the radial growth of *F. oxysporum* f. sp. *momordicae.*

Raj and Kapoor (1996) assessed groundnut, mustard, sesame and cotton seed oil cakes for their ability to reduce wilt of tomato caused by *F. oxysporum* f. sp. *lycopersici* and also reported that groundnut and mustard oilcakes of 2 per cent concentration in soil (w/w) were most effective in reducing the pathogen population and disease incidence. However, groundnut oil cake was superior to mustard as it recorded higher reduction in disease index (77.1 per cent). Padmodaya and Reddy (1999) reported that FYM and Neem cake were found most effective against *F. oxysporum* f. sp. *lycopersici* causing seedling disease of tomato under glass house conditions.

Antifungal effects of organic amendments against *F. oxysporum* f. sp. *cubense* revealed that neem cake was most inhibitory to mycelial growth of the pathogen followed by groundnut cake. In glass house experiment also, neem cake exhibited maximum reduction in rhizosphere population of the pathogen and vascular discoloration index (Sarvanan *et al.,* 2004). *Trichoderma harzianum* effectively suppressed fusarium wilt of pigeonpea. Mandhare and Suryawanshi (2005) reported that application of *Trichoderma* spp. as seed and soil application was found effective showing 63.25 per cent wilt reduction.

Sterility Mosaic of Pigeonpea

Sterility mosaic was first reported from Pusa (Bihar) by Mitra (1931). It is also called "Green plague of pigeonpea" and it is one of the most devastating disease of pigeonpea growing states of the country. The disease is mainly confined to Asia and has been reported from Bangladesh, Nepal, Thailand, Myanmar and Sri Lanka. In India, it is a serious problem in Uttar Pradesh, Bihar, Gujarat, Karnataka, Tamil Nadu and Andhra Pradesh. In India, annual loss of 2.05 L tonnes of pigionpea grains (worth Rs. 75 crores) is reported by several workers.

Symptoms

The disease affected plants remain green even after maturity of the crop. The affected crop looks from distance as a green forage crop standing in the field. Disease affected plants appear bushy with yellowish green foliage, light and dark green mosaic pattern of reduced size and suppression of flowers and pods. Severe stunting, reduction in leaf size, increased number of secondary, tertiary branches arising from the leaf axils and complete or partial cessation of development of the reproductive structures hence it is called Sterility mosaic disease. Mainly three types of symptoms are observed:

- ☆ Light and dark mosaic pattern on leaflets: Plants do not produce flowers and pods.

- ☆ Ring spot, where there is no sterility; it is characterized by green islands surrounded by a chlorotic halo on leaflets. The rings disappear as plants mature.
- ☆ Mild mosaic with partial sterility.

Causal Organism

Most of the workers assumed that sterility mosaic is caused by Pigeonpea Sterility Mosaic Virus (PSMV) simply on the basis of symptomatology. Transmission of the pathogen is mainly by eriophyid mites (*Aceria cajani*). There is no information about morphology or properties of the pathogen *in vitro* condition.

Biology of Mite Vectors

Eriophyid mite (*Aceria cajani*) is a worm-like, microscopic vector which is about 200-250 μm in length. The mite has two pairs of legs and does not possess wings and eyes. It has hairy structure on hind portion of the body by which it sticks with lower surface of leaves and their dispersal in field is mainly by wind currents. The mite has short life cycle which completes in two weeks. Maximum number of mites present in under surface of leaves on lower portion of the plant in comparisons to upper portion. Mites feed with puncturing and sucking types of mouthparts which consists of slender stylets.

Disease Cycle

Perennial, wild and ratooned pigeonpeas infected with disease seem to be a source of Pigeonpea Sterility Mosaic Virus (PSMV) and also for the vector. *A. cajani* is the only vector of PSMV. Besides the pigeonpea, these mites have been observed onto common weeds such as *Oxalis circulate* and *Cannabis sativa*. However the role of these two weeds in survival of vector or virus is obscure.

Management

- ☆ Seed treatment with 10 per cent aldicarb @ 3 g/kg seed.
- ☆ Cultural practices – select a field away from perennial or ratoon pigeonpeas
- ☆ Destroy source of infection *i.e.*, perennial or ratooned pigeonpeas
- ☆ Destroy infected plant in early stage of disease development.
- ☆ Crop rotation with non leguminous crop.
- ☆ Uses of acaricide or insecticides like Kelthane or metasystox @ 0.1 per cent to control the mite vector in early stages of plant growth.
- ☆ Resistant varieties: Host resistance is the best solution against the disease. Some pigeonpea varieties show resistance to the sterility mosaic like Bahar, DA-11, 33, ICPL306, Hy3C, Pusa9, Pusa885, Asha, Sharad (DA-11), Amar, Azad, Narendra-Arhar-1, Malviya arhar 6, Malviya arhar13.

Phytophthora Blight

It is also known as stem blight, stem canker and stem rot. It is recently recognized disease of pigeonpea. Phytophthora blight was first observed at IARI, New Delhi,

India, in 1966 by Williams, Grewal and Amin. A survey was conducted between 1975 and 1980 indicated that Phytophthora blight to be widespread with an average incidence of 2.6 per cent but its incidence was very high (26.30 per cent) in West Bengal. The disease affects crop at any stage of growth and development when environmental conditions are conducive for the pathogen and disease development.

Symptoms

Symptoms depend upon the age and affected part of the plant. Pigeonpea seedlings may be infected with Phytophthora blight as soon as they emerge. Young seedlings are killed within 3 days. The seedling shows crown rot symptoms which topple over and die. When seedlings are about 1 month old, first symptom appear as water-soaked lesions on the primary and trifoliate leaves which turn purple to dark brown necrotic within 5 days. The leaflet lesions are circular to irregular in shape and can be large as 1 cm in diameter. The whole foliage can become blighted within a week. The symptoms of stem usually appear later on the main stem, branches and petioles as brown to dark brown lesions which distinctly different from the healthy green portions. During favorable weather condition upper portion of the mature stems and branches may be badly affected.

Causal Organism

The disease is caused by *Phytophthora drechsleri* f. sp. *cajani*. The fungus produces aerial white mycelium on culture medium. The hyphae are hyaline, cottony, coenocytic, branched, smooth, and slender with measuring 3-6 μm in diameter. Irregular swellings with tubular projections are present on the hyphae. Sporangiospores are hypha-like and swollen at tip portion. The sporangia are ovate to pyriform, rarely spherical and with a minute papilla on some substrates. The sporangia measures 41-78 x 28-45 μm. Zoospores mature within the sporangia and are release individually after dissolution of the apical portion of the sporangium. Zoospores are biflagellate, hyaline, ovoid to reniform, tapering slightly at the anterior end. Oogonia are hyaline and purple to brown in colour. Antheridia are simple, hyaline, amphigynous and measure 12.5-19 x 10-17 μm. Oospores are spherical to globose and chlamydospores are also formed.

Disease Cycle

The survival of the pathogen occurs in the soil and plant debris as oospores. They may germinate to form sporangia or mycelium under favourable conditions. The disease is most severe in rainy season (July to September) on both seedlings and two months old plants. In rainy season oospores germinate by sporangia directly through germ tube and infection occurs in young seedlings.

Secondary inoculum comes from primary infection as a large number of sporangia which are produced on the mycelium. Wind, movement of water and raindrop splashes are sources of dispersal of secondary inoculum. The optimum temperature for growth and development of sporangia and zoospore germination is around 25°-30°C. In the absence of potassium (K) and high doses of nitrogen (N) inhances disease incidence. Application of K decreases disease incidence regardless presence of N or P in the soil. According to ICRISAT disease development is faster when day and night temperatures are more or less the same, *i.e.*, between 20 and 25°C, cloudy weather and relative humidity 70-80 per cent.

Disease Management

Cultural practices are most effective method in the management of the disease *viz.* Select fields free from blight previously, timely sowing, avoid sowing in low-lying land prone to waterlogging, grow on raised seedbeds having good drainage, summer solarization and deep summer ploughing. Use resistant varieties/lines/germplasms of pigeonpea have been found resistant or tolerant against Phytophthora blight like; Pusa A-3, Pant A-83-14, METH 12, COMP-1, ESR-6, AS-3, ICPL 161 and 366. *Trichoderma viridae, T. harzianum, Bacillius subtilis* and *Pseudomonas fluorescence* are antagonistic against pathogen. Ridomil MZ @ 3g/kg seed as seed treatment is most effective fungicide against the disease. Two foliar sprays of Ridomil MZ at 15 day intervals starting from 15 days after germination is effective against the disease.

Alternaria Blight

The leaf spot disease is reported only from India and Alternaria blight caused by *Alternaria alternata* has also been reported to cause a similar leaf spot. In severe infestation alternaria blight may cause 40-50 per cent reduction in yield.It is prevelant in most pigeonpea growing states of India (Kushwaha *et al.*, 2010).

Symptoms

Symptoms appear as blighten of leaves, severe defoliation and drying of infected branches. Brown spots are present on the leaves with concentric rings. The lesions appear on all aerial parts including pods. Initially small necrotic spots appear on the leaves which gradually increase in size. Lesions with dark and light brown concentric rings have a wavy outline and purple margin. As infection progresses, the lesions enlarge and coalesce. The disease is mostly confined to older leaves in adult plants, but may infect new leaves of young plants particularly in the post rainy-season.

Causal Organism

The genus *Alternaria* was established with *A. alternata* (originally *A. tenuis*) as the type isolate in 1867. Because of absence of an identified sexual stage for the majority of *Alternaria* species, this genus was classified into division of mitosporic fungi or the phylum Fungi imperfecti. *Alternaria* produces large, multicellular, dark-coloured (melanized) conidia with longitudinal as well as transverse septa. These conidia are broadest near the base and gradually taper to an elongated beak, providing a club-like appearance. They are produced in single or branched chains on short and erect conidiophores. *Alternaria* forms conidia which arise as protrusions of the protoplast through pores in the conidiophores's cell wall. At the onset of conidial development, the apex of conidiophore thickens and a ring-shaped electron-transparent structure is deposited at the apical dome.

Disease Cycle

Alternaria species are mainly saprophytic fungi. However, some species have acquired pathogenic capacities which collectively causing disease over a wide host range. The fungus sporulates well under warm, humid conditions. Late sown or post rainy season crop favours disease development. *Alternaria* has no known sexual stage or overwintering spores, but the fungus can survive as mycelium or spores on decaying plant debris for a considerable time or induce latent infection in seeds.

In seed-borne, the fungus can attack the seedling once the seed has germinated. In other cases, once the spores are produced they are mainly spread by wind on to plant surfaces where infection can occur. Typically weakened tissues, either due to stresses, senescence or wounding are more susceptible to *Alternaria* infection than healthy tissues. The observation of saprophytic *Alternaria* species can become parasitic when they meet a weakened host which illustrates that the distinction between saprophytic and parasitic behavior is not always evident.

Disease Management

- ☆ Use resistant varieties ICPL 366 and DA 2.
- ☆ Avoid fields close to perennial pigeon pea.
- ☆ Select seeds from healthy plant and early sowing of pigeonpea is effective against the disease.
- ☆ Seed treatment with carbendazim 2.0 g/kg seed or thiram 3.0 g/kg seed.
- ☆ Spray the crop with mixture of carbendazim and mancozeb @ 2gm/litre water at 15 day intervals.

Collar Rot

Collar rot is a minor disease of pigeonpea but becomes a serious problem when it is grown on undecomposed organic matter in the soil. Plant debris remains in the fields also favours disease development.

Symptoms

The disease appears in early stage of plant growth within a month of sowing. Patches of dead seedlings found scattered in the field at seedling stage.Seedlings become chlorotic before they die.The characteristic symptom is rotting in the collar portion which is covered with white mycelial growth; this symptom differentiates from other diseases. Affected seedlings with collar rot can be uprooted easily and the lower portion of root remains in soil. Sometimes white or brown sclerotial bodies can be attached to the collar region of seedling or soil around it.

Causal Organism

Collar rot disease is caused by a soil born fungus *Sclerotium rolfsii*.High soil moisture at the time of sowing and temperature around 30° C predispose seedlings to infection. Early sown crop during month of June shows more infestation than later sown crop.

Disease Management

- ☆ Select field free from undecomposed organic matter
- ☆ Collect rice stubbles from the field and destroy them before sowing pigeonpea.

- Seed treatment with carbendazim @ 2 gm / kg seed.
- Spraying carbendazim 2 gm / litre water at 15 day intervals when disease appears.
- Use recommended varieties.

Refrances

Agrios, G.N. 2005. *Plant Pathology*. 5th ed. Elsevier Academic Press, London.

Bart, P.H. and Thomma, J. 2003. *Alternaria* spp.: from general saprophyte to specific parasite Molecular Plant Pathology. Laboratory of Phytopathology, Wageningen University, Binnenhaven 5, Netherlands.

Common Names of Diseases, The American Phytopathological Society Retrieved from"http: / / en.wikipedia.org / w / index.php?title=List_of_pigeonpea_diseases and oldid=581610763".

DAC. 2011. *Fourth advance estimates of production of Food grains for 2010-11*. Agricultural statistics division, Directorate of economics and statistics, Department of agriculture and cooperation, Government of India, New Delhi (http: / / eands.dacnet.nic.in / advance_estimate / 3rd advance_estimates_2010-11(english).

Ingole, M.N., Ghawade, R.S., Raut, B.T. and Shinde, V.B. 2005. Management of Pigeonpea wilt caused by *Fusarium udum* Butler", *rop Protection and Productivity*, 1 (2): 67-69.

Kannaiyan, J., Nene, Y.L., Reddy, M.V., Ryan, J.G. and Raju, T.N. 1984. Prevalence of pigeonpea diseases and associated crop losses in Asia, Africa and the Americas. *Trop. Pest Management*. 30: 62.

Kushwaha, A., Srivastava, A., Nigam, R. and Srivastava, N. 2010. Management of alternaria blight of pigeonpea crop through chemicals. *Inter. J. Pl. Prot.* 3(2): 313-315.

Mandhare, V.K. and Suryawanshi, A.V. 2005. Application of *Trichoderma* species against pigeonpea wilt", *JNKVV Research Journal*, 32 (2): 99-100.

Mandhare, V.K. and Suryawanshi, A.V. 2005. Application of *Trichoderma* species against pigeonpea wilt. *JNKVV Research Journal*, 38: 99-100.

Mori, P.K. 2003. Investigations on *fusarium* wilt of bitter gourd (*Momordica charantia* L.) and its management. M. Sc. (Agri.) thesis (un-published) submitted to Gujarat Agricultural University, Junagadh, p97.

Nene, Y.L., Sheeila, V.K. and Sharma, S.B. 1996. A world list of chickpea and pigeonpea pathogens. Fifth Edition, ICRISAT, Patancheru, Andhra Pradesh, India, p. 19-20.

Padmodaya, B. and Reddy, H.R. 1999. Effect of organic amendments on seedling disease of tomato caused by *F. oxysporum* f. sp. *lycopersici*. *J. Mycol. Pl. Pathol.*, 29: 38-41.

Rai, B. and Upadhyay, R.S. 1982. *Gibberella indica,* the perfect stage of *Fusarium udum. Mycologia,* 74: 343.

Raj, H. and Kapoor, I.J. 1996. Effect of oil cake amendment of soil on tomato wilt caused by *Fusarium oxysporum* f. sp. *lycopersici*. *Indian Phytopath.,* 49: 355-61.

Reddy, M.V., Sharma, S.B. and Nene, Y.L. 1990. Pigeonpea: Disease management. In: *The Pigeonpea* (Eds. Y.L. Nene, S.D. Hall and V.K. Sheila), CAB International, Wallingford, U.K., p303-347.

Sarvanan, T., Muthusamy, M., Loganathan, M. and Prabakar, K. 2004. Antifungal effects of organic amendments against *fusarium* wilt pathogen in banana. *Madras Agric. J.,* 91: 526-529.

Sharma, P.D. 2006. *Plant Pathology*. Narosa Publishing House Pvt. Ltd., New Delhi.

Singh, R.P. 2008. *Plant Pathology*. 8th ed. Kalyani Publisher, New Delhi.

Singh, R.S. 2005. *Plant Diseases*. 8th ed. Oxford and IBH Publishing Co., New Delhi.

2016, Diseases of Pulse Crops and their Sustainable Management 159–191
Editors: Samir Kumar Biswas, Santosh Kumar and Gireesh Chand
Published by: BIOTECH BOOKS, NEW DELHI

Chapter 11

Important Diseases of French Bean (*Phaseolus vulgaris* L.) and their Management Strategies for Sustainable Agriculture

J.N. Srivaastava[1]*, *Gireesh Chand*[1], *Santosh Kumar*[1] *and Prabhat Kumar*[2]

[1]*Department of Plant Pathology, Bihar Agricultural University, Sabour, Bhagalpur, Bihar*
[2]*Betelvine Research Centre, Islampur, Nalanda – 801 303 (B.A.U., Sabour)*

Introduction

Rajmash/French bean (*Phaseolus vulgaris* L.), popularity known as rajmashalso known as common bean, haricot bean, kidney bean, string bean, salad bean, runner bean, snap bean, dry bean, pinto bean, black bean, green beans, wax bean is an important leguminous vegetable crop in India. It is consumed as tender pods, shelled green beans and dry beans. It is nutritious vegetable and grown in hilly areas of Himachal Pradesh, Jammu and Kashmir and North Eastern states during winter and as an autumn crop in part of Uttar Pradesh, Maharashtra, Karnataka and Andhra Pradesh. In India French bean contribute a major part in total pulse production scenario.

* Corresponding Author: E-mail: j.n.srivastava1971@gmail.com

During the last few years, there has been a sharp decline in availability of pulses from 70g per capita per day. The major constrain of pulses as well as Rajmash/ French bean production in India is diseases. Various diseases are known to attack Rajmash/French bean during their growing period and cause huge yield losses (Butler,1918). In this chapter important diseases of Rajmash/French bean has been discussed alongwith suitable management of strategies.

Bean Root Rots

Causal Organism

Pythium spp, *Rhizoctonia solani, Fusarium solani f. sp. phaseoli, Sclerotium rolfsii* and *Macrophomina phaseolina*.

Economic Importance

Root rot is caused by different soil-borne fungi like *Pythium spp, Fusarium solani f. sp. phaseoli, Rhizoctonia solani, Sclerotium rolfsii* and *Macrophomina phaseolina*. The disease is widespread in Africa. They are very common in bean crops that are under stress, *i.e.*, low soil fertility, high humidity, warm to high temperatures, high or low soil moisture, compacted soils, drought, acid soils or soils fertilized with ammonium fertilizers. This is very common especially in over cultivated soils. Root rot of French bean was first reported during the year 1924 from New Jersey (Cook, 1924). In India root rot french bean was first reported during the year 1980 from Bangalore (Sharma and Sohi, 1981). The disease causes huge amount of loss but the intensity of loss varies from location to location and even year to year.

Symptom

Symptoms vary depending on causative organisms and on environmental conditions. However, root rots should be suspected whenever leaves turn yellow or drop off. In general plants wilt and die, plants may become stunted or pods are malformed with under-sized seeds.

Pythium Root Rot: *Pythium* spp.

Pythium root rot causes seedling and post-emergence damping-off. Symptoms appear as elongated water-soaked areas on hypocotyls and roots at 1–3 weeks after emergence and extend several centimetres above or below the soil level (Figure 11.1). Lesions become dry and turn brown with a slightly sunken surface. The fungus also attacks lateral roots causing plant wilt and death. The pathogen can extensively prune roots (Figure 11.2) and destroy much of the hypocotyls and main root system. Pods in contact with moist soil may also become infected and exhibit a watery soft rot and mass of white fungal mycelia (Schwartz 2005).

Rhizoctonia Root Rot: *Rhizoctonia solani*

Initial symptoms on the hypocotyls and tap roots are dark circular to oblong sunken cankers delimited by brown margins (Figure 11.4). Later the cankers enlarge and become red, rough, dry and pithy. The cankers girdle the stem resulting in retarded plant growth (Figure 11.3). Minute brown sclerotia may develop on or in the cankers (Schwartz,2005).

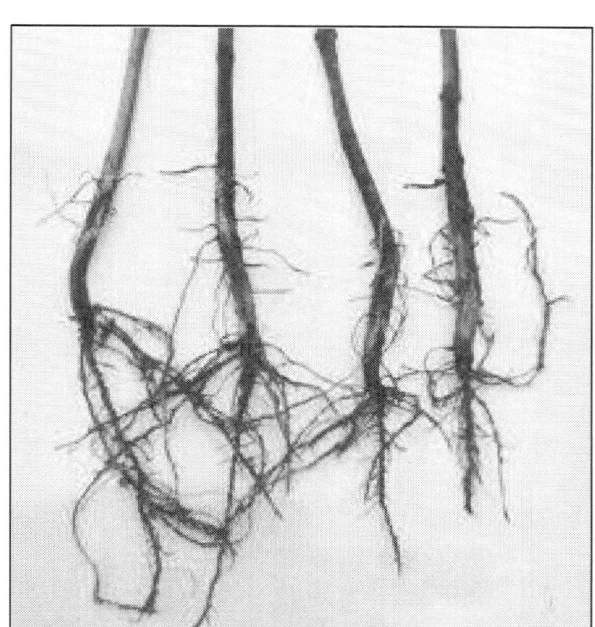

Figure 11.5: Lesions on Roots.

Figure 11.6: Adventitious Roots Formed in Response to *Fusarium* or *Rhizoctonia* Root Rot Damage.

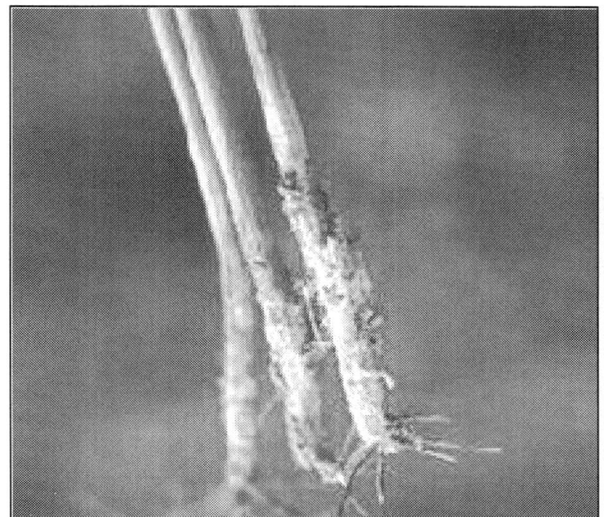

Figure 11.7: Sclerotium Root Rot.

Figure 11.8: Pods which Touch Soil may become Infected.

Disease Cycle

The main source of infection for bean root rots is infected soil. Other sources of infection include plant debris and infected or contaminated seed.

The disease is spread by irrigation water, wind-blown, soil particles, farm tools and machinery (Schwartz, 2006).

Disease Management

A combination of the following preventive measures may be used:

- ☆ Seed and soil treatments: *e.g. Pythium* species can be managed by seed treatment with Ceresan or Captan @ 2.5g / kgof seed. Drench the seed bed with Captan @ 2g / litre water.
- ☆ *Rhizoctonia solani* can be managed by seed treatment with Ceresan or Captan @ 2.5g / kgof seed. Drench the seed bed with captan @2 g / litre water.
- ☆ Brassicol @ 1.5g / litre water.
- ☆ *Fusarium* rot can be managed by seed treatment with Thiram or Emisan @ 2.5g / kg of seed. Apply Thiram in ferrows at time of planting. *Sclerotium* root rot may be managed with Terraclor 75 WP.
- ☆ Application of farm yard manure or other soil amendments such as Green manure or inorganic fertilizer will be helpful in reducing the disease.
- ☆ Planting in raised beds or ridges will facilitate rapid plant growth.
- ☆ Hilling up soil around the stem before flowering may stimulate development of adventitious roots (*Rhizoctonia* and *Fusarium* infected plant).
- ☆ Practicing 2–3 year rotation with non-host crops such as cereals.

Charcoal Rot/Ashy Stem Blight

Causal Organism

Macrophomina phaseolina

Economic Importance

The disease is also known as ashy stem blight, root rot, ashy grey stem and Macrophomina rot. The disease has a worldwide distribution and occurs on large number of plants. The disease is found throughout the tropics and subtropics and has a wide host range. It is most damaging in areas of unreliable rainfall and high temperature.

Symptoms

The disease infects the cotyledons and hypocotyls at soil level producing black, sunken cankers which have a sharp margin and concentric rings. The growing tip of plant may be killed and the stem may break at the cankers point. The ash-like wound symptoms are also develop on stem of young plant (Figure 11.9), thus the name "ashy stem blight". Black charcoal-like flakes also occur on old grey wounds on stems, pods and seeds (Sikora *et al.*, 2008).

Figure 11.9: Ashy Stem Blight. **Figure 11.10**: Leaf Wilting and Defoliation.

Infection on older plants may cause stunting, leaf chlorosis, wilting, premature defoliation, rotting of the stem and roots and plant death (Figure 11.10). The infection is more pronounced on one side of the plant. Damaged seeds shrivel and lose colour.

Disease Cycle

The pathogen is soil borne in nature which produced sclerotia presence in soil or embedded in the host tissues. However, Survival is longer in dead host tissues than in free condition the soil. Dry soils favour longer survival of these sclerotia. The pathogen is also seed borne in nature (Gupta *et al.*, 1991).

Disease Management

The following measures may be uses alone or in combination to manage the disease.

- ☆ Sowing clean disease-free seed.
- ☆ Seeds dressing with fungicides such as Thiram, Murtano, etc.
- ☆ Rouging and burning infected bean plants
- ☆ Sowing resistant varieties if available
- ☆ Practicing a 2–3 year rotation with non-hosts crops like cereals
- ☆ Fumigating the soil with Chloropicrin, methyl bromide, etc.
- ☆ Flooding of the soil if possible to suffocate the pathogen.
- ☆ Apply or Thiram @ 8kg/ha in furrow at the time of planting which will reduce the disease incidence.

Bean Anthracnose

Causal Organism

Colletotrichum lindemuthianum

Economic Importance

Anthracnose is of the most important diseases of bean worldwide. It is especially common in regions with frequent rainfall and more destructive under cool to moderate temperatures with high relative humidity. The pathogen was collected from France in 1843 by Great Mycologist Desmazieres (1943), but the first description of bean anthracnose was done in 1975 based on specimens originating in Bonn, Germany (Zaumeyer and Thomas, 1957). Its Occurrence in India was first noticed in Nilgiri hills by Hutchinson and Ram Ayyar (1915). In Himachal Pradesh, the disease incidence ranged from 5.0 to 65.0 per cent in different location (Sharma, *et al.*, 1994).

Symptoms

The disease affects all part of the plant, depending upon time of infection and source of inoculum. The initial symptoms of anthracnose appear as a dark brown to black lesion along the veins on the underside of the leaves (Figure 11.12). Leaf petioles

Figure 11.12: Bean Field Affected by Anthracnose Disease.

Figure 11.13: Anthracnose Lesions on Bean Stem, Pods and Underside of Bean Leaf.

and even stems may also show this symptom. Sunken cankers also occur on the developing hypocotyls.

The most striking symptoms of bean anthracnose appear on pods (Figure 11.13). Pod lesions are typically sunken and are encircled by a slightly raised black ring surrounded by a reddish border. Under severe infection, young pods may shrivel and dry prematurely. The fungus may penetrate through the pod causing discoloration and distortion of the seed. The fungus may also penetrate the seed coat and become firmly established within the seed. Such infected seed, when planted, serve as the source of infection to the succeeding crop (Tu, 1994).

Disease Cycle

The main sources of infection are through infected seeds and crop debris. The disease is spread by rain splash, wind-blown rain and movement of insects, animals and man, especially when plant foliage is moist.

Disease Management

Alone or in combination of the following method may help reduce infection and spread of the disease:

- ☆ Best control may be achieved by using disease free seed
- ☆ Crop rotation with non-hosts (such as cereals) for at least 2 years
- ☆ Removal or burial of plant debris from fields soon after harvest
- ☆ Sowing resistant varieties
- ☆ Dressing seeds with Thiram, Ziram, Cerasan @ 2.5 g/kg seed, Bavistin or Benlet @ 2g/kg seed help control infection in the seed coat
- ☆ Overhead irrigation discouraged to avoid spread of infection.
- ☆ Cleaning of seed storage facilities to avoid contamination.
- ☆ Spraying the crop with protectant fungicides or systemic fungicides *e.g.*, Bavistin or Benlet Zineb @ 1g/litre water in standing crop at 7 days interval after the first appearance of disease.

Alternaria Leaf Spot (Figure 11.14a)

Causal Organism

Alternaria alternata

Economic Importance

The disease has caused problems in processing beans in the USA and in Europe, and is particularly destructive if pod lesions develop resulted in blemished produce.

Symptoms

The disease is likely to occur on the older trifoliate leaves of beans, especially during prolonged periods of wet cool weather. Small, brown, irregular-shaped spotting occurs on the leaf surface, and the spots develop into larger grey-brown, round lesions containing concentric rings. The spotting tends to occur between the

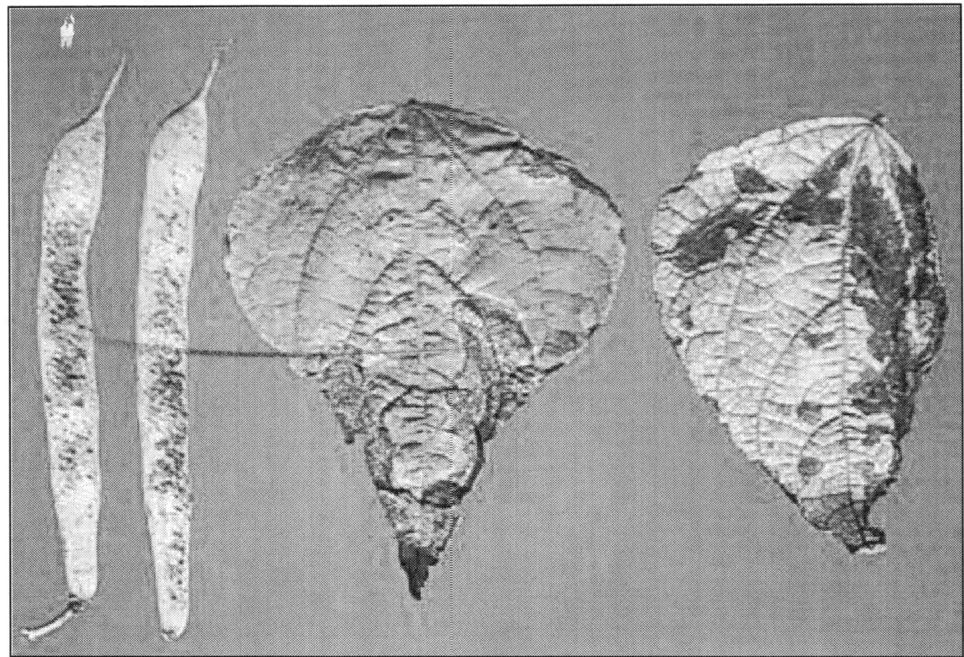

Figure 11.14a: Alternaria Leaf Spot.

major leaf veins. As the disease progress, the lesions become more angular and then coalesce causing larger areas of the leaf to die back. Pod spotting can occur in the form of red-brown flecking over the surface (Tu, 1983).

Disease Cycle

The principal pathogen, *Alternaria alternata,* is common in all situations. Spores are produced on the surface of diseased lesions and can be spread easily by rain. However, the leaf surface needs to be wet for at least 24 hours before the spores are able to germinate and infection to develop. The disease is not seed transmitted.

Disease Management

- ☆ Follow clean cultivation and field sanitation
- ☆ Dressing seeds with Thiram, Captan @ 3g/kg of seed
- ☆ Spray the plant with Blitox-50/Fyolan @ 3g/litre of water or Dithane Z-78/Dithane M-45 @ 2g/litre of water immediately after the appearance of disease

Angular Leaf Spot

Causal Organism

Phaseoriopsis griseola

Economic Importance

Angular leaf spot is a serious disease of beans that is widely distributed in Africa. Infection and disease development are favored by humid conditions and moderate temperatures (20-25°C). The disease was first reported in year 1878 from Italy (Saccardo, 1878). In India, the disease was reported first by Srinivasan (1953) from Nilgiri hills and subsequently from Himachal Pradesh, Uttar Pradesh and Karnataka. The disease has been appearing regularly in varying intensities in all bean growing areas of the country. The losses caused by the disease has been estimated about 40 to 70 per cent (Singh and Saini, 1980)

Symptoms

The disease may affect above ground all parts of the bean plant (Figure 11.14b). Symptoms are more prominent during the late flowering and early pod formation stages. The disease forms round lesions which usually appear as brown spots with a tan or silvery centre (Mackie,1945).

Figure 11.14b: Angular Leaf Spot Affecting all above-Ground Part of Plant.

The spots are initially confined to tissue between major veins, giving it an angular appearance (Figure 11.15). On some varieties a yellow halo may surround the lesions and eventually the entire leaf becomes yellow before aborting prematurely.

Figure 11.15: Spots Initially Confined to Tissue between Veins.

Figure 11.16: Lesions as seen on Upper Leaf Surface.

Figure 11.17: Lesions as seen on Underside of Leaf.

Figure 11.18: Pod Lesions.

Figure 11.19: Shrivelled or Discoloured Seeds.

Lesions can be observed on the underside of the leaf and appear slightly paler than those on the upper leaf surface (Figures 11.16 and 11.17). Lesions on stems and petioles appear elongated and dark brown. Lesions on pods are sunken irregular to circular black spots with reddish-brown centers and may be similar to those caused by anthracnose (Figure 11.18). Infected pods may bear poorly developed or shrivelled or discoloured seeds (Figure 11.19).

Disease Cycle

The main sources of infection are plant debris or residue and infected volunteer plants or may be infected seeds. The disease is spread by contaminated equipment and clothing, splashes of main water and windblown rain (Singh, 1980).

Disease Management

The disease can be managed by alone or in combination of the following strategies:

- Sowing resistant cultivars
- Crop rotation with a non host crop *e.g.*, maize for at least 2 years
- Removal or deep ploughing of crop debris or residues
- Seed dressing with suitable fungicides such as Mancozeb, Molybdenum (Mo) sprays, Benzimidazole, Thiabendozole, Trifloxystrobin and Azoxystrobin
- On the appearance of disease, spray chlorothalanil/propineb/quantaf/ cabendazim @ 0.1 per cent.

Ascochyta Leaf Spot

Causal Organism

Phoma exigua var. exigua, Ascochyta phaseolorum

Economic Importance

The disease occurs widely throughout Africa, particularly under cool and humid conditions.

Symptoms

Symptoms appear first on leaves as large dark grey to black spots. Later the infected area becomes zonate with concentric rings around the spot containing black pycnidia (Figures 11.20 and 11.21).

A blackening of nodes is characteristic of infection on stems (Figure 11.22); it can girdle the stem and kill the plant. Under severe infection, extensive blight and premature leaf drop may occur. Flower infection can lead to end rot of pods and cause extensive cankers (Figure 11.23). Pod infection often results in seed infection which can be transmitted to the next crop (Dixon,1981).

Figure 11.20: First Appearance as Large, Dark Grey to Black Spots.

Figure 11.21: Concentric Rings around Areas of Infection.

Figure 11.22: Severe Infection.

Figure 11.23: Pod Rot and Extensive Cankers.

Disease Cycle

The major sources of infection are infected seed and crop debris. The disease is spread by rain splash and contaminated equipment and clothing.

Disease Management Strategies

- ✰ Crop rotation with cereals
- ✰ Deep ploughing of infected plant debris or residue
- ✰ Planting of pathogen / disease-free seed
- ✰ Wide spacing to improve aeration and reduce disease development
- ✰ Mulching the crop to reduce water splash and spread of inoculum
- ✰ Spray the crop with fungicides such as Zeneb, Benomyl, Dithane M 45, or Mancozeb @ 2.5g / litre of water after first appearance of symptom on plant.

Collar Rot or Web Blight

Causal Organism

Rhizoctonia solani

Economic Importance

Web blight is sporadic disease in Africa. It occurs mainly in humid lowland tropical regions characterized by high to moderate temperatures. It is caused by the aerial forms of the fungus that causes *Rhizoctonia* root rot. It is a major disease of bean in hot and humid parts of Ethiopia, Madagascar.Web blight was first describe by Matz (1917). In India the disease was fist time recorded from Bagalore (Sharma and Sohi, 1981). Yield loss due to the disease varied from 8.45 to 64.68 per cent (Sharma and Sohi, 1980)

Symptoms

The general characteristic symptom is that the leaves appear to have been scalded by hot water and may appear grey-greenish to dark brown (Figure 11.24). Small brown necrotic, water-soaked spots first appear on primary leaves. Under favorable conditions the spots expand rapidly and irregularly. Finally the spots cover the whole leaf and make the plant appear burnt (Figures 11.25 and 11.26). The necrotic

Figure 11.24: Leaves Appearing as if Scalded, Grey-Greenish to Dark Brown0

Figure 11.25: Bean Plants Appearing Burnt due to Web Blight.

Figure 11.26: Leaves Appearing Burnt Due to Web Blight.

leaf tissue becomes covered with brown sclerotia and light-brown hyphae which form a web of mould over the plant and lead to defoliation. Young pod infection appears as light brown, irregular-shaped lesions, which may coalesce and kill the pod. Lesions on older pods are dark brown, circular, lightly zonate and sunken with a dark border. The pod is usually not killed unless the peduncle is destroyed.

Disease Cycle

The major sources of infection are crop debris and infected soil. Web blight has a host range of over 200 species including, beans, tomatoes, cucumber, cabbage, water melon as well as foliage and fruit of uncultivated plants and weeds.

The disease is spread by wind and rain, running water and movement of animals, man or agricultural implements through the field.

Disease Management Strategies

- ✰ Crop rotation with non-host crops such as tobacco, maize, or grass
- ✰ Deep ploughing of infected crop debris and residue

- Planting pathogen or disease free seed
- Wide spacing to maximize air circulation and change microclimate that inhibit disease development
- Mulching, where possible, to avoid rain splash and spread of inoculum onto other plants
- Treat seed with Thiram @ 0.3 per cent or Carbendazim
- Apply PCNB or Thiram @ 8 kg/ha in furrows at the time of planting which will reduce the disease incidence in seedling stage. (Singh,1985)
- Spraying with chemicals such as Maneb, Carbendazim, Benomyl, and Captafol @ 2g of water.

Fusarium Wilt/Fusarium Yellows

Causal Organism

Fusarium oxysporum f. sp. *phaseoli*

Economic Importance

The disease is widespread but unevenly distributed in Africa. It is favoured by relatively low humidity and high temperatures during drought periods as well as when plants are wounded.

Symptoms

The characteristics symptom of the disease may confused with those of Fusarium root rot. However in seedlings, infection causes stunting, wilting and death. The fungus can also cause water-soaked lesions on pods. The pathogen penetrates the vascular tissue of the root and hypocotyl producing a reddish discoloration throughout the root, stem, petioles and peduncles. Infection causes yellowing of lower leaves and may progress to the upper leaves, causing premature defoliation.

Figure 11.27: Fusarium Wilt.

Disease Cycle

Infected soil is the most common source of infection. Other sources include infected plant residues, debris and contaminated seed. Spread of disease is through irrigation, wind-blown soil particles and contaminated farm equipment.

Disease Management Strategies

- ☆ Crop rotation with non-host crops, *e.g.*, maize or sorghum
- ☆ Deep ploughing of infected bean residues and plant debris
- ☆ Use of organic amendments such as farmyard manure, green manure, etc. to improve the soil fertility
- ☆ Use plant resistant varieties
- ☆ Seed dressing with fungicides, *e.g.*, Cerasan, Thiram, Benomyl @ 2.5 g/kg seed

Powdery Mildew

Causal Organism

Erysiphe polygoni

Economic Importance

This disease is sparsely distributed in Africa and is of minor importance. It's mainly favoured by warm temperatures (20-24°C), low humidity and shade.

Symptoms

Powdery mildew can be confused with Floury leaf spot caused by *Mycovellosiella phaseoli*, however, white mould is confined to the upper surface of the leaf unlike floury leaf spot where mould is confined to the lower surface. Symptoms first appear on the upper leaf surface as small, circular, white powdery mould (Figure 11.28). The area under the mould becomes dark brown (Figure 11.29) and subsequently the white mould may cover the entire upper surface of the leaf. The plant may become stunted, malformed and yellow and leaves may fall prematurely. Pods may also become

Figure 11.28: Small Circular Moldy Spots on Upper Leaf Surface.

Figure 11.29: Later Stage of Infection.

stunted, malformed or killed. Seeds get infected through pod infection and may carry the infection to the next crop if planted (Howard *et. al.*,1994).

Disease

The major source of infection is contaminated seed and infected plant debris. The disease is spread by wind, rain and insects.

Disease Management

- ✰ Crop rotation with non-hosts such as, maize and other cereals
- ✰ Deep ploughing of residue from diseased plants after harvest
- ✰ Use of clean, disease free seed
- ✰ Wide spacing of plants to increase aeration
- ✰ Spraying the crop with sulphur, lime-sulphur, Dinocap, Benomyl, Triforine (SAPROL) etc.

Floury Leaf Spot

Causal Organism

Mycovellosiella phaseoli (*Drummond*)

Syn:Ramularia pasiolina (*Petrak*)

Economic Importance

This disease is widespread in eastern Africa in mid- to high-altitude areas with cool to moderate temperature and high relative humidity. Floury leaf spot also known as Farinose spot is an important disease of Rajmas / French bean in relatively high altitudes in tropics where moderate temperatures along with plentiful moisture in the form of rain or dew is prevalent. The occurrence of this disease was first reported from Ecuador in 1950 (Petrak, 1950). In India, this disease appears in an epidemic form in the North Western Himalayan region and was first reported from Solan (H. P.) during the year 1964 (Sohi *et al.*, 1965).

Symptoms

Initial symptoms are round, white spots of mould on the lower surface of older leaves. The spots coalesce, become irregularly shaped and powdery. On the under surface, white fluffy growth of the fungal conidiophores can be seen which give the appearance of leaf sprinkled with coarse flour, hence the disease is named as floury leaf spot. On the upper leaf surface, the symptoms appear as yellow or green and brown spots which seldom contain mould (Figures 11.30 and 11.31). Severe infections may cause premature defoliation. Chlorosis occurs on the upper leaf surface corresponding to lower leaf lesion.

Disease Cycle

Very little information is available in literature regarding the disease cycle of this pathogen. The initial occurrence of lesions on lower most leaves of plants that

Figure 11.30: Symptoms of Floury Leaf Spot on Upper Leaf Surface.

Figure 11.31: Symptoms of Floury Leaf Spot on Lower Leaf Surface.

indicate soil or derbies may act as source of primary infection. Overwintering conidia may be splashed by rain to lower leaves. Conidia produced on primary spots are blown by wind and caused secondary infection.

Disease Management Strategies

- Seed dressing with fungicides, *e.g.* Benomyl or Carbendazim or Thiophanate @2g/kg seed.
- Timely application of the same fungicides to the crop@ 1g/l of water

White Mould

Causal Organism

Sclerotinia sclerotiorum

Economic Importance

Widespread but only seasonally important under cool, moist conditions at intermediate altitudes.

Symptoms

The disease attacks old plant tissue such as flowers, cotyledons, pods, seeds, and leaves or injured plant tissue. Symptoms appear initially as greyish-green water-soaked lesions, followed by a white / cotton-like mould growth, accompanied by a watery soft rot on the affected plant part (Figures 11.32 and 11.33). Black sclerotia form in and on infected tissue. The infected tissue then becomes dry and light coloured, appearing as if bleached. Wilting may be observed within the plant canopy. Seeds are damaged through pod infection (Figure 11.34) (Tu, 1983).

Disease Cycle

The major sources of infection are through infected soil, infected or contaminated seed, and infected plant debris / residue. Alternate hosts include sugar beet, tomato, cauliflower, lettuce, as well as grasses. The disease is spread through furrow irrigation and use of irrigation runoff water (Gabrielson, 1971).

Figure 11.32: Bean Pod and Stem Damaged by White Mold.

Figure 11.33: Symptoms of White Mold on Bean Pod.

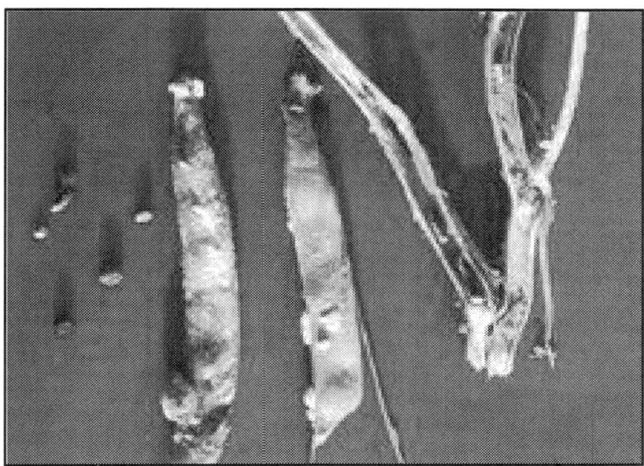

Figure 11.34: Mature Pods and Bean Stem Killed by White Mold.

Disease Management

- ☆ Burning of infected crop and other plant debris / residues
- ☆ Wide spacing of plants to increase aeration
- ☆ Mulching to avoid contact of pods with the soil
- ☆ Post harvest flooding to suffocate the pathogen
- ☆ Timely application of fungicides such as, Benomyl, Dicloran, or PCNB (Natti,1979; Ormrod *et al.*, 1993)
- ☆ Exposing diseased plant debris to the sun to desiccate and kill sclerotia (Sweeney, *et al.*, 1983)

Leaf Rust

Causal Organism

Uromyces phaseoli

Economic Importance

Rust occurs wherever beans are grown. It is favoured by cool to moderate temperatures with moist conditions that result in prolonged periods of free water on the leaf surface for more than 10 hours. Plants infected during the early stages suffer more yield losses. Any factor that delays plant maturity, such as late planting, herbicide damage, excess nitrogen or hail damage, may increase the potential for significant yield loss. Bean rust first describe from Germany (Person,1975) and now it is distributed worldwide. In India, the disease was first reported by Butler (1981). The disease can cause serious economic losses in favorable environmental conditions. The losses in green pod yield has been reported from 4.7 to 69.0 per cent, depending upon the cultivars, age of the crop at the time of infection and per cent disease intensity(Sharma, 1989).

Symptoms

Rust can be distinguished from other leaf spots. The spores from the leaf spots rub off onto your fingers, while blights do not. These leaf spots enlarge to form reddish brown or rust coloured pustules (Figures 11.35–11.37) and give a rusty appearance

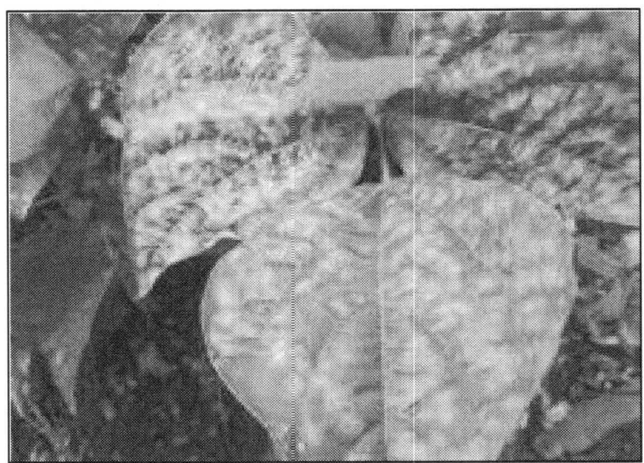

Figure 11.35: Leaf Rust, Upper Leaf Surface.

Figure 11.36: Leaf Rust, Lower Leaf Surface.

Figure 11.37: Spots become Reddish Brown or Rust Coloured.

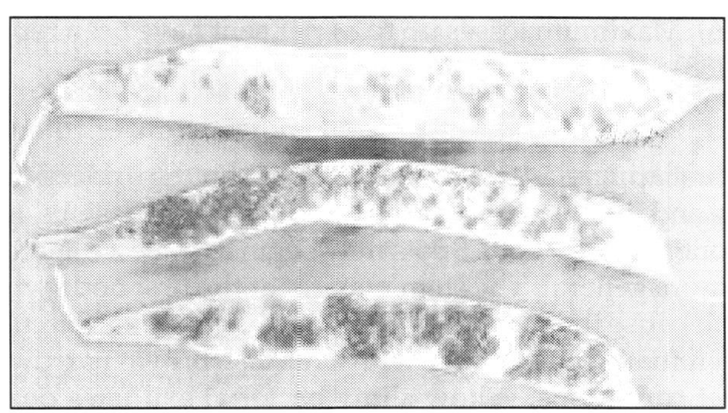

Figure 11.38: Infected Pods.

to anything they contact. Severe infection may cause leaves to turn brown, curl upwards, dry, and abort prematurely. Green pods and occasionally stems and branches also may become infected and develop typical rust pustules. Pod set, pod fill and seed size can be reduced if early infection is severe. A severely damaged bean field looks as scorched (Stavely, *et. al.*,1984).

Disease Cycle

Major sources of infection are infected bean debris and infected volunteer plants. The disease is mainly spread by wind.

Disease Management

- ☆ Practicing a 2-3 year rotation with non-hosts crops like cereals
- ☆ Practicing deep ploughing of infected plant debris
- ☆ Removal of potential sources of rust spores *e.g.*, alternate hosts and volunteer plants
- ☆ Monitoring plants frequently during flowering and early pod development for initial rust signs, then applying chemicals such as those listed below
- ☆ Timely use of fungicides such as Chlorothalonil and Maneb formulations.
- ☆ *T. viride and T. koningii* as 2 per cent foliar spraya were effective both in French bean and other bean.

Common Bacterial Blight (CBB)

Causal Organism

Xanthomonas phaseoli

Economic Importance

The disease is widespread and is favored by warm to high temperatures and high humidity. Common blight affects the foliage and pods of beans and causes significant losses in both yield and seed quality. The disease was first reported from New York, USA (Beach, 1892). In India, this disease was first reported from Bombay

Symptoms

Symptoms first appear as small water-soaked pinprick spots on the underside of the leaves. The tissues surrounding the spots gradually become yellow green (Figure 11.42) resembling a halo, thus the name is halo blight. Infected branches and petioles bear greasy spots which develop a reddish discolouration. The stem may rot at the first node where cotyledons were attached and cause the plant to break. The spots soon turn reddish brown (Figure 11.43). Infected pods exhibit water-soaked lesions which also develop a reddish discolouration (Figure 11.44). Under humid conditions, whitish to yellow bacterial exudates appear on the lesions. Seed may become shrivelled and discoloured or rotten (Figure 11.45). But sometimes no visible symptoms may be seen at all.

Figure 11.43: Spots Turn Reddish Brown.

Figure 11.44: Infected Pods Exhibit Water-Soaked Lesions.

Figure 11.45: Seed may become Shrivelled and Discolored or Rotten.

Disease Cycle

Major sources of infection are infected seed and plant debris.

The disease is spread by water splash, windblown rain and contact among plants due to movement of equipment and/or workers, especially when the foliage is wet.

Disease Management

- ☆ Crop rotation with a non-host crop for at least for 2-3 years
- ☆ Deep ploughing to bury and destroy the pathogen infected debris
- ☆ Use of clean seed
- ☆ Avoiding movement of workers in the field when wet
- ☆ Removal of all infected seedlings from the field immediately when sighted
- ☆ Suppression by spraying bean crops with registered copper based fungicides such as Kocide 4.5LF

Viral Disease

Two important viral diseases of bean are bean common mosaic and bean common mosaic necrosis. These two are the most important viral diseases affecting beans in Africa.

Symptoms

The typical symptoms of both BCMV (Figure 11.45) and BCMNV (Figure 11.46) are a light green or yellow and dark green mosaic pattern on leaves (Figure 11.47) usually accompanied by puckering, distortion and rolling of the leaves. Other symptoms seen on susceptible hosts include mottling, curling and malformation of leaves, as well as general stunting of the plant (Figure 11.48) (Hampton,1975).

Severe infection may cause leaf and flower distortions, blistering, stunting, and small pods. Plants infected early in the growing season or grown from infected seed may suffer a delay in maturity and have fewer pods and fewer seeds per pod than healthy plants. BCMNV causes the black root disease in susceptible varieties which occurs as systemic necrosis appearing first on young trifoliate leaves as diffuse reddish-brown areas with reddish-brown to black veins. Stems, pods, seeds and roots may be darkened and later turn apical wilting, die-back and necrosis (Davis *et al.*, 2001).

Figure 11.45: BCMV Infection.

Figure 11.46: BCMNV Infection.

Figure 11.47: Mosaic Pattern on Leaves.

Figure 11.48: Mosaic Pattern on Leaves.

Transmission

The major source of infection is infected seed. Other legumes crop and weed hosts also serve as the main inoculums of the virus. Infection is spread by several species of migrant aphids. The virus is also transmitted by mechanical sap inoculation, pollen and infected seed. Several species of aphid vector *viz. Aphis gossypii, A. medicaginis, A. rumicis,A.spiraecola, Brevicoryne brassicae, Macrosiphum pisi, M. gei and Myzus persicae transmit the virus under natural condition.*

Disease Management

- ✰ Planting clean disease free seeds
- ✰ Rogue infected plants from field
- ✰ Remove weed hosts from around the bean field
- ✰ Planting early in the season to help escape the high aphid population period
- ✰ Use resistant French bean cvs. GM 108, 1140, Silver, Pi217179, Sinilac, Sea Farer and Idle Refugi (Sastry *et al.*, 1981)
- ✰ Spray the crop with dimethoate (Rogar 30EC) or Monocrotophos (Nuvacron 40 EC) or 1 per cent Mineral oil for vector control in case of heavy infection.

References

Beach, S.A. 1872. Some beae diseases. New York State Agric. *Expt. Stn. Bull.* **48:** 305-333.

Boomstra, A.G. 1977. New sources of *Fusarium* root rot resistance in *Phaseolus vulgaris* L. *J. Amer. Soc. Hort. Sci.* **102:** 182-185.

Butler, E.J. 1918. Fungi and Diseases in plants. Thacker Spink and Co, Calcutta. 547 p.

Cook, M.T. 1924. Common diseases of beans and Pea, *Agric Exp. Stn. Circ.* **142:** 8.

Davis, R.M., Hall, A.E. and Gilbertson, R.L. 2001. Beans: bean common mosaic. University of California Pest Management Guidelines, UC IPM Online: http: / / www.ipm.ucdavis.edu / PMG / r52101611.html.

Desmazieres, J.B.H.J. 1843. Dixieme Notice Sur Quelques Plantes Cryptogames La Plupart Inedites Recemment Decouvertes En France, Et qui Vont Paraitre en Nature Dans La Collection Publiee Par L; Auteur. Ann. Des Sci. *Nat.* (Ser 2) **19:** 335-375.

Dubin, H.J. and Ciampi L.R. 1974. *Pseudomonas phaseolicola* in Chile. *Frutopatologia,* **9:** 91-92

Dixon, G.R. 1981. "Pathogens of legume crops" *in* Vegetable Crop Diseases. AVI Publishing Company, Inc. Westport, Connecticut.

Gabrielson, R.L. 1971. Field control of white and gray molds of beans in Western Washington. *Plant Dis. Rep.,* **55:** 234-238.

Hampton, R.O. 1975. The nature of bean yield reduction by bean yellow and bean common mosaic viruses. *Phytopathology*, **65:** 1342-1346.

Howard, R.J., Garland, J.A. and Seaman, W.L. (Eds.). 1994. Diseases and Pests of Vegetable Crops in Canada. The Canadian Phytopathological and Entomological Society of Canada, p. 554.

Hutchinson, C.M. and Ram Ayyar, C.S. 1915. The Indian rice beer ferment. *Mem. Dep. Agric. Indian Bact. Ser.,* **1:** 137-168.

Kishun, R. 1988. Loss in yield of French beans affected with common blight. P. 29-31. In: Advances in Research on plant pathogenic Bacteria. (Gnanamanickan S S and Mahadevan A Eds.). Today and Tomorrow's Printers and publishers, New Delhi.

Mackie, W.W. *et al.*, 1945. Production in California of Snap bean seed free from blight and antharacnose. *Calif. Agric. Exp. Sta., Bull.* **689:** 1-23.

Maier, C.R. 1968. Influence of nitrogen nutrition on Fusarium root rot of pinto beans and on its suppression by barley straw. *Phytopathology,* **58:** 620-625.

Matz, J. 1917. A Rhizoctonia of fig. *Phytopathology,***28:** 202-205.

McNab, A. 2007. Beans: root rot diseases. Penn State University Vegetable Disease Identification Website: http: //www.ppath.cas.psu.edu/EXTENSION/VEGDIS/VegDisases/Identification_files/bean_rrot.html

Miller, D.E. and Burke, D.W. 1974. Influence of soil bulk density and water potential on *Fusarium solani* f. sp. *phaseoli* root rot of beans. *Phytopathology,* **64:** 526-529.

Natti, J.J. 1979. Epidemiology and control of bean white mold. *Phytopathology,* **61:** 669-674.

Ormrod, D.J., Sweeney, M.E. and Brookes, V.R. 1993. Efficacy of single applications of fungicides at 10 per cent bloom against gray and white mold of Snapbeans. *Pesticide Research Report. ECPUA Ottawa.*

Persoon, C.H., 1795. Observations Mycologicae: *In usteri, P., Ann. Bot.* **3:** 1-43

Petrak, F. 1950. Beitrage zur Pilsflora von Ekuador. *Sydowia,* **4:** 450-587.

Saccardo, P.A. 1878. Fungi veneti nove vel critici, Vel mycological venetae addendi, *Isariopsis griseola* Sacc. *Michelia,* **1:** 273.

Sastry, K.S., Satyanarayan, A., Singh, S.J. and Rajendra, R., 1981. Screening of different bean cultivar against commen bean masaic virua, *Pulse Crops Newl.,***1:** 30.

Singh, A.K. and Saini, S.S. 1980. Inheritance of resistance to angular leaf spot (*Isariopsis griseola*) *in French bean* (*Phaseolus vulgaris* L.) *Euphytica, 29: 175-176.*

Singh, R.S. 1985. Diseases of Vegetable Crops. Oxford and IBH Publishuing Co., New Delhi.

Schwartz, H.F. 2005. *Compendium of Bean Diseases.* APS Press, St. Paul, Mn., USA. 109pp.

Schwartz, H.F. 2006. Root rots of dry beans. Colorado State University Cooperative Extension. Fact Sheet no. 2. 938

Sharma, P.N.J., Kumar, A., Sharma, O.P. and Tyagi, P.D. 1994. Virulence pattern of *Colletotrichum lindemuthianum* on *Phasiolus vulgaris* in Himachal Pradesh. *National Symposium on current trend in the managementof plant disease.CCSHAU, Hissar,* Nov. 10-11, 1994.

Sharma, S.R. and Sohi, H.S. 1980. Assessment of losses in French bean caused by *Rhizoctonia solani* Kuhn. *Indian Phytopath.***33:** 366-369.

Sharma, S.R. and Sohi, H.S. 1981. Effect of different fungicides against *Rhizoctonia* rot of French bean (*Phasiolus vulgaris*). *Indian J. Mycol. Plant Pathol.* **11:** 216-220.

Sharma S. 1989. Transmission of Rhizoctonia solani Kuhn. In seeds of been (Phaseolus vulgaris L.) *Curr. Sci.* **58:** 972.974.

Sherf, A.F. and MacNab, A.A. 1986. *Vegetable Diseases and Their Control.* John Wiley and Sons, Inc. New York. 728pp.

Sikora, E., Kemble, J. and Bauske, E. 2008. Common diseases of snap and lima beans. Alabama Cooperative Extension System. Factsheet ANR-1024.

Sohi, H.S., Sharma, S.L., Verma, B.R. and Sachdeva, K.B. 1965. Floury leaf spot of bean caused by *R. phasiolina. Indian Pthtopath,***18:** 384.

Srinivasan, K.V. 1953. *Isariopsis griseola* Sacc. On *Phaseolus vulgaris* L. *Curr. Sci.* **22:** 20.

Stavely, J.R. 1984. Pathogenic specialization in *Uromyces phaseoli* in the United States and rust resistance in beans. *Plant Dis.* **68:** 95-99.

Sweeney, M.E. and Ormrod, D.J. 1983. Fungicide applications for the control of grey and white molds of Snapbeans. *Pesticide Research Report. ECPUA, Ottawa.* P. 229.

Tu, J.C. 1983. Efficacy of iprodione against alternaria black pod and white mold of white beans. *Can. J. Plant Pathol.* **5:** 133-135.

Tu, J.C. 1994. Occurrence and characterization of the alpha-Brazil race of bean anthracnose (*Colletotrichum lindemuthianum*) in Ontario. *Can. J. Plant Pathol.* **16:** 129-131.

Uppal, B.N, Patel, M.K. and Nigam, B.G. 1946. Bacterial blight of French bean in Bombay Province. *Proc. Natl. Inst. Sci.* **12:** 351-359.

Webster, D.M. 1983. Bacterial blights of Kidney beans and their control. *Plant Dis.,* **67:** 935-940.

Wallen, V.R. and Galway, D.A. 1979. Effective management of bacterial blight of field beans in Ontario. *Can. J. Plant Pathol.,* **1:** 42-46.

Zaumeyer, W.J. and Meiners, J.P. 1975. Disease resistance in beans. *Ann. Rev. Phytopathol.* **13:** 313-334.

Zaumeyer, W.J. and Thomas, H.R. 1957. A monographic study of bean diseases and methods for their control. *U.S. Dep. Agric., Bull.* **868**: 90-107.

2016, Diseases of Pulse Crops and their Sustainable Management 193–223
Editors: Samir Kumar Biswas, Santosh Kumar and Gireesh Chand
Published by: BIOTECH BOOKS, NEW DELHI

Chapter 12

Major Diseases of Pea and their Integrated Managements

Mohd Ali[1]*, Deepak Singh[2], Ashish Kumar[3] and A.K. Singh[4]

[1]Department of Plant Pathology, S.V.P. University of Agriculture and Technology, Meerut – 250 110, U.P.
[2]SMS (Plant Protection), Krishi Vigyan Kendra, Sitamarhi – 843 330
[3]Senior Scientist (Plant Pathology) IARI Regional Station, Pusa, Samastipur – 848 125, Bihar
[4]Zonal Project Director, Zonal Project Directorate, Zone-2, Kolkata, W.B.

Introduction

Pea (*Pisum sativum* L.) also known as Mater is an important Leguminous crop belonging to the family "*Leguminoceae*" sub family "Papilionaceae. It is one of the most important multipurpose pulse crop. Pea is the second most important food legume in the world after pigeonpea and native of the plant is central Asia. The crop is widely grown in Asia, Europe, North America, etc. In India Pea is grown mainly as a *Rabi* crop, sown in October and November and harvested in February and March. It is consumed in various form, such as green peas, vegetables, dal, chhola, besun and soup. Green pea are dehydrates, frozen and preserved. The crop is also used for

* Corresponding Author: E-mail: mohdali.patho7220@gmail.com

grazing purposes in some parts of the world. Green plants along with pods are useful for our milk and draft animals. The foliage and hulls of the plant are used as fodder for cattle.

Table 12.1: Nutritional Composition of Raw Green Pea per 100 gram

Sl.No.	*Constituents*	*Content*
1.	Energy	81 k cal
2.	Carbohydrate	14.5 g
3.	Sugars	5.7 g
4.	Dietary fiber	5.1 g
5.	Fat	0.4 g
6.	Protein	5.4 g
7.	Vitamins:	
	Vitamins A equiv.	38 µg
	Beta carotene	449 µg
	Lutwin and zeaxanthin	2593 µg
	Thiamine (Vit. B_1)	0.3 mg
	Riboflavin (Vit. B_2)	0.1 mg
	Niacin (Vit. B_3)	2.1 mg
	Pantothenic acid (B_5)	0.1 mg
	Vitamin B_6	0.2 mg
	Folate (Vit. B_9)	65 µg
	Vitamin C	40 mg
8.	Minerals:	
	Calcium	25 mg
	Iron	1.5 mg
	Magnesium	33 mg
	Phosphorus	108 mg
	Potassium	244 mg
	Zinc	1.2 mg

Source: USDA Nutrient database.

Being a legume, it is capable of fixing atmospheric nitrogen in the soil. Inspite of high yielding varieties and new improved agronomic practices, per hectare production of pea is low in this state. Among various factors, disease constitute a major barrier in obtaining good and stable yield of pea crop. Diseases of pulse crop have received comparatively little attention as compared to the disease of other field crop like cereals, millets, oilseeds and potato. During recent years improved varieties comparatively protected crop against the major disease of pea caused by fungi, bacteria, viruses, and nematode have been reported from India. Among the fungal diseases wilt (*Fusarium oxysporum* f.sp. *pisi*), powdery mildew (*Erysiphe polygoni*), Downy mildew

(*Peronospora pisi*) and Rust (*Uromyces fabae*) are the most important disease of field pea in India (Sharma, 1998).

Fusarium Wilt of Pea

Introduction

Wilt of pea caused by *Fusarium oxysporum* f.sp. *pisi.* (linford) snyder and Hansen, is one of the most important disease. The Fusarium wilt of pea was first recognized during 1918 by Bisby in Minnesota (Chupps and Sherf, 1960). In India, first report of the occurrence of pea wilt was made available by Patel *et al.* (1949) from Bombay. Peshne (1966) made a comparative study on morphology, physiology and pathogenicity of *Fusarium oxysporum* f. sp. *pisi.* Wilt of pea or mater causes serious losses worldwide and is one of the major yield reducers in India. The pea is prone to attack by various diseases. The soil borne disease like root rot and wilt of pea caused by *Fusarium oxysporum* f. sp. *pisi* (Linford) snyder and Hansen is one of the most important diseases. Maheshawari *et al.* (1983) conducted a survey of wilt and root rot complex of pea in the various pea growing areas in Northern India and reported 13.9 to 95 per cent loss in Hosiarpur district of Punjab where disease intensity was 25-100 per cent. They also reported that early sowing of pea crop reduced incidence of disease. Sharma *et al.* (1998) found that soil borne disease like root rot and wilt are the limiting factors in producing early crop of pea.

Symptoms

Symptoms of pea wilt have been described by different workers in different ways from time to time. Linford (1928) observed distortion and wilting of leaf-lets by sudden collapse of the plants, sometimes accompanied by water soaking of the collapsed parts and extensive cortical decay of the roots. The first and the foremost characteristics symptoms is a downward incurving of the margins of younger leaves and stipules, accompanied by a slight yellowing of the leaves and a superficial grayness suggesting an excessive development of waxy bloom. The lower internodes increase in diameter and the entire stem becomes rigid (Schroedar and Walker, 1942). Buxton (1955) reported that pea wilt symptoms were similar to symptoms of the plant grown in alkali soil. He also reported that initial symptoms of wilt in the field are difficult to be distinguished. He reported that discoloration of vascular tissue may range from pale yellow bright orange to orange brown. Sukapure *et al.* (1957) reported symptoms of pea wilt as rolling of leaves, the upper parts of the plants may be pale, the growth of terminal bud is checked. Then stem and upper leaves may become more rigid than normal and the roots crisp and brittle, while the lower leaves turn pale and commence to wither, sometimes, the entire plant becomes yellow and the lower leaves wither progressively upwards, characteristically, however, after the collapse of a few basal leaves, the upper part of the plant wilts abruptly and may become dry while still green in colour. After such wilting the stem, shrivels downwards from the tip to basal basal internode, which remains firm and turgid till the end discoloration of vascular system. Partial wilting and sterility are characteristic symptoms of fusarium wilt. If pods are formed, they contain only a few shrunken, immature seeds which soon dry up. The disease as it's occurs in Maharashtra state resembles with the fusarium wilt

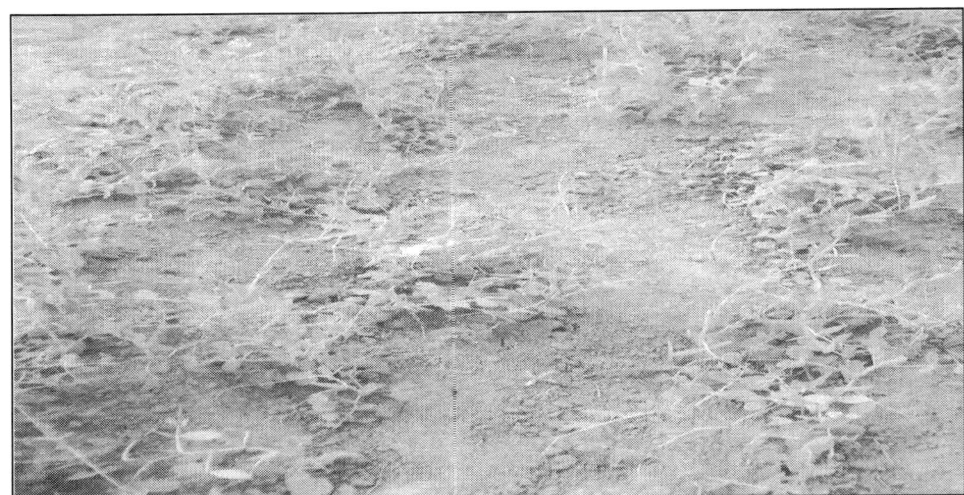

Figure 12.1: Pea Plants Infected with *Fusarium oxysporum* f. sp. *pisi.*

Note: the downward curl of the leaves on the plant that is still green, and the progression of the yellowing of leaves from the plant base toward the tip on the plant with the more advanced disease.

of peas occurred in U.S.A. Nyvall and Haglund (1972) found symptoms of pea wilt is appeared in 7-14 days old plant and infection takes place either through wounds or by direct penetration through epidermis. They further believed that the severity of symptoms was directly related to the number of infection sites.

Causal Organism

The disease, root rot and wilt of pea caused by *Fusarium oxysporum* f. sp. *pisi* (Linford) snyder and Hansen is one of the most important diseases. The fungus belong to kingdom Fungi, Phyllum-Ascomycota, order Tuberculariales and Family *Tuberculariaceae*. The fungus hyphae is septate, delicated white to peach coloured, usually with a purple tinge. Microconidia are borne on simple phialades arising laterally on hyphae or from short less branched conidiophores. These are oval ellipsoid to cylindrical or curved and measures 5-12 x 2.2 -3.5 micrometer. Macroconidia are borne on elaborately branched conidiophores or on the surface of sporodochia. They are thin walled, 3-5 septate, fusoid–subulates pointed at both ends and measure 27-46 x 3-4.5 micrometer. Chlamydospores are both terminal and intercalary.

Disease Cycle

Fusarium oxysporum is a soil-borne pathogen that is commonly found in soil and plants. *Fusarium oxysporum* f. sp. *pisi* only affects peas. Anwar *et al.* (1994) isolated the pathogen from both ungerminated seed and abnormal seedlings. The fungus causes infection on the fibrous roots or epicotyls region, grows both inter and intercellular in the cortex and ultimately concentrates in the xylem vessels. The mycelium may grow systemically through the vascular system and reach seed causing seed borne infections. The infected seeds germinate resulting in production of abnormal seedlings. After

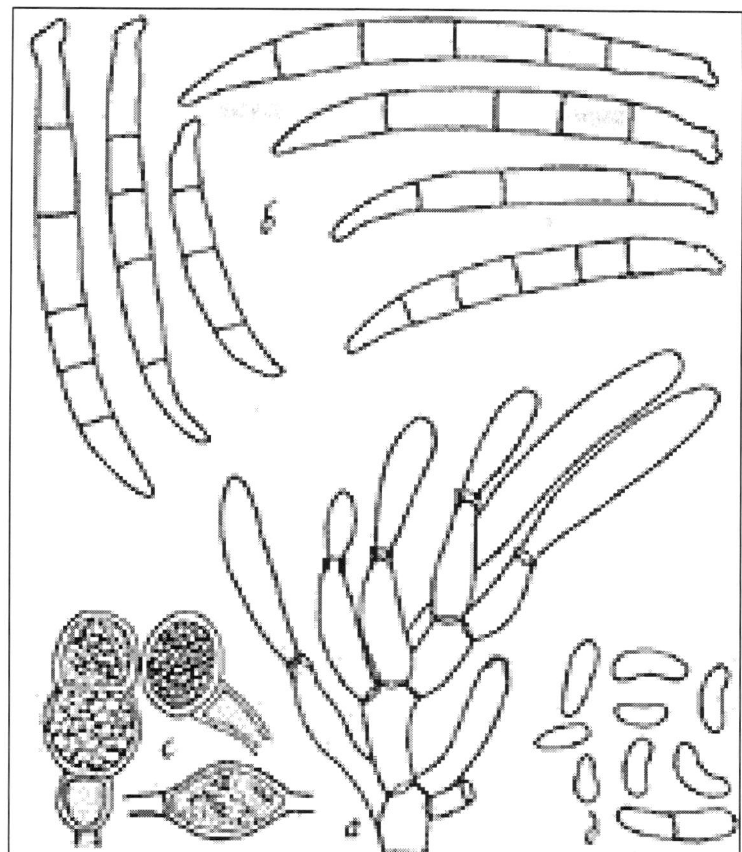

Figure 12.2: *Fusarium oxysporum* f. sp. *pisi* as seen Under a High-Power Microscope: (a) Conidiophores bearing young macroconidia at their tips; (b) Macroconidia; (c) Chlamydospores; (d) Microconidia (drawing by Lenore Gray).

the death of the plant, the fungus continues to grow and sporulate on the stem cortex, resulting in production of soil borne inoculum. The disease is thus monocyclic in nature without any secondary spread.

Epidemiology

The fungus grows optimally in culture medium at 24-28 °C while the disease intensity is maximum at 21°C. At higher temperatures, it may result in root decay. Condition favorable for plant growth does not allow the disease to progress rapidly. The pathogen establishes in some areas quickly than others and is serious under low moisture conditions. The disease however does not appear in very wet soils favour the disease while the acid soils retarted the same. The reduction in shoot length could be used to supplement the visual severity rating for fusarium wilt in field pea (Neumann and Xu, 2003). Optimum temperature for growth of race 2 of the fungus in culture medium is 28°C and it causes maximum disease at temperature between 24-28°C. Wet soils also favors near wilt development.

Management

Cultural

Use of disease free seed, field sanitation and long crop rotations are important in reducing the primary inoculums of the pathogen. Early planting of pea crop has also been suggested to lower the disease levels. Soil amendment and repeated cropping with different non-host crops such as wheat, oats, maize and sorghum reduced the population of *F. o.* sp. *pisi* and *Fusarium solani* f. sp. *pisi* in the soil and the severity of pea root rot (Vinay and Sugha, 2004).

Resistant Variety

Shukla *et al.* (1982) tested 602 varieties / cultivars under field condition in Kanpur but none varieties was found complete resistance. 11 varieties namely P-43, P-124, P-180, P-185, 450 B, 6113, 6234, Kalamukhi, 6583, 6587 and 6588 were found moderately resistant (5.1 to 10 present mortality). Datar (1983) tested 36 cultivars against *Fusarium oxysporum* f. sp. *pisi* in field condition, out of which five cultivars namely BR-12, Khapera Khada, 15/1, 11, 4-3 and 479-13 were found resistant and rest of them were found susceptible. Kumar and Kohli (2001) reported that sixteen cultivars of pea (*Pisum sativum*) were screened against. *F. o.* f. sp. *Pisi.* Arkel suited for autumn sowing was most susceptible, whereas, accession DPR-3 was most resistant. Matter Ageta-6 exhibited relatively low disease incidence.

Chemical Control

Utikar *et al.* (1978) reported that seed treatment of pea with Bavistin and Benlate gave complete control of *F. o.* f. sp. *pisi*. Shukla *et al.* (1979) tested the efficacy of some seed dressing fungicides for the control of pea wilt and reported that Bavistin have the best result in improving germination, reducing mortality and giving significant higher yield. Maheshwari *et al.* (1981) reported that seed dressing with Benomyl and Dithane M-45 and soil treatment with Phorate + Captan reduced plant mortality against *F.o.* f. sp. *pisi* and increased yield. The pathogen was also controlled by soaking pea seeds in a 1:1 combines suspension of Captafol (0.1 per cent) and Captan (0.2 per cent) (Gangopadhyay and Kapoor, 1979). Sinha and Upadhyay (1990) tested 11 compounds and found that pathogen growth was completely inhibited by Emisan-6 and Sulfex (80 per cent) at all concentrates tested. Sumitha and Gaikwad (1995) observed that the radial growth of pathogen in culture was completely inhibited by bavistin, Topsin M-70 and Thiram, each at 0.1 per cent; Captan at 0.15 per cent and Dithane Z-78 at 0.3 per cent. Wang *et al.* (1995) reported that seed treatment with fungicide like, thiram or carbendazim is known to reduce seed borne inoculums of pathogen and has been recommended for control of disease. Pandey and Upadhyay (1999) reported Bavistin as highly effective, while *T. viride* and *T. harzianum* best among all bio-control agents. Integration of bio-agents with bavistin was not beneficial. Bioagent + Thiram was highly effective to minimise the disese. Verma and Dohroo (2002) found that seed treatment with bavistin germination, lowest pre emergence rot and highest yield with the lowest wilt incidence. Agarwal *et al.* (2003) recommended seed treatment with 3 g carbendazim + thiram (1:2)/kg of seed reduced the incidence of wilt (*Fusarium udum*). Sinha *et al.* (2003) reported that bavistin, carbendazim and raxil (tebuconazole) were effective in controlling wilt in pigeonpea. Maheswari *et al.* (2008) reported that Carbendazim proved most effective fungitoxicant for checking the mycelial growth of *F.o.*f. sp. *lentis* (5.6 mm) followed by captan (9.9 mm), hexaconazole (12.5 mm) and diniconazole (16.44 mm).

Plant Extracts

Singh *et al.* (2000) studied fungitoxic effect of different plant extracts against *F. udum in vitro* and *in vivo*. Leaf extract of *Citrus medica* and *Azadirachta indica* rhizome extract of *Cucurma longa* and *Zinziber afficinale* and bulb extract of garlic completely inhibited mycelial growth at higher concentration. Kamalkanhan *et al.* (2001) showed that out of 20 plant leaf extracts tested against *Pyricularia oryzae, Prosophis juliflora, Zizyphus jujuba* and *Abutilon indicum* significantly inhibited mycelial growth and toxin production. Sharma *et al.* (2003) reported the antifungal activity of different plant extracts against *F. o.* f. sp. *pisi in vitro*. Leaf extracts of *Datura stramonium* and *Azadirachta indica* had superior antifungal activity. Verma and Dohroo (2003) studied fungitoxic effect of different plant extracts against *Fusarium oxysporum* f. sp. *pisi in vitro*. Leaf extract of garlic resulted 100 per cent inhibition followed by ocimum extract. Devi and Paul (2003, 2005) the fungitoxic activity of extract of 10 plant species was evaluated against the pea wilt caused by *F. o.* f. sp. *pisi*. The *Ranuculus muricatus* extract completely inhibited the growth of the pathogen followed by Lantana, *Ocimum* and *Datura*. Karande *et al.* (2007) reported that leaf extracts of tulsi, sadafuli were quite effective in controlling *Colletotrichum gloeosporioides* and *Fusarium oxysporum*, followed by bougainvillea, onion, neem and glyricidia. Sahni and Saxena (2009) the antifungal activity of ethanolic extracts of medicinal plants were evaluated against *F. o.* f. sp. *pisi* by "Modified disc technique". Various plant extracts resulted inhibition of growth of mycelium however bark of *Euphorbia nerifolia* exhibited absolute toxicity against the test fungus (98.0 per cent).

Essential Oil

Sharma *et al.* (2003) reported the antifungal activity of different plant oils were evaluated against *F. o.* f. sp. *pisi in vitro*. The oils of *Datura stramonium* and *Azadirachta indica* considerable antifungal activity against the pathogen. Abo-El Seoud *et al.* (2005) reported essential oils of fennel, peppermint, caraway, eucalyptus, geranium and lemon were tested for their antimicrobial activities against *Fusarium oxysporum*. Essential oils of fennel, peppermint were selected as an active ingredient for the formulation of biocides due to their efficiency in controlling the tested *Fusarium oxysporum*. Singh and Sumbali (2007) reported that essential oils and Mentha and *Ocimum* species tested at three concentration (1.0, 0.5 and 0.25 per cent) caused by significant reduction against *Penicillium expansum* causes post harvest Rot of Apples. Sitara *et al.* (2008) reported essential oils extracted from the seed of neem, mustard, black cumin and asafoetida were evaluated for their antifungal activity against *F. o., F. moniliforme, F. nivale* and *F. semitectum*. All the oils extracted except mustard, showed fungicidal activity of varying degree against test species.

Biological Control

Velikanov *et al.* (1994) reported that *T. aureoviride, T. harzianum, T. viride* and *G. virens* were found to be hyperparasitic on both *F. oxysporum* and *F. solani*. Gourdar *et al.* (2000) tested antagonistic nature of *T. viride, T. harzianum, Aspergillus niger, A. flavus, B. subtilis, P. fluorescens, Penicillium* sp. and *Streptomyces* sp. against *Fusarium udum*. Among all these, *T. viride* exhibited maximum inhibition of fungal growth, followed by *T. harzianum*. Gholve and Kurundkar (2002) evaluated the efficacy of *T. viride* and *P. fluorescens* isolates against wilt of pigeonpea *in vitro* with ICP2376 @ 20 g/kg seeds. All the treatments except seed treatment with *T. viride* alone increased seedling emergence and reduced wilt incidence. The greatest reduction in wilt incidence was recorded from seed treatment with *T. viride* + *P. fluorescens* isolate from chilli. Sawant *et al.* (2003) evaluated different products of biological control agents (*Trichoderma* spp.) as seed and soil applicants and reported that seed treatment with Phule Trichokill, (8 g/kg seed), gave lowest wilt incidence and next best treatments was seed treatment with Trieco (@ 4 g/kg seed) Verma and Dohroo (2003) studied the efficacy of the fungal antagonists *T. harzianum* and *T. viride* against fusarium wilt of pea caused by *F. of.* sp. *pisi in vitro. T. harzianum* and *T. viride* showed the maximum growth inhibition of the pathogen. Rajib *et al.* (2003) reported that antagonistic potential of *T. harzianum* and *G. virens* was observed as these overgrew and suppressed the growth of *F. o.* f. sp. *lentis* in dual culture with in 120 hrs of co-inocualtion. Prasad *et al.* (2003) reported that *T. harzianum* (44.1 per cent) isolate was found superior than the *T. viride* (39.1 per cent) isolate in reducing the colony diameter of *Sclerotium rolfsii* tested by dual culture technique. Chaudhary and Prajapati (2004) evaluated biological agents against *Fusarium udum* and reported that maximum growth inhibition of *F. udum* was recorded from *G. virens* the *T. viride* while, in culture filtration of biological agents, the maximum inhibition was recorded for *Aspergillus niger* followed by *T. viride*. Vaidya *et al.* (2004) reported that antagonistic potential of bio-control agents in dual culture showed maximum inhibition of mycelial growth of *F. o.* f. sp. *dianthi* and *Fusarium oxysporum* f. sp. *gladioli* strains of *Streptomyces* sp. followed by *Trichoderma* sp. and *P. fluorescens* respectively.

Rust of Pea

The rust fungi comprise more than 100 genera and around 7000 species (Maier *et al.*, 2003). *Puccinia* represents the largest genus with about 4000 species, followed by the genus *Uromyces* with about 600 species (Maier *et al.*, 2003). Maier *et al.* (2003) also suggests that these two genera are polyphyletic. This implies that although they are morphologically similar but might be differences at the molecular level. Mainly five species of rust fungi have served as model organisms in the laboratory. *Melampsora lini* and its host *Linum usitatissimum*, for example, were used by Flor (1956) to demonstrate the gene-for-gene hypothesis. *Uromyces appendiculatus* and *Puccinia graminis* have been used in a number of cytological and physiological studies (Zhou *et al.*, 1991; Leonard and Szabo, 2005). Today molecular analyses of rust fungi mainly focus on *P. triticina* (Thara *et al.*, 2003), *M. lini* (Catanzariti *et al.*, 2006), and *Uromyces fabae* (Jakupovic *et al.*, 2006). The very limited informations are available regarding to the epidemiology and management of field pea rust in India and other countries.

Historical Background

Two species of *Uromyces* have been reported to cause rust of pea. One of them *Uromyces pisi* (Persoon) de Bary, has been reported from several European countries (Deutelmoser, 1926; Jorstad, 1984). It is a heteroecious species having its aecial stage on *Euphorbia cyparissias* and rarely occurs in India, other species *U. viciae-fabae* (Pers.) de Bary has been found to cause pea rust in India (Butler, 1918; Prasada and Verma, 1984; Kapooria and Sinha, 1966).

Symptoms

The minute raised yellow rust pustules appear on above ground parts of the plants and most preferably on underside of the leaves, less abundant on the pods and stems. All the four stages develop on every green part of the host including the pods (Figure 12.2). The first symptoms appeared with the development of aecia. The yellow spots have aecia appear first on the under surface of the leaves, stems and petioles and persist for longer time The formation of aecial stage is preceded by a slight yellowing which gradually turns brown. The uredopustules are powdery, light brown in appearance. Later in season dark brown teleutopustules develop on the leaves occurs in the same sours as the uredia and develop from the same mycelium.

Pathogen

The rust fungus *Uromyces viciae fabae* (Pers.) de Bary is one of the important pathogens of field pea (*Pisum sativum* L.). It is worldwide distributed and also reported from faba bean (*Vicia faba* L.), lentil (*Lens culinaris* Medic.) and sweet pea (*Lathyrus sativus* L.) (Chung *et al.*, 2004; Emeran *et al.*, 2005 and 2008; Kushwaha *et al.*, 2006; Shroff and Chand, 2010). The fungus is an autoecious with aeciospores, urediospores and teliospores on the surface of host plant and completes its life cycle on the same host. However, initial inoculum has not been clearly established (Gaumann, 1998). In India, under field condition, urediospores are converted to teliospores in the month of March due to the higher temperature. Urediospores are short-lived, but the teliospores can survive in plant debris from one season to another (Hebblethwaite, 1983). However, urediospores can be preserve for two years under freezing condition has been documented by Cunningham, (1973), Davison *et al.* (1963). Germination of teliospores takes place between 17 to 22 °C temperature and at the start of next season producing basidiospores which initiated new infection cycle (Joseph and Hering 1997 and Joshi and Tripathi 2012). The basidial stage and initiation of primary infection are not fully understood. The disease is favoured by high humidity, cloudy and rainy weather condition. Disease development in field is favoured between 20°C to 22°C (Webb and Hawtin, 1981; Kushwaha *et. al*, 2006). On pea the fungus starts its life cycle with the formation of pycnia and aecia. The uredia and telia follow these stages. The yellow spots having aecia in round or elongated clusters are the earliest visible symptoms of the disease on the leaves. Pycnia occurs in small groups associated with the aecia. The aecia are cupulate, 0.3 to 0.4 mm in diameter. The peridium is short and whitish. The aeciospores are round to angular or elliptical with the hyaline wall 1 micron thick and verrucose. The aeciospores measures 14 to 22 μm in diameter. The uredia develop on both the sides of the leaves and on other parts of the plant. They represent a powdery light brown appearance. The urediospores

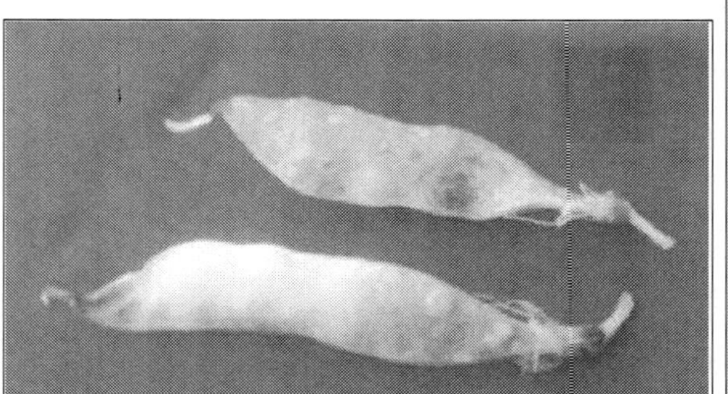

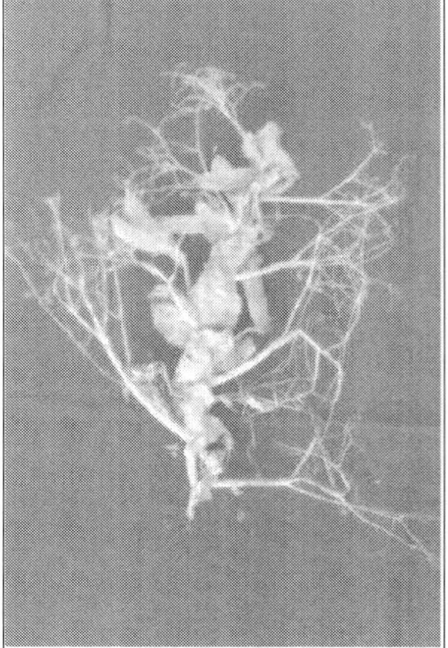

Figure 12.2: Infected Leaves, Pods and Entire Plant with Rust.

are round to ovate, light-brown, echinulate, with 3-4 equatorial germ pores and measure 20-30 (22-28) × 16-26 (19.22) μm. On host the aeciospores are known to act in place of urediospores as repeating spores. The telia occur in the same sorus as the uredia are developed from the same mycelium. They are dark-brown or black. The teliospores are subglobose, ovate or elliptical, with rounded or flattened apex which is considerably thickened and appears papillate. These spores are smooth, brown and measure 25-38 (40) × 18-27 μm. The wall is 1-2 μm thick at one side and 5-12 μm thick at another. The pedicel is persistent, yellowish-brown, thick and up to 100 μm long. The mycelium in intercellular, branched, septate having yellowish or orange-red oil drops in the cytoplasm, knob-like haustoria are formed in the host cell. The fungus does not produce basidiocarps. Caryogamy occurs in the thick walled teliospores which may be simple (unicellular). It originates at the tip of a binucleate hyphae. Young teleutospores are binucleate (dicaryotic) but after caryogamy, it contains a single diploid nucleus. The teliospores represent the sexual stage of the fungus and on germination give rise to a promycelium from each cell. The diploid nucleus passes into the promycelium, under goes meiosis and four haploid nuclei are formed. These nuclei may than be separated by three septa in the promycelium. Each nucleated cell of the promycelium then produces a short sterigmata at the tip of which swells to from a globular basidiospore in which the single nucleus form the cell of the promycelium moves. The basidium in the uredinales thus consists of two stages: the probasidium or hypobasidium (teliospore and epibasidium (promycelium). Polymorphism is the *Uromyces* spp. exhibited by development of following types of spore in the sequence listed:

Stage 0

Pycnia or spermogonia bearing pycniospores or spermatia. The pycnia are produced on a haploid thallus resulting from infection by basidiospores. This is the only monocaryotic stage of the rust fungus on the host. They contain a palisade of sporogenous cells which cut off at their tips one-celled, uninucleate pycniospores or spermatia in a sweet nectar. The spore laden nectar in exuded from the pycnia and carried by insects. The spermatia affect fertilization by fusion with receptive or flexuous hyphae present at the mouth of pycnia of the opposite mating type. The pycnia are highly variable in shape, being globose, conical, hemispherical, lens shape or undefined shape without proper delimiting boundary. On the host, they may be subcuticular or subepidermal.

Stage I

Aecia bearing aeciospores. The aecia are formed as a result of dicaryotization of the monocaryotic mycelium. Thus, they are associated with pycnia, however, they also function as repeating spores or uredia (uredial aecia). The aecia contain a palisade of binucleate sporogenous cells at their base. These cells invariably bear unicellular, binucleate, hyaline aeciospores in chains. They normally germinate to produce a dicaryotic mycelium initiating the dicaryon which bears the uredinia and telia.

Stage II

Uredia or uredinia bearing urediospores or urediniospores. The uredinia are sori produced by binucleate thallus and bearing one-celled urediniospores singly on pedicel. Paraphyses may also present in these sori. The urediniospores are binucleate with hyaline or coloured walls. The latter are perforated by conspicuous pores (germ pores). The wall is echinulated with pointed conical spines.

Stage III

Telia bearing teliospores or teleutospores. Telia and teleutospores vary enormously. The spores may be stalked or sessile embedded in the host or free and uni or multicellular. Mostly they are smooth, young teliospore is binucleate but at maturity it has a single diploid nucleus representing the diplophase in the life cycle.

Stage IV

Basidium or promycelium bearing basidiospores. Its represents the transition of diplophase to haplophase by germination of teliospore, meiosis of diploid nucleus and formation of haploid basidiospores.

Life Cycle

Uromyces viciae fabae is a macrocyclic rust fungus, it exhibits all five spore forms belonging to order Uredinales. It is also autoecious, as all spore forms are produced on a single host (Mendgen, 1997). The *U. viciae fabae* perpetuate overwintering on residual plant material, diploid teliospores germinate in the spring with a metabasidium. After meiosis, the latter produces four haploid basidiospores with two different mating types. These are ejected from the metabasidium and after landing on a leaf of a host, germinate and produce infection structures. Pycnia are produced which contain pycniospores and receptive hyphae. Pycniospores are exchanged between pycnia of different mating types and after spermatization, dikaryotization occurs in aecial primordia. An aecium differentiates and dicaryotic aeciospores are produced. These aeciospores germinate and form infection structures from which uredia which produce urediospores are formed. Urediospores are the major asexual

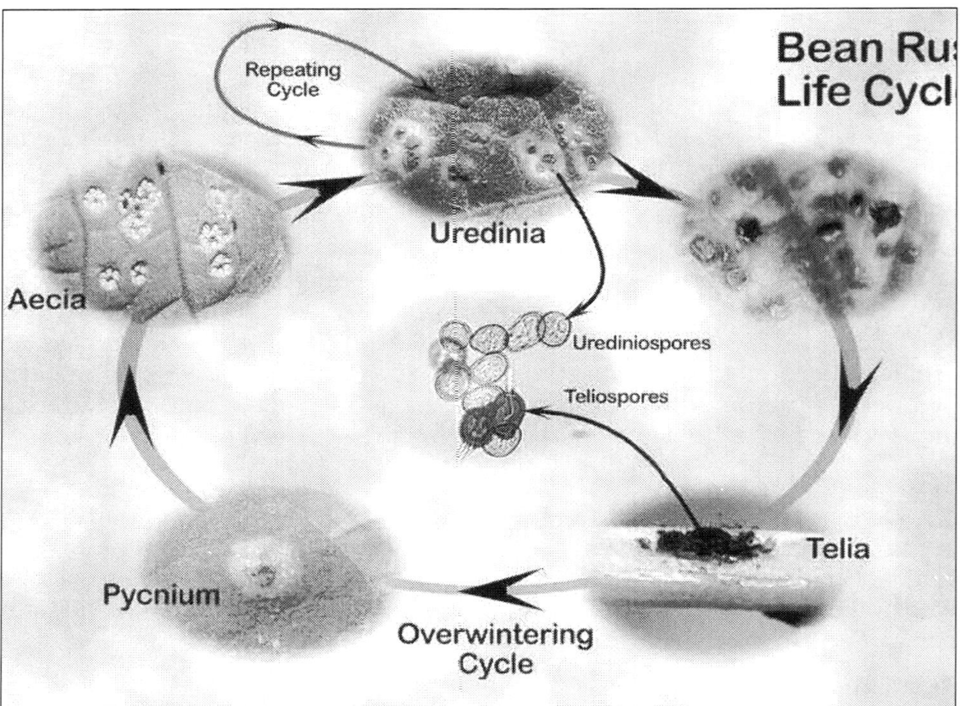

Figure 12.3c: Life Cycle of *Uromyces viciae-fabae.*

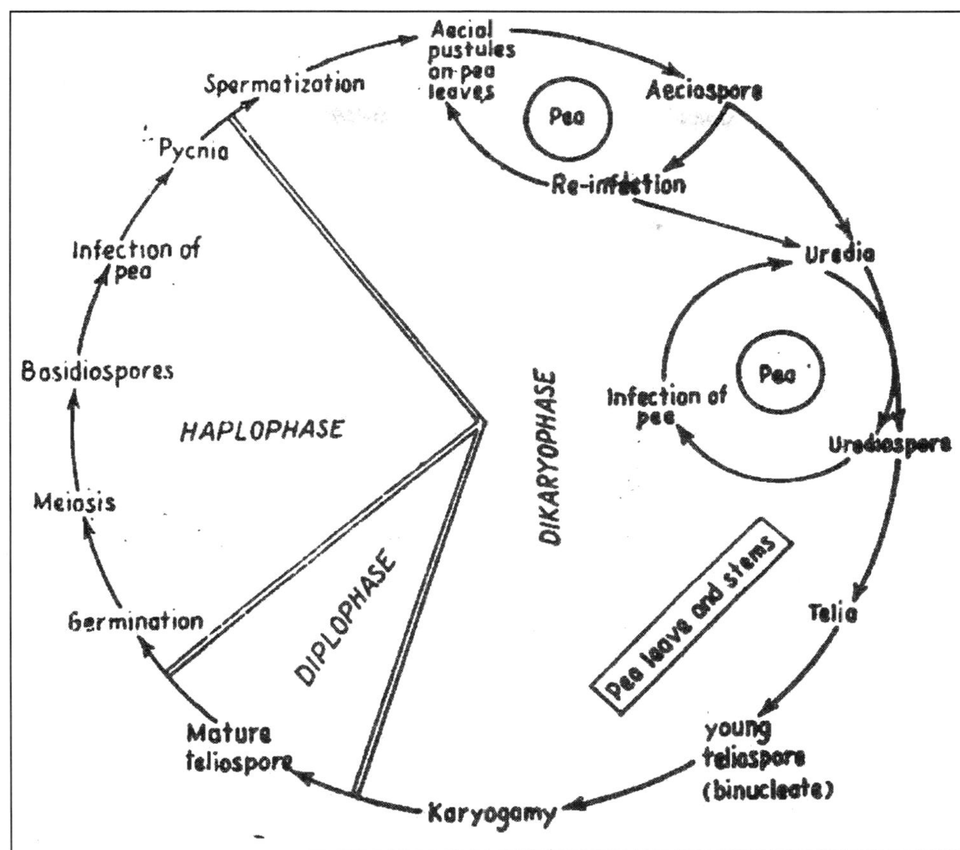

Figure 12.3b: Life Cycle of *Uromyces viciae-fabae.*

spore form of rust fungi in massive amounts through repeated infection of host plants during summer. Urediospores are aerially dispersed and can travel thousands of kilometers carried by the wind (Brown and Hovmoller, 2002). In the fall, uredia differentiate into telia, the nuclei fuse during sporogensis and single-celled diploid teliospores develop for the winter.

Uromyces pisi is heteroecious rust with 0 and I stage on the ornamental plants *Euphorbia cyparrisias* but is not common in India. The uredial and telial stages occurs on pea, broad bean and other legume.

Epidemiology

Effect of Temperature on Survival of Aeciospore/Urediospore

Effect of temperature on the perpetuation of fungus (*Uromyces viciaefabae*) in nature have been reported by various workers. Prasada and Verma (1948) reported that at relatively low temperature 17-22°C results in formation of secondary aecia while at 25°C the infection causes development of uredia. No infection by aeciospores occurs at 30°C. These spores remain viable for 8, 6, 4, 3 and 2 weeks at temperatures of 3-8°C, 10-12°C, 17-18°C and 30°C, respectively. No viability is retained after 6 weeks showing that the aeciospores do not survive during the off season for the crop. Singh (1998) reported that optimum temperature for germination of uredospores is 16-25°C. No germination occurs at 28-29°C. in the plain of North India, where warm season sets towards the end of March. The rust spores from lentil were found to remain viable for 16-17 week, when stored at 3-8°C and only for 2 weeks at 36-37°C. Which is indicated that the spores do not survive in the hot summer interning two successive crop of lentil. The teliospores of the lentil rust fungus have been found to have no dormancy and can germinate at 12-22°C soon after their formation. Batra and Stavely (1994) on reported that the urediospore of *Uromyces appendiculatus* (Pers.) germinate at 15°C to 24°C while the teliospores in crop debris germinate at 10°C to 15°C under favourable conditions. The spore complete the infection cycle within next 5-10 days. Rust of chickpea caused by *Uromyces ciceris-arietini* (Gregnon), the urediospore lose viability is often within 2-4 weeks but can survive for longer period if kept in sealed vials at 6°C (Singh 1998). Shroff and Chand (2010) reported that infection process of *Uromyces viciae fabae* started after deposition of aeciospores on the surface of pea leaves at a temperature 25-30°C and relative humidity of 99-100 per cent. Percentage germination of aeciospores on host surface after 24 hrs was 5.35 percent at increased up to 6.30 per cent after 48 hrs and 6.06 percent after 72 hrs of inoculation. Joseph and Hering (1997) reported that urediospores of *U. viciae fabae* (Broad bean rust) germinated well in the range of 5-26°C, fasted germination at 20°C, exposure to 30° C gave poor germination and damage the spores and at 20°C some infection occurred within 4 hrs leaf wetness, but longer wetness periods up to 24 hrs gave increased infection. Kushwaha *et al.* (2006) observed that aeciospores in *Uromyces fabae* were found to be repeating spores and play an important role in pea rust outbreaks in the North Eastern Plain Zone (NEPZ) of India. Infection of pea plants was occurred either by aeciospores or urediospores. Production of aeciospores was observed at a temperature range of 10-25°C with a maximum at 25± 2°C. Among the different growth stages of pea, the pod formation stage was highly susceptible and production the maximum number (744) of aecidia/leaf at 20-25°C. Urediospore production mainly coincided with the senescence of the pea plants.

Host Range

The pathogen has wide host range. It can infects broad bean (*Vicia faba*) and other species of *Vicia* (Deutelmoser, 1926; Hiratsuka, 1933), lentil (*Lens esculanta*) (Nattrass, 1932; Patil, 1933) and *Lathyrus* species (Hiratsuka, 1933; Patil, 1933; Dupias, 1943; Anderson, 1952). *Uromyces fabae* on lentil are applicable to the fungus on pea also. Aeciospores form on lentil has been found to infect pea and *Vicia fabae* and those from pea could infect lentil (Prasada and Verma, 1948). Magsi (2000) observed that Lucerne (*Madicago sativa* L.) is suitable host for the causes of rust disease by *Uromyces striatus*. The *U. viciaefabae* infects many other species of *Vicia*. They also found that *Pisum* is the more favourable host than *Lathyrus* (Kapporia and Sinha, 1966). Abbasi (2000) reported that *Uromyces viciae-craccae*, infecting *Viciae variabilis* (*V. caracce* sub sp. *stenopylla*) first time in Iran. Mahanta *et al.* (2000) observed that *Uromyces appendiculatus* infect the rajmash (*Phaseolus vulgaris* L.) in north plateau zone of Orissa State of India.

Environmental Factors

Environmental factors play important role in the development of the disease. The environmental variables *viz.* temperature, humidity, rainfall and sunshine hours are the most crucial. Since, disease development as a result of host or host pathogen interaction. So for disease prediction, quantitative relationship between disease development and weather variables has been established in several host pathogen systems.

In the literature, information pertaining to severity of infection by *Uromyces viciae-fabae* and pustules per plant increased progressively with an increase in the duration of leaf wetness up to 24 h, but did not increase further significantly. It was observed that relationship between severity of pea rust and duration of leaf wetness at 20°C and above temperature may be useful in predicting disease out break if initial inoculum is present (Chauhan and Singh, 1994). Sharma (1998) observed that free moisture (rain, dew) or high relative humidity (96 per cent plus) are essential for infection of *Uromyces appendiculatus,* when moisture and temperature are favourable for more than eight hours and spore are present, infection occurs. Sharma (1998) reported that the occurrence of rust disease (*Uromyces appendiculatus*) in french bean in mid-hills of Uttar Pradesh is erratic. Analysis of meteorological data for eight year (1986-1993be reported that exclusive precipitation is deleterious to disease development, which explains the occurrence of severe infection during period of low rainfall. Atmospheric temperatures between 5–20°C with high RH (60-70 per cent mean weekly) and light shower or drizzle favour for *U. viciae-fabae* development and spread (Mittal, 1997) reported that maximum temperature 25°C and minimum 7-8°C and rains are unfavourable for rust spread.

Management

Cultural Practices

Cultural practices mostly affect the environmental conditions favorable for growth and buildup of inoculum. These can be utilized alone or in combination as a means of plant disease control. Age old methods of manipulation of environment as to make it unfavourable to pest and pathogens and also favourable to host and natural enemies have been used to suppress the plant diseases. The cultural practices is very important tool in avoiding or reducing the inoculum/disease without any unwanted side effects like pollution of environment.

Sowing Time

Delay in sowing after 5 October, increased the severity of rust infestations and decreased grain yield. Cultivar 'khaparkheda' gave the highest seed yield (1.54 t/ha) and had the lowest severity (Sangar and Singh, 1994). Bhardwarj and Sharma (1996) reported that plants from 15 October sowing were taller, produced the highest number of marketable pods and highest green pod yield (4.74 t/ha). Among the genotype, Arkel and VL-7 were earlier to flower and recorded the higher green pod yield (4.34

and 4.22 t/ha, respectively). The percentage disease index of rust (*Uromyces viciae-fabae*) was lowest in 15 October sowing followed by 30 September. Singh *et al.* (1996) found that severity of rust (*Uromyces viciae-fabae*) increased as sowing become delayed. Bhardwaj and Sharma (1996) reported that percent disease index of rust was maximum (30.6) in early (25 September) sown crop in comparison to late sowing. However, sowing of pea on 18, 23 and 13 October gave maximum pod yield of 90.6, 84.2 and 82.3 kg/ha, where yield losses were estimated to the tune of 17.81, 23.56 and 26.03 per cent, respectively.

In early sown crop low disease severity may be due to unfavourable environmental conditions (*i.e.* low temperature, relative humidity and soil moisture etc.) for disease development. In early sown crop, in relation to crop disease has one principal aim to reduce minimum beside over infective agents (propagules, vectors) meets susceptible stage of crop.

Chemical Control

Suppression of rust disease by application of chemical is sometimes spectacularly effective, to manage plant diseases. In some cases chemicals are applied many times, leading to unwanted environmental problems, but in other cases, they are applied one thus acceptable. Till today chemical pesticides had been the backbone of disease control of rust in cereals and pulses crops. Spraying of fungicides alone or in combination has been considered necessary to provide adequate protection of the crop from rust. Upadhyay and Gupta (1994) reported that Tridimefon, maneb + tridemorph were effective against rust disease under field conditions. Systemic sterol biosynthesis inhibiting fungicides are effective against *Uromyces viciae-fabae* (Fuzi, 1995). Khaled *et al.* (1995) reported that fungicides like benomyl carboxin, metalaxyl, oxycarboxin, thiram, triadimefon and triforine alone or with Dithane M-45 (mencozeb) gave good production against rust. Tilt gave the best control, reducing rust intensity and increased pod yield. Folicur and calixin were also effective against rust (Ayub *et al.*, 1996). However, Istran (1996) observed that opus (expoxyconazole) was most effective against *Uromyces viciae-fabae* and *Erysiphe pisi* and increased grain yield, while several formulation combining polyram DF (Metiram) with Altro combi (cyproconazole + carbendazim) or kumulus-s (Sutfer) found non-effective. Gupta *et al.* (1998) observed that hexaconazole (0.1 per cent) and difenoconazole (0.01 per cent) were best against rust and increased yield. Mohy *et al.* (1999) reported that efficacy of mancozeb, benomyl, captan and thiophanate-methyl as seed dressing and foliar sprays, with oxycarboxin is effctive for the control of lential rust caused by *Uromyces viciae-fabae.* Gupta and Shaym (2000) reported that seven ergostral biosynthesis inhibiting fungicides like cyperconazole, flusilezole, penconazole and hexacanozole completely inhibited rust incidence and rust severity on leaves. Singh and Tripathi (2004) observed that Baycor (0.1 per cent), 2 to 3 prophylactic sprays at 15 days interval was found most effective in reducing the disease severity and appreciable increase in grain yield.

Host Resistance

Envolvement of rust resistant varieties seems to be the most effective but there is need for screening of existing lines/germplasm/cultivars against the above disease, because most of the cultivated pea are either susceptible or highly susceptible to rust. However, in general KFP 106, DMR 11, HUP 8603, type 163 and KPMR 22 showed good level of resistance, being conditioned by a number of genes with small effects in more desirable (Singh and Tripathi, 2004). Xue and Warkentin (2002) observed that Tara and Century were the most resistant to both UF-1 and UF-3 isolates. Victoria and Topper were the most resistant to UF-2 only. Barilli *et al.* (2009) collected 2759 pea accessions and screened for resistance against *Uromyces pisi* (Pers.) Wint. All accessions displayed a complete interaction (high infection type) both in adult plants under field condition and in seedlings under growth chamber conditions, but with varying levels of disease reduction. The identified resistance was based on reduction of disease severity with no associated host cell necrosis, which fit the definition of partial resistance. No complete resistance or incomplete resistance based on hypersensitive was observed.

Induced Resistance

Induced resistance seems to a new approach for the control of pea rust. The mechanism resistant has been found that the accumulation of Baily phaseolin and other phytoalexins in host cell resulted resistance of plant bean cells expressed resistance to against *Uromyces appendiculatus* and Ingham (1971); Elanghy and Heitefuss (1976); Hoppe *et al.* (1980) observed that localized induction of resistance in bean against *U. appendiculatus* by an amino acid derivative have also been noticed by several workers. The most recent development on systemic induced resistance in the bean system comes from the availability of 2, 6-dichloro isonicotinc acid (CGA-41369). 20 $\mu g\ m^{-1}$ leaves against *U. appendiculatus,* Walter and Murray (1992) found that treatment of the first two leaves with either 10 m^M tripotassium phosphate or $5m^M$ EDTA in place of rust inoculation also caused significant increase in resistance of the upper leaves to challenge inoculation with the rust fungus.

Future Needs

Introduction

Powdery mildew of pea is an economically important disease of pea in semi-arid region of India a well as in the world. The pathogen is air borne in nature. It is particularly damaging in late sowings or in late maturing varieties of pea. The disease is caused by *Erysiphe pisi,* although other fungi such as *Erysiphe trifolii* and *Erysiphe baeumleri* have also been reported causing this disease on pea. The disease attains a serious status causing huge losses in the pod quality and total yield losses. The disease can cause 25–50 per cent yield losses, reducing total yield biomass, number of pods per plant, number of seeds per pod, plant height and number of nodes. The disease also affects green pea quality. Severe infections can reduce yield by 10–20 per cent. Powdery mildew is most prevalent late in the season. Crops sown late are more likely to be affected by powdery mildew than early sown crops. Severe pod infection can lead to poor seed quality. Control measures of the disease include early planting, the use of fungicides and resistant cultivars. Chemical control is feasible with a choice of protective and systemic fungicides. However, public attitude and environmental concern towards the use of pesticides as well as the development of powdery mildew resistant strains to different fungicides have reduced the appeal of chemicals and have led to the search of alternative control methods. The current control practices and highlights the future challenges for efficient and sustainable strategies for management of powdery mildew. Non-fungicided products, such as soluble silicon, oils, salts and plant extracts are under study but are not fully ready yet for commercial application. Attempts have also been made to control powdery mildews with mycolytic bacteria, mycophagous arthropods, fungi, yeasts and other possible non-fungal biological control agents, but more efforts are still needed to prove the efficacy of these methods in agricultural practice.

Symptoms

First symptoms appear on the upper surface of the foliage as very small, slightly discolored spots. These soon give rise to white powdery areas which continue to enlarge. As they do so, the white color and powdery appearance become more pronounced. Tissue beneath these areas may turn purplish, then brown. Multiple infections may cover the entire above ground plant. Severely infected plants are unthrifty and have poor yield and quality. Small, oval, black fruiting structures may form in mature lesions (Gupta and Paul, 2001).

Causal Organism

It is particularly damaging in late sowings or in late maturing varieties. It is caused by *Erysiphe pisi*, although other fungi such as *Erysiphe trifolii* and *Erysiphe baeumleri* have also been reported causing this disease on pea. The fungus, *Erysiphe pisi* perpetuate overwinters on infected plant debris and on alternate hosts. Air currents, spread the fungus locally and over long distance.

Figure 12.4: Whitish Color and Powdery Appearance on Leaves, Stems and Pods.

Disease Cycle and Epidemiology

Pea powdery mildew is an air-borne disease of worldwide distribution. The absence of rain and presence of at least slight dew favor disease development. Rain controls the disease by washing off the spores and causing them to burst instead of germinates. Epidemics of polycyclic plant diseases such as powdery mildews frequently start from foci of infection. Spatial distribution of fungal plant pathogens is determined by components of the disease cycle such as pathogen survival, source of primary inoculums and mode and amount of inoculums dispersal. Host plant genotypes with quantitative resistance may have less disease because attacking pathogens have reduced infection efficiency, longer latent period or reduced propagule production. One or more of these components can reduce the disease progress (temporal increase) and/or spread (spatial increase) in the field. These aspects of fungal plant pathogens have been modeled to differentiate amounts of plant resistance to pathogens. This approach may enable the presence of quantitative resistance in the field to be confirmed, and parents and progeny with this attribute to be selected for breeding programs. Powdery mildew develops quickly in warm (15 -25°C), humid conditions (over 70 per cent) for 4–5 days, particularly at flowering and after canopy closure. As with downy mildew, rainfall will wash spores off of the plants. Free moisture on plants will also restrict germination of the spores and does not promote the epidemic. If infection occurs earlier than four weeks from maturity, yield losses due to powdery mildew arise from the infection covering stems, leaves and pods, which will lead to shriveled seeds (Yarwood *et al.*, 1954).

Management

Cultural

Cultural practices should be utilised to minimise conditions favourable for infection and spread of powdery mildew. Pratibha and Amin (1991) reported that sowing of pea crop from late September to early October late November or early December show more powdery mildew and greater give reduced yields. The adjustment of date of sowing could be important in avoiding or reducing powdery mildew infection. The disease does not develop fast under sprinkler irrigation system (Hagedorn, 1984), thus such arrangements for cash crop could be used. Plant in sunny areas as much as possible, provide good air circulation, and avoid applying excess fertilizer. A good alternative is to use a slow-release fertilizer. Overhead sprinkling may help reduce powdery mildew because spores are washed off the plant. However, overhead sprinklers are not usually recommended as a control method in vegetables because their use may contribute to other pest problems.

Resistance

To minimize the chance of resistance developing, rotate chemical groups frequently. Resistant cultivars are available and are being more developed. Genetic resistance is acknowledged as the most effective, economic and environmentally friendly method of control. However, only three genes (*er1*, *er2* and *Er3*) have been described so far in *Pisum* germplasm and only *er1* has been widely used in breeding programmes, which is very risky also. Expansion of cultivation areas of pea varieties harboring the same resistance gene could promote the occurrence of new races of the pathogen that would lead to a breakdown of the resistance. The use of polygenic resistance or combining several major genes could enhance the durability of the resistance.

Biological

Introduction of pea with pea non pathogenic powdry mildews (*Oidium* sp., *Phyllactinia corylea*) induced resistance against a subsequent infection with *Erysiphe pisi* and resulted in reduced conidial germination, appesorium formation and secondry branch development up to 12 days after inoculation. Singh *et al.* (2003) reported aqueous extracta of vermicompost (AVC) to inhibit spore germination and development of powdery mildews on pea (*Pisum sativum*) in the field at very low concentration (0.1-0.5 per cent).

Plant Extract

Maurya *et al.* (2004) reported that methanol extract (NBM) of neem bark and ethyl acetate extract (CRE) of motha (*Cyperus rotundus*) rhizome significantly reduced disease intensity. Ginger, tulsi, mahua and cashew nut extracts were also effective against the disease at 2000 ppm *in vivo.*

Chemical

Current powdery mildew control methods are early planting, the use of fungicides and use of resistant cultivars. Chemical control is feasible with a choice of protective and systemic fungicides. However, public attitude and environmental concerns towards the use of pesticides as well as the development of powdery mildew resistant strains to different fungicides have reduced the appeal of chemicals and have led to the search of alternative control methods. Use sulfur spray or dust or benomyl can be used to manage the disease. Seed treatment can lesser or prevent seedling infection. A range of fungicides was assessed on late-sown green peas in south-east Queensland and north-west Tasmania for the control of powdery mildew caused by *Oldium* sp. Triadimefon (Bayleton 125 EC) at 500 mL/ha applied at early flowering gave excellent control under high disease pressure in both States, In Tasmania, single sprays with tebuconazole, propiconazole or fenpropimorph also reduced leaf and pod infections significantly with corresponding increases in yield. Wettable sulphur also use to control the disease in both States. It has a significant chemical cost advantage over systemic fungicides and is acceptable to biodynamic/organic growers.

Downy Mildew of Pea

Introduction

Downy mildew of pea is an important disease under cool and foggy weather conditions. The disease develops systemic, local leaf infections and also pod infections. Downy mildew of peas is widely distributed in all over the world (Dixon, 1981). In India, the disease is prevalent thought the Indo- Gangetic plains. Systemic infection of seedlings can cause considerable yield loss. Olofsson (1966) and Biddle *et al.* (1988) reported yield losses of 30 per cent in Sweden and 45 per cent in UK, respectively.

Symptomatology

Downy mildew causes different types of symptoms on pea plants. Three different infection types with different symptoms can be recognized during a crop cycle growing stages. Systemic infection in seedlings causes stunted growth with conidia sporulation, which often covers a major part of the plant surface and this is caused by oospores in the soil which infect germinating seeds. These infections can seriously reduce the plant population. Germinating seedlings cannot be infected by inoculating the roots. However, Ryan (1971) found that 90 per cent systemic infection caused by placing oospore inoculum around or slightly above the seed. A lower frequency of infection (50 per cent) was obtained when placed 3 cm above the seed level and infection was even more reduced (1 per cent) when the oospores were placed 3 cm below the seeds.

Mence and Pegg (1971) induced systemic infection by inoculating conidia into the apical bud of young plants, or onto the epicotyl or hypocotyl, but not by inoculating the roots of germinating seedlings. Taylor *et al.* (1990) showed that systemic infection could also originate from leaf infection. Top systemic infection is the result of direct infection of the top meristem. The type of infection is found more frequently in varieties with reduced stipule size, determined by the gene st (Taylor *et al.,* 1990). In these varieties, the top meristem is not protected by the stipules, which wrap around the apex in varieties with normally sized stipules. Following infection, the mycelium develops in the intercellular spaces penetrating the stem the leaf stalks and even the pods through the veins (Kosevskii and Kirik, 1979).

Local foliar and tendril lesions with conidial sporulation on the abaxial foliar surface is a typical symptom of the disease. Local infections on leaves or tendrils develop from conidia present on the plant sur-face. Germ tubes penetrate the cuticle at an epidermal cell but did not find appressoria (Mence and Pegg, 1971). Mycelium grows irregularly in the spongy parenchyma of the leaf and under high relative humidity and often develops between the upper epidermis and the palisade parenchyma. Haustoria are most frequently found in the leaf mesophyll and filiform in epidermal cells; oogonia are terminal, isolated, spherical when mature and pyriform when young (Kosevskii and Kirik, 1979).

Pod infection causes yellow lesions on the pod surface and epithelial proliferations on the endocarp (Snyder, 1934). Pod infection develops from conidia

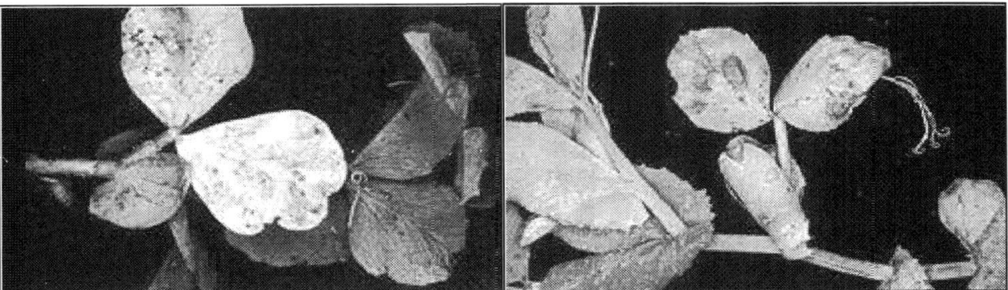

Figure 12.5: Thick Grey Fungal Growth on the Lower Leaf Surface, is Typical of Downy Mildew and an Older Plant also Affected by Downy Mildew.

deposited on young pods rather than by mycelial growth through the peduncle and pedicel (Mence and Pegg, 1971). Oospores are formed within the yellow lesions. Pod infection often causes distorted pods, seed abortion and brown discoloured small peas with a bitter taste. Pod infections directly affect pea quality and are therefore a serious expression of the disease.

Causal Organism

Peronospora viciae (Berk) Casp f sp *pisi* (Sydow) Boerema and Verhoeven (syn *P pisi* Sydow) (Perono-sporacea; Oomycetes) causes downy mildew on peas. The pathogen belong to kingdom Chromista, Phylum Oomycota, Class Oomycetes, Order Peronosporales and Family Peronosporaceae.

Species of *Peronospora* produce conidia that lack modification in the apical region, the operculum, do not contain zoospores, and germinate by germ tubes (Shaw, 1981). *P viciae* also produces oospores, which have a typical reticulate pattern of the exosporium. The sexual breeding system for *P. viciae* has not been described in literature but may be similar to *Parasitica* and *Bremia lactucae* Regel. These species are also capable of regular and predictable production of oospores in large numbers. Both heterothallic and homothallic isolates of these 2 species have been found. Sexual reproduction is probably important for the adaptation of the fungus to various host genotypes by recombination of virulence genes. The vegetative stage is probably diploid like in other species of Peronospora (Tommerup, 1981).

The fungus is an obligate parasite which can only grow on living plant tissue. Forma specialis pisi can only infect Pisum species and not species of the genus *Vicia* within the tribe Vicieae (Campbell, 1935).

Disease Cycle and Epidemiology

The disease is both seed and soil borne in nature. In the seed the fungus may be present as contaminant or the seed infection may extend beneath the seed, coat. Such infected seeds might act as the source of inoculums. However the main sources of primary inoculums are the oospores in the diseased crop debris where these can survive up to 2 years. Oospores on germination produce a germ tube which infects the seedlings systemically. Oospores also in the soil are the primary inoculums early in the season. The oospores can survive for a long time in the soil. Infections are common in south Sweden where a 6 year crop rotation is common practice. Oospores survive for 10-15 years in the soil (Olofsson, 1966). Oospores of *P. destructor* (Berk) Casp in onion debris showed good viability after 25 years of outdoor storage (McKay, 1957). Under favorable temperatures for sporangial germination, four hours leaf wetness is required for infection to take place. High humidity also helps in the dissemination of sporangia. Cool nights coupled with foggy weather or presence of dew favor the disease development. Optimum temperature for infection to occur is 16°C, though it can take plce between 0-22°C. The plants are predisposed to infection by other pathogens like *Uromyces viciae*.

Management

Cultural

Proper drainage and crop spacing create micro conditions which are unfavorable for disease development. Field sanitation, destruction of infected crop debris, 3-4 years crop rotation with non host crops are important in reducing the primary inoculums. Since the pathogen is also seed borne in nature, it is always recommended to use disease free seed. Seed treatment in hot water at 50°C for 25 minutes has also been found effective to eradicate the seed borne inoculums. Pick off and dispose of (by burial, burning or consigning to the council green waste) affected leaves as soon as symptoms are seen in the field. Avoid dense planting and control weeds, so that there is good air circulation around the plants. In glasshouses, try to avoid prolonged leaf wetness or periods of high humidity. Avoid overhead watering where possible. Open the doors and vents, when conditions allow, encouraging air movement, avoid watering plants in the evening, as this can lead to high humidity or leaf wetness that persists throughout the night. Watering early in the morning so that leaf surfaces dry out rapidly. To avoid infection from soil-borne resting spores, practice crop rotation for vegetables, and avoid re-planting with the same host for at least a year where an ornamental plant has been affected (Stegmark, 1994).

Resistance

Pea varieties are known to differ in their reaction to downy mildew disease. Bains *et al.* (1993) reported pea varieties PWR-3 and Bonneville to have resistance under Punjab condition. Variation in resistance between pea cultivars has been reported by Stegmark (1988). Some cultivars are completely resistant to some isolates but are fully susceptible to others. However, there are also pea genotypes that have stable partial resistance, never complete, to different isolates (Stegmark, 1990). Race-specific resistance was found in several cultivars, but there is no pea genotype with complete resistance to all known pathogenic races (Matthews and Dow, 1983). The cultivar 'Dark Skin Perfection' (DSP) is more resistant to downy mildew than some other cultivars used for the production of peas for canning and freezing (Stegmark, 1988).

Chemicals

Seed borne inoculums should be eradicated through seed dressing with systemic acylalanine fungicides (*e.g*, metalaxyl) which is very effective against systemic seedling infections (Brokenshire, 1980).However, later in the season, the pod infection can still be severed and there is no real effective fungicidal treatment against pod infection in peas grown for freezing. In the long run, the current acylalanine fungicides may become ineffective due to development of tolerance by the pathogen. The secondry spread of the disease should be checked by spraying fungicides like captan + fosetyl aluminium + thiobendazole, carbendazi + chlorothaonil, maneb, zineb etc.

References

Aanwar, S.A., Bhutta, A.R., Raut, C.A. and Khan, M.S.A. 1994. Seed borne fungi of pea and their role in poor germination of pea seed. *Pakistan J. Phytopathology*, 6: 135-139.

Abbasi, M. 2000. *Uromyces viciae*- creaccae a new rust fungus in Iran. *Rostaniha,* 1(1-4): 75-76.

Abo-El Seoud M.A.; Sarhan, M.M.; Omar, A.E. and Helal, M.M. 2005. Biocides formulation of essential oils having antimicrobial activity. *Ar. of Phytopath. and Pl. Protec.,* 38: 175-184.

Agarwal, S.C.; Singh, K. J. and Tripathi, A. K. 2003. Integrated pest management in pigeon pea (*Cajanus cajan*). *Indian J. Agri. Sci.,* **73(5)**: 291-293.

Anderson, J.P. 1952. The uredinales of Alaska and adjacent parts of Canada. *Iowa St. Call J. Sci.,* 26(4): 507-526.

Ayub, A., Rahaman, M.Z., Ali, S. and Khatun, A. 1996. Fungicidal spray to control leaf rust of lentil. *Bangladesh J. Plant Pathol.,* **12(1-2)**: 61-62.

Bailey, J.A. and Ingham, J.L. 1971. Phaseollin accumulation in bean (*Phaseolus vulgaris*) in response to infection by tobacco necrosis virus and the rust *Uromyces appendiculatus. Physiological Plant Pathology,* **1**: 415-456.

Bains, S.S., Dhiman, J.S. Singh, H. and Singh, H. 1993. Occurence of *Peronospora pisi* with *Uromyces viciae fabae* on Leaves on pea genotypes. *Indian Phytopathol,* **48**: 365-366.

Barilli E., Sillero J.C., Aparicio M.F. and Rubiales D. 2009. Identification of Resistance to Uromyces pisi (Pers.) Wint. In Pisum spp.germplasm. *Field Crop Research,* **117(2)**: 198-203.

Batra, L.R. and Stavely, R. 1994. Attraction of two spotted spider mites to bean rust uredinia. *Plant Dis.,* **78**: 282-284.

Biddle, A.J., Knott, C.M. and Gent, G.P. 1988. Pea Growing Handbook. Processors and Growers Research Organization, Peterborough, UK, p. 264.

Brokenshire, T. 1980. Control of pea downy mildew with seed treatments and foliar sprays. In: Test of Agrochemicals and Cultivars., *Suppl Annals of Applied Biology,* **64(1)**: 34-35.

Brown, J.K. and Hovmller, M.S. 2002. Arial dispersal of pathogen on the global and continental scales and its impact on plant disease. *Science* **297**: 537–541.

Butler, E.J. 1918. *Fungi and Diseases in Plants.* Thacker Spink and Co., Calcutta, India.

Buxton, E.W. 1955. Fusaium disease of Gladiolus. *Trans. Brit. Mycol. Soci.,* **38:** 193.

Campbell, L. 1935. Downy mildew of peas caused by *Peronospora pisi* (De Bary) Syd. *Washington Agric Exp Sta Bull* **318,** 42.

Catanzariti A.M., Dodds P.N., Lawrence G.J., Ayliffe M.A. and Ellis J.G. 2006. Haustorially expressed secreted proteins from flax rust are highly enriched for avirulence elicitors. *Plant Cell* **18**: 243–256.

Chaudhary, R. G and Prajapati, R. K. 2004. Comparative efficacy of fungal biological agents against *Fusarium udum. Ann. Pl. Protec. Sci.,* **12(1)**: 75-79.

Chauhan, R.S. and Singh, B.M. (1994). Effect of different duration of leaf wetness on pea rust development. *Pl. Dis. Res.,* **9(2)**: 200-201.

Chung W.H., Tsukiboshi T.O.Y., Kakishma M. 2004. Phylogenic analyses of *Uromyces viciae-fabae* and its varieties on Vicia, Lathyrus and Pisum in Japan. *Mycoscience* **45**: 1-8.

Chupp, C. and Sherf, A.F. 1960. Vegetable disease and their control. The Ronald Press Company, New York, p. 668.

Cunningham, J.L. 1973. Longevity of rust spores in liquid nitrogen. *Plant Dis. Rep.* **57**: 736-737.

Datar, V.V. 1983. Reaction of pea wilt (*Pisum sativum* L.) varieties to wilt caused by *Fusarium oxysporum* f. sp. *pisi*. *Indian J. Mycol. and Pl. Pathol.*, **13**: 88-89.

Davison, A., Vaugham, E.K., 1963. Longevity of urediospores of race 33 of *Uromyces phaseoli* var. *phaseoli* in storage. *Phytopathology* **53**: 736-737.

Deutelmoser, E. 1926. Plant protection measures in vegetable culture. I. Measures against fungus pests. Obst.-U. Gemusebu I. XXII, **19**: 291-292. (*Rev. Appl. Mycol.*, 1927: 137).

Devi, Meena and Paul, Y.S. 2003, 2005. Management of pea wilt / root rot complex by integrating plant extracts and biocontrol agents. Integrated Plant Disease Management. Challenging problems in horticultural and forest pathology, Solan, India, p. 101-105.

Dixon, G.R. 1981. Downy mildews on peas and beans.In: The Downy Mildews (DM Spencer, ed), 87-154, Academic Press, New York, USA, p. 636.

Dupias, G. 1943. Chemical control of bean rust (Uromyces phaseoli tynica). Investigaciones Agropecuario, **2:** 23-27(Es, enltab) Exp. Agric., La Matena, Peru.

Ebbels, D.L. 1967. Effect of soil temperature on Fusarium wilt and nodulation of peas (*Pisum sativum* L.). *Ann. Appl. Biol.*, **60:** 391-398.

Elnaghy, M. A. and Heitefuss, R. 1976. Permeability changes and production of antifungal compounds in Phaseolus vulgaris infected with *Uromyces phaseoli*. II. Role of phytoalexins. *Physiological Plant Pathology*. **8**: 269-277.

Emeran A.A., Roma N.B., Sillero J.C., Satovic Z. and Rubiales D. 2008. Genetic variation among and within *Uromyces* species infecting legumes. *J. Phytopathol*, **156**: 419-424.

Emeran A.A., Sillero J.C., Niks, R.E. and Rubiales D. 2005. Infection structure of host-specialized isolates of *Uromyces viciae-fabae* and of other species of *Uromyces* infecting leguminous crops. *Plant Dis.* **89**: 17-22.

Flor H.H. 1956. The complementary genetic systems in flax and flax rust. *Adv. Gen.*, **8**: 29–54.

Fuzi, I. 1995). Fungicides against diseases of pea, *Pisum sativum* L., in Hungary. *Pesticides Science*, **45(3)**: 292-295.

Gangopadhayay, S. and Kapoor, K.S. 1979. Wet seed treatment to control Fusarium wilt of pea. *Veg. Sci.*, **3**: 74-78.

Gaumann, E.A., 1998. *Comparative morphology of fungi*, Translated by Caroll William Dodge, Biotech Book, Delhi. p. 563.

Gholve, V.K. and Kurundkar, B.P. 2002. Biological control of pigeonpea wilt with *Pseudomonas fluorescens* and *Trichoderma viride*. *Indian J. Pulses Res.*, **15(2)**: 174-176.

Gourdar, S.B.; Kulkarni, Srikant and Kulkarni, S 2000. Bioassay of antagonists against *Fusarium udum*, the causal agent of pigeonpea wilt. *Karnataka J. Agric. Sci.*, **13 (1)**: 64-67.

Gupta, S.K. and Shyam, K.R. 1998. Control of powdery mildew and rust of pea by fungicide. *Indian Phytopathology*, **51(2)**: 184-186.

Gupta, S.K. and Shyam, K.R. 2000. Post infection activity of ergosterol biosynthesis inhibiting fungicides against pea rust. *J. Mycol. Pl. Pathol.*, **30(3)**: 414-415.

Gupta,V.K. and Paul,Y.S. 2008. *Disease of vegetable crops*. Kalyani publishers,New Delhi, India, p. 74-79.

Hagedorn, D.J. 1984. Root rot of peas.In Compendium of pea diseases. The American Phytopatholgical Society, Minnesote, USA.

Hebblethwaite, P.D. 1983. The faba bean. Butterworths London, U.K., p. 573.

Hiratsuka, N. (1933). Studies on *Uromyces fabae* its related species. *Jap. J. Bot.* **6**: 329-379. (*Rev. Appl. Mycol.*, 1933: 596-597)

Hoppe, H.H., Humme, B and Heitefuss, R.1980. Elecitor induced accumulation of phytoalexins in healthy and rust infected leaves of *Phaseolus vulgaris*. *Phytopathologische Zeitschrift* **97**: 85-88.

Istran, F. 1996. Control fungicides (IS) against fungal diseases of peas. *Novenyvedelem*, **32(3)**: 144-146.

Jakupovic, M., Heintz, M., Reichmann P., Mendgen K and Hahn, M. 2006. Microarray analysis of expressed sequence tags from haustoria of the rust fungus *Uromyces fabae*. *Fungal Genet B*, **43**: 8–19.

Jorstad, I. 1948. Erterust, *Uromyces pisi* (D.C.) Fuck., 1. Norge (Pea rust: *Uromyces pisi* (D.C.) Fuck., in Narway). *Nord, Jordbr-Forskn.*, **7-8**: 198-207. (*Rev. Appl. Mycol.*, 1950: 242-243).

Joseph, M.E., Hering T.F. 1997. Effect of environment on spore germination and infection by broad bean rust (Uromyces *viciae fabae*). *The J. of Agri. Science*. **128**: 73-78.

Joshi A. and Tripathi, H.S. 2012. Studies on epidemiology of lentil rust (*Uromyces viciae fabae*). *Indian Phytopathology*, **65(1)**: 67-70.

Kamalkanhan, A.; Shanmugam, V.; Surendran, M. and Srinivasan, R. 2001. Evaluation of plant extracts against *Pyricularia oryzae*. *Ann. Pl. Protec. Sci.*, **9 (1)**: 68-72.

Kapooria, R.G. and Sinha, S. 1966. Studies on the host range of *Uromyces fabae* (Pers.) de Bary. *Indian Phytopathol.*, **19**: 229-230.

Khaled, A.A., EL-Moity, SMHA, and Omar, SAM. 1995. Chemical control of some faba bean diseases with fungicides. *Egyptian J. Agri. Res.*, **73(1)**: 45-56.

Kosevkii II and Kirik, N.N. 1979. Features of the develop-ment of mycelium of *Peronospora pisi* Syd in pea tissue, Osobennosti razvitiya mitseliya *Peronospora pisi* Syd v tkanyak gorokha. *Mikilogiya I Fitopatologiya* **13**: 46-48.

Kumar, Ashok and Kohli, U.K. 2001. Evaluation of garden pea genotypes for horticultural.

Kushwaha C., Chand R., Srivastava C.P. 2006. Role of aeciospores in outbreak of pea (*Pisum sativum*) rust (*Uromyces fabae*). *Eur.J.Plant Ptho.* **115**: 323-330.

Leonard, K.J. and Szabo, L.J. 2005. Stem rust of small grains and grasses caused by *Puccinia graminis.Molecular plant pathology,* **6**: 99-111.

Linford, M.B. 1928. A Fusarium wilt of peas in Wisconsin. *Bull. Wis. Agric. Exp. Sta.,* **85**: 1-44.

Magsi, M.R. 2000. Studies on physiology of *Uromyces striatus* Schroet. *Pakistan J. Phytopathol.,* **12(2)**: 130-133.

Mahanta, I.C., Dhal, A., Mohanty, A.K. and Mohapatra, S.S. 2000. Integrated management of diseases of kidney bean or rajmash (*Phaseolus vulgaris* L.) in North central plateau zone of Orissa. *Orissa J. Hort.,* **28(2)**: 25-28.

Maheshwari, S.K., Jhooty, U.S. and Gupta, J.S. 1981. Studies on the wilt and root rot of pea. Efficiency of various chemicals in controlling the wilt and root rot complex of pea. *Agri. Sci. Dig.,* **1**: 37-38.

Maheshwari, S.K., Jhooty, U.S. and Gupta, J.S. 1983. Survey of wilt and root rot complex of pea in northern India and assessment of losses. *Agric. Sci. Dig. India,* **3**: 139-141.

Maheshwari, S.K., Nazir, A. Bhat; Masoodi, S.D. and Beig, M.A. 2008. Chemical control of lentil wilt caused by *Fusarium oxysporum* f. sp. *lentis. Ann. Pl. Protec.,* **16(2)**: 419-421.

Maie,R. W., Begerow, D., Weib M, and Oberwinkler, F. 2003. Phylogeny of the rust fungi: an approach using nuclear large subunit ribosomal DNA sequence. *Can J. Bot.,* **81**: 12-23.

Matthews, P. and Dow, K.P. 1983. Disease resistance improvement in dried peas. In: 72nd Report 1981-1982, John Innes Institute, Norwich, UK, p. 195.

Maurya,S., Singh, D.P., Srivastava, J.S. and Singh, U.P. 2004 Effect of some plant extracts on pea powdery mildew (*Erysiphe pisi*). *Annals of Plant Protection Sciences,* **12**: 296-300.

McKay, R. 1957. The longevity of the oospores of onion downy mildew Peronospora destructor (Berk) Casp. *Sci Proc R Dublin Soc Nat Sci.,* **27**: 295-307.

Mence M.J. and Pegg, G.F. 1971. The biology of Peronospora viciae on pea: factors affecting the susceptibility of plants to local infection and systemic colonization. *Ann Appl Biol.,* **67**: 297-308.

Mendgen, K. 1997. The Uredinales. The Mycota Vol. V. Plant Relationships Part B(Esser K and Lemke PA,eds), p. 79-94. Springer-Verlag, Berlin.

Mittal, R.K. 1997. Effect of sowing dates and diseased development in lentil as sole and mixed crop with wheat. *J. Mycol. Pl. Pathol.*, **27(2)**: 203-209.

Mohy, Ud. Din. G.; Kahn, M.A. and Khan, S.M.C. 1999. Evaluation of seed and foliar applied fungicides to control lentil rust. *Pakistan J. Phytopathol.*, **11(1)**: 77-80.

Nattrass, R.M. (1932). Annual Report of the Mycologist for 1931. *Ann. Rept. Dept. of Agric.* Cyprus for the year 1931, p. 56-64. (*Rev. Appl. Mycol.*, 1957: 530).

Neumann, S. and Xue, A.G. 2003. Reaction of field pea cultivars to four race of *Fusarium oxysporum* f. sp. *pisi*. *Canadian Journal of Plant Science*, **83**: 377-379.

Nyvall, R.F. and Haglund, W.A. 1972. Sites of infection of *Fusarium oxysporum* f. sp. *pisi* race on peas. *Phytopathology*, **62:** 1419-1424.

Olofsson, J. 1966. Downy mildew of peas in western Europe. *Plant. Dis Rep*, **50**: 257-261.

Pandey, K.K and Upadhyay, J.P. 1999. Comparative study of chemical, biological and integrated approach for management of Fusarium wilt of pigeonpea. *J. Mycol. Pl. Pathol.*, **29(2)**: 214-216.

Patel, M.K.; Kamath, M.N. and Bhide, V.P. 1949. Fungi of Bombay supplement I. *Indian Phytopath.*, **2**: 142-155.

Patil, M.K. 1933. India: diseases in the Bombay Presidency. *Int. Bull. Pl. Prot.* VII, **11**: 246 (*Rev. Appl. Mycol.*, 1934: 215).

Peshne, N.L. 1966. A comparative study of morphology, physiology and pathogenicity of Nagpur and Poona isolates of *Fusarium oxysporum* f. sp. *pisi*. *Ann. Agril. Res. Abst. Grad. Res.*, W.K. 1960-65. *Agric. Coll. Mg.* 1966, *Space Res.*, 174-175.

Prasad, C.S.; Gupta, Vishal; Tyagi, Ashish and Pathak, S. 2003. Biological control of *Sclerotium rolfsii* Sacc., The incitant of cauliflower collar rot. *Ann. Pl. Protec. Sci.*, **11(1)**: 61-63.

Prasada, R. and Verma, U.N. 1948. Studies on lentil rust, *Uromyces fabae*. *Indian Phytopathol.*, **1**: 142-146.

Pratibha Shukla and Amin, K.S. 1991. Integrated disease management of powdery mildew of field peas (*Pisum sativum L.*) caused by *Erysiphe polygoni* D.C. *Legume Res.* **14**: 59-63

Rajib, J. D., Dwivedi, R.P. and Narain,U. 2003. Biological control of *Sclerotium rolfsii* Sacc., the incident of cauliflower Collar Rot. *Ann. Pl. Protec. Sci.*, **11(1)**: 61-63.

Ramphal and Choudhary, B. 1978. Screening of garden peas for resistance to Fusarium wilt. *Indian J. Agric. Sci.*, **48:** 407-410.

Ryan, E.W. 1971 Two methods of infecting peas systemically with *Peronospora pisi*, and their application in screening cultivars for resistance. *Ir J Agric Res* **10**: 315-322.

Sahani, R.K. and Saxena, A.R. 2009. Fungitoxic properties of medicinal and aromatic plants against *Fusarium oxysporum* f. sp. *pisi. Ann. Pl. Protec. Sci.,* **17:** 146-148.

Sangar, R.B. and Singh, V.K. 1994. Effect of sowing dates and pea varieties on the, severity of rust, powdery mildew and yield. *Indian J. Pul. Res.,* **7(1)**: 88-89.

Sara, F. and Diego, R. 2011. Powdery mildew control of pea. A review. *Agronomy Sus Developm.*

Sawant, D.M.; Kolase, S.V. and Bachkar, C.B. 2003. Efficacy of bio control agents against wilt of pigeonpea. *J. Maharashtra Agric. Univ.,* **28 (3)**: 303-304.

Schroeder, W.T. and Walker, J.C. 1942. Influence of controlled environment and nutritive on the resistance of garden pea to Fusarium wilt. *J. Agric. Res.,* **65**: 221-248.

Sharma, A.K. 1998. Epidemialogy and management of ruts disease of french bean. *Vegetable Science,* **25(1)**: 85-88.

Sharma, Pankaj; Singh, S.D. and Rawal, P. 2003. Antifungal activity of some plant extracts oils against seed borne pathogen of pea. *Plant Disease Research,* Ludhiana, **18 (1)**: 16-20.

Sharma, R.R. Rajiv, Sharma, K. and Munshi, A.D. 1998. Breeding for Fusarium wilt resistance in pea (*Pisum sativum*). *Ann. Pl. Protec. Sc.,* **6 (1)**: 1-10.

Shaw, C.G. 1981. Taxonomy and evolution. In: The Downy Mildews (DM Spencer, ed), 17-29, Academic Press, New York, USA, p. 636.

Shroff, S. and Chand, R. 2010. Pre-infection Biology of aeciospores of *Uromyces fabae. Inter J Curr. Trends Sci. Tech.* **1(2)**: 1-10.

Shukla, P., Singh, R.P. and Mishra, A.N. 1979. Efficacy of seed dressing fungicides for the control of pea wilt. *Pesticide,* **13**: 40-47.

Shukla, P., Singh, R.P. and Mishra, A.N. 1982. Evaluation of varieties of strains of pea against wilt. *Indian J. Mycol. Pl. Pathol.,* **12**: 93-94.

Shukla, P. and Amin, K.S. 1991. Integrated management of powdery mildew of field peas (*Pisum sativum* L.) caused by *Erysiphe polygoni* DC. *Legume Res.* **14**: 59-63.

Singh D, and Tripathi H.S. 2004. Epidemiology and Management of field pea rust. *J.Mycol.Pl. Pathol* **34(2)**: 675-679.

Singh, P.P., Amit Wadhwa and Sandhu, K.S. 2003. Influence of post-inoculation diurnal conditions on development of Ascochyta blight of peas.*Indian Phytopatholgy*. **56**: 446-447.

Singh, R.S. 1998. Plant diseases. Oxford and IBH, New Delhi, p. 396.

Singh, R.S. 1998. Plant diseases. Oxford and IBH, New Delhi, .p 512.

Singh, Rajesh, Rai, Bharat, Singh, R. and Rai B. 2000. Antifungal potential of some higher plants against *Fusarium udum* causing wilt disease of *Cajanus cajan. Microbios,* **102**: 165-173

Singh, V.K., Sangar, R.B.S. and Singh, R.N. 1996. Effect of varieties and sowing dates on disease incidence and productivity of fieldpea (*Pisum sativum*). *Indian J. Agron.*, **41(3)**: 451-453.

Singh, Y.P. and Sumbali, Geeta 2007. Efficacy of leaf extracts and essential oils of some plant species against *Penicillium expansum* Rot of Apples. *Ann. Pl. Protec. Sci.*, **15 (1)**: 135-139.

Sinha, P., Biswas, S.K. and Singh, P. 2003. Integrated management of wilt (*Fusarium udum*) in pigeonpea *Farm Sci. J.*, **12 (2)**: 107-109.

Sinha, V. and Upadhyay, R.S. 1990. Fungitoxicity of some pesticides against *Fusarium udum*, the wilt causing organism of pigeonpea. *Hindustan Antibiotic Bulletin*, **32**: 1-2, 36-38.

Sitara,U.; Niaz, I., Naseem, J. and Sultana, N. 2008. Antifungal effect of essential oils on *in vitro* growth of pathogenic fungi. *Pak. J. Bot.*, **40**: 409-414

Stegmark, R. 1988 Downy mildew resistance of vari-ous pea genotypes. *Acta Agric Scand.*, **38**: 373-379.

Stegmark, R. 1990 Variation for virulence among Scaninavian isolates of *Peronospora viciae* f sp *pisi* (pea downy mildew) and response of pea geno-types. *Plant Pathol* **39**: 118-124.

Stegmark, R. 1994. Downy mildew on peas (*Peronospora viciae* fsp*pisi*), Plant pathology (review). *Agronomie*, **14**: 641-647.

Sukapure, R.S.; Bhide, V.P. and Patel, M.K. 1957. Fusarium wilt of garden peas (*Pisum sativum* L.). In Bambay State. *Indian Phytopath.*, **10**: 11-17.

Sumitha, R. and Gaikwad, S.J. 1995. Efficacy of some fungicides against *Fusarium udum* in pigeonpea. *J. Soils and Crops*, **5 (2)**: 137-140.

Taylor, P.N., Lewis, B.G. and Matthews, P. 1990 Factors affecting systemic infection of *Pisum sativum* by *Peronospora viciae*. *Mycol Res* **94**: 179-181.

Thara, V.K., Fellers J.P. and Zhou, J.M. 2003. In planta induced genes of *Puccinia triticina*. *Mol Plant Pathol* **4**: 51-56.

Tommerup, I.C. 1981. Cytology and genetics of downy mildews. In: *The Downy Mildews* (DM Spencer, ed), Academic Press, New York, USA, p. 121-142.

Upadhyay, A.L. and Gupta, R.P. 1994. Fungicidal evaluation against powdery mildew and rust of pea (*Pisum sativum* L.). *Ann. Agric. Res.*, **15(1)**: 114-116.

Utikar, P.G., Gade, U.A. and More, B.B. 1978. Seed treatment to control of *Fusarium* wilt of pea. *Pesticides*, **12**: 29-30.

Vaidya, Manita; Shanmugan, V. and Gulati, Arvind 2004. Evaluation of biocontrol agents against Fusarium isolates infecting carnation and gladiolus. *Ann. Pl. Protec. Sci.*, **12(2)**: 314 –320.

Velikanov. L.L., Cukhonosenko, E.Yu, Nikolaeva, S.I. and Zavlisko, I.A. 1994. Comparison of hyperparasitic and antibiotic activity of the genus *Trichoderma* Per: Fr and *Gliocladium virens* Miller Giddens et Foster isolates towards the pathogen causing root rot of pea. *Mikologiya-I- Fitopatologiya*, **28**: 52-56.

Verma, S. and Dohroo, N.P. 2002. Evaluation of fungicides against *Fusarium oxysporum* f. sp. *pisi* causing wilt of autumn pea in Himachal Pradesh. *Plant Disease Research,* **17**: 261-268.

Verma, Shalini and Dohroo, N.P., 2003. Comparative efficacy of biocontrol agents against *Fusarium* wilt of pea. Integrated plant disease management. Challenging problem in horticultural and forest pathology, Solan, India, p. 93-99.

Vinay Sagar and Sugha, S.K., 2004. Role of soil amendments and repeated cropping in the management of pea root rot. *Tropical Science*. **44**: 1-5.

Virin, W.J. and Walker, J.C., 1939. Relation of temperature and moisture to near wilt of pea. *J. Agric. Res.,* **59**: 591-600.

Virin, W.J. and Walker, J.C., 1940. Relation on near wilt fungus to the pea plant. *J. Agric. Res.,* **60**: 241-248.

Walters, D.R. and Murray, D.C., 1992. Induction of systemic resistance to rust in *viciae faba* by phosphate and EDTA: effects of calcium. *Plant Pathology,* **41**: 444–448.

Wang, K., Zhan,Z.S.,Chen, J. N., Fan, S.N. B.,Wang, K.L., Zhang, Z.S., Chen, J.W., Fan, Z.O. and Niu, B.S., 1995. A study on the occurrence of root rot in pea. (*Pisum sativum*) and the techniques for its integrated control. *NinKia, J. Agric. Forestry, Sci. and Technol.,* **3**: 1-5.

Webb, C. and Hawtin, G., 1981. Lentils. Commonwealth Agricultural Bureaux, London,U.K., p. 216.

Xue, A.G. and Warkentin, T.D., 2002. Reaction of fieldpea varieties to three isolates of *Uromyces fabae*. *Canadian J. Plant Sci.,* **82(1)**: 253-255.

Yarwood, C.E., Sidky, S.,Cohen,M. and Santilli, V., 1954. Temperature relations of powdery mildews. *Hilgradia* **22**: 603-622.

Zhou XL, Stumpf, M.A., Hoch H. C. and Kung, C., 1991. A mechanosensitive channel in whole cells and in membrane patches of the fungus Uromyces. *Science*. 253: 1415.

2016, Diseases of Pulse Crops and their Sustainable Management **225–242**
Editors: **Samir Kumar Biswas, Santosh Kumar and Gireesh Chand**
Published by: **BIOTECH BOOKS, NEW DELHI**

Chapter 13

Diseases of Soybean and their Management

Sunil Kumar[1], Sanjeev Kumar[2] and Gireesh Chand[2]*

[1]School of Agricultural Sciences and Rural Development, Nagaland University, Medziphema – 797 106, Nagaland
[2]Department of Plant Pathology, Bihar Agricultural University, Sabour, Bhagalpur, Bihar

Introduction

Soybean (*Glycine max* L. Merrill) is a leguminous crop; it belongs to the family Leguminocae. It is rich in high quality protein (40-42 per cent), oil (18-20 per cent) and other nutrients like calcium, iron and glycines. It is a good source of isoflavones. Soybean helps in preventing heart diseases, cancer, HIV etc (Kumar, 2007). Soybean protein is rich in the valuable amino acid lysine (5 per cent) in which most of the cereals is deficient. In addition, it contains good amount of minerals, salts and vitamins (thiamine and riboflavin). The sprouting grains contain a considerable amount of vitamin C, minerals, salts and vitamins (thiamine and riboflavin) (Singh *et al.*, 2003).Soybean is the richest, cheapest and easiest source of best quality protein and fat. Hence, it is called as vegetarian meat and wonder crop. The crop is severely affected by a number of diseases and causes huge amount of yield losses.

* Corresponding Author: E-mail: gireesh_76@rediffmail.com

Fungal Diseases

Collar Rot

Causal Organism

The causal organism of this disease is *Sclerotium rolfsii. S. rolfsii* s a well known polyphagous and most destructive soil borne fungus. It was first reported by Rolfs (1892) as a cause of tomato blight in Florida. Later, Saccardo (1911) named the fungus as *Sclerotium rolfsii*. In India, Shaw and Ajrekar (1915) isolated the fungus from rotted potatoes and identified as *Rhizoctonia destruens* Tassi. However, later, studies showed that, the fungus involved was *S. rolfsii* (Ramakrishnan, 1930). Higgins (1927) worked detail physiology and parasitism of *S. rolfsii*. However, its perfect stage was first studied by Cruzi (1931) and proposed generic name as *Corticium*. Mundkur (1934) successfully isolated the perfect stage of *S. rolfsii*. Sclerotium is soil inhabitant fungai, produces abundant white fluffy, branched mycelium that forms numerous sclerotia but is usually sterile (does not produce spores) and cause serious diseases on many hosts by affecting the roots, stems, tubers, corns and other plant parts that develop in or on the ground. The perfect stage of the fungus is *Aethalium rolfsii*. In fact, the fungus' growth is so fast, Rolfs mentioned that "if the temperature is 80-90°F, in 48 hours you will have a growth thathttp://www.cals.ncsu.edu/course/pp728/Sclerotium/cultlnk.htm will in appearance rival swan down." Both in culture and in plant tissue, a fan-shaped mycelial expanse may be observed growing outward and branching acutely. The fungus produces two types of hyphae. Coarse, straight, large cells (2-9 um x 150-250 um) have two clamp connections at each septation, but may exhibit branching in place of one of the clamps. Branching is common in the slender hyphae (1.5-2.5um in diameter) which tend to grow irregularly and lack clamp connections. Slender hyphae are often observed penetrating the substrate. Sclerotia (0.5-2.0mm diameter) begin to develop after 4-7 days of mycelial growth. Initially a felty white appearance sclerotia quickly melanise to a dark brown coloration. Townsend and Willetts (1954) recognize four zones in the mature Sclerotium: i) thick skin, ii) rind of thickened cells, iii) cortex of thin walled cells, and iv) medulla containing filamentous hyphae. Sclerotia forming on a host tend to have a smooth texture, whereas those produced in culture may be pitted or folded. Serving as a protective structure, sclerotia contain viable hyphae and serve as primary inoculum for disease development.

Symptoms

The infected plants gradually lose their colour and turn pale, followed by drooping. The affected roots, particularly the collar portion turn yellowish-brown. Affected plants can be easily pulled out from the soil. White to tan-brown mustard seed like sclerotia are seen around the infected roots. The symptoms may be extended on stem, causing shrivelling of the stem (the fungus can also be seen naturally causing water-soaked spots on leaves) and finally result in the death of the plants (Kolte, 1985).

Disease Cycle

The fungus overwinters mainly as sclerotia. Pathogen is spread by contaminated tools, infected transplants seedling, moving water, infested soil, infected vegetables and fruits and in some hosts as sclerotia mixed with the seed. The fungus attacks tissues directly. However, the mass of mycelium it produces secretes oxalic acid and also pectinolytic, cellulolytic and other enzymes and it kills and disintegrates tissues before it actually penetrates the host. Fungus once establishes in the plants, advances and produces mycelium and sclerotia quite rapidly, especially at high moisture and high temperature from 30 to 35 C°.

Disease Management

Management of collar rot disease is difficult. Crop rotation provides only partial control. Cultural practices *i.e.* deep summer ploughing to bury the fungal sclerotia in surface debris, ammonia fertilizations, and calcium compounds application are effective in controlling the diseases. Soil solarisation is very effective for controlling of this disease. The control is attributed to the hydrolysis products of glucosinolates in to allyl and butenyl isothicyanates which are toxic to *Pythium aphanidermatum, Sclerotium rolfsii, Sclerotinia sclerotiorum* and *Phytophthora capsici* (Singh, 2009).

Cercospora Leaf Spot

Causal Organism

Cercospora leaf spot is caused by *Cercospora kikuchii* (Teleomorph - *Mycosphaerella*). The fungus produces long, slender and colourless to dark, straight to slightly curved, multicellular conidia on short dark conidiophores. Conidiophores arise from the plant surface in the clusters through stomata and form conidia successively on new growing tips. Conidia are detached easily and are often blown long distances by the wind. The fungus is favoured by high temperatures and therefore is most destructive in the summer months and in warmer climates. Fungus produces non specific toxin cercosporin which acts as a photosensitizing agent in the plant cells *i.e.* it kills cells only in light. The pathogen remains over seasons in or on seed and as small black stomata in plant debris.

Symptoms

Foliar symptoms usually are seen at the beginning of seed set and occur in the uppermost canopy on leaves exposed to the sun. Affected leaves are discoloured, with symptoms ranging from light purple, pinpoint spots to larger, irregularly shaped patches typically only on the upper leaf surface. As disease develops, affected leaves may become leathery and dark purple with bronze highlights. Symptoms may be confused with sunburn. Discoloration may extend to the upper stems, petioles and pods. Infection of petioles and severe symptoms may lead to defoliation of the uppermost leaves and give the appearance of a maturing crop. However, petioles of fallen leaves remain attached to the stem, and lower leaves of the plant remain green. Symptoms of purple seed stain are distinct pink to dark purple discolorations of seed. Discoloured areas vary in size from small spots to the entire surface of the seed coat; however, infected seeds may not show symptoms.

Disease Cycle

The fungus survives winter in infected crop residue and infected seed. Mostly early season infections do not cause symptoms but contribute to infection of foliage and pods later in the season. Warm and wet weather is favourable for infection.

Foliar symptoms are the result of an interaction between a toxin produced by the fungus and sunlight. Weather conditions during flowering and plant maturity will affect the incidence of purple seed stain. Despite being caused by the same organism, there is no consistent relationship between the occurrence of Cercospora leaf blight and purple seed stain.

Management

Use the disease free seeds and resistant varieties to control this disease. Seed treatment is essential to eliminate the seed-borne inoculum. Disinfection of seeds by dip in 0.5 per cent copper sulphate solution for 30 minutes. Foliar application of fungicides namely hexaconazole @ (0.025 per cent), bavistin (0.025 per cent) and chlorothalonil (0.2 per cent) are economic and effective to control this disease. Applications made during pod-filling stages can reduce the incidence of purple seed stain, but may not affect soybean yield. Rotation to non-host crops such as alfalfa, corn and small grains and tillage to bury infested crop residue will reduce pathogen levels. If considering tillage, use proven conservation practices to maintain soil quality.

Downy Mildew

Causal Organism

The causal organism of this disease is *Peronospora manshurica*. Downy mildew is a very common foliar disease of soybeans, but it seldom causes serious yield loss. The pathogen may also infect seed and reduce seed quality. Diseased plants are usually widespread within a field.

Symptoms

Seedlings that are infected from oospores on the seed can develop large chlorotic areas on the first and second pairs of true leaves. The disease is more common in late vegetative and reproductive growth stages. Lesions occur on upper surfaces of leaves as irregularly shaped, pale green to light yellow spots that enlarge into pale to bright yellow spots. Older lesions turn brown with yellow-green margins. Young leaves are more susceptible than older leaves, so disease is often found in the upper canopy. Lesion size varies with the age of the leaf affected. On the underside of the leaf, fuzzy, gray tufts may be seen growing from each lesion, particularly when humidity is high or leaves are wet, for example, early in the morning. Infected pods show no external symptoms, but the inside of the pod and seed may be covered with a dried, whitish fungal mass that appears crusty and contains spores. Infected seed can be smaller, appear dull white and have cracks in the seed coat.

Disease Cycle

The pathogen is primarily soil borne through oospores lying in the diseased plant debris. *Peronospora manshurica* survives in leaves and on the surface of seed. Extended periods of leaf wetness are favourable for movement of the pathogen. High humidity and moderate temperatures favour the infection. The increased resistance of older leaves and higher temperatures midseason usually stop disease development before extensive damage occurs.

Management

Use only resistant and certified seeds for sowing. However, many races of the pathogen have been identified, and varieties that are resistant to all known races have not yet been developed. Crop rotation and burial of infested crop residue using conservation tillage practices can reduce pathogen levels. Two to three foliar spray of fungicide such as sulphur fungicide should be done at the disease initiation and after that 15 days interval.

Frogeye Leaf Spot

Causal Organism

Frogeye leaf spot has become more prevalent in north hills zone of India. The causal organism of this disease is *Cercospora sojina*. It is especially problematic in continuous soybean fields. Diseased plants are usually widespread within a field.

Symptoms

Early season infections from infected seed result in stunted seedlings. On leaves, lesions are small, irregular to circular and gray with reddish-brown borders that most commonly occur on the upper leaf surface. Lesions start as dark, water-soaked spots that vary in size, and as lesions age, the central area becomes gray to light brown with dark, red-brown margins. In severe cases, disease can cause premature leaf drop and will spread to stems and pods. Symptoms on stems are not as common or distinctive as foliar symptoms and appear as narrow, red brown lesions that turn light gray with dark margins as they mature. Lesions on pods are circular or oval shaped and are initially red-brown and turn to light gray with a dark brown margin. Seed close to lesions on pods can be infected. Infected seeds have light to dark gray discoloured blotches that vary in size and cover the entire seed in severe cases. The seed coat often cracks.

Disease Cycle

The fungus survives in infested crop residue and infected seed. Early season infections contribute to infection of foliage and pods later in the season. Warm, humid weather promotes spore production, infection and disease development. Young leaves are more susceptible to infection than older leaves, but visible lesions are not seen on young, expanding leaves because the lesions take two weeks to develop after infection. It is common for disease to be layered within the canopy. This is a result of little to no infection during dry periods and higher levels of infection during wet or humid weather.

Management

Resistant varieties are available and should be used where disease is a potential problem. Several races of the pathogen have been identified, and varieties with resistance to all known races are available. Crop rotation and tillage will reduce survival of *Cercospora sojina*. Crops not susceptible to this pathogen are alfalfa, corn and small grains. If tillage is considered to promote decay of crop residue, great care should be taken to minimize soil erosion and maintain soil quality. Foliar fungicides applied during late flowering and early pod set to pod-filling stages can reduce the incidence of frogeye leaf spot and improve seed quality and yield.

Septoria Brown Spot

Causal Organism

Brown spot is the most common foliar disease of soybean. The pathogen of this disease is *Septoria glycines*. Disease develops soon after planting and is usually present throughout the growing season. Yield losses depend on how far up the canopy the disease progresses during grain fill. Diseased plants are usually widespread within a field.

Symptoms

Symptoms are typically mild during vegetative growth stages of the crop and progress upward from lower leaves during grain fill. Infected young plants have purple lesions on the unifoliate leaves. Lesions on later leaves are small, irregularly shaped and dark brown, and are found on both leaf surfaces. Adjacent lesions can grow together and form larger, irregularly shaped blotches. Infected leaves quickly turn yellow and drop. Disease starts in the lower canopy and, if favourable conditions continue, will progress to the upper canopy. Lesions on stems, petioles and pods are not as common, but appear as brown, irregularly shaped spots ranging from small specks to 1/2 inch in diameter.

Disease Cycle

Warm and wet weather favours the disease development. The fungus survives on infected leaf and stem residue. Disease usually stops developing during hot and dry weather, but may become active again near maturity or when conditions are more favourable.

Management

Use of resistant variety is good source of managing this disease but there are no known sources of resistance, but differences in susceptibility occur among soybean varieties. The host range of *Septoria glycines* includes other legume species and common weeds such as velvet leaf. Crop rotation with non-host crops such as alfalfa, corn and small grains and incorporation of infested crop residue into the soil will reduce the survival of *Septoria glycines*. If tillage is an option, use conservation tillage practices to maintain soil quality. Foliar fungicides labelled for brown spot control are available. Applications made after appearance of the disease may slow the rate of disease development into the middle and upper canopy and protect yield.

Soybean Rust

Causal Organism

The causal organism of soybean rust is *Phakopsora pachyrhizi*. Soybean rust is an aggressive disease capable of causing defoliation and significant yield loss. Soybean rust is an endemic to India and found in most soybean growing areas of the world.

Symptoms

Soybean plants are susceptible at any stage of development, but symptoms are most common after flowering. Early symptoms of rust infection begin on lower leaves. Lesions begin to form on lower leaf surfaces, starting as small, gray spots and changing to tan or reddish-brown. Lesions are scattered within yellow areas that appear translucent if the affected leaves are held up to the sun. Mature lesions contain one to more small pustules that usually occur on lower leaf surfaces. These pustules produce uredospores and spore production may continue for weeks. Premature defoliation and early maturity occurs while an infection is severe.

Disease Cycle

The rust pathogen can only survive on green tissue; thus, the pathogen is unable to survive in areas where killing frosts eliminate susceptible hosts. The movement of rust depends on rust spores increasing at sites where the pathogen has survived the winter, dispersal of the spores to new areas and establishment of the disease in those areas. These steps need to be repeated several times within a growing season in order for rust to cause an epidemic in the country. When spores land in new areas, infection takes place only when prolonged periods of leaf wetness (6 to 12 hours) and moderate temperatures occur in those areas. Cool, wet weather or high humidity favour soybean rust epidemics. Dense canopies also can provide ideal conditions that encourage disease development. Infection can spread rapidly to middle and upper leaves once the canopy closes.

Management

A limited number of resistant breeding lines have been identified; however, there is currently some commercially available soybean rust resistant variety (DSb 21) in India. Resistant varieties have been released in other countries, but none are resistant to all known races of the pathogen. Currently, foliar fungicides are the only viable option for managing soybean rust. To manage the disease effectively and profitably, fungicides need to be sprayed prior to infection or, at the latest, very soon after initial infection *i.e.* hexaconazole @ 0.1 per cent or propiconazole @ 0.1 per cent (Singh, 2009). National and local spread of soybean rust can be tracked to help gauge if/when to start scouting or initiate fungicide applications.

Charcoal Rot

Causal Organism

Charcoal rot can be an important disease and is most yield-limiting when weather conditions are hot and dry. This disease is caused by *Macrophomina phaseolina*. This disease is more common is southern and North Eastern part of the India and causes huge losses.

Symptoms

Symptoms of charcoal rot usually appear after flowering. Initial symptoms are patches of stunted or wilted plants. Leaves remain attached after plant death. The lower stem and taproots of these plants are discoloured light gray or silver. When stems are split, black streaks are evident in the woody portion of the stem. In addition, the fungus produces numerous tiny, black fungal structures called microsclerotia that are scattered throughout the pith and on the surface of taproots and lower stems. These microsclerotia give the tissue a charcoal-like appearance. Infected seed either show no symptoms or having microsclerotia embedded in seed coat cracks or on the seed surface. Infected seed have lower germination, and if seed germinates, the seedlings usually die within a few days.

Disease Cycle

The fungus survives in soil or soybean residue as microsclerotia. Microsclerotia infect roots of soybean plants, sometimes very early in the season. Environmental factors like temperature, humidity, rainfall etc. affect microsclerotia survival, root infection and disease development. The fungus is more abundant in soil when pH is very acidic or alkaline. Charcoal rot is most prevalent during hot, dry weather, especially when it occurs during the flowering / pod formation stages.

Management

In summer crops, irrigation lowers soil temperature and increases soil moisture. These conditions are unfavourable for the disease. Most efforts on control of *M. phaseolina* involve management of populations of microsclerotia. Growing small grains, such as wheat or barley, can reduce microsclerotia numbers. Corn is also a host of *M. phaseolina* so it will not reduce levels of the fungus when planted in rotation with soybeans. The fungus is less damaging to corn than to soybeans. Fields with minimal or no tillage may have fewer symptoms because of lower soil temperatures and greater water-holding capacity. Avoid excessive seeding rates so that plants do not compete for moisture, which increases disease risk during a dry season.

Fusarium Wilt and Root Rot

Causal Organism

Fusarium is a very common soil fungus, and more than 10 different species are known to infect soybean roots and cause root rot. The species *Fusarium oxysporum* is responsible for causing Fusarium wilt. Although Fusarium root rot is a widespread disease in the country, the economic impact on yield is not well documented.

Symptoms

Symptoms of Fusarium wilt are more noticeable under reduced moisture and hot conditions and are often misdiagnosed as those of Phytophthora root rot. Infected plants have brown vascular tissue in the roots and stems and show wilting of the stem tips. However, external decay or stem lesions are not seen above the soil line. Foliar symptoms include scorching of the upper leaves, while middle and lower canopy leaves can turn chlorotic and later wither and drop from the plant. Young plants are at the greatest risk to root rots caused by *Fusarium* species. Infected plants may exhibit poor or slow emergence, and seedlings are often stunted and weak. Seedlings with root rot have reddish-brown to dark brown discoloured roots. Infected plants may have poor root systems and poor nodulation, which may cause the plants to wilt and finally die.

Disease Cycle

The fungus survives in the soil either as spores or as mycelium in plant residue. Certain weeds may serve as hosts to some pathogenic *Fusarium* species. The fungi can infect plants at any stage of soybean development but infection is particularly favoured when plants are weakened. Stresses such as herbicide injury, high soil pH, iron chlorosis, nematode feeding and nutritional disorders can all predispose plants to infection. After infection, damage to plants can be worsened if soil moisture is limited because of the compromised root systems.

Management

Varieties have varying levels of susceptibility, but no resistant varieties have been described. Reducing or eliminating stress factors, such as use of herbicides that cause injury to soybeans, wet soils and soybean cyst nematode, can help reduce root rot problems. Growing of tolerant varieties to iron deficiency chlorosis should be considered if the root rot seems associated with iron deficiency chlorosis. If *Fusarium* is a problem in a field, seed treatments with bavistin @ 2g/kg seed may protect seedlings in subsequent years.

Powdery Mildew

Causal Organism

The powdery of soybean is caused by *Microsphaera diffusa*. The disease is more prevalent in cooler than normal seasons. While this disease is uncommon, when it does show up in fields, there can be noticeable yield loss.

Symptoms

The most common and characteristic sign of powdery mildew is white, powdery fungal growth appeared on aboveground plant parts, particularly the upper surface of leaves. Powdery mildew usually does not appear until mid- to late reproductive stages. Initially, small fungal colonies form and grow together as they enlarge. Eventually, entire surfaces of infected plant parts are covered with white fungal growth. Advanced symptoms include yellowing of plant tissues and premature defoliation.

Disease Cycle

Microsphaera diffusa is a biotrophic parasite. The fungus survives in infested crop residue. In general belief had been that the pathogen survives between crop seasons through cleistothecia in soil. The favourable conditions for the disease development are cool, cloudy weather and low humidity. Powdery mildew of soybean is severely affect when crop sown in late season.

Management

Planting of resistant varieties to minimise the disease and early sowing to escape the disease. Chemicals such as sulphur fungicide effectively manage the powdery mildew; however, there are limited situations where fungicide use will be profitable. Efficacy of some plant extracts and plant products against the pathogen has been experimentally demonstrated Nemadole (a neem product) and *Allium cepa* (onion), *Allium sativum* (garlic), rhizome of ginger and neem leaves (*Azadiracta indica*) are non phytotoxic but fungicidal and at par with Karathane in the suspension of powdery mildew of pea. Several fungi such as *Ampelomyces*, *Tilletiopsis* and *Verticillium* and insects (*Thrips tabaci*) are natural biocontrol agents of the powdery mildew.

Anthracnose Stem Blight

Colletotrichum truncatum (hemibiotrophoc fungus) causes the anthracnose stem blight of soybean. Anthracnose is generally a late season disease that is prevalent on maturing soybean stems throughout the world. Soybean, however, is susceptible to infection throughout the growing season. Diseased plants are usually widespread within a field.

Symptoms

Infected seed may or may not show symptoms. When seed symptoms do occur, they appear as brown discoloration or small gray areas with black specks. Foliar symptoms include reddish veins, leaf rolling and premature defoliation. On stems and petioles, symptoms typically appear as irregularly shaped red to dark brown blotches during early reproductive stages. Damping off may occur if infected seed is planted. Leaves, pods and stems may also be infected without showing symptoms. Petiole infection may result in a shepherd's crook. Early infection of leaf petioles may cause premature defoliation and yield loss. Infection of young pods results in seedless pods at maturity while pods infected later contain seeds that are infected. Near maturity, black fungal bodies that produce small, black spines and spores are evident on infected stems, petioles and pods.

Disease Cycle

The fungus overwinters as mycelium in crop residue or infected seed. Although plant stand may be affected by early season infection, most infection occurs during the reproductive stage of the crop. Spores produced by the fungus are sensitive to drying; thus, free moisture for 12 hours or longer is necessary for successful infection. Warm, wet weather favours infection and disease development. The most important factors affecting the infection are temperature and moisture. Moderate temperatures between 13° and 26°C favour infection. No infection occurs at temperatures above 27° and at 13°C also the disease is considerably reduced. A relative humidity of above 92 per cent is necessary for infection, the optimum being close to 100 per cent. A 10 hour wet period is reported to be necessary for conidial infection and new lesions usually appear in 3 -7 days depending on prevailing temperature.

Management

There are no known sources of resistance to anthracnose, but soybean varieties differ in susceptibility. The seed must be disease free hence it should be collected from only healthy pods. Usually, seed produced in dry areas or free from infection. Crop rotation and tillage will reduce survival of *Colletotrichum* species. Non-legume crops such as corn are not susceptible to this pathogen. If tillage is considered, great care should be taken to minimize soil erosion and maintain soil quality. Foliar fungicides labelled for anthracnose are available. Benlate, Ziram, Vitavax, Ferbam and lime sulphur, in order listed had been recommended for foliar sprays. Bavistin, Vitavax and Agroson GN were recommended for seed treatment. Applications should be made during the early to mid-reproductive growth stages of the crop, although there are limited situations where fungicide use will be profitable. There are many reports of biological control of the anthracnose of bean through seed bacterization and through inoculation with avirulent strains of the pathogen (Sticher *et al.*, 1997; Van Loon *et al.*, 1998).

Phytophthora Root and Stem Rot

Causal Organism

Phytophthora sojae causes the Phytophthora root and stem rot of soybean. Phytophthora root and stem rot is an economically important disease of soybeans that is most severe in poorly drained soils. Diseased plants often occur singly or in patches in low-lying areas of the field that are prone to flooding.

Symptoms

The most characteristic symptom of Phytophthora root rot, however, is a dark brown lesion on the lower stem that extends up from the taproot of the plant. *Phytophthora sojae* can infect soybeans at any growth stage from seed to maturity. Early season symptoms include seed rot and pre- and post-emergence damping off. Stems of infected seedlings appear water-soaked, while leaves may become chlorotic and plants may wilt and die. On older plants, symptoms vary depending on the variety. For susceptible plants, leaves become chlorotic between the veins and plants wilt and die, with the withered leaves remaining attached. Varieties that are not fully susceptible may appear stunted, but plants are typically not killed. The lesion often reaches as high as several nodes and will girdle the stem and stunt or kill the plant.

Disease Cycle

Phytophthora sojae survives on crop residue or in the soil as oospores. Optimum soil moisture is 15 to 20 per cent is needed for oospores germinate to produce structures that release swimming spores, called zoospores, under saturated soil conditions. The zoospores are attracted to soybean roots. Infection occurs via the roots, and from there the pathogen colonizes the roots and stems. Disease is most common in poorly drained soils, but may occur in other soils as well.

Management

Management of Phytophthora root rot is by planting resistant varieties. Many race-specific resistance genes (called *Rps* genes) to *Phytophthora sojae* have been identified in soybean breeding lines. Some of these genes have been incorporated in commercial soybean varieties; thus, there are soybean varieties available that have complete resistance to a specific race of *Phytophthora sojae*. There are numerous races (now called pathotypes) of *Phytophthora sojae*, and many pathotypes can exist in a single field. Furthermore, new pathotypes can develop that can infect varieties with specific *Rps* genes. Partial resistance is available to *Phytophthora sojae*. Partial resistance is effective against all races of *Phytophthora sojae*; however, it is only expressed after the first true leaves emerge, not in very young seedlings. Continuous soybean production may increase disease severity. But rotation to non-hosts may reduce disease severity because oospores can survive in soil for long periods of time. Disease is more severe in no-till fields because these fields can be wetter. If tillage is considered to improve drainage, use proven conservation tillage practices to maintain soil quality. Where *Phytophthora sojae* is a serious problem, seed treatments with metalaxyl as an active ingredient can provide some protection. Seed treatments are especially helpful with poor quality seed and in fields with a history of this problem.

Pod and Stem Blight and Phomopsis Seed Decay

Causal Organism

The causal organism of these diseases is *Diaporthe phaseolorum* var. *sojae* and *Phomopsis longicolla*. Pod and stem blight is one of three diseases that make up the

Diaporthe-Phomopsis complex. Other diseases in this complex include seed decay and stem canker. Stems, petioles, pods and seeds are severely affected by this disease.

Symptoms

The most characteristic symptoms of pod and stem blight is linear rows of black specks on mature stems of soybeans. The specks, which are flask-shaped fruiting structures of the fungus known as pycnidia, can be seen during the season on prematurely killed petioles or stems. Poor seed quality may result from infection. Seed infection occurs only if pods become infected. Pod infection can occur from flowering onwards, but extensive seed infection does not occur until plants have pods that are beginning to mature. Insect damage to pods favours development of seed infections. *Phomopsis*-infected seed are cracked and shrivelled and are often covered with chalky, white mold. Infected seedlings have reddish-brown, pinpoint lesions on the cotyledons or reddish-brown streaks on the stem near the soil line. If infected seeds are planted, emergence may be low due to seed rot or seedling blight.

Disease Cycle

The fungi survive winter in infected seed and infested crop residue. Certain weeds may serve as hosts to some pathogenic *Diaporthe* and *Phomopsis* species. Infection can occur early in the growing season without causing symptoms. Disease is favoured by warm, humid weather, when soybean plants are maturing. Also, disease is more severe if harvest is delayed.

Management

Sources of resistance have been identified, and variation in seed infection has been reported among commercial soybean varieties. Unfortunately, there currently are no resistant varieties or lists of seed reactions of current varieties available. Varieties with an earlier relative maturity for a region are at greater risk of Phomopsis seed decay and pod and stem blight than fuller-season varieties. Do not plant seed with a high incidence of infection. Crop rotation and tillage will reduce survival of *Diaporthe* and *Phomopsis* species. Non-host crops include corn. If tillage is considered to promote decay of pathogen-infested residue, be careful to minimize soil erosion and maintain soil quality. Application of foliar fungicides near R5 stage can protect seed quality, but may not affect yield. Use of early maturity varieties, lower the incidence of seed rot. Fungicidal seed treatments with Thiram, Ziram and Apron are effective against *Phomopsis* species. Treating *Phomopsis*-infected seed lots may increase germination and improve plant establishment.

Pythium Root Rot

Causal Organism

Several species of *Pythium* are reported to cause this disease. Early planting dates increase the risk of disease in the major soybean growing areas. Diseased plants often occur singly or in small patches in low-lying areas of the field that are prone to flooding.

Symptoms

Pythium species cause pre- or post-emergence damping off. Infected seed appear rotted and soil sticks to them. Infected seedlings have water-soaked lesions on the hypocotyl or cotyledons that develop into a brown soft rot. Diseased plants are easily pulled from the soil because of rotted roots. Older plants become resistant to soft rot, but root rot may cause plants to become yellow, stunted or wilted if infection is severe.

Disease Cycle

The pathogen survives either in plant residue or in soil as oospores. Severity of disease depends on the amount of the pathogen in the soil, plant age and environmental conditions at the time of infection. Saturated soil is critical for infection for all *Pythium* species. As *Phytophthora, Pythium* produces zoospores that swim in free water and infect the roots of plants. In general, *Pythium* species that are prevalent in the north at lower temperatures (10 to 15°C), and *Pythium* species in the south at warmer temperatures (30 to 35°F), although there are exceptions.

Management

Planting in cold, wet soils should be avoided to reduce infection by *Pythium* species that infect at low temperatures. Seed treatments with Apron, metalaxyl or strobilurins are effective for management of disease. Resistance to metalaxyl/ mefenoxam has been accepted; however, they are generally considered more effective than strobilurins. Soil application of metalxyl at transplanting time followed by weekly sprays of potassium phosphonate (1g/L) plus acibenzolar-S-methyl (0.025 g/L) also significantly reduced root rot infection. No-till soils often have higher soil moisture and lower soil temperatures, factors that increase the risk of *Pythium* infection. If tillage is considered to improve drainage, use conservation tillage practices to maintain soil quality.

Rhizoctonia Root Rot

Causal Organism

The pathogen of Rhizoctonia root rot is *Rhizoctonia solani.* Rhizoctonia root rot is one of the most common soil borne diseases of soybeans. Diseased plants usually occur singly or in patches in the field. Disease is typically more common on the slopes of fields.

Symptoms

Rhizoctonia infects young seedlings, causing pre- and post-emergence damping off. Infected seedlings have reddish-brown lesions on the hypocotyls at the soil line. These lesions are sunken, remain firm and dry and are limited to the outer layer of tissue. If seedlings survive the damping off phase, infections may expand to the root system, causing a root rot. The root rot phase may persist into late vegetative to early reproductive growth stages. Older infected plants may be stunted, yellow and have poor root systems.

Disease Cycle

The fungus survives on plant residue or in soils as sclerotia. When soils warm, the fungus becomes active and infection may occur soon after seed is planted. The fungus grows better in aerated soils; thus, disease is more severe on light and sandy soils. Symptoms may disappear if infected plants grow out of the root rot problems although plants may remain stunted.

Management

Resistance has been reported in some varieties; however, there are no varieties being developed for resistance to Rhizoctonia root rot. Unfortunately, many strains of *Rhizoctonia* can infect corn, alfalfa, dry bean and some cereal crops. Eliminating stress factors, such as use of herbicides that cause injury to soybean roots, can help reduce root rot problems. Most fungicide seed treatments such as Bavistin or Benlate (2 g/kg) are effective against *Rhizoctonia* and same fungicide can be used as foliar sprays 2 – 3 times gives good control.

Bacterial Disease

Bacterial Pustule

Causal Organism

This disease caused by *Xanthomonas axonopodis* pv. *glycines*. Bacterial pustule occurs mid to late season when temperatures are warmer and more favourable for disease development. Symptoms may be mistaken for bacterial blight, septoria brown spot or soybean rust. Diseased plants are usually widespread within a field.

Symptoms

Lesions are found on outer leaves in the mid to upper canopy. Lesions start as small, pale green specks with elevated centres and develop into large, irregularly shaped infected areas. Unlike bacterial blight, no water soaking is associated with lesions, but each lesion is surrounded by a greenish-yellow halo. A pustule may form in the centre of some lesions, usually on the lower leaf surfaces. Pustules crack open and release bacteria. Bacterial pustule will not cause leaves to tatter like bacterial blight.

Disease Cycle

Bacteria survive winter in crop residue and seeds and are spread by rain and wind. Infection occurs through leaf stomata or wounds. Rainy weather favours disease development. Unlike bacterial blight, high temperatures do not slow disease development.

Management

Avoid planting extremely susceptible varieties. Some varieties are marketed as resistant to this disease. Rotation and tillage reduce survival of *Xanthomonas axonopodis* pv. *glycines*. Other legume crops may be hosts; non-hosts include alfalfa, corn and small grains. If tillage is considered, use proven conservation tillage practices to maintain soil quality.

Viral Disease

Soybean Yellow Mosaic

It is the most destructive disease of soybean in India. It was first reported in 1960 and is now known to occur throughout the country. The loss of yield depends upon the stage at which the crop is infected. If the infection is early in the season there may be total loss of seed yield.

Causal Organism

Four viruses causing yellow mosaic disease of legumes across the South Asia have been identified as bipartite begomoviruses (genus *Begomovirus*, family *Geminiviridae*). The soybean strain of MYMV occurring in north India is distinct from the strain occurring in southern and western India (Usharani *et al.*, 2004). A strain of MYMIV, designated as MYMIV-Cp causes golden mosaic of cowpea. It has restricted host range and transmission by *Bemisia tabaci*. These viruses have evolved independently of the begomoviruses in plant species of other families. The paired particles of the virus measure 30 × 18 nm. The particle contains two circular ssDNA molecules which account for 20 per cent of the particle weight. The coat protein contains one polypeptide with MW of 28.5 kDa.

Symptoms

Disease appeared in the field when the crop is about one month old. Two types of symptoms appeared depending upon the host response. The general pattern of development of both symptoms is the same. The first visible sign of the disease is the appearance of yellow spots scattered on the lamina surface. They are mostly round in the shape. In yellow mottle, the spots are diffuse and expand rapidly. The leaves show yellow patches alternating with green areas that also turn yellow. Such completely yellow leaves gradually change to a whitish shade and ultimately become necrotic. These colour changes of affected plants are so conspicuous that the disease can be spotted in the field from a distance. In case of necrotic mottle, the centre of yellow spots develops necrosis which is demarcated by finer veins. The virus becomes systemic in the plant and all newly formed leaves show signs of mottle from the very beginning. Number of size of spots per plant and seeds per pod are greatly reduced.

Management

Certified and healthy seeds use for sowing. Cultivar PK 21-22 of soybean is tolerant to the disease. Control of the disease through prevention of population build up of the vector has also been recommended. Sprays of 0.1 per cent metasystox, starting when the crop is about a month old or as soon as a single diseased plant is seen in the field, can give relief from severe incidence of the disease. Anthio is effective at 0.2 per cent when used as spray 3 times.

Nematode Disease

Soybean Cyst Nematode

In India the most important pathogen of soybean is soybean cyst nematode (SCN). In high-yielding production fields or during years when soil moisture is plentiful, damage from SCN may not be obvious. However, yield losses up to 40 percent on susceptible varieties are still possible. When symptoms are associated with damage, infected plants usually occur in patches within a field.

Causal Organism

The soybean cyst nematode that causes the disease is known as *Heterodera glycines*. The body of females is swollen, pearly white and lemon shaped and usually varies between 0.6 to 0.8 mm in length and 0.3 to 0.5 mm in diameter. The male is wormlike about 1.3 mm long and 30 – 40 *u*m in diameter. The males remain in the root for a few days during which they may or may not fertilize the females and then they move into the soil and soon die. Cysts are typically lemon shaped. Mature cysts of Indian populations measures 470 – 1010 × 370 – 730 μm. Each females produces 300 to 600 eggs most of which remain inside her body when the females die. Eggs in the gelatinous matrix may hatch immediately and the emerging second stage juveniles may cause new infestation.

Symptoms

Obvious symptoms may not develop, even though yield loss occurs. Noticeable symptoms of SCN include stunting, slow or no canopy closure and chlorotic foliage. Infected plants have poorly developed root systems. Soybean cyst nematode infection also may reduce the number of nodules formed by the beneficial nitrogen-fixing bacteria necessary for optimum soybean growth. Signs of SCN include white females that are most readily seen in the field starting about six weeks after crop emergence. To see them, roots must be dug and soil carefully removed. However, the only way to get a reliable diagnosis as to the amount of SCN in the soil is through analysis of a properly collected soil sample by a diagnostic laboratory. Plant damage is not just limited to direct and indirect effects of feeding by the nematodes. Wounds caused by infecting nematodes and by maturing females serve as entry points for other soil borne pathogens. Diseases such as brown stem rot, Rhizoctonia root rot, sudden death syndrome and charcoal rot are more severe in the presence of SCN.

Disease Cycle

SCN survives in the soil as eggs within dead females called cysts. These eggs can survive several years in the absence of a soybean crop. The second-stage juvenile (J2) hatches from the eggs and infects soybean plants. After infection, these juveniles migrate to the vascular system before setting up specialized feeding cells within the root. As they feed, the nematodes become immobile. The juveniles moult three more times before maturing into adults, with females becoming so large they burst through the outer surface of the roots. A female will produce 200 to 300 eggs that are deposited in an external egg mass or are retained within her body. Soybean cyst nematode can complete four or more generations during the growing season, depending on planting date, soil temperature and length of the growing season, host suitability, geographic location and maturity group of the soybeans. Conditions that favour soybean growth are also favourable for SCN development. High soil pH may be used to predict where SCN is more problematic. Areas of fields with soil pH levels of 7.0 to 8.0 typically have more SCN compared to areas with soil pH 5.9 to 6.5.

Management

The number of SCN in a field can be greatly reduced through proper management, but it is impossible to eliminate SCN from a field once it is established. Soil tests are recommended prior to every third or fourth soybean crop to monitor SCN population densities (numbers). Resistant varieties are available to manage SCN. The three most common sources of resistance are PI 88788 (most common), PI 548402 (Peking) and PI 437654 (also referred to as Hart wig or PUSCN-14). Resistant varieties are not resistant to all SCN populations. Most resistant varieties contain only one source of genetic resistance. Rotating sources of SCN resistance may help prevent the development of more damaging SCN populations. SCN-resistant varieties, even high-yielding varieties, can vary considerably in how well they control nematode population densities. Greater SCN reproduction will result in a higher SCN egg population in the soil at the end of the growing season, and consequently, higher numbers of SCN in subsequent seasons. Thus, growers must consider how SCN-resistant soybean varieties affect SCN population densities, in addition to how well the varieties yield, to maintain the long-term productivity of the land for soybean production. If SCN is a problem, rotation should include non-host crops (usually corn) and resistant soybean varieties. Years of non-host crops may decrease SCN numbers by as much as 90 percent south, but only 10 to 40 percent in the north. Maintaining adequate soil fertility, breaking hardpans, irrigation and controlling weeds, diseases and insects improves soybean plant health. These practices help plants compensate for damage by SCN, but do not decrease SCN numbers. Zero tillage practices may slow SCN movement and lower population densities. Soil that remains on tillage and harvest equipment can move SCN and should be removed before equipment is relocated from an infested to a non-infested field. Seed treatments labelled for use on SCN may provide early season protection. A limited number of nematicides labelled for use on SCN can be applied at planting.

References

Cruzi, M. 1931. Aleumi csidi 'Canorena Pedale' da sclerotium observation in Italia. *Atti Academia Nazionale des* Lincei. Rendiconti, **14**: 233-236.

Higgins, B.B. 1927. Physiology and parasitism of *Sclerotium rolfsii* Sacc. *Phytopath,* **17(7)**: 417-448.

Kolte, S.J. 1985. Diseases of annual edible oilseed crops. Vol. 3. Rapeseed-Mustard and Sesame Diseases. CRC Press, Inc. Boca Raton, Florida. USA, p. 97.

Kumar. 2007. A study of consumer attitudes and acceptability of soy food in Ludhiana. MBA research project report, Department of business management, Punjub Agricultural University, Ludhiana, Punjab.

Mundkur, B.B. 1934. Perfect stage of *Sclerotium rolfsii* Sacc. in pure culture. *J. of Agric. Sci.*, **4**: 779-781.

Ramakrishnan, T.S. 1930. A wilt of zinnia caused by *Sclerotium rolfsii. Madras Agric. J.*, **16**: 511-519.

Rolfs, P.H. 1892. Tomato blight some hints. *Bulletin of Florida Agricultural Experimental Station*, p. 18.

Saccardo, P.A. 1911. Notae mycologicae. *Annales Mycologici*, **9**: 249-257.

Singh Chhidda, Singh Prem, Singh Rajbir. 2003. *Modern Techniques of Raising Field crops, Oxford* and IBH Publishing Co. Pvt. Ltd. p. 273.

Singh, R.S. 2009. Plant Diseases. Oxford and IBH Publishing Co. Pvt. Ltd. New Delhi.

Sticher, L.B. Mauch-Mani and Metroux, 1997. Systemic acquired resistance. *Annual Review of Phytopathology*, **35**: 235.

Townsend, B.B., and Willetts, H.J., 1954. The development of sclerotia of certain fungi. *Ann. Bot.* **21**: 153-166.

Usharani, K.S., Surendranath, B., Haq, Q.M.R. and Malathi, V.G. 2004. Yellow mosaic virus infecting soybean in northern India is distinct from the species infecting soybean in southern and western India. *Current Science,* **86(6)**: 845.

Van Loon, L.C., Bakker, P.A.H.M. and Pieterse, C.M.J. 1988. Systemic resistance induced by rhizosphere bacteria. *Annual Review of Phytopathology,* **36**: 453.

2016, Diseases of Pulse Crops and their Sustainable Management 243–267
Editors: **Samir Kumar Biswas, Santosh Kumar and Gireesh Chand**
Published by: **BIOTECH BOOKS, NEW DELHI**

Chapter 14

Biological Control in Crop Disease Management (CDM): A Gateway to Environmental Security and Sustainable Agriculture

Deepak Kumar Verma[1]*, Balaram Mohapatra[2], Diganggana Talukdar[3], Shikha Srivastava[4] and Bavita Asthir[5]

[1]*Department of Agricultural and Food Engineering,*
[2]*Environmental Molecular Microbiology Lab, Department of Biotechnology,*
Indian Institute of Technology, Kharagpur – 721 302, West Bengal
[3]*Department of Plant Pathology,*
Assam Agricultural University, Jorhat – 785 013, Assam
[4]*Department of Botany, Deen Dyal Upadhyay Gorakhpur University,*
Gorakhpur – 273 009, Uttar Pradesh
[5]*Department of Biochemistry, Punjab Agricultural University,*
Ludhiana – 141 004, Punjab

Introduction

In modern agriculture, the fundamental knowledge in biology has been a gateway to solution for many practical problems and this can be achieved by the better understanding of the surrounding disease-retardant interactions between the

* Corresponding Author: E-mail: deepak.verma@agfe.iitkgp.ernet.in, rajadkv@rediffmail.com

microbes and plants with their knowledge in complex ecology. Biological control is the emerging and important tools in modern biology for the crop disease management (CDM) under the commercialization of agriculture and changing food habits (Manczinger *et al.*, 2002) which afford effective means in increasing sophisticated measures to analyse the multichannel dialogue depicted in Figure 14.1 which are expected to attention the environmental security and to create the possibilities of increasing the yields of agricultural crops and food production by disease reduction and its suppression in a significant level to fulfil of the world's food demand, and also supplying chemical residue-free farm produce to the consumer (Narayanasamy, 2013) and to realize the practical potential of biological agents for CDM to improved agricultural production.

The term "biological control" in agriculture may be defined as the suppression of pathogenic populations by the actions of their native or introduced enemies to

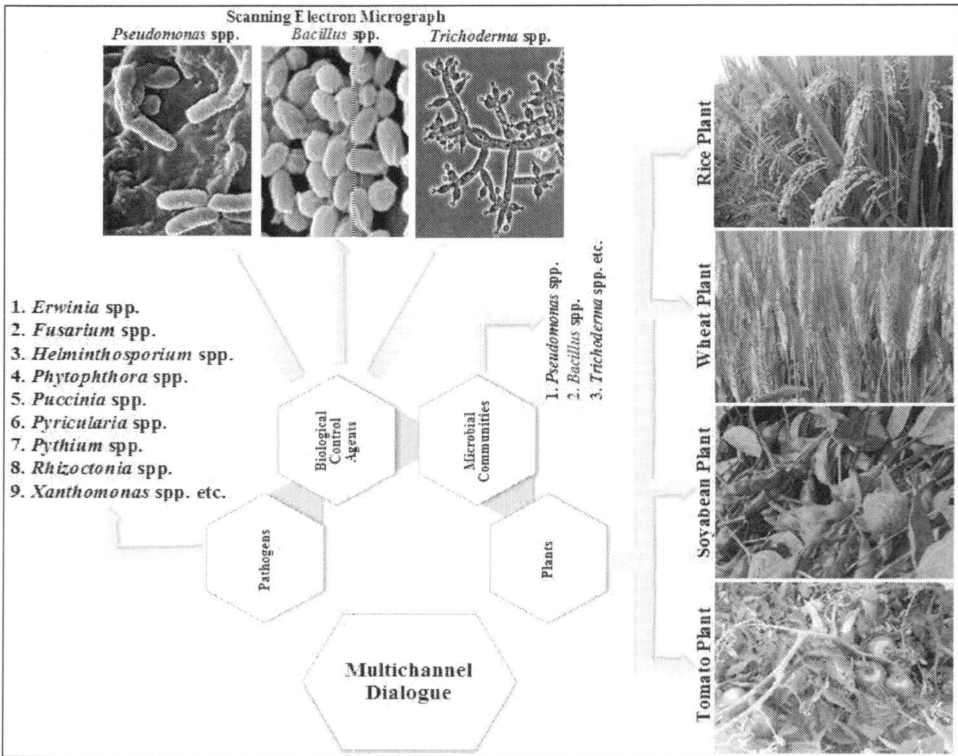

Figure 14.1: Multichannel Dialogue of Biological Control is Consequence of the Interactions between the Plant, Pathogens, Microbial Community and Biological Agent, Concept Adopted from Handelsman (2002).

Picture Courtesy: *Pseudomonas* spp. (http://www.health.qld.gov.au/endoscopereprocessing/module_1/1_3d.asp); *Bacillus* spp. (http://www.extension.org/pages/13225/bacillus#.U6rNKfmSy8c): *Trichoderma* spp. (http://edrosenthal.com/2011/11/ask-ed-marijuana-grow-tip-36-trichodermaharzianum.html)

afford greater levels of protection and sustainability to the agricultural yields. It will be imperative to unite knowledge of mechanistic interactions with an appreciation of the complexity of the agro-ecosystem. In narrow sense, biological control indicates to control one organism by another organism (Beirner, 1967). Biological control has many different definitions (Bull, 2001). The definition proposed by Baker and Cook (1974): "Biological control is the reduction of inoculum density or disease producing activity of a pathogen or parasite in its active or dormant state, by one or more organisms, accomplished naturally or through manipulation of the environment, host or antagonist, or by mass introduction or one or more antagonists". Monte and Liobell (2003) defined as the use of natural organisms or genetically modified, genes or gene products the effects of undesirable organisms to favour organisms useful to human, such as crop, trees, animals and beneficial microorganisms. In other hand, Saba *et al.* (2012) defined the biological control as the use of natural organisms, or genetically modified genes or gene products, to reduce the effects of undesirable organisms to favour organisms useful to human, such as crops, trees, animals and beneficial microorganisms. More narrowly, biological control refers to the purposeful utilization of introduced or resident living organisms, other than disease resistant host plants, to suppress the activities and populations of one or more plant pathogens (Pal and Gardener, 2006; Ouda, 2014). By most definitions, biological control has been recognised as an applied field of research involving the human use of biological agents as microorganism to enhance crop productivity (Handelsman, 2002). The research efforts identify the various eco-friendly approaches in addition of microbes uses for crops diseases management (Narayanasamy, 2013), they are; 1) to mitigate the ill effects of infection by microbial plant pathogens, 2) to reduce or replace the use of synthetic chemicals, and 3) to integrate the compatible and synergistic strategies for enhancing the effectiveness of disease suppression. In addition, many advantage of the microbes in biological control are not limited, some of them are; 1) Avoid adverse effects on the beneficial microbes including antagonists in soil; 2) Avoid pollution of soil, air and water; 3) Cost effective (less expensive); 4) Avoid the development of resistant strains in pathogen.

As these research avenues pursued reveals significant efforts on biological control of diseases, with the majority of effort focused on new principles, community function and new strategies of biological agents will emerge which recognise as a very vital, eco-friendly and cost-effective for agricultural crop's disease management in developing countries, where the chemical treatment methods are very costly and can be prohibitive.

Biological Control in Crop Disease Management (CDM)

Aims

Crop diseases have been concerned with the mankind since our agriculture began and also played a crucial role in the natural resources destruction which is contributing global crop production losses about 13-20 per cent (Anonymous, 1993). For CDM, biological control is an alternative to the use of chemicals (Cook, 2000). The crop diseases occurrence and their counterattack for self-defence might have evolved simultaneously. Thus the aims of the biological control for crop diseases management are; 1) Selection of the most effective biotic agents, 2) Identification of the abiotic agents that can act individually or in combination of biotic agents, 3) Assessing the effects of biotic agents on growth promotion in treated plants, and 4) Examination of the possible measures to reduce the uses or replace the uses of the chemicals without any compromising the effectiveness of disease control (Narayanasamy, 2013).

Concepts

Garrett (1956) defined biological control of plant diseases as "any condition under which or practice whereby survival and activity (excluding the role of the host and man) with the result that there is a reduction in the incidence of the disease caused by the pathogen". Keeping in view the functional relationship between the pathogen, the host and the environment in the development of the disease, Baker and Cook (1974) have given a broader definition of biological control. According to them "Biological control is the reduction of inoculum density or disease producing activity of a pathogen or parasite in its active or dormant state, by one or more organisms, accomplished naturally or through manipulation of the environment, host or antagonist, or by mass introduction or one or more antagonists". Disease producing activity involves growth, infectivity, aggressiveness, virulence and other qualities of pathogen or process that controls infections, symptom development and reproduction. According to includes Ukey and Meshram (2009) organism having the biological control potential:

1. A virulent or hypovirulent individuals or population with the pathogenic species itself.
2. The host plant is manipulated genetically by cultural practices or with microorganisms towards greater or more effective resistance to the pathogens
3. Antagonist of the pathogen as microorganisms that interfere with the survival or disease producing activities of the pathogens.

Principle

According to Gnanamanickam *et al.* (2002), the current approaches in principle of biological control to implement in CDM are considered as; 1) Introduction of Microbes in the Phylloplane, Rhizosphere, or Soil; 2) Stimulating Indigenous Antagonists; 2) Induced Resistance; and 4) Bio-rational Approaches.

Biological Agents used Against Plant Pathogens

Biological agents are widely regarded by the general public as natural and therefore non-threatening products (Monte and Liobell, 2003). They are a new tools and one of the most promising means to achieve the goal of crop disease control from their pathogen alone, or to integrate with reduced doses of chemicals with minimal impact on the environment (Chet and Inbar, 1994; Harman and Kubicek, 1998), although risk assessments must clearly be carried out on their effects on non-target organisms and plants (Monte and Liobell, 2003). Biological agents act against plant pathogens through different mechanisms / modes of action (Cook and Baker. 1983). They have been considered as a potential control strategy for plant pathogens in recent years and search for these biological agents is increasing day by day (Kumar and Mukerji, 1996). Moreover, knowledge concerning the behaviour of such antagonists is essential for their effective use (Monte and Liobell, 2003). Antagonistic behaviour is interactions that can lead to biological control (Cook and Baker. 1983) which include;

- ✰ Antibiosis: A biological interaction between two or more organisms in which at least one of them is detrimental and have direct inhibitory effect due to production of antibiotics or toxic metabolites. The example includes relationship between antibiotics and bacteria and the relationship between plants and disease causing organism (Howell, 1998).
- ✰ Competition: It occurs when two or more microorganisms *e.g.* biological agents and the pathogen which cause disease which require the same resources in excess of their supply. These resources can include space, nutrients and oxygen. In the biological control system, pathogens as out-competes are the less efficient competitor while in-competes the more efficient one, *i.e.* the biological agents (Bandyopadhyay and Cardwel, 2003).
- ✰ Hyperparasitism: Also known as predation *i.e.* results from biotrophic or necrotrophic interactions which lead to the parasitism of the plant pathogen by the biological agent. Some microorganisms, particularly those in soil, can reduce damage from diseases by promoting plant growth or by inducing host resistance against a variety of pathogenic agents (Hutchinson, 1998; Cook, 2000: Kerry, 2000).

In addition of antibiosis, competition and hyperparasitism, efficient biological agents often express more than one mode of action for suppressing the plant pathogens in CDM. Recent studies have shown that they are opportunistic, avirulent plant symbionts, as well as the parasites on other microbes cause crop disease. Biological agents includes to the use of beneficial microbes, their genes, and / or products, such as metabolites, that reduce to the negative effects of plant pathogens and to promote positive responses by the plant (Vinale *et al.*, 2008). The use of biological agents against plant pathogens are the fundamental matter of eco-friendly management of the organism's community. In the nature, diverse groups of microbes such as bacteria, fungi, viruses etc. are existing among them biological agents are not limited to any specific group but very few majority of them (other than the pathogenic agents or damaged hosts which causing the disease) can be used for various crop disease depicted in Tables 14.1 and 14.2. The disease suppression is the consequence of the interactions between the plant, pathogens, and the microbial community which is mediated by biological agents and the success of these agents dependants upon the complex interactions between these beneficial microbes and with pathogens and plants should be established in the ecosystem shown in Figure 14.1 (Vinale *et al.*, 2008). Indeed, they cannot perform their function with the lack of any one of the components among them. In biological control, they are receiving enormous attention and are globally acclaiming as ideal candidates for obvious reasons.

Table 14.1: Biological Agents from Bacterial Genera against Pathogens in Crop Disease Managements

Sl.No.	Bacterial Genera	Reference
1.	**Agrobacterium* spp.	
2.	**Bacillus* spp. *B. amyloliquefaciens, B. cereus, B. laterosporus, B. penetrans, B. polymyxa, B. pumilus, B. subtilis*	Stirling, 1984; Brown *et al.*, 1985; Aspiras and de la Cruz, 1986; Oka *et al.*, 1993; Rosales *et al.*, 1993; Asaka and Shoda, 1996; Mari *et al.*, 1996; Sharga and Lyon, 1998; Giffel and Beumer, 1999)
3.	*Burkholderia* spp. *B. cepacia*	Heydari and Misaghi, 1998; Zaki *et al.*, 1998
4.	*Enterobacter* spp. *E. cloacae*	Van Dijk and Nelson, 1998
5.	*Erwinia* spp. *E. herbicola*	Yuen *et al.*, 1994; Vanneste and Yu, 1996
6.	*Pantoea* spp. *P. agglomerans, P. dispersa*	Zhang and Birch, 1996
7.	*Pasteuria* spp. *P. nishizawa, P. penetrans*	Dube and Smart Jr, 1987; Chen *et al.*, 1994
8.	**Pseudomonas* spp. *P. aeruginosa, P. aurefaciens, P. cepacia, P. chitinolytica, P. corrugate, P. fluorescens, P. lilacinus, P. lindbergii, P. putida, P. solanacearum, P. syringae*	Aspiras and de la Cruz, 1986; Dube and Smart Jr, 1987; Sakthivel and Gnanamanickam, 1987; Sakthivel, 1987; Janisiewicz and Roitman, 1988; Keel *et al.*, 1989; Trigalet and Trigalet-Demery, 1990; Spiegel *et al.*, 1991; De La Cruz *et al.*, 1992; Elangovan and Gnanamanickam, 1992; Sabet *et al.*, 1992; Ghewande *et al.*, 1993; Rosales *et al.*, 1993; Chun and Shetty, 1994; Valasubramanian, 1994; Vanneste and Yu, 1996; Duffy and De´fago, 1997; Khan and Saxena, 1997; M'Piga *et al.*, 1997; Shamim *et al.*, 1997; Ganesan, 1999; Mondal *et al.*, 1999;
9.	*Rhizobacteria* spp.	Podile *et al.*, 1988; Racke and Sikora, 1992; Mondal *et al.*, 1999
10.	*Rhizobium* spp. *R. etli*	Reitz *et al.*, 2000
11.	*Serratia* spp. *S. marcescens*	Press *et al.*, 1997
12.	**Streptomyces* spp.	Podile and Kishore, 2002

* Species registered and are available as commercial products (Vinale *et al.*, 2008).

Table 14.2: Biological Agents from Fungal Genera against Pathogens in Crop Disease Managements

Sl.No.	Fungal Genera	Reference
1.	**Ampelomyces* spp. *A. quisqualis*	Grove and Boal, 1997
2.	*Aspergillus* spp. *A. carbonarium, A. terreus, A. niger*	Bandyopadhyay and Cardwel, 2003
3.	**Candida* spp.	
4.	*Chaetomium* spp. *C. globosum*	Singh, 1994; Pereira and Dhingra, 1997
5.	**Coniothyrium* spp. *C. minitans*	Sabet *et al.*, 1992; McLaren *et al.*, 1996
6.	*Cryptococcus* spp. *C. laurentii*	Roberts, 1990
7.	*Dicyma* spp. *D. pulvinata*	Mitchell and Taber, 1986
8.	*Fusarium* spp. *F. chlamydosporum*	Mathivanan *et al.*, 1997
9.	**Gliocladium* spp. *G. roseum, G. virens*	Keinath *et al.*, 1991; Ristaino *et al.*, 1991; Lewis and Papavizas, 1991
10.	*Glomus* spp. *G. fasciculatum, G. microcarpum, G. intradices*	Datnoff *et al.*, 1995
11.	*Paecilomyces* spp. *P. lilacinus*	Dube and Smart Jr, 1987; Khan and Saxena, 1997; Shamim *et al.*, 1997
12.	*Penicillium* spp. *P. islandicum, P. steckii*	Podile and Kishore, 2002
13.	*Sporidesmium* spp. (*Teratosperma* spp.) *S. sclerotivorum* (*T. sclerotivora*), *T. oligocladum*	Adams, 1981; Adams and Ayers 1981; Rio *et al.*, 1998
14.	*Talaromyces* spp. *T. flavus*	Fravel *et al.*, 1986; Spink and Rowe, 1989; McLaren *et al.*, 1996
15.	**Trichoderma* spp. *T. aureoviridae, T. hamatum, T. harzianum, T. koningii, T. roseum, T. viride*	Cole and Zvenyika, 1988; Truong *et al.*, 1988; Brahmbhatt *et al.*, 1989; Lashin *et al.*, 1989; Knudsen *et al.*, 1991; Latunde-Dada, 1991; Devaki *et al.*, 1992; Datnoff *et al.*, 1995; Inbar *et al.*, 1996; Karthikeyan, 1996; O'Neill *et al.*, 1996; Thennarasu, 1997; De Meyer *et al.*, 1998; Elad *et al.*, 1998; Menendez and Godeas, 1998; Elad, 2000; Sivan *et al.*, 1984; 1987; Singh, 1991; 1994;
16.	*Verticillium* spp. *V. biguttatum, V. lecanii*	Subrahmanyam *et al.*, 1990; Morris *et al.*, 1995; Meyer and Meyer, 1996

* Species registered and are available as commercial products (Vinale *et al.*, 2008).

However, biological agents as microbes are typically have a relatively very complex spectrum of their activities if compared with synthetic chemicals and exhibit very limited commercial use in agriculture for suppression of crop disease (Mathre *et al.*, 1999; Chaube *et al.*, 2002). Researchers are on the way to overcome this limitation for improving the commercialization of biological agents and their product formulations with increasing effect spectrum. To date, several naturally occurring microorganisms have been identified as biological agents to control of plant pathogens and among them a number of biological agents have been registered and are available as commercial products in markets, including strains belonging to bacterial genera (Table 14.1) and fungal genera (Table 14.2). Here, some of them bacterial genera and fungal genera are discussed.

Bacillus spp.

Bacillus is a genus of Gram-positive (low G+C), endospore-forming, rod-shaped (bacillus), bacteria and a member of the phylum Firmicutes (Harwood, 1989; Arnesen *et al.*, 2008). The *Bacillus* genus was first described and classified in 1872 by a German biologist Ferdinand Julius Cohn (Soule, 1932). Representative of this genus are ubiquitous in nature and widely distributed in soil, air, and water (most found in soil and water) worldwide where they are involved in chemical transformations that rival those of the *Actinomycetes* and *Pseudomonads* (Harwood, 1989; Ruger, 1989; Arnesen *et al.*, 2008). Taxonomical identification of species within the *Bacillus* genus has changed over time as differentiation methods have improved (Griffiths, 2010). The genus *Bacillus* is large, encompassing more than 60 distinct species have been described with a great genetic diversity (Priest, 1993; Giffel and Beumer, 1999). They may be divided into five or six groups (groups I–VI), based on 16S rRNA phylogeny or phenotypic features, respectively (Priest, 1993). *Bacillus* spp. includes both free-living (non-parasitic) and parasitic pathogenic species (Madigan and Martinko, 2005), most of which are considered non-pathogenic. *Bacillus* species can be obligate aerobes (primarily aerobic: oxygen reliant) (Harwood, 1989), to/or facultative anaerobes (facultatively anaerobic saprophytes: having the ability to be aerobic or anaerobic) (Turnbull, 1996; Arnesen *et al.*, 2008). They will test positive for the enzyme catalase when there has been oxygen used or present (Turnbull, 1996). The *Bacillus* genus includes range of the most diverse and commercially useful species of microorganisms of human interest (Harwood, 1989) which proving to be dominant bacterial workhorses industrial microorganisms (Schallmey *et al.*, 2004) and among the species of *Bacillus* genus, only few *viz. B. amyloliquefaciens, B. cereus, B. laterosporus, B. penetrans, B. pumilus, B. pumilus B. subtilis, B. subtilis, B. subtilis* etc. are leading as the representatives of this group being used in wide range of CDM for sustained agriculture.

Pseudomonas spp.

The genus *Pseudomonas* is very much interesting microbe due to virtue of its ability to use as biological agent against the pathogen of crop plants which cause disease and reduce the agriculture production. *Pseudomonas* is a genus of gram-negative, rod-shaped, non-sporing, motile, with a single polar flagellum (Migula, 1900; 1894). They belong to the family *Pseudomonadaceae* which contain 191 validly described species (Figure 14.2) (Stanier *et al.*, 1966; Euzéby, 1997). The members of this genus demonstrate a great deal of metabolic diversity and consequently are able to colonise a wide range of niches (http://en.wikipedia.org/wiki/Pseudomonas). The taxonomy of the genus is controversial (Stanier *et al.*, 1966). Some of them are well defined, particularly the fluorescent pseudomonads including *P. aeruginosa, P. putida,* and *P. fluorescens* (Stanier *et al.*, 1966). The certain members of the *Pseudomonas* genus *viz. P. aeruginosa, P. aurefaciens, P. cepacia, P. chitinolytica, P. corrugate, P. fluorescens, P. lindbergii, P. putida, P. solanacearum, P. syringae* etc. have been applied since the mid-

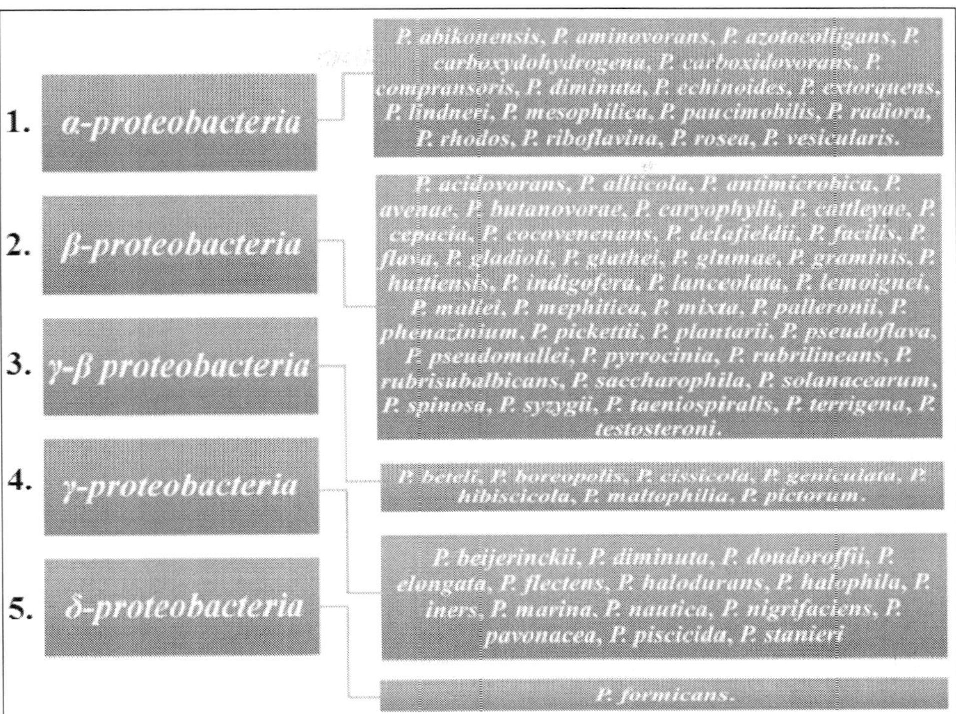

Figure 14.2: Classification of *Pseudomonas* genus (Adopted from http://en.wikipedia.org/wiki/Pseudomonas).

1980s as biological agents to preventing the growth or establishment of crop plant pathogens (http://en.wikipedia.org/wiki/Pseudomonas).

Trichoderma spp.

Trichoderma is a genus of free living, asexually reproducing, anamorphic and filamentous fungi (Kubicek *et al.*, 2002; Monte and Llobell, 2003; Harman *et al.*, 2004) generally characterized by rapid growth, mostly bright green conidia and a repetitively branched conidiophore structure (Saba *et al.*, 2012). In the genus of *Trichoderma* spp., there are 89 species (Samuels, 2006) and today, about 104 (internationally) and 13 (from India) have been recorded and isolated from various substrates and locations with the high levels of diversity (www.isth.info.in; Pandya *et al.*, 2011). The genus *of this* fungal species was first proposed and described in Germany as a genus by Persoon (1794) over 200 years ago in the end of 17th century and in India, it was first time isolated by Thakur and Norris during the year 1928 from Madras (Pandya *et al.*, 2011).

Trichoderma spp., are soil-borne, green-spored ascomycetes, ubiquitous in nature with worldwide occurrence (Saba *et al.*, 2012), are commonly associated and highly interactive with soil ecosystems, root ecosystems and other forms of plant organic debris (Ranasingh *et al.*, 2006) and also easily isolated from them (Howell, 2003). The

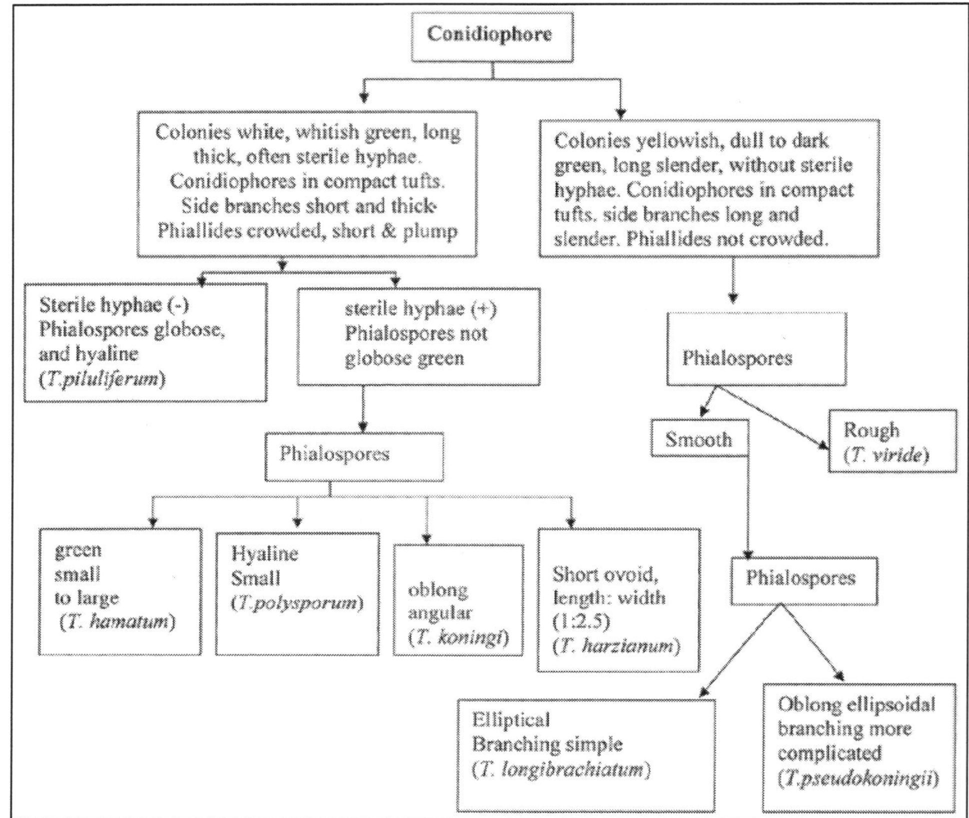

Figure 14.3: Classification of *Trichoderma* spp. Based on Morphology of Conidia and Conidiophores.

Source: Ranasingh *et al.* (2006).

review of Harman *et al.* (2004) report that *Trichoderma* spp. are opportunistic, avirulent plant symbionts, as well as being parasites of other fungi with direct confrontation. Primarily, they are isolated from soil and decomposing organic matter (Monte and Llobell, 2003) but often considered as most frequently isolated soil fungi and presented nearly in all types of soils *viz.* agriculture, forest, prairie, salt marsh and desert soils etc. of temperate and tropical climatic zones (Danielson and Davey, 1973; Domsch *et al.*, 1980; Harman *et al.*, 2004).

Trichoderma spp. has been found effective against aerial, root and soil pathogens (Weller, 1988; Whipps *et al.*, 1993; Elad *et al.*, 1998; Elad, 2000; Chaube *et al.*, 2002; Harman *et al.*, 2004) and have long been recognized as biological agents by many researchers (Papavizas, 1985; Chet, 1987; Latunde-Dada, 1991; Shamim *et al.*, 1997; Harman, 1998; Adekunle *et al.*, 2001; Howell, 2003; Ranasingh *et al.*, 2006; Saba *et al.*, 2012) for the control of crop diseases due to their effective antagonists against the pathogens (Papavizas, 1985, Chet, 1987; Kumar and Mukerji, 1996; Ha, 2010), for their ability to increase plant root growth and development, crop productivity, abiotic

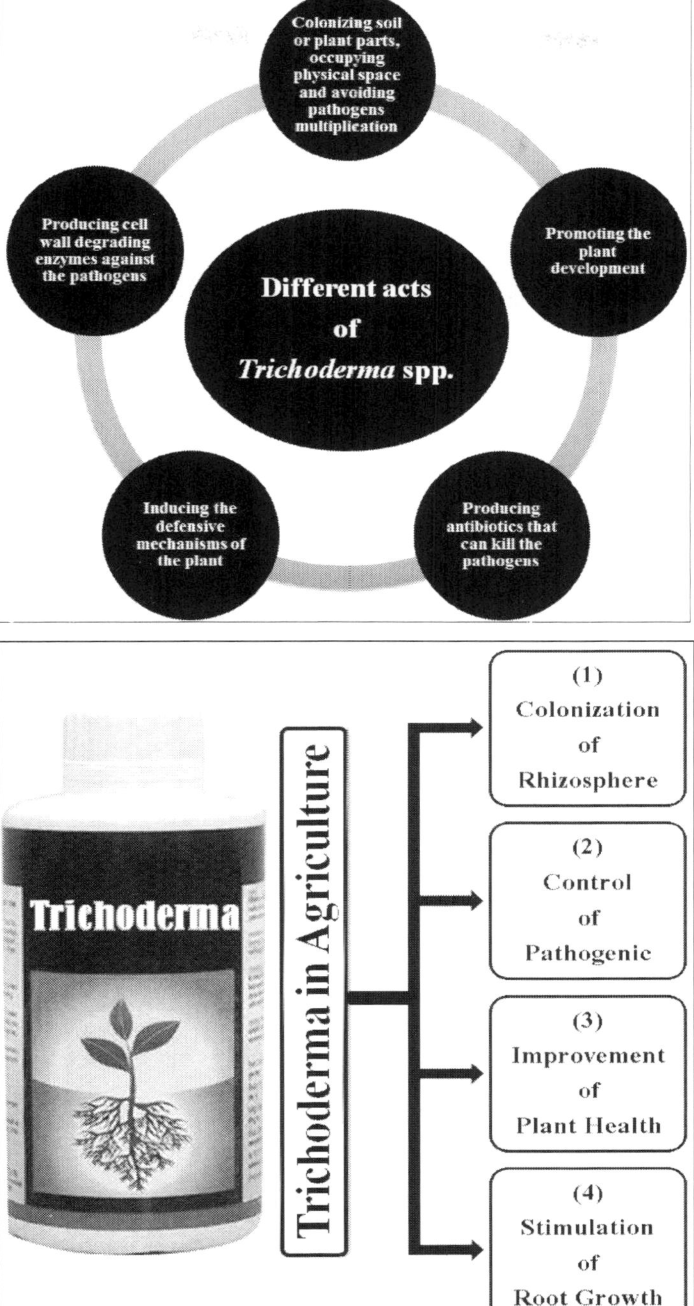

Figure 14.4: *Trichoderma* spp. as Biological Agents Due to their different Acts (Monte and Llobell, 2003) and Advantages Use in Agriculture (Harman *et al.,* 2004).

stresses resistance, and uptake and use of nutrients (Monte and Llobell, 2003; Harman *et al.*, 2004; Ranasingh *et al.*, 2006; Saba *et al.*, 2012) and much more indicated in Figure 14.3. These spices became popular biological agents due to their potential to protect crops against their pathogens all over the world (Howell, 2003; Ha, 2010) and are becoming widely used in agriculture and horticulture (Howell, 2003; Harman *et al.*, 2004), since the first recognized application in the early of 1930, and have also been controlled and added many diseases to the list in subsequent years (Howell, 2003).

The biological control agent, *Trichoderma* spp. are known as the most commonly used (Papavizas, 1985, Chet, 1987; Kumar and Mukerji, 1996) as an alternative to the chemical control against a wide range of fungal plant pathogens (Chet, 1987; Harman and Björkman 1998; Harman, 2006). Today, there are numerous strains of these fungi are *viz. T. harzianum, T. viride, T. hamatum* etc. have been developed and extensive studies on still progress to made these microbes versatile model organisms for research (Saba *et al.*, 2012). *Trichoderma* spp. have been culminated in the commercial production in the United States, India, Israel, New Zealand, and Sweden for the protection and growth enhancement of a number of crops (McSpadden Gardener *et al.*, 2002) due to their well-known opportunistic plant symbionts nature and direct confrontation mechanisms with effective mycoparasitism against plant pathogenic agents (Papavizas, 1985; Harman and Kubicek, 1998; Howell, 2003; Saba *et al.*, 2012). Mycoparasitism describes the complex process in which several events are *viz.* recognition of the host, attack and subsequent penetration and killing are involved (Vinale *et al.*, 2008). *Trichoderma* spp. releases or produces a variety of chemical compounds that induce to systemic resistance responses in plants (Ranasingh *et al.*, 2006) but have no harmful effects on humans, wildlife and other beneficial organisms which contribute towards the control of plant pathogen, in addition of moderate effects on soil balance. They are really very effective and safe in both natural and controlled environments and never accumulate in the food chain (Saba *et al.*, 2012).

Biological Control and Sustainable Agriculture

Sustainable agriculture means keep in existence; keep up; maintain or prolong etc. many goals encompasses in literally meaning. It is the act of farming which can understand an ecosystem approach to agriculture by using ecology principles and the study of relationships between organisms and their environment (Altieri, 1995). Simply defined, sustainable agriculture is a gateway to agriculture that contributes to the livelihood of communities and does not degrade to the environment during producing food. In words of Lewandowski *et al.* (1999) sustainable agriculture can fulfil to the significant ecological, economic and social functions at the different levels *viz.* local, national and global and also never harm to other ecosystems in addition of the agricultural because it focussed on to maintains biological diversity, agricultural productivity, regeneration capacity, vitality and ability to function etc. through mannerly management and utilisation of the agricultural ecosystem (Rodrigues *et al.*, 2003).

Today, more food and income are needed for growing population especially from agriculture and agricultural related activities have been the most significant challenges faced by our sustainable agriculture (Dordas, 2009). According to the United Nations (UNs) and Food and Agricultural Organization (FAO), often asserted in the next few decades about 9 billion people will approach from the global population where is a need for 70–100 per cent more food (Godfray *et al.*, 2010). In such a way, significant losses are increasingly being affected by various problems such as diseases, pests, droughts, decreased soil fertility due to use of hazardous chemical pesticides, pollution and global warming in agricultural yield and quality of crops due to the traditional agricultural practices worldwide. In addition of all, plant diseases continue to play a major limiting role in agricultural production which cannot fulfil to that particular global demand. Thus, there is a need for some eco-friendly management of crop disease through biological control agents for sound plant health because it is key to sustainable agriculture and may help to resolve some of these problems. In recent years the importance of biological control has risen to become one of the most important issues in sustainable agriculture. Due to nature-friendly and ecological approach, the use of specific microorganisms have dictated the need for alternative CDM techniques that interfere with plant pathogens and also made to solve and overcome to the serious concerns about food safety, environmental quality and standard chemical methods of CDM (Harman *et al.*, 2004). To achieve the goal of double food production from global agriculture by 2050 in order to feed the worldwide growing population and at the same time reduce its reliance on synthetic chemicals, there is an urgent need of biological control to harness the multiple beneficial interactions occurs between plants and microorganisms to solve CDM and global food demands.

Environmental Security and Biological Control

The value of nature's services to agricultural production and also the importance of maintaining their viable environment have long considered by ecologists and environmental scientist not only, although plant pathologists has been stressed environmental influences with emphasize of disease triangle (Figure 14.5) in their study of CDM through biological control (Krupinsky *et al.*, 2002; Garrett, 2008; Grulke, 2011). In the context of environmental security, if biological agents are very much essential and urgently needed to their biological and ecological study due to continue ensue of safety problems and ecological disruptions. Toward the environment and environmental security, considerable effort has been directed by biological control with the help of different discipline of sciences such as modern chemistry, molecular biology, biotechnology, genetic engineering etc. as the new technology which has been implemented to replace hazardous chemicals and or nontoxic to environments and others organism including humans but still emerging.

Future Perspectives, Opportunities and Challenges

In the biological control of CDM, the biological agents are fundamentally very much different from synthetic chemical used against crop pathogen and as they grow and proliferate effectively. Therefore, the effective antagonist of biological agents under favourable conditions in crop ecosystem is to establish the active action against targeted pathogens (Caldwell, 1958; Lewis and Papavizas, 1984). The biological control emerging tools in which the agents are receiving much more attention due to

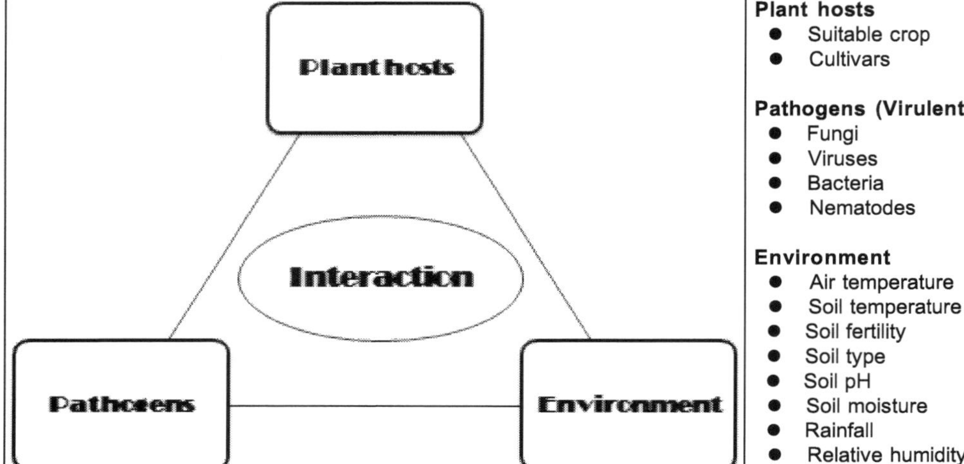

Figure 14.5: Disease Triangle Demonstrates how Interactions between Plant Hosts, Pathogens and Environment Cause Disease.

Concept adopted from Krupinsky *et al.* (2002)

the importance of their biological control strength of plant pathogens and for environmental protection of natural ecosystems. The biological control holds promise with the future to provide solutions for both, disease problems affecting worldwide agricultural production and to needs for environmental security in natural ecosystems. In addition of future prospective and opportunities, there are several challenges rising and facing, and they are not allaying to the general public concerns, not satisfying to the regulatory agencies and also not promoting to the commercial acceptance in CDM. Some of them are:

1. To select suitable naturally occurring strains of biological control agents or create them through genetic engineering by incorporating useful genes to made them ecologically sustainable as well as socially acceptable (Kerr, 1987).
2. Modifications in farm equipments and practices to accommodate biological treatment as one component in CDM.
3. The quality control of commercial biological control agents should be strictly enforced which ensure that their bio-formulation products are not treating as well as hazards chemicals.
4. For considering the environmental, social, economic as well as sustainable values, there are no any easier, earlier, and fast techniques for plant pathologists have been developed to detect the pathogens and also to postulate appropriate management practices.
5. Biological control move beyond the suppression of plant disease pathogen which include the impacts on human health and disasters because of no safety of biological control agents, as well as their efficacy.

Conclusions

The new era in the agriculture has begun with the implementation of biological control for CDM. Owing to inconsistent performance of biological agents in the field developed thus far has plagued researchers and their efforts to exploit them for commercial application. Biological control is rapidly developing as an important component of integrated disease management (IDM) because of its cost-effectiveness and the focus on environmental security. Biological control for CDM involves the participation of biological agents against crop pathogen. Commonly used biological agents for CDM are microbes (like bacteria, fungi, virus etc.) or their genes or their products. In the last 10 years our understanding regarding biological control mechanisms and plant responses has increased tremendously. The choice of the correct biological candidate is indeed as one of the most important factors governing the success of biological control programs on the commercial basis. *Bacillus* spp., *Pseudomonas* spp. and *Trichoderma* spp. are reliable candidates as they are highly effective against crop pathogen to limit the yield loss. This emerging tool is currently receiving inadequate attention because of its significant role in sustainable agriculture. As it is illustrated by numerous research studies that biological control has given new dimensions in the field of agriculture as it is eco-friendly and complementary to environmental sustainability. The research towards biological control improvement and implementation is gaining momentum. There is a need of enthusiastic attempt of disseminating knowledge about biological control to increase production efficiency. It is assumed cost-effective formulations of biological agents that perform consistently in their fields when made available, either by themselves or as part of an integrated disease-management (IDM) package, will be revolutionary innovation in the field of agriculture than never before. Further research is needed into the basic biology of biological control agents and plant pathogens in order to understand their ecology, physiology, biochemistry, and genetics necessary to predict the behaviour of these agents under different environmental conditions and to develop reliable biological control systems.

References

Adams, P.B., 1989. Comparison of antagonists of *Sclerotinia* species. *Phytopathol.*, **79**: 1345–1347.

Adams, P.B. and Ayers, W.A., 1981. *Sporidesmium sclerotivorum*: Distribution and function in natural biological control of sclerotial fungi. *Phytopathol.*, **71**: 90–93.

Adekunle, A.T., Cardwell, K.E., Florini, D.A. and Ikotun, T., 2001. Seed treatment with *Trichoderma* species for control of damping-off of cowpea caused by *Macrophomina phaseolina*. *Biocont. Sci. Technol.*, **11**: 449-457.

Altieri, M.A., 1995. *Agroecology: The Science of Sustainable Agriculture.* Westview Press, Boulder, CO.

Anonymous, 1993. *Production Year-Book,* Food and Agriculture Organization (FAO), Rome.

Arnesen, L., Stenfors, P., Fagerlund, A., and Granum, P.E., 2008. From soil to gut: *Bacillus cereus* and its food poisoning toxins. *FEMS Microbiol. Rev.*, **32**: 579-606.

Asaka, O. and Shoda, M., 1996. Biocontrol of *Rhizoctonia solani* damping-off of tomato with *Bacillus subtilis* RB14. *Appl. Environ. Biol.*, **62**: 4081–4085.

Aspiras, R.B. and de la Cruz, 1986. Biocontrol of bacterial wilt in tomato and potato through pre emptive colonization using *Bacillus polymyxa* FU- and *Pseudomonas fluorescens*. *Philipp. J. Crop Sci.*, **11**: 1–4.

Baker, K.F. and Cook, R.J., 1974. Biological control of plant pathogen, Willey Freeman, San Francisco, reprinted Edn. Am. Phytopathol. Soc., St. Paul, St. Paul, Minnesota, USA., p. 433.

Bandyopadhyay, R. and Cardwel K.F., 2003. Species of *Trichoderma* and *Aspergillus* as Biological Control Agents against Plant Diseases in Africa. *In*: Biological Control in IPM Systems in Africa (eds Neuenschwander P., Borgemeister C. and Langewald J.). CAB Inte., p. 193-206.

Beirner, B.P., 1967. Biological control and its potential. *World Rev. Pest Cont.*, **6**: 7–20.

Brahmbhatt, A.B., Mukhopadhyay, A.N. and Patel, K.K., 1989. *Trichoderma harzianum*, a potential bio-control agent for tobacco damping-off. *PKV Res. J.*, **13**: 170–172.

Brown, S.M., Kepner, J.L. and Smart, Jr G.C., 1985. Increased crop yields following application of *Bacillus penetrans* to field plots infested with *Meloidogyne incognita*. *Soil Biol Biochem.*, **17**: 483–486.

Bull, C.T., 2001. Biological control. *In*: *Encyclopedia of Plant Pathology* (eds. Malloy, O.C. and Murray, T.D.). Vol.-1, J. Wiley and Sons, New York, p. 128–135.

Caldwell, R., 1958. Fate of spores of *Trichoderma viride* Pers. Ex. Ft. introduced in the soil. *Natu.*, **181**: 1144 -1145.

Chaube, H.S., Mishra, D.S., Varshney, S. and Singh, U.S., 2002. Biological control of plant pathogens by fungal antagonistic: Historical background, present status and future prospects. *Annu. Rev. Plant Pathol.*, **2**: 1-42.

Chen, S., Dickson, D.W. and Whitty, E.B., 1994. Response of *Meloidogyne* spp. to *Pasteuria penetrans*, fungi, and cultural practices in tobacco. *J. Nematol.* **26**: 620–625.

Chet, I., 1987. *Trichoderma:* Application, mode of action and potential as a biocontrol agent of soil borne plant pathogenic fungi. *In*: Innovative approaches to plant disease control (ed. Chet I.). John Wiley and Sons, New York. p. 137-160.

Chet, I. and Inbar, J., 1994. Biological control of fungal pathogens. *Appl. Biochem. Biotechn.*, **48**: 37–43.

Chun, W.W.C. and Shetty, K.K., 1994. Control of the silverscurf disease of potatoes caused by *Helminthosporium solani* Dur. and Mont. with *Pseudomonas corrugata* (abstr). *Phytopathol.*, **84**: 1090.

Cole, J.S. and Zvenyika, J.S., 1988. Integrated control of *Rhizoctonia solani* and *Fusarium solani* in tobacco transplants with *Trichoderma harzianum* and triadimenol. *Plant Pathol.*, **37**: 271–277.

Cook, R.J., 2000. Advances in plant health management in the 20th century. *Ann. Rev. Phytopathol.*, **38**: 95-116.

Cook, R. J. and Baker, K.F., 1983. *The Nature and Practice of Biological Control of Plant Pathogens*. Am. Phytopathol. Soc., St. Paul, St. Paul, Minnesota, APS Press. USA.

Danielson, R.M. and Davey, C.B., 1973. The abundance of *Trichoderma* propagules and the distribution of species in forest soils. *Soil Biol. Biochem.*, **5**: 485-494.

Datnoff, L.E., Nemec, S. and Pernezny, K., 1995. Biological control of Fusarium crown and root rot of tomato in Florida using *Trichoderma harzianum* and *Glomus intradices*. *Biol. Control*, **5**: 427–431.

De La Cruz, A.R., Poplawsky, A.R. and Wiese, M.V., 1992. Biological suppression of potato ring rot by fluorescent pseudomonads. *Appl. Environ. Microbiol.*, **58**: 1986–1991.

De Meyer, G., Bigirimana, J., Elad, Y. and Hofte, M., 1998. Induced systemic resistance in *Trichoderma harzianum* T39 biocontrol of *Botrytis cinerea*. *Eur. J. Plant Pathol.*, **104**: 279–286.

Devaki, N.S., Bhat, S.S., Bhat, S.G. and Manjunatha, K.R., 1992. Antagonistic activities of *Trichoderma harzianum* against *Phythium aphanidermatum* and *Pythium myriotylum* on tobacco. *J. Phytopathol.*, 136: 82–87.

Domsch, K.H., Gams, W. and Anderson, T.H., 1980. *Compendium of Soil Fungi*. Acade. Press, London, p. 794-809.

Dordas, C., 2009. Role of Nutrients in Controlling Plant Diseases in Sustainable Agriculture. In: *Sustainable Agriculture* (eds. Lichtfouse E. *et al.*). Springer Science + Business Media, New York, p. 443-460.

Dube, B., and Smart, Jr. G.C., 1987. Biological control of *Meloidogyne incognita* by *Paecilomyces lilacinus* and *Pasteuria penetrans*. *J. Nematol.*, **19**: 222–227.

Duffy, B.K. and De´fago, G., 1997. Zinc improves biocontrol of Fusarium crown and root rot of tomato by *Pseudomonas fluorescens* and represses the production of pathogen metabolites inhibitory to bacterial antibiotic biosynthesis. *Phytopathol.*, **87**: 1250–1257.

Elad, Y., David, D.R., Levi, T., Kapat, A., Krishner, B., Guvrin, E. and Levine, A., 1998. Trichoderma harzianum, Trichoderma-39 mechanisms of biocontrol of foliar pathogens. In: *Modern fungicides and antifungal compounds II*. H. Lyr, Edn. Intercepts Ltd. Andover, Hampshire, UK p. 459-467.

Elad, Y., 2000. Biological control foliar pathogens by means of Trichoderma harzianum and potential modes of action. *Crop Protect.*, **19**: 709-714.

Elangovan, C. and Gnanamanickam, S.S., 1992. Incidence of *Pseudomonas fluorescens* in the rhizosphere of rice and their antagonism towards *Sclerotium oryzae*. *Indian Phytopathol.*, **45**: 358–361.

Euzéby, J.P., 1997. List of Bacterial Names with Standing in Nomenclature: a folder available on the Internet. *Inte. J. Syst. Bacteriol.*, **47(2)**: 590–592.

Fravel, D.R., Davis, J.R. and Sorenson, L.H., 1986. Effect of *Talaromyces flavus* and metham on Verticillium wilt incidence and potato yield, 1984–1985. *Biol. Cult. Cont. Plant Dis.*, **1**: 17.

Ganesan, P., 1999. Biological control of fusarium wilts of cotton and tomato with bacterial strains of *Pseudomonas fluorescens* and *P. putida*. *Ph.D. dissertation,* University of Madras.

Garrett, K.A., 2008. Climate change and plant disease risk. *In: Global climate change and extreme weather events: understanding the contributions to infectious disease emergence* (eds. Relman D. A., Hamburg M. A., Choffnes E. R. and Mack A.). National Acade. Press, Washington, DC. p. 143–155

Ghewande, M.P., Desai, S., Narayan, P. and Ingle, A.P., 1993. Integrated management of foliar diseases of groundnut (*Arachis hypogea* L.) in India. *Inte. J. Pest Manage.,* **39**: 375–378.

Giffel, M.C. te and Beumer, R.R., 1999. *Bacillus cereus:* a review. *J. Food Technol. Afri.,* **4**: 7-13.

Gnanamanickam, S.S., Vasudevan, P., Reddy, M.S., Kloepper, J.W. and Defago, G., 2002. Principles of Biological Control. *In: Biological Control of Crop Diseases* (ed. Gnanamanickam, S.S.). Marcel Dekker, Inc., New York, NY.

Godfray, H.C.J., Beddington, J.R., Crute, J.I., Haddad, L., Lawrence, D., Muir, J.F., Pretty, J., Robinson, S., Thomas, S. and Toulmin, C., 2010. Food security: The challenge of feeding 9 billion people. *Sci.,* **327**: 812–818.

Griffiths, M.W., 2010. Pathogens and toxins in foods: Challenges and interventions. ASM Press, Washington, DC. p. 1-19.

Grove, G.G. and Boal, R.J., 1997. Apple powdery mildew control trials using the mycoparasite *Ampelomyces quisqualis* (AQ10). *F and N Tests,* **52**: 7.

Grulke, N.E., 2011. The nexus of host and pathogen phenology: understanding the disease triangle with climate change. *New Phytologist,* **189**: 8–11.

Ha, T.N., 2010. Using *Trichoderma* species for biological control of plant pathogens in Vietnam. *J. ISSAAS,* **16**: 17-21.

Handelsman, J., 2002. Future Trends in Biocontrol. *In: Biological Control of Crop Diseases* (ed. Gnanamanickam, S.S.). Marcel Dekker, Inc., New York, NY.

Harman, G.E., 2006. Overview of mechanisms and uses of *Trichoderma* spp. *Phytopathol.,* **96**: 190–194.

Harman, G.E. and Björkman, T., 1998. Potential and existing uses of *Trichoderma* and *Gliocladium* for plant disease control and plant growth enhancement. *In: Trichoderma* and *Gliocladium* (eds. Kubicek, C.P. and Harman, G.E.). vol.-2., Taylor and Francis, London, p. 229-265.

Harman, G.E. and Kubicek, C.P., 1998. *Trichoderma* and *Gliocladium*. Taylor and Francis, London, p. 278.

Harman, G.E., Howell, C.R., Viterbo, A., Chet, I. and Lorito, M., 2004. *Trichoderma* species-Opportunistic, avirulent plant symbionts. *Nat. Rev. Microbio.,* **2**: 43–56.

Harwood, C.R., 1989. *Bacillus. In: Introduction to the Biotechnology of Bacillus* (ed. Harwood C. R.). vol.-2, Biotechnol. Handbooks, Plenum Press, New York, p. 1-4.

Heydari, A. and Misaghi, I.J. 1998. Biocontrol activity of *Burkholderia cepacia* against *Rhizoctonia solani* in herbicide-treated soils. *Plant Soil*, **202**: 109–116.

Howell, C.R., 1998. The role of antibiosis in biocontrol. *In*: *Trichoderma* and *Gliocladium*. Enzymes, Biological Control and Commercial Application (eds. Harman, G.E. and Kubicek, C.P.). vol. 2. Taylor and Francis Ltd., London, p. 173–183.

Howell, C.R., 2003. Mechanisms employed by *Trichoderma* species in the biological control of plant diseases: The history and evolution of current concepts. *Plant Dis.*, **87**: 4–10.

Hutchinson, S.W., 1998. Current concepts of act we defence in plants. *Ann. Rev. Phytopath.*, **36**: 59-90.

Inbar, J., Menendez, A. and Chet, I., 1996. Hyphal interaction between *Trichoderma harzianum* and *Sclerotinia sclerotiorum* and its role in biological control. *Soil Biol. Biochem.*, **28**: 757–763.

Janisiewicz, W.J. and Roitman, J., 1988. Biological control of blue mold and gray mold on apple and pear with *Pseudomonas cepacia*. *Phytopathol.*, **78**: 1697–1700.

Karthikeyan, A., 1996. Effect of organic amendments, antagonist *Trichoderma viride* and fungicides on seed and collar rot of groundnut. *Plant Dis. Res.*, **11**: 72–74.

Keel, C., Voisard, C., Berling, C.H., Kahr, G. and Defago, G., 1989. Iron sufficiency, a prerequisite for the suppression of tobacco black root rot by *Pseudomonas fluorescens* strain CHA0 under gnotobiotic conditions. *Phytopathol.*, **79**: 584–589.

Keinath, A.P., Fravel, D.R. and Papavizas G. C. 1991. Potential of *Gliocladium roseum* for biocontrol of *Verticillium dahliae*. Phytopathol., 81: 644–648.

Kerr, A., 1987. The impact of molecular genetics on plant pathology. *Annu. Rev. Phytopathol.*, **25**: 87–110.

Kerry, B.R., 2000. Rhizosphere Interactions and the Exploitation of Microbial Agents for the Biological Control of Plant-Parasitic Nematodes. *Ann. Rev. Phytopath.*, **38**: 423-441.

Khan, T.A. and Saxena, S.K., 1997. Integrated management of root knot nematode *Meloidogyne javanica* infecting tomato using organic materials and *Paecilomyces lilacinus*. *Bioresource Technol.*, **61**: 247–250.

Knudsen, G.R., Eschen, D.J., Dandurand, L.M. and Bin, L., 1991. Potential for biocontrol of *Sclerotinia sclerotiorum* through colonization of sclerotia by *Trichoderma harzianum*. *Plant Dis.*, **75**: 466–470.

Krupinsky, J.M., Bailey, K.L., McMullen, M.P., Gossen, B.D. and Turkington, T.K. 2002. Managing plant disease risk in diversified cropping systems. *Agron. J.*, **94**: 198-209.

Kubicek, C.P., Bissett, J., Druzfinina, L., Kulling, G. and Szakacs, G., 2002. Genetic and metablic diversity of *Trichoderma:* a case study on southeast Asian isolates. *Fungal Genet. Biol.*, **38**: 310-319.

Kumar, R.N. and Mukerji, K.G., 1996. Integrated disease management future perspectives. *In: Advances in Botany* (eds. Mukerji, K.G., Mathur, B., Chamala, B.P. and Chitralekha, C.). APH Publishing Corporation, New Delhi, p. 335-347.

Lashin, S.M., El-Nasr, H.I.S., El-Nagar, M.A.A. and Nofal, M.A., 1989. Biological control of *Aspergillus niger* the causal organism of peanut crown rot by *Trichoderma harzianum*. *Ann. Agric. Sci.*, **34**: 795–803.

Latunde-Dada, A.O., 1991. The use of *Trichoderma koningii* in the control of web blight disease caused by *Rhizoctonia solani* in the foliage of cowpea (*Vigna unguiculata*). *J. Phytopathol.*, **133**: 247-254.

Lewandowski, I., Hardtlein, M. and Kaltschmitt, M., 1999. Sustainable crop production: definition and methodological approach for assessing and implementing sustainability. *Crop Sci.*, **39**: 184–193.

Lewis, J.A. and Papavizas, G.C., 1984. Chlamydospores formation by *Trichoderma* spp. In natural substrates. *Can. J. Microbiol.*, **30**: 1-7.

Lewis, J.A. and Papavizas, G.C., 1991. Biocontrol of cotton damping-off caused by *Rhizoctonia solani* in the field with formulation of *Trichoderma* spp. and *Gliocladium virens*. *Crop Prot.*, **10**: 396–402.

M'Piga, P., Belanger, R.R., Paulitz, T.C. and Benhamou, N., 1997. Increased resistance to *Fusarium oxysporum* f. sp. *radicis lycopersici* in tomato plants treated with the endophytic bacterium *Pseudomonas fluorescens* strain 6328. *Physiol. Mol. Plant Pathol.*, **50**: 301–320.

Madigan, M. and Martinko, J., 2005. *Brock Biology of Microorganisms* (eds Madigan M. and Martinko J.). 11th Ed., Prentice Hall.

Manczinger, L., Antal, Z. and Kredics, L., 2002. Ecophysiology and breeding of mycoparacitic *Trichoderma* strains (a review). *Acta Microbiolo. Immunolo.* Hunga., **49**: 1-14.

Mari, M., Guizzardi, M., Brunelli, M. and Folchi, A., 1996. Postharvest biological control of grey mould (*Botrytis cinerea* Pers.: Fr.) on fresh-market tomatoes with *Bacillus amyloliquefaciens*. *Crop Prot.*, **15**: 699–705.

Mathivanan, N., Kabilan, V. and Murugesan, K., 1997. Production of chitinase by *Fusarium chlamydosporum*, a mycoparasite to groundnut rust, *Puccinia arachidis*. *Indian J. Exp Biol.*, **35**: 890–893.

Mathre, D.E., Cook, R.J. and Callan, N.W., 1999. From discovery to use: Traversing the world of commercializing biocontrol agents for plant disease control. *Phytopathol.*, **83**: 972-983.

McLaren, D.L., Huang, H.C. and Rimmer, S.R., 1996. Control of apothecial production of *Sclerotinia sclerotiorum* by *Coniothyrium minitans* and *Talaromyces flavus*. *Plant Dis.*, **80**: 1373–1378.

McSpadden Gardener, B.B. and Fravel, D.R., 2002. Biological control of plant pathogens: Research, commercialization, and application in the USA (http: / / naldc.nal.usda.gov / download / 12141 / PDF). Online, Plant Health Progress.

Menendez, A.B. and Godeas, A., 1998. Biological control of *Sclerotinia sclerotiorum* attacking soybean plants. Degradation of the cell walls of this pathogen by *Trichoderma harzianum* (BAFC 742). *Mycopathologia*, **142**: 153–160.

Meyer, S.L.F. and Meyer, R.J., 1996. Greenhouse studies comparing strains of the fungus *Verticillium lecanii* for activity against the nematode *Heterodera glycines*. *Fund. Appl. Nematol.*, **19**: 305–308.

Migula, W., 1894. Über ein neues System der Bakterien. Arb. Bakteriol. Inst. Karlsruhe., **1**: 235–328.

Migula, W., 1900. System der Bakterien. vol.-2. Jena, Germany: Gustav Fischer.

Mitchell, J.K. and Taber, R.A., 1986. Factors affecting the biological control of *Cercosporidium* leaf spot of peanuts by *Dicyma pulvinata*. *Phytopathol.*, **76**: 990–994.

Mondal, K.K., Singh, R.P. and Verma, J.P., 1999. Beneficial effects of indigenous cotton rhizobacteria on seed germinability, growth promotion and suppression of bacterial blight disease. *Indian Phytopathol.*, **52**: 228–235.

Monte, E. and Llobell, A., 2003. *Trichoderma* in Organic Agriculture. *In*: Proceedings V World Avocado Congress (Actas V Congreso Mundial del Aguacate), 10-14 Des., p. 725-733.

Morris, R.A.C., Coley-Smith, J.R. and Whipps, J.M., 1995. The ability of the mycoparasite *Verticillium biguttatum* to infect *Rhizoctonia solani* and other plant pathogenic fungi. *Mycol. Res.*, **99**: 997–1003.

Narayanasamy, P., 2013. Introduction. *In*: *Biological Management of Diseases of Crops* (ed. Narayanasamy P.). vol.-2, Integration of Biological Control Strategies with Crop Disease Management Systems. Springer Science + Business Media, New York, p. 1-6.

O'Neill, T.M., Niv, A., Elad, Y. and Shtienberg, D., 1996. Biological control of *Botrytis cinerea* on tomato stem wounds with *Trichoderma harzianum*. *Eur. J. Plant Pathol.*, **102**: 635–643.

Oka, Y., Chet, I. and Spiegel, Y., 1993. Control of the root knot nematode *Meloidogyne javanica* by *Bacillus cereus*. *Biocontrol. Sci. Technol.*, **3**: 115–126.

Ouda, S.M., 2014. Biological control by microorganisms and ionizing radiation. *Inte. J. Adv. Res.*, **2**: 314-356.

Pal, K.K. and Gardener, B.M., 2006. Biological Control of Plant Pathogens (https: / / www.apsnet.org / edcenter / advanced / topics / Documents / PHI-BiologicalControl.pdf). *The Plant Health Instructor*, p. 1-25.

Pandya, J.R., Sabalpara, A.N. and Chawda, S.K., 2011. *Trichoderma*: a particular weapon for biological control of phytopathogens. *J. Agric. Technol.*, **7(5)**: 1187-1191.

Papavizas, G.C., 1985. *Trichoderma* and *Gliocladium*: biology, ecology, and potential for biocontrol. *Ann. Rev. Phytopathol.*, **23**: 23–54.

Pereira, J. and Dhingra, O.D., 1997. Suppression of *Diaporthe phaseolorum* f. sp. *Meridionalis* in soybean stems by *Chaetomium globosum*. *Plant Pathol.*, **46**: 216–223.

Persoon, C.H., 1794. Neuer Veersuch einer systematischen Eintheilung der Schwamme (*Dispositio methodica fungorum*). *Romer's News Mag. Bot.*, **1**: 63-128.

Podile, A.R. and Kishore, G.K., 2002. Biological Control of Peanut Diseases. *In*: *Biological Control of Crop Diseases* (ed. Gnanamanickam S. S.). Marcel Dekker, Inc., New York, NY.

Podile, A.R., Kumar, B.S.D. and Dube, H.C., 1988. Antibiosis of rhizobacteria against some plant pathogens. *Indian J. Microbiol.*, **28**: 108–111.

Press, C.M., Wilson, M., Tuzun, S. and Kloepper, J.W., 1997. Salicylic acid produced by *Serratia marcescens* 90–166 is not the primary determinant of induced systemic resistance in cucumber or tobacco. *Mol. Plant-Microbe Interact.*, **10**: 761–768.

Priest, F.G., 1993. Systematics and ecology of Bacillus. *In*: Bacillus subtilis and other Gram positive bacteria - Biochemistry, physiology, and molecular genetics (eds. Sonenshein A. L., Hoch J. A., Losick R.). ASM press, Ame. Soci. Microbiol., Washington, D.C.

Racke, J. and Sikora, R.A., 1992. Isolation, formulation and antagonistic activity of rhizobacteria towards potato cyst nematode *Globodera pallida*. *Soil Biol. Biochem.*, **24**: 521–526.

Ranasingh, N., Saurabh, A. and Nedunchezhiyan, M., 2006. Use of *Trichoderma* in Disease Management (http: / /orissa.gov.in / e-magazine / Orissareview / sept-oct2006 / engpdf / 68-70.pdf). *Orissa Rev.*, p. 68-70.

Reitz, M., Rudolph, K., Schroder, I., Hoffmann-Hergarten, S., Hallmann, J. and Sikora, R.A., 2000. Lipopolysaccharides of *Rhizobium etli* strain G12 act in potato roots as an inducing agent of systemic resistance to infection by the cyst nematode *Globodera pallida*. *Appl. Environ. Microbiol.*, **66**: 3515–3518.

Rio, L.D., Martinson, C.A. and Yang, X.B., 1998. Control of Sclerotinia stem rot of soybeans with *Sporidesmium sclerotivorum*. Proceedings of the 1998 Inte. Sclerotinia Workshop, Fargo, ND, p. 64–65.

Ristaino, J.B., Perry, K.B. and Lumsden, R.D., 1991. Effect of solarization and *Gliocladium virens* on sclerotia of *Sclerotium rolfsii*, soil microbiota, and the incidence of southern blight of tomato. *Phytopathol.*, **91**: 1117–1124.

Roberts, R.G., 1990. Postharvest biological control of gray mold of apple by *Cryptococcus laurentii*. *Phytopathol.*, **80**: 526–530.

Rodrigues, G.S., Campanhola, C. and Kitamura, P.C., 2003. An environmental impact assessment system for agricultural R&D. Environ. *Impact Assess. Rev.*, **23**: 219–244.

Rosales, A.M., Vantomme, R., Swings, J., De Ley, J. and Mew, T.W., 1993. Identification of some bacteria from paddy antagonistic to several rice fungal pathogens. *J. Phytopathol.*, **138**: 189–208.

Ruger, H.J., 1989. Benthic studies of the northwest African upwelling region of psychrophilic bacterial communities from areas with different upwelling region psychrophilic bacterial communities from areas with different upwelling intensities. *Marine Ecol. Progr. Seri.*, **57**: 45-52.

Saba, H., Vibhash, D., Manisha, M., Prashant, K.S., Farhan, H. and Tauseef, A., 2012. *Trichoderma* - a promising plant growth stimulator and biocontrol agent (http: / / mycosphere.org / pdfs / MC3_4_No14.pdf). *Mycosphere*, **3**: 524–531.

Sabet, K.K., Mostafa, M.A., El-Bana, O.H.I. and El-Sherif, E.M., 1992. *Pseudomonas lindbergii* and *Coniothyrium minitans* as biocontrol agents effective against some soil fungi pathogenic to peanut. *Egypt J. Agric. Res.*, **70**: 403–414.

Sakthivel, N., 1987. Biological control of *Sarocladium oryzae* (Sawada) Gams and Hawksworth, sheath-rot pathogen of rice by bacterization with *Pseudomonas fluorescens* migula. *Ph.D. Dissertation*, University of Madras.

Sakthivel, N. and Gnanamanickam, S.S., 1987. Evaluation of *Pseudomonas fluorescens* for suppression of sheath rot disease and for enhancement of grain yields in rice, *Oryza sativa* L. *Appl. Environ. Microbiol.*, **53**: 2056–2059.

Samuels, G.J., 2006. *Trichoderma*: Systematics, the Sexual State, and Ecology. *Phytopathol.*, **96**: 195–206.

Schallmey, M., Singh, A. and Ward, O.P., 2004. Developments in the use of *Bacillus* species for industrial production. *Can. J. Microbiol.*, **50**: 1–17.

Shamim, S., Ahmad, N., Rahaman, A., Haque, S.E. and Chaffer, A., 1997. Efficacy of *Pseudomonas aeruginosa* and other biocontrol agents in the control of root rot infection in cotton. *Acta Agrobotan.*, **50**: 5–10.

Sharga, B.M. and Lyon, G.D., 1998. *Bacillus subtilis* BS 107 as an antagonist of potato blackleg and soft rot bacteria. *Can. J. Microbiol.*, **44**: 777–783.

Singh, D., 1991. Biocontrol of *Sclerotinia sclerotiorum* (Lib.) de Bary by *Trichoderma harzianum*. *Trop. Pest Manage.*, **37**: 374–378.

Singh, N., 1994. *Trichoderma harzianum* and *Chaetomium* sp. as potential biocontrol fungi in management of red rot disease of sugarcane. *J. Biol. Cont.*, **8**: 65–67.

Sivan, A., Elad, Y. and Chet, I., 1984. Biological control of *Pythium aphanidermatum* by a new isolate of *Trichoderma harzianum*. *Phytopathol.*, **74**: 498–501.

Sivan, A., Ucko, O. and Chet, I., 1987. Biological control of Fusarium crown rot of tomato by *Trichoderma harzianum* under field conditions. *Plant Dis.*, **71**: 587–592.

Spiegel, Y., Cohn, E., Galper, S., Sharon, E. and Chet, I., 1991. Evaluation of a newly isolated bacterium *Pseudomonas chitinolytica* sp. nov., for controlling the root-knot nematode *Meloidogyne javanica*. *Biocon. Sci. Technol.*, **1**: 115–125.

Spink, D.S. and Rowe, R.C., 1989. Evaluation of *Talaromyces flavus* as a biological control agent against *Verticillium dahliae* in potato. *Plant Dis.*, **73**: 230–236.

Stanier, R.Y., Palleroni, N.J. and Doudoroff, M., 1966. The aerobic pseudomonads. *J. Gen. Microbiol.*, **43**: 159-271.

Stirling, G.R., 1984. Biological control of *Meloidogyne javanica* with *Bacillus penetrans*. *Phytopath.*, **74**: 55–60.

Subrahmanyam, P., Reddy, P.M. and McDonald, D., 1990. Parasitism of rust, early and late leaf spot pathogens of peanut by *Verticillium lecanii*. *Peanut Sci.*, **17**: 1–4.

Thennarasu, R., 1997. Biological control of sett rot (*Ceratocyctis paradoxa*) of sugarcane. *M.Sc. (Agri.) dissertation,* Tamil Nadu Agricultural University, Coimbatore.

Trigalet, A. and Trigalet-Demery, D., 1990. Use of avirulent mutants of *Pseudomonas solanacearum* for the biological control of bacterial wilt of tomato plants. *Physiol. Mol. Plant Pathol.*, **36**: 27–38.

Truong, H.X., Salinas, M.D., Obien, A.S. and Carasi, R.C., 1988. Biological control of *Sclerotium rolfsii* and *Pythium aphanidermatum* damping off on tobacco with *Trichoderma harzianum* culture. *Brighton Crop Protection Conference: Pests and Diseases*, Brighton, p. 1189–1194.

Turnbull, P.C.B., 1996. *Bacillus.* In: *Barron's Medical Microbiology* (eds. Baron S. *et al.*,) (http: / / www.ncbi.nlm.nih.gov / books / bv.fcgi?rid=mmed.section.925). 4th Ed., University of Texas, Medical Branch.

Ukey, R.C. and Meshram, M.K., 2009. Biological Control in Plant Disease Management. *In*: Role of Biocontrol Agents for Disease Management in Sustainable Agriculture. Res. India Pub., p. 388–391.

Valasubramanian, R., 1994. Biological control of rice blast with *Pseudomonas fluorescens* Migula: role of antifungal antibiotics in disease suppression. *Ph.D. Dissertation,* University of Madras.

Van Dijk, K. and Nelson, E.B., 1998. Inactivation of seed exudate stimulants of *Pythium ultimum* sprorangium germination by biocontrol strains of *Enterobacter cloacae* and other seed-associated bacteria. *Soil Biol. Biochem.*, **30**: 183–192.

Vanneste, J.L., and Yu, J., 1996. Biological control of fire blight using *Erwinia herbicola* Eh252 and *Pseudomonas fluorescens* A506 separately or in combination. *Acta Hortic.*, **411**: 351–353.

Vinale, F., Sivasithamparamb, K., Ghisalbertic, E.L., Marraa, R., Wooa, S.L. and Moritoa, L., 2008. *Trichoderma*: Plant-pathogen interactions. *Soil Biol. Biochem.*, 40: 1–10.

Weller, D.M., 1988. Biological control of soil borne plant pathogens in the rhizosphere with bacteria. *Annu. Rev. Phytopathol.*, **26**: 379-469.

Whipps, J.M., McQquilken, M.P. and Budge, S.P., 1993. Use of fungal antagonists for biocontrol of damping-off and Sclerotinia disease. *Pestic. Sci.*, **37**: 309-313.

Yuen, G.Y., Craig, M.L., Kerr, E.D. and Steadman, J.R., 1994. Influences of antagonist population levels, blossom development stage, and canopy temperature on the inhibition of *Sclerotinia sclerotiorum* on dry edible bean by *Erwinia herbicola*. *Phytopathol.*, **84**: 495–501.

Zaki, K., Misaghi, I.J. and Heydari, A., 1998. Control of cotton seedling damping-off in the field by *Burkholderia* (*Pseudomonas*) *cepacia*. *Plant Dis.*, **82**: 291–293.

Zhang, L. and Birch, R.G., 1996. Biocontrol of sugarcane leaf scald disease by an isolate of *Pantoea dispersa* which detoxifies albicidin phytotoxin. *Lett. Appl. Microbiol.*, **22**: 132–136.

2016, Diseases of Pulse Crops and their Sustainable Management 269–286
Editors: Samir Kumar Biswas, Santosh Kumar and Gireesh Chand
Published by: BIOTECH BOOKS, NEW DELHI

Chapter 15

Aerial Blight of Soybean Caused by *Rhizoctonia solani* and Biological Management

Lalan Sharma[1]*, *Santosh Kumar*[2], *D.T. Nagrale*[1], *S.K. Singh*[3] *and Sanjeev Kumar*[4]

[1]*National Bureau of Agriculturally Important Microorganisms, Kusmaur, Maunath Bhanjan – 275101, U.P.*
[2]*Department of Plant Pathology, Bihar Agricultural University, Sabour, Bhagalpur – 813 210, Bihar*
[3]*Department of Plant Pathology, Anand Agricultural University, Anand, Gujarat*
[4]*Subject Matter Specialist (Plant Pathology), Krishi Vigyan Kendra, Sabour, Bhagalpur – 813 210, Bihar*

Introduction

Soybean (*Glycine max*. (L.) Merrill) is being a leguminous crop, considered to be a pulse crop but due to high content of oil (20 per cent), protein about 40 per cent and greater response to applied nitrogen levels it has now been placed in oil seed category. In fact, soybean is one of the nature's most efficient protein producers. It produces about three times more protein than rice, wheat and maize on the basis of per hectare production. It prefers warm moist climate and excessive dry, hot and cold weather is

* Corresponding Author: E-mail: drsharmanbaim@rediffmail.com

not congenial for the crop. It may be grown on a variety of soil type but a fertile, friable, well drained soil of neutral soil reaction is found to be better because high salinity or alkalinity inhibit the germination of the seeds. Thus loam soils to black cotton soils are used for crop cultivation. The seed-bed should be well pulverized, free from clods and perennial weeds, well leveled and should have sufficient moisture. It is mainly a kharif season crop but it may be grown during spring season, thus its sowing is done in June / July as kharif crop, February / March as spring crop on plains and in May / June on hills. The spring planting is common in southern states of India.

Soybean was domesticated by farmers in the plains of Northern China in about 500 to 1000 years ago. It was I[st] introduced into the United States in 1765, which is the leading country in soybean production. Now, it is widely planted and grown in more than 39 countries. The soybean has served as a stable diet for thousands of years in china. To the Chinese, Soybean was known as "the meat of the fields". In United States, due content of high nutritive value and various forms of uses, it is also known as "Gold of America". In India, it was introduced in 1960's and has emerged as an important oilseed crop. The major soybean growing states in India are Madhya Pradesh, Maharashtra, Rajasthan, Karnataka and Andhra Pradesh. Madhya Pradesh contributes about 65 per cent of the total area and around 57 per cent of total production in the country and therefore, it is known as "Soya State". During seventies, In India, soybean was considered to be a rather disease and pest free crop. But, with the rapid area expansion, year after year and season after season, continuous mono-cropping of the same cultivars in the same fields without any rotation and indiscriminate movement of seed materials etc, the scenario has altogether changed and at present mostly around hundred pathogens have been reported occur on soybean on different parts of country. Some of these have established themselves in different regions and has assumed of economic significance in telling into serious yield loss.

In India, during the year 2005 – 06, 22 diseases were reported across the country. Among them 9 diseases were of wide occurrence and rest was limited to some locations (NRCS, 2005 – 06). Among the major diseases Aerial Blight / Web blight of Soybean caused by *Rhizoctonia solani* Kuhn is a serious problem in Soybean production. The main characteristic symptoms of aerial blight of soybean are appearance as small water soaked spots on leaves.The spot become greenish to reddish brown and later turn brown to black in colour. In severe form whole leaf may be blighted. Due to fungus attacks aerial parts of the Soybean plant, losses 30-80 per cent chlorophyll formation and reduce photosynthesis process and ultimately go for reduction in yield and biomass production.

Occurrence and Distribution

Aerial blight of soybean was first reported by Sawada (1919) from Taiwan (Formosa) and was described the causal fungus as *Hypochnus sasakii*. Later on Matsumoto *et al.* (1932) described the physiology and parasitology of the fungus from Japan under the name *Hypochnus sasakii* Shrirai. Crall (1950) in Iowa recorded the disease in 1949, 1950 and 1951. Atkins and Lewis (1952) reported it from Batan Rouge, Louisiana (U.S.A.). They found that there was no resistant material and the susceptible cultivars varied in their degree of susceptibility. In 1954, they identified the causal organism as *Rhizoctonia microsclerotia* (Matz.) Weber. Cripsin and Gallegos (1963) reported the disease from Mexico naming it web blight. The fungus *R. microsclerotia* was originally described by Matz (1917) on fig plant. The synonyms of fungus are *Corticium microsclerotia* (Matz) Weber and *Pellicularia filamentosa* (Pat.) *Rogers and Weber (1939) and Rogers (1943). In 1987, natural infection of seedlings by *R. solani* AG-1, which produce only sasakii type sclerotia, was observed from Louisiana (U.S.A.) by *Yang *et al.* (1988). Jogoe *et al.* (1952) reported *Corticium solani* as the perfect stage of *Rhizoctonia solani* from Malaya for the first time on soybean. *Voelcker (1953) reported that *Corticium solani* caused leaf rot of soybean which was a new record for Malayasia. *Zevada *et al.* (1955) also reported the occurrence of *Pellicularia filamentosa,* the synonyms of the fungus is *Corticium microsclerotia,* on soybean in Mexico.

In India, the disease and the pathogen was first reported by Uppal *et al.* (1935) from Mirpurkhas (Bombay). Later on *Verma and Thapliyal (1976) reported aerial blight of soybean (*Rhizoctonia solani*) is an important disease during *kharif* season in warm humid areas like Pantnagar. Collar rot symptom on soybean due to *Thanatephorus cucumeris* described by Mathew and Nair, (1980) in Kerala. The disease assumed severe proportion in Rajasthan (Goyal and Ahmed, 1988). Srivastava and Gupta (1989) reported that *R. solani* was responsible for causing aerial blight on *Vigna mungo, V. radiata, Macrotyloma uniforum, Phaseolus vulgaris,* groundnut and soybean. Shukla *et al.* (1993) first time recorded severe leaf infections on *Coleus forskohlii* in experimental plantations in Lucknow, during July to September 1990 and 1991. *R. solani* was isolated from infected leaf and was confirmed pathogenicity by test. Dubey and Mishra (1994) reported that the pathogen causing web blight of sweet potatoes in Bihar was isolated and identified as *T. cucumeris.* Mehrotra (1998) recorded first time, the occurrence of leaf blight caused by *Rhizoctonia solani* (the anamorph of *Thanatephorus cucumeris*), as a new disease of *Cassia fistila, Bauhinia variegata, Dalbergia sissoo* and *Populus deltoides* in the nurseries in western Uttar Pradesh.

Economic Importance and Disease Severity

Soybean plants are infected by the pathogen at any stage of development, which causes very rapid defoliation and frequently crop failure (Crispin and Gallegoes, 1963). Fenille *et al.* (2002) reported 31-60 per cent yield losses due to foliar blight of soybean in north and northeast Brazil. Upto 50 per cent loss in yield is reported from United States (Tachibana *et al.*, 1971; Horn and Fontenot, 1980; Joye, 1986). Crop losses and epidemiology of *R. solani* on soybean were investigated in Madhya Pradesh, India. Disease reduced shoot length, pod and seed formation, total number of seeds and seed weight (Patel and Bhargava, 1998). The symptoms of aerial blight of soybean caused by *Rhizoctonia microsclerotia,* as leaf and pod spots, leaf blight, defoliation, stem and petiole lesions, cob web like mycelium and sclerotia developed over infected leaves were described by Atkins and Lewis (1954). The necrotic areas varied in size, earlier smaller spots and then covering entire leaflet. Plants get infected on stem near the soil surface. Lesions so produced were reddish brown coloured, oval to linear in shape. The pods attacked in young stages, developed brown necrotic areas or turned completely necrotic. The microsclerotia were oval to irregular, 159.9

μm (range 99.9-166.5 μm) in size, in aggregates consisting of chains or masses, produced in abundance along the petiole and few one on the diseased lesions of the infected leaves. Lesions produced by pathogen on leaf initially appeared as water soaked at the base of blade and petioles, subsequently, spread in a fan shaped manner to the rest of the leaf blade. Lesions later become necrotic and brown to tan in colours with thin reddish brown margin. Affected leaves curl and dry out (Verma and Thapliyal, 1976; O'Neill *et al.*, 1977).

The Pathogen

The causal agent of the disease has been reported in the literature to be *Corticium sasakii* (Shirai) Matsumota, *C. vagum* Berk and Curt, *Sclerotium irrgulare*, I. Miyake, *Hypochnus sasakii, Pellicularia sasakii* (Shirai) S.I to and *Rhizoctonia solani* Kühn. Researchers now accept *R. solani*, perfect stage *Thanatephorus cucumeris* (Frank) Donk, in the AG1 anastomosis series as the pathogen (O'neill *et al.*, 1977). The genus *Rhizoctonia* was first described by de Candole (1815). Since than, more than one hundred species of this fungus have been described. These species are distinguished from one another primarily on the basis of their morphology and pathogenicity. Akino and Ogoshi (1995) characterized genus *Rhizoctonia* as follows:

1. Branching near the distal septum of cells in young vegetative hypae
2. Formation of a septum in the branch near the point of origin
3. Constriction of the branch
4. Dolipore septum
5. No clamp connection
6. No conidium except moniloid cells
7. Sclerotium not differentiated into rind and medulla
8. No rhizomorph

Rhizoctonia isolates were differentiated into three groups. One is the multinucleate *Rhizoctonia*, which has three or more nuclei per cell, larger hyphae (6-10 μm diameter) and telomorph of the genus; *Thanatephorus* Donk. Another is the binucleate *Rhizoctonia*, which has only 2 nucleus per cell (rarely one or three), smaller hyphae (4-7 μm diameter) and telomorph on the genus *Ceratobasidium* Rogers. The third group includes *R. oryzae* and *R. zeae* which are multinucleate and have telomorphs in the genus *Waitea* Warcup and Tolbot (Ogoshi, 1987).

Rhizoctonia solani Kühn was reported in 1958 by German scientist Kühn on diseased potato tubers (Parmeter, 1970). Since then, this fungus has gained the reputation of being wide spread, destructive and versatile plant pathogen. It is capable of attacking a wide range of host plants, causing seed decay, damping off, stem cankers, root rots, aerial blights and fruit decay. It is the combination of its competitive saprophytic ability and high pathogenic potential that make *R. solani* a persistent and destructive plant pathogen. As the description by Kühn was insufficient and it was often difficult to determine whether or not a particular isolate was *R. solani*. Until, Duggar (1915) reported that *R. solani* was characterized as (a) the diameter of vegetative hyphae 8-12 μm; (b) constriction at the point of branching; and (c) right angle branching of mature hyphae.

The systematic positions of the sclerotial and basidial stages are as follows:

Sclerotial Stage

Subdivision	:	Deuteromycotina
Form-class	:	Deuteromycetes
Form-subclass	:	Agonomycetales
Genus	:	*Rhizoctonia*
Species	:	*solani* (Kühn) (Alexopoulus *et al.*, 1996)

Basidial Stage

Sub division	:	Basidiomycotina
Class	:	Basidiomycetes
Sub-class	:	Holobasidiomycitide (Hypomycetes)
Order	:	Tulasnellales
Family	:	Ceratobasidiaceae
Genus	:	*Thanetephorus*
Species	:	*cucumeris* (Frank) Donk

(Alexopoulus *et al.*, 1996)

Dasgupta *et al.* (1992) described the mycelium in culture initially as silvery becoming yellow and brown with maturity, 1-2 μ broad and infrequently septate. Three types of mycelium have been observed.

1. **Runner Mycelium**: Straight, creeping tropic, non-infectious hyphae or some times thick and flattened.
2. **Lobate Mycelium**: Branching out from the former as short, swollen, much branched, single or multiple lobate appressoria and penetration peg, which later becomes intracellular all through the lesions. The lobate state associated with the infection process may be counter part of the sporogenous moniloid state (Gangopadhyay, 1983).
3. **Moniloid Mycelium**: Forming the sclerotia. Ou (1985) reported that mycelium aggregates and forms sclerotia which are spherical, dark brown to black, 4-5 mm in diameter.

The sclerotia of *R. solani* are brown to black composed of clusters of melanin encrusted, thick walled cells, rich in nutrients, formed by repeated branching from short, thick, lateral hyphae, when produced on plant parts it is difficult to separate the sclerotia from their surrounding embeded sclerotia (Gangopadhyay and Chakraborti, 1982). Sclerotia are 4-5 mm in diameter, spherical but flattened when pressed between leaf sheath and culm. Five phases of sclerotial morphogenesis have been recognized in *R. solani*.

1. Repeated hyphal branching resulting in short, thick and lateral monoiloid hyphae.
2. Hyphal aggregation as clusters of thick walled cells and network formation.
3. Formation of sclerotial initial.
4. Formation of whitish immature sclerotia.
5. Maturation and pigmentation by mycelium encrustation (Morozimoto *et al.*, 1980).

Gangopadhyay and Chakrabarti (1982) reported the size of basidia as 11-15 μm × 8 - 9 μm, sterigmata as 7-10 μm × 2-3 μm and basidiospores as 9-12 μm × 5-7 μm. Dasgupta (1992) reported that basidiospores are 2-4 terminal in imperfect cymose or racemose clusters formed by branching of short celled ascending hyphae. Ogoshi (1987) first time reported anastomosis reactions to differentiate strain of *R. solani*. Four main groups, each a non-inter breeding population, were recognized by Talbot (1970). *Yokoyama and Ogoshi (1984) conducted an experiment to show that when isolates of *R. solani* are paried 2-3 cm apart on a medium (usually 2 per cent water agar) in a petridish, their mycelia grow and overlap, which can be observed under a light microscope at low magnification. If hyphal fusion occurs these isolates belongs to the same anastomosis group and often, attraction of hyphae and death of fused cells are observed. If fusion, attraction and hyphal death do not occur, the isolates belong to different anastomosis grouping (AGs). *Wang and Hsieh (1993) observed anastomosis behaviour of 12 isolates of *Rhizoctonia* spp. obtained from diseased stem of turf grass in Central and Southern parts of Taiwan. Ten isolates were binucleate and fell into six anastomosis groups (AG-Ba, Ag-C, Ag-F, AG-G, AG-L and AG-Q). The other two isolates were multinucleate and belonged to AG-4 and WAG-O. The major amastomosis groups were; rice sheath blight, *R. solani* AG-11A; maize sheath blight, AG-11 A and AG-4; sheath blight of wheat and barley binucleate *Rhizoctonia*, damping off of solanaceous vegetable seedlings, AG-4 ground nut leaves, root and shotos, Ag-4, Sheath Blight of soybean, AG-1 1B; potatoes, AG-3; crucifers, AG-2-1 in winters and AG-4 during other season, Vigna, Phaseolus, cucurbitaceae and solanaceae, usually AG-4 (Chen *et al.*, 1990). Nine fungus belonging to the anastomosis group AG-1, 1A of *R. solani* is the major cause of sheath blight on rice in Venezuela. Nuclear staining with HCL-Giemsa reveled that somatic cells were multinucleate. In distinct terminal and penullimate cells, the number of nuclei varied between strains CBS 250-84 of the anastomosis group AG-1-1A (Cenedo *et al.*, 1996). By hyphal anastomosis, 10 anastomosis group (AGs) have been reported in *R. solani* all represented by the same teleomorph. Each AG represents a monitor breeding population and a genetically independent entry. Again, within type 10 AGS of *R. solani*, by pathogenic variation and by DNA-DNA homology, interspecific group (ISGS) have been recognized (Kuninga and Yokosawa, 1982-85).

Ogoshi (1978) divided *R. solani* into 13 ISGs that differ in anastomosis behaviour, cultural appearance and pathogencity. They are AG-1 1A, AG-1 1B, AG-1 1C, AG-2-1, AG-2-2 IIIB, AG-2-2 IV, AG-3, AG-4, AG-5, AG-6, AG-7, AG-8 and AG-BI. The AG system now considered as the most useful grouping system for *R. solani*, however, the behaviour of this fungus can not be completely under stood solely in terms of AGs.

For example isolates of AG-1 cause not only arial / web blight but also sheath blight of rice or leaf blight on many hosts (Kuninaga, 1986). Muyolo *et al.* (1993) reported that anastomosis group AG-2-2-IIIB and AG-4 of *R. solani* were responsible for causing hypocotyl / root rot in case of *Phaseolus vulgaris* and soybean. Naito and Itoh (1999) reported that soybean root rot disease caused by *R. solani* AG-2-3 in Japan. Fenille *et al.* (2002) also reported the *R. solani* causes pre and post emergence damping-off, root and hypocotyl rot and foliar blight in soybean, resulted yield losses in Brazil. About 73 isolates of *R. solani* associated with soybean in Brazil were examined. Six were binucleate and 67 were multinucleate. Among the all isolates AG-1-IA caused foliar blight in adult soybean plants cv. Xingle (Fenile *et al.*, 2002). Botton rot caused by *R. solani* is an increasing problem in field grown lettuce in Germany. Ninty three isolates were identified and categorized in groups (AG)1-1B, one is AG-1-1C and one is AG-1-1C and one is AG-2-1, pathogenic potential at six AG-1-1b isolates was determined on 14 plant species (Grosch *et al.*, 2004). A new sub-group of *R. solani* proposed (AG-1-1D) which foliar disease on coffee in phillipines (Priyatmojo *et al.*, 2001). The fungus classified by nine anastomosis groups, designated AG, that differ in pathological and cultural characteristics (Anderson, 1982), AG-1-1b causes web blight (Yang *et al.*, 1990). The primary cause of disease is *R. solani* anastomosis group one *i.e.* AG-1 (Jones, *et al.*, 1989). Sclerotia of *R. solani* AG-1 have been extracted from soil in Japan (Naiki *et al.*, 1981).

Biological Control

There is growing concern in the entire world, to reduce the amount of chemicals being released in the environment and their persistence and accumulation over period of time in the environment (especially in soil and aquatic ecosystems) and to reduce their inherent hazards to plants and animal life. To increase world food production and feed the ever-increasing population, agriculture produce can be augmented with biological control instead of pesticides.

The micro-organisms which are used in controlling plant diseases are called as biocontrol agents. The biocontrol agents act by

1. Reduction of population of pathogen
2. Preventing the pathogen to infect the plants
3. Limiting disease development after inoculation

Biological control is more economical and often more effective method of disease control as it introduces the antagonist with the planting material (Cook and Baker, 1983). This approach may prove to be the most successful means of biological control of seedborne as well as soilborne pathogens. Biological control is the use to living agent to control plant pathogens (Mukhopadhyay and Chandra, 1996). Baker and Cook (1974) reported that majority of antagonists are saprophytes. Saprophytes are more broadly adapted to the environment and are able to use a greater range of nutrients than the plant parasites, a situation favorable for the biological control.

Successful use of fungal and bacterial biocontrol agents like *Trichoderma* and *Pseudomonas* spp for the control of soil borne diseases caused by pathogens like,

Rhizoctonia, Sclerotium, Fusarium, Pythium, and *Phytophthora* in several crops have been reported (Cook and Baker 1983). To date, a number of BCAs have been registered and are available as commercial products, including strains belonging to fungal genera such as *Gliocladium, Trichoderma, Ampelomyces, Candida* and *Coniothyrium and* bacterial genera such as *Agrobacterium, Pseudomonas, Streptomyces* and Bacillus. Biological control remains a very vital disease-management strategy for developing countries, where the costs of chemical treatments can be prohibitive.

A Case Study, Biological Disease Management

Experimental Design for Fungal and Bacterial Antagonists and Disease Occurrence

Experimental plot was prepared by giving one deep ploughing followed by three harrowing and leveling. Finally fine and well pulverized soil was prepared. Randomized block design (RBD) replicated three times of each treat was followed. There were 10 treatments including check. The plot size used was 3x3 m. To avoid effect of other treatment 50cm and 100cm distance were spaced between plot and replications respectively. In this method fungal and bacterial antagonist used as soil treatment, seed treatment and foliar spray separately or added with other material and without treated plot served as check.

T_1: Seed Treatment with *T. harzianum* @ 10g/Kg seed.
T_2: Seed Treatment with *P. fluoresens* @ 10g/Kg seed.
T_1: Soil Treatment with Pant Bioagent 3 mixed with FYM (1:100) @ 50g/plot.
T_4: $T_1 + T_3$
T_5: $T_2 + T_3$
T_6: T_4 + Foliar spray at *T. harzianum* @ 0.25 per cent
T_7: T_5 + Foliar spray at *T. harzianum* @ 0.25 per cent
T_8: T_4 + Foliar spray at *P. fluoresens* @ 0.25 per cent
T_9: T_5 + Foliar spray at *P. fluoresens* @ 0.25 per cent
T_{10}: Control

The germination of seedlings was recorded 15-20 days after sowing by counting the total number of emerged seedlings in all rows. Percent emergence for each plot was calculated by the following formula.

$$\text{Percent emergence} = \frac{\text{Total no. of emerged seedlings}}{\text{Total no. of seeds planted}} \times 100$$

The disease severity was recorded on 5 plants selected randomly from each plot. Three observations were recorded at fortnight interval starting from 01-10-2005 – 31-10-2005. Individual selected plants were tagged and divided in to bottom, middle and top and leaves (3 – 5) from each portion were graded as per the rating scale of 0 to 9 (Annual report on soybean, 2004 – 05). The rating scale is described as further:

Scale Grade	Description
0	No lesion/spot
0.1–1	Spots visible on the few plants upon very careful examination(1 per cent leaf area covered) only on new borne leaves.
1.1–3	Spots visible on the few plants upon careful examination (2-10 per cent leaf area covered), no spots on the stem.
3.1–5	Spots easily visible on many plants (11-25 per cent leaf area covered), no defoliation, little damage.
5.1–7	Spots present on almost all the plant parts (26-50 per cent leaf area covered), some leaves drop, death of few plants, damage conspicuous.
7.1–9	Spots very common on all plants (>50 per cent leaf area covered), defoliation common, death of plants common, damage directly reduce severe yield loss.

The numerical values were further used for the calculation of PDI (Percent Disease Index) using the formula:

$$\text{PDI} = \frac{\text{Sum of individual rating}}{\text{No. of leaves examined}} \times \frac{100}{\text{Maximum disease rating}}$$

For statistical analysis PDI was converted to angular transformed value.

The yield was recorded after 4-5 days of sun-drying of the soybean grain and the grain samples, from the produce of each plot were taken to determine 1000 grain weight. The yield data was analyzed statistically. In this, five randomly selected plants from each treatment were digged out and thoroughly washed the root with tap water. Plant height, weight of nodules, no of nodules, fresh root and shoot weight was measured.

Biological Control of Rhizoctonia Aerial Blight (RAB) through Introduced Fungal and Bacterial Antagonists

The experiment was performed in RBD replicated three times at each treatment. There were 10 treatments including one check. The bioagents *T. harzianum, Pseudomonas fluorescencs* and Pant bioagent-3 were used of seed treatment @ 10g/kg seed as well as foliar spray @ 0.25 per cent and pant bioagents (*T. harzianum* + *P. fluorescens*) mixed with FYM (1: 1000) @ 50 g/ha as soil application The crop (cv, PK-262) was planted on 15.07.2005 and terminated 12.11.2005. The spraying was done at 30 and 45 days after sowing. The minimum seeding mortality (0.46 per cent) was recorded in treatment 8 closely followed by 1.97 in treatment No 3. Other treatments also reduced the seedlings mortality as compared to check plots but the differences between treatment No. 6 while minimum germination (42.50 per cent) in treatment No 2. All the treated plots significantly reduced the disease severity and increased the yield as compared to check plot. However, treatment No. 8 found to be most effective by reducing maximum disease severity (PDI 6.17) as compare to other treatment. Maximum yield 1.90 kg/plot was achieved in treatment no 3 followed by treatment No. 6.

The maximum plant height was recorded in treatment No. 4 and 7 superior to others and increased no. of nodules was recorded in all treatments in comparison to check. Weight of nodules in all the isolate was recorded non-significantly different. The maximum fresh root weight was recorded in the treatment No. 6 (7.55 g) closely followed by treatment No. 4 (7.09g) along with maximum fresh shoot weight was recorded in treatment No. 9 (191.79) closely followed by treatment No. 6 (191.24 g) and treatment No. 2 (191.21 g). The weight of 1000 grains was recorded and observed that treatment No. 3 shown maximum (106.32 g) closely followed by treatment no 6 (104.51 g) and treatment No. 4 (104.48 g).

Conclusion

Soybean (*Glycine max* (L.) Merrill), the wonder crop contains approximately 40 per cent protein and 20 per cent low cholesterol free edible oil. The crop suffers from various diseases, among them aerial blight incited by *R. solani* Kuhn has assumed serious proportion in warm and humid agro-climate zones of the country.

In the present investigation, isolatic variations that occur under natural ecosystem has been studied. Incidentally this soil inhabitant attacks underground plant parts but in soybean all aerial parts (stem, branches, leaves, pods) are attacked. The attacked plant parts show scorched and blighted patches covered with aerial mycelial mats and sclerotial bodies. Two bioagents namely *fungal and bacterial bioagents* were tested against rhizoctonia solani. In totality, the results indicate that there are differences in pathogenic potential of the in the natural crop ecosystem and disease severity can be managed by judicious application of biocontrol agents. This fact must be taken into account when developing integrated disease management schedule.

References

Abdul Rahim, A.M. and Abu Surrieh, A.A. 1989. Biological control of *Rhizoctonia solani* the causal agent of seedling blight in Okra. *Arab J. Pl. Protec.*, **7**: 167-171.

Akino, S. and Ogoshi, A. 1995. Pathogenicity and host specificity in *Rhizoctonia solani.* In: *Pathogenesis and Host Specificity in Plant Disease.* Vol. IInd (ed. Keisuke Kohmoto, Uma, S. Singh and Rudra, P. Singh) Programon Science Ltd. Publication, U.K. p. 37-46.

Alexopoulus, C.J.; Mims, C.W. and Blackwell, M. 1996. Introductory Mycology, Wiley Eastern, New Delhi, p. 869.

Anderson, N. A. 1982. The genetics and pathology of *Rhizoctonia solani. Ann. Rev. Phytopathol.*, **20**: 329 – 347.

Atkins, J.G. and Lewis, W.D. 1952. Rhizoctonia aerial blight of soybean in Louisiana. *Phytopath*ology., **44**: 215-218.

Atkins, J.G. and Lewis, W.D. 1954. Rhizoctonia aerial blight of soybean in Louisiana. *Phytopathology*, **44**: 1.

Baker, K.F. and Cook, R.J. 1974. Biological control of plant pathogens W.H. Freeman and Co. San Francisco, p. 433.

Bora, K., Das, B.C. and Roy, A.K. 1999. Integrated management of sheath blight disease of rice with *Trichoderma harzianum* and chemical. *Oryza*, **36**: 238-240.

Bradley, C.A., Hartman, G.L., Mueller, D.S., Haffman, D.D., Nickell, A.D. and Pederson, W.L., 2005. Genetic analysis of resistance to Rhizoctonia solani in Soybean cultivar 'Savoy'. *Canadian Journal of Plant Pathology*, **27**: 137 – 142.

Carling, D.E. Helim, D.J. and Leiner, R.H. 1990. *In vitro* sensitivity of *Rhizoctonia solani* and other multinucleate and binucleate *Rhizoctonia* to selected fungicides. *Plant Dis.*, **74**: 860-863.

Chang Senk, P., Paulitz, T.C. and Baker, R., 1988. Biocontrol of Fusarium wilt of cucumber resulting from interactions between *Pseudomonas putida* and non pathogenic isolates of *Fusarium oxysoprium*. *Phytopathology*, **78**: 190-194.

Chen, J. S. Ge, G.X. and Zhang, B.X., 1990. Identification of Rhizoctonia solani on crops and in the related soil. *Acta Agric. Univ. Zhenjiangensis*, **16**: 219-224.

Chet, I., Elad, Y., Kalfon, A., Hader, Y. and Katan, J. 1982. Integrated control of soilborne and bulbborne pathogens in Iris. *Phytoparasitica*, **10**: 229-231.

Cook, R.J. and Baker, K.F., 1983. The nature and practice of biological control of plant pathogens. *St. Paul. Amer. Phytopathol. Soc.*, p. 359.

Cotterill, P.J. and Mclean, L.K., 1992. Evaluation of fungicides to control take all and rhizoctonia root rot of wheat. *Pl. Prot. Qucrt.*, **75**: 51-54.

Crall, J.M. 1950. Soybean diseases in Iowa in 1949. *Pl. Dis. Reptr.*, **34**: 96-97.

Crispin, A. and Gallegos, C.C., 1963. Web blight: A severe disease of beans and soybeans in Mexico. *Plant Dis. Rep.*, **47**: 1010-1011.

Cudom, M.A.; Mazza, S.M. and Gutierrez, S.A. 2003. Selection of *Trichoderma* spp. isolates against *Rhizoctonia solani*. *Spanish J. Agric. Res.*, **1(4)**: 79-82.

Dasgupta, M.K. 1992. Rice sheath blight: The challenge continues *In*: *Plant Diseases of International Importance, Disease of Cereals and Pulses*. Vol. I (eds. U.S. Singh, A.N. Mukhopadhyay, J. Kumar and H.S. Chaube) Prentice Hall. Engle Wood Cliffs, New Jersey. p. 130-157.

De Candolle, A.P. 1815. Memoir Swiles rhizoctones, houvean gnre de champignons quialtaqueless sicines, des plantes et in particular celle de la lazurene cultivee. *Mem Mus Natl. Hist. Nat. Ser.* **2**: 209-216.

Dhingra, O.D. and Sinclair, J.B. 1986. *Basic plant pathology methods*. CRC Press Inc. Boca Raton, Florida. p. 280-339.

Diaz-Polanco, C.; Maurezutt, P. and Salas-de-Diaz, G. 1978. *Rhizoctonia solani* pathogen on cowpea (*Vigna unguiculata*) in *Venezuela. Agronomica Tropical*. **28(4)**: 409-418.

Donk. M.A. 1956. Notes on resurpinate hymenomycetes. The Tulasnelloid Fungi. Reinwardita. **3**: 363-379.

Dubey, S.C. 2000. Biological management of web blight of groundnut (*Rhizoctonia solani*). *J. Mycol. Pl. Pathol.*, **30**: 89-90.

Dubey, S.C. 2002. Evaluation of *Gliocladium virens* and *Trichoderma viride* as foliar spray against web blight of urd and mung. *J. Mycol. Pl. Pathol.*, **32(2)**: 236-237.

Dubey, S.C. 2003. Integrated management of web blight of urd and mung bean. *Indian Phytopathol.*, **56(4)**: 413-417.

Dubey, S.C. and Dwivedi, R.P. 1988. Chemical control of web blight of black gram. *Farm Sci. J.*, **3**: 182-183.

Dubey, S.C. and Dwivedi, R.P. 1992. *Agric. Sci. Digest.*, **12**: 90-92.

Dubey, S.C. and Mishra, B. 1994. Web blight of sweet potato caused by *Thanatephorus cucumeris* (Fr.) Donk. *Indian J. Mycol. Pl. Pathol.*, **24**: 155.

Dubey, S.C. and Patel, B. 2001. Evaluation of antagonists against *Thanatephorus cucumeris* causing web blight of urd and mung bean. *Indian Phytopath.*, **54**: 206-209.

Dubey, S.C. and Prasad, R.B. 1997. Evaluation of different fungitoxicants against web blight of horse gram. *Indian J. Agric. Res.*, **31**: 161-166.

Duggar, B.M. 1915. *Rhizoctonia crocorum* (Pers.) D.C. *Rhizoctonia solani* Kuhn (*Corticium vagum* B and C) with notes on other species. *Ann. Missouri Botan. Garden*, **2**: 403-458.

Ehteshamul, H. S. and Ghaffer, A. 1995. Role of *Brady rhizobium japonicum* and *Trichoderma spp.* in the control of root rot disease of soybean. *Acta Mycologica*, **30(1)**: 35-40.

Elad, Y. and Chet, I. 1987. Possible role of competition for nutrients in biocontrol of *Pythium* damping-off by bacteria. *Phytopathology*, **77**: 190-195.

Elad, Y.I. Chet and Katan, J. 1980. *Trichoderma harzianum* biocontrol agent effective against *Sclerotium rolfsii* and *Rhizoctonia solani*. *Phytopathology*, **70**: 119-121.

Elliot, M.L., Jardine, E.A., Batson, W.E., Caceres, J. Jr., Caceres, P.M., Brannen, C.R., Howell, D.M., Benson, K.E., Conway, C.S., Rothrock, R.W., Schnieder, B.H., Ownle, C.H., Canaday, A.P., Keinath, D.M., Huber, D.R., Sumnber, C.E., Motsenbocker, P.M., Thaxton, M.A., Cubeta, P.D., Adams, P.A., Backman, J., Fajardo, M.A., Newman, R.M., Pereira and E.A., Jardine-des., 2001. Viability and stability of biocontrol agents on cotton and snap bean seeds. *Pest Mgt. Sci.*, **57(8)**: 695-706.

Escobar, P., Montealegre, J. and Herrera, R. 2004. *In vitro* response of *Trichoderma harzianum* strain to Fe^{3+}, salinity, pH and temperature, in order to be used in the biological control of *Rhizoctonia solani* and *Fusarium solani* in tomato. *Boletin Micologico*, **19**: 95-102.

Fadl, F.A. and Hessein, A.M. 1978. Root-rot disease of soybean in Egypt, Causal organisms and varietal resistance. *Agri. Res. Rev.*, **56**: 87-93.

Farzana, A.; Ghaffar, A. and Ali, F. 1991. Effect of seed treatment with biological antagonists on rhizosphere mycoflora and root infecting fungi of soybean. *Pakistan J. Bot.*, **23**: 183-188.

Fenille, R.C.; Souza, N.L.-De and Kuramae, E.E. 2002. Characterization of *Rhizoctonia solani* associated with soybean in Brazil. *European J. Pl. Pathol.*, **108**: 783-792.

Flentje, N. T. and H. M. Stretton (1964). Mechanism of variation in *Thanatephorus cucumeris* and *T. praticolus*. *Auot. J. Biol. Sci.*, **12**: 686–704.

Ganacharya, N.M. 1979. Effect of fungicidal seed treatment on emergence, nodulation and grain yield of soybean in Marathwada. *J. Maharashtra Agric. Univ.*, **4**: 112-113.

Gangopadhyay, S. 1983. Etiology and perpetuation of *Rhizoctonia solani* in India. *Curr. Sci.*, **114**: 852-855.

Gangopadhyay, S. and Chakravarti, N.K. 1982. Sheath blight of rice. *Rev. Pl. Path.*, **61**: 451-460.

Gautam, S.R. 1981. Studies on Bayleton and Bayton: In vitro stability systemicity and response in soybean, *Glycine max* (L.) Merill. Ph.D. Thesis, G.B.P.U.A.T., Pantnagar, p. 159.

Gomes, A.M.A., Peixoto, A.R., Mariono, R.L.R. and Michoreff, S.J. 1996. Effect of bean seed treatment with fluorescent *Pseudomonas* spp. on *Rhizoctonia solani* control. *Arquivos de Biologia Technolgia*, **39**: 537-545.

Goyal, J.P. and Ahmed, S.R. 1988. Rhizoctonia aerial blight of soybean- a new record from Rajasthan. *Indian J. Mycol. Pl. Pathol.*, **18**: 219.

Grosch, R. and Kofoet, A. 2003. Influence of temperature, pH and inoculum density on bottom rot on letuce caused by *Rhizoctonia solani*. *Zeitschrift-fur-Pflanzenkrankheiten-und-Pflanzenschutzt*, **110(4)**: 366-378.

Grosch, R., Schneider, J.H. and Kofoet, A. 2004. Characterization of *Rhizoctonia solani* anastomasis groups causing bottom rot in field grown letuce in Germany. *European J. Pl. Pathol.*, **110(1)**: 53-62.

Grover, R. K. and J. D. Moore (1966). Toximetric studies of fungicides against brown rot organisms *Sclerotinia fructicola* and *Sclerotinia lexa*. *Phytopathology*, **52**: 876–880.

Hader, Y., Harman, G.E. and Taylor, A.G. 1984. Evaluation of *Trichoderma koningii* and *T. harzianum* from New York soil for biological control of seed rot caused by *Pythium* spp. *Phytopathology*, **74**: 106-110.

Harman, G.E., Chet, I. And Baker, R. 1980. Factors affecting *Trichoderma hamatum* applied to seed as a biological agent. *Phytopathology*, **71**: 569-572.

Horn, N.L. and Fontenot, M.F. 1980. Aerial blight of soybean in Locisiana. *Proc. South Soybean Dis. Workers Conf.* **9**: 117.

Jain, R.K. and Thapliyal, P.N. 1980. Toxic metabolites from *Rhizoctonia solani* Kuhn: Production and possible role in pathogenesis. *Indian J. Expt. Biol.*, **18**: 316-318.

Jogoe, R.B. and Allen, E.F. 1952. Notes on current investigations, July to September 1952. *Malayan Agric. J.*, **35**: 218-227.

John Tuite. 1969. Plant pathological methods, fungi and bacteria. *Burgress Publishing Company*, p. 239.

Jones, R.K. and Belmer, S.B. 1989. Characterization and pathogenicity of *Rhizoctonia* spp. isolated from rice, soybean and other crops grown in rotation with rice in Texas. *Plant Dis.*, **73**: 1004-1010.

Joye, G.F. 1986. Management of Rhizoctonia aerial blight of soybean and biology of sclerotia of *Rhizoctonia solani* Kuhn. Ph.D. Thesis, Louisiana State Univ. Baton Rouge. p. 91.

Khan, A.A. and Sinha, A.P. 2005a. Comparative antagonisitc potential of some biocontrol agents against sheath blight of rice. *Indian Phytopath.,* **58(1)**: 41-45.

Khan, A.A. and Sinha, A.P. 2005b. Influence of different factors on the affectivity of fungal bioagents to manage rice sheath blight in nursery. *Indian Phytopath.,* **58(3)**: 289-293.

Khan, T.A. and Husain, S.I. 1991. *In vitro* studies on the toxicity of culture filterates of different fungi on the growth of *Rhizoctonia solani. New Agriculturist,* **1**: 107-110.

King, E. O., Wood, M. K. and Raney, D. E. 1954. Two simple media on the demonstration on Pyocianin and fluoricin. *J. Lab. Clinical Med.,* **44**: 301 – 307.

Kuhn, J.G. 1858. Die Krankheiten der Kulturegewachse, ihre ursachen und thre verhutung. Gustav Bosselmann, Berlin, p. 312.

Kumar, M. and Dubey, S.C. 2002. Relationship of disease intensity with weather and management of web blight of winged bean. *Indian Phytopath.,* **55(2)**: 152-157.

Kuninaga, S. 1986. Differentiation of *R.solani* by physiochemical characteristics. *Pl. Prot. Jpn.* p. 137–142.

Kuninaga, S. and Yokosama, R. 1984. DNA base sequence homology in R. solani kuhn IV, Genetic relatedness with in AG-4 *Ann. Phytopath. Soc. Japan,* **34**: 322-330.

Lewis, J.A. and Papavizas, G.C. 1987. Reduction of inoculum of *Rhizoctonia solani* in soil by germlings of *Trichoderma hamatum. Soil Biol. Biochem.,* **19**: 195-201.

Lim, T.K., Ng, C.C. and Chin, C.L. 1987. Etiology and control of durian foliar blight and die back caused by *Rhizoctonia solani. Ann. Appl. Biol.,* **11**: 301-307.

Maier, C.R. 1962. Response of selected *Rhizoctonia solani* isolates to different soil chemical tests. *Phytopathology,* **52**: 19.

Malik, G., Dawar, S., Saltar, A. and Dawar, A. 2005. Efficacy of *Trichoderma harzianum* after multiplication on different substrates in the control of root rot fungi. *Int. J. Biol. Biotech.,* **2(1)**: 237-242.

Manibhusan ao, K.; Sreenivasaprasad, S.; Babu, O.I. and Joe, Y. 1989. Susceptibility of rice sheath blight pathogen to mycoparasistes. *Curr. Sci.,* **58**: 515-518.

Mathew, A. and Gupta, S.K. 1996. Studies on web blight of french bean caused by *Rhizoctonia solani* and its management. *Indian J. Mycol. Pl. Pathol.,* **26(2)**: 171-177.

Mathew, A.V. and Nair, M.C. 1980. Collar rot of soybean-a new report form India (*Thanatephorus cucumeris*). *Curr. Sci.,* **49**: 158.

Matsumoto, T., Yamamoto, W. and Hirane, S. 1932. Physiology and parasitology of the fungi, generally referred to as *Hypochnus sasakii* Shirai. I. Differential media *Soc. Trop. Agric. (Taiwan). J.* **4**: 370-388.

Matsumoto, T., Yamamoto, W. and Hirane, S. 1932. Physiology and parasitology of the fungi generally referred to as *Hypochnus sasakii* (Shirai). II. Temperature and Humidity Relations. *Ibid.* **5**: 332-345.

Matz, J. 1997. Rhizoctonia of the fig. *Phytopathology*, **7**: 110-117

Mayer, C.R. 1962. Response of selected *Rhizoctonia solani* isolates to different soil chemical tests. *Phytopathology*, **52**: 19.

Mckinney, H. H. 1923. Influence of soil temperature and moisture on infection wheat seedling by *Halminthosporium sativum*. *J. Agric. Res.* **26**: 195 – 210.

Meena, P.D. and Chattopadhyay, C. 2002. Effect of some physical factors, fungicides on growth of *Rhizoctonia solani* Kuhn and fungicidal treatment on groundnut seed germination. *Indian J. Plant Prot.*, **30(2)**: 172-176.

Mehrotra, M.D. 1998. Rhizoctonia aerial blight- a destructive nursery disease and its management *Indian Forester*, **124(8)**: 637-645.

Mishra, D.K. and Narain, A. 1994. *Gliocladium virens and Streptoverticillium* as sources of biocontrol of few phytopathogenic fungi. *Indian Phytopath.*, **47**: 236-240.

Mishra, D.S. and Sinha, A.P. 1999. Laboratory evaluation of fungicides against *Rhizoctonia solani* Khun, the cause of sheath blight of rice. *Agric. Sci. Dig. Karnal.*, **19**: 21-213.

Moromizoto, Z., Matsugama, N. and Wakirnoto, S. 1980. Developmental process of sclerotium and its inhibition with several amino acids. *Annals of the phytopathologic J. Soc. of Japan*, **46**: 21-25.

Morton, D.J. and Stronbe, W.H. 1955. Antagonistic and stimulatory effect of soil microorganisms upon *Sclerotium*. *Phytopathology*, **45**: 417-420.

Mousa, A.A., Abe-El-Sayed, W.M. and El-Khali, M.M.A. 1997. Biological control of *Rhizoctonia* damping–off sugarbeet by *Pseudomonas fluorescens*. *Arab-Univ. J. of Agric. Sci.*, **5**: 433-447.

Mukherjee, P.K. and Mukhopadhyay, A.N. 1995. *In situ* mycoparasitism of *Gliocladium virens* on *Rhizoctonia solani*. *Indian Phytopathol*, **48**: 101-102.

Mukhopadhyay, A.N. and Chandra, I. 1996. Biological control of sugarbeat and tobacco damping-off by *Trichoderma harzianum*. Proc. *Malaysian Pl. Prot. Soc. Geneting Thailand, Malaysia*, p. 60-65.

Mukhopadhyay, A.N. and Mukherjee, P.K. 1991. Innovative approaches in biological control of soilborne diseases in chickpea. *In*: Fourth Intl. *Trichoderma* and *Gliocladium* Workshop. Belgrade, Italy, July 17-20, 1991, Petria. 1: 146.

Muyolo, N.G., Lipps, P.E. and Schmitthenner, A.F. 1993. Anastomosis groupings and variation in virulence among isolates of *Rhizoctonia solani* associated with dry bean and soybean in Onion and Zaire. *Phytopathology*, **83**: 438.

Naiki, J., and Ui, J. 1981. *Rhizoctonia solani* Kuhn root rot of bean, soybean and adzuki bean seedlings. *Mem. Fac. Agric. Hokkaido Univ.*, **12**: 262 – 269.

Naito, S. and Itoh, N. 1999. Occurrence of soybean root rot disease caused by *Rhizoctonia solani*, AG-2-3 in Japan. *Ann. Rep. Soc. Pl. Prot. North Japan*, **50**: 54-57.

Nene, Y.L. and Thapliyal, P.N. 1993. Evaluation of fungicides. *In*: Fungicides in Plant Disease Control (3rd ed.). Oxford and IBH Publishing Co. Pvt. Ltd., New Delhi. p. 531.

O'Neill, N.R., Rush, M.C.; Horn, N.L. and Carver, R.B. 1977. Aerial blight of soybeans caused by *Rhizoctonia solani*. *Plant Dis. Rept.*, **61(69)**: 713-717.

Ogoshi, A. (1978). Studies on the grouping of *R. solani* Kuhn with hyphal anastomosis and on the perfect stage of the groups. *Bull. Nat. Inst. Agric. Sci. Ser. C. Pl. Pathol. Entomol.* **30**: 1–63.

Ogoshi, A. 1987. Ecology in pathoenicity of anastomosis and inter specific groups of *Rhizoctonia solani* Khun. *Ann. Rev. Phytopath.*, **25**: 125.

Ou, S.H. 1985. Rice Disease (2nd ed.) CAB International Mycological Institute, Kew, Surrey, U.K. p. 272.

Pandey, S. and Srivastava, S.N. 1990. Evaluation of fungitoxicants against *Rhizoctonia solani* causing seedling disease of sugarbeet. *Indian Phytopath.*, **43**: 80-84.

Parmeter, J.R. Jr. and Whitney, H.S. 1970. *Rhizoctonia solani*: *Biology and Pathology* (ed. Parameter, J.R. Jr., Berkeley), Univ. California Press.

Patel, B.L. and Bhargava, P.K. 1998. Aerial blight of soybean (*Glycine max.*) caused by *Rhizoctonia solani*. *Indian J. Agric. Sci.*, **68(5)**: 277-278.

Polonenko, D.R., Scher, F.M., Kloepper, J.W., Singhleton, C.A., Caliberte, M. and Zales Ka, I. 1987. Effects of root colonizing bacteria on nodulation of soybean roots by *Bradyrhizobium japonicum*. *Canadian J. Microbiol.*, **33**: 498-503.

Priyatmojo, A., Escopalao, V.E., Tangonan, N.G., Pascual, C.B., Suga, H., Kageyama, J. and Hyakumachi, M. 2001. Characterization of a new subgroup of *Rhizoctonia solani* anastomosis group 1 (AG-1-ID) causal agent of a necrotic leaf spot on coffee. *Phytopathology*, **91(11)**: 1054-1061.

Rahman, M.H., Agarwal, V.K., Thapliyal, P.N. and Singh, R.A. 1995. Effect of date of sowing and seed treatment on seedling emergence and diease incidence on soybean. *Bangladesh J. Plant Pathol.*, **11**: 19-22.

Rogers, D.P. 1943. The genus *Pellicularia* (Thelephoraceae). *Farlowia*. **1**: 95-118.

Ruppel, *E.G.*, Baker, R., Harman, G.E., Hubbard, J.P., Hecker, R.J. and Chet, I. 1983. Field tests of *Trichoderma harzianum* rafai aggr as a biocontrol agent of seedling disease in several crops and Rhizoctonia root rot of sugarbeet. *Crop Prot.*, **2**: 399-408.

Safrankova, I. 2002. Influence of temperature and light on formation of the basidial stage of some isolates of *Rhizoctonia solani* Kuhn. *Acta Universitatis Agricultural et Silviculturae Mendelianae-Burnensis*, **50(5)**: 97-101.

Sawada, K. 1919. Descriptive catalogue of the Formosan fungi part-I (in Japanese) *Agricultural Expt. Sta. Govt. of Formosa Special Bull.*, **19**: 1-695.

Shanmugam, V., Vishwanathan, R., Raguchander, T., Balasubramanian, P. and Samiyappan, R. 2002. Immunology of the pathogen virulence and phytotoxin production in relation to disease severity: A case study in sheath blight of rice. *Folia Microbiologica*, **47(5)**: 551-558.

Sharma, J. 1999. Studies on epidemiology and management of web blight of urdbean caused by *Rhizoctonia solani* Kuhn. Ph.D. Thesis, G.B. Pant University of Agriculture and Technology, Pantnagar. 185 p.

Sharma, M. and Gupta, S.K. 2003. Ecofriendly methods for the management of root rot and web blight (*Rhizoctonia solani*) of french bean. *J. Mycol. Pl. Pathol.*, **33(3)**: 345-361.

Sherwood, R.T. and Lindberg, C.G. 1962. Production of a phytotoxin by *Rhizoctonia solani*. *Phytopathology*, **52**: 586-587.

Shukla, R.S., Kumar, S., Singh, H.N. and Singh, K.D. 1993. First report of aerial blight of *Coleus forskohlii* by *Rhizoctonia solani* in India. *Plant Dis.*, **77**: 429.

Singh, A., Malhotra, S.K. and Singh, A. 1996. Effect of temperature and pH on growth and sclerotia formation of *Rhizoctonia solani* causing web blight of winged bean. *Agric. Sci. Dig. Karnal*. **14**: 141-142.

Singh, L.P. and Singh, S. 2002. *In vitro* growth of fungus *Rhizoctonia solani* in relation to heavy metals treatment. *Bionotes*, **4(4)**: 101.

Singh, R., Shukla, T.N., Dwivedi, R.P., Shukla, H.P. and Singh, P.N. 1974. Studies on the soybean blight caused by *Rhizoctonia solani*. *Indian J. Mycol. Pl. Path.*, **4**: 101-103.

Srivastava, L.S. and Gupta, D.K. 1989. Aerial blight of French bean, groundnut, soybean, black gram and horse gram- a new record from India. *Pl. Dis. Res.*, **4**: 163-164.

Sunder, S. and Dodan, D.S. 1993. Growth, sensitivity to fungi toxicants and pathogenic behaviour of binucleate *Rhizoctonia* isolates. *Pl. Dis. Res.*, **8**: 11-18.

Tachibana, H., Jowett, D. and Fehr, W.R. 1971. Determination of losses in soybean caused by *Rhizoctonia solani*. *Phytopathology*, **61**: 1444-1446.

Talbot, P.H.B. 1970. Taxonomy and nomeculature of the perfect state. *In*: J.R. Parameter (Jr.) *Rhizoctonia solani* Biology and Pathology, p. 255. Univ. California Press Berkeley. p. 20-31.

Tiwari, A. and Khare, M.N. 2002. Conditions inducing imperfect and perfect stages in *Rhizoctonia solani* causing diseases of mungbean. *J. Mycol. Pl. Path.*, **32(2)**: 176-180.

Tiwari, R.K.S. and Singh, A. 2004. Efficacy of fungicides on *Rhizoctonia solani* and *Sclerotium rolfsii* and their effect on *Trichoderma harzianum and Rhizobium leguminosarum*. *J. Mycol. Pl. Pathol.*, **34(2)**: 482-485.

Uppal, B.N., Patil, M.K. and Kamal, M.N. 1935. The fungi of Bombay. *Bombay Deptt. Agr. Bull*. **176**: 56.

Verma, H.S. 1973. Rhizoctonia aerial blight of soybean symptomotology, etiology and chemical control. M.Sc. Thesis, G.B.P.U.A. and T., Pantnagar, p. 58.

Verma, H.S. and Thapliyal, P.N. 1976. Rhizoctonia aerial blight of soybean. *Indian Phytopah.*, **29**: 389-391.

Voelcker, O.J. 1953. Annual reports of the department of agriculture, Malaya for the year 1952, p. 65.

Wang, T. C. and Hsieh, Y. 1993. *Rhizoctonia* sp. Causing turfgrass disease and their anastomosis groups in Taiwan. *Plant Pahol. Bull.*, **2**: 221-224.

Weber, C.F. 1939. Web blight: a disease of beans caused by *Corticium microsclerotia. Phytopathology*, **29**: 559-575.

Wu, W.S. 1980. Biological and chemical seed treatment of soybean. Memories of the College of Agriculture, National Taiwan University. **20**: 1-16.

Yang, X. B., Berggren, G. T., and Snow, J. P. 1990. Types of Rhizoctonia foliar blight on soybean in Louisiana. *Plant Dis.*, **74**: 501 – 504.

Yang, X.B., Berggsen, G.T. and Show, J.P. 1988. Seedling infection of soybean by *Rhizoctonia solani* AG-1, causal agent of aerial blight. *Plant Dis.*, **72**: 644.

Yokogama K. and Ogoshi, A. 1984. Studies on hyphal anastomosis of Rhizoctonia solani IV: observation of imperfect fusion. *Ann. Phytopath. Soc. Japan*, **50**: 398-401.

Yoyogi, K. 1927. On the *Hypochnus* disease of soybean and its comparison with that of rice plants (in Japanese). *J. Pl. Prot.*, **14**: 146-158.

Zaidi, S.F.A. 2003. Inoculation with *Bradyrhizobium japonicum* fluorescent *Pseudomonas* to control *Rhizoctonia solani* in soybean (*Glycine max*). *Ann. Agric. Res.*, **24(1)**: 151-153.

Zevada, M.Z., Yer kes, W.D. and Nei Derihauser, J.S. 1955. First list of fungi of Mexico arranged by hosts. (in Spanish). *Mex. Ofic. De E stud Espec. Fol. Tec.***14**: 43.

Zhou, E., Yang, M. and Chen, Y. 2002. The effects of soil environmental factors on the saprophytic colonization of *Rhizoctonia solani*. *Acta Phytopath. Sinica*, **32(3)**: 214-218.

2016, Diseases of Pulse Crops and their Sustainable Management *287–302*
Editors: **Samir Kumar Biswas, Santosh Kumar and Gireesh Chand**
Published by: **BIOTECH BOOKS, NEW DELHI**

Chapter 16

Botrytis Grey Mould of Chickpea (*Cicer arietinum* L.) and their Management

Usha Suyal[1]*, Santosh Kumar[2], Amarendra Kumar[2] and H.S. Tripathi[1]

[1]*Centre of Advance Studies in Plant Pathology, G.B. Pant University of Agriculture and Technology, Pantnagar – 263 145, Uttarakhand*
[2]*Department of Plant Pathology, Bihar Agricultural University, Sabour, Bhagalpur – 813 210, Bihar*

Introduction

Chickpea (*Cicer arietinum* L.) commonly known as gram is a versatile crop among the legumes and ranks first among the pulses both in acreage and production. It is an important pulse crop in over 45 countries of Asia, Africa, Americas and Oceania, with an annual production of 8.62 million tones from 11.12 million hectares (FAO, 2005). Chickpea is used as an important source of protein in human nutrition and cattle feed, and also to improve soil fertility by biological nitrogen fixation. Chickpea usually required few inputs other than labour, insecticides and seed. The major constraints to production include insect-pests, diseases, environmental stresses, drought and poor crop management. Among the different diseases of chickpea,

* Corresponding Author: E-mail: u.suyal@gmail.com

botrytis grey mould (BGM) account about 20 per cent losses in the world and their cost is estimated about 10-100 billion euros per year (Genoscope, 2008). The fertile land and frequent rain during crop growth are responsible for dense canopy creating micro-environment highly variable to growth and development of the pathogen. Several epidemics of BGM causing complete crop loss in the major chickpea producing countries in the world have been reported. Botrytis grey mould (BGM) caused by *B. cinerea* Pers. Ex. Fr., is the second most potentially important disease of chickpea after Ascochyta blight caused by *Ascochyta rabiei* [Pass] Lab. BGM can devastate chickpea, resulting in complete yield loss in years of extensive winter rains and high humidity (Reddy *et al.*, 1993; Pande *et al.*, 1982).

Geographical Distribution, Economic Importance and Losses

The occurrence of botrytis grey mould of chickpea was first reported by Shaw and Ajrekar in 1915. Joshi and Singh (1969) observed that BGM of chickpea in epiphytotic form in Nainital, *Tarai* area of Uttar Pradesh, now part of Uttarakhand State. Since 1967-68, *Botrytis cinerea* has caused vast devastation in in parts of West Bengal, Bihar, Uttar Pradesh, Rajasthan, Haryana, Punjab and Himachal Pradesh (Singh, 1997). The disease was responsible for heavy losses in the Indo- Gangetic plains of India during 1979-1982 (Grewal and Laha, 1982) and caused 70-100 per cent losses in yield at Central State farm Hissar and several parts of Punjab. During 1978-79, an epidemic of BGM destroyed chickpea crop completely over an area of 20,000 ha in the states of Punjab, Haryana, Uttar Pradesh and Bihar (Singh *et al.*, 1982; Grewal and Laha, 1982). In Bangladesh, the damage caused by the disease was estimated to be 70-80 per cent in 1989 (Bakr *et al.*, 1997). Mahmood and Sinha (1990) reported more than 60 per cent losses in grain yield. Reddy *et al.* (1993) observed an epiphytotic of grey mould in *Tarai* region of Nepal, where the yield losses were recorded up to 100 per cent. In Nepal the disease occurs almost every year and an estimated loss of 66 per cent in the experimental field and about 15 per cent in the farmers' field (Joshi, 1992). The effects of BGM on pod yield depend on the onset of the disease in relation to crop growth and disease severity, both of which depend largely on weather conditions and inoculums level of the pathogen.

Causal Organism

BGM of chickpea is caused by *Botrytis cinerea* Pers. Ex. Fr. The asexual stage of the necrotrophic fungus, *B. cinerea* (Moniliaceae, Hyphales) is dominant on chickpea crops. *B. cinerea* grown on potato dextrose agar (PDA) has a white, cottony appearance, which turns light grey with age. The mycelium is septate, brown and 8-10μm wide. Young hyphae are thin and hyaline. Conidia and conidiophores are not in pycnidia or acervuli. Conidiophores lighter brown than hyphae with hyaline tip, septate, 8-24μm wide. Tips of conidiophores or their branches are slightly enlarged and bear small pointed sterigmata. Conidia are hyaline, one- celled oval or globose or short cylindrical and borne in clusters at the tips of conidiophores branches.

Conidia from chickpea host measures 4-24 x 4-18μm (average 14.9 x 8.4μm) and from potato dextrose agar 4.16 x 4-10μm (average 7.4 X 6.1μm). Conidia readily germinate in water or even on the host surface under conditions of high humidity by producing 1-3 thin, hyaline germ tubes. The sporodochia on the host surface are fairly large, 0.5- 5μm in diameter consisting of densely interwoven brown, septate hyphae and produce round to oval unicellular conidia measuring 4-8μm (av. 5.26μm) diameter. These conidia do not germinate, the sporodochia soon become, non-soporiferous and changed into hard sclerotia measuring 2-11 x 2-7 (5-6 x 3-8) μm (Joshi and Singh, 1969). However, no white hyphal matrix is produced and these sclerotial structures resemble those of *Sclerotinia* spp. The perfect stage of *Botrytis cinerea* is *Botryotinia fuckeliana* (Groves and Loveland, 1953). However, there is no report of the occurrence of perfect stage of the fungus on chickpea. Joshi and Singh (1969) found that sometimes only sclerotia are found in old cultures of *B. cinerea* and sporodochia were not seen. These may or may not be surrounded by white mycelial growth characteristics of *Sclerotinia* sp. Two week old cultures formed an abundance of sporodochia and sclerotia (Joshi and Singh, 1969). The sexual stage germinates from fertilized sclerotia by the emergence of apothecia that release sexually produced ascospores (Faretra and Grindle 1992). Apothecia formation requires either two sexually compatible isolates (MAT 1-1 and Mat 1-2) or a pseudo homothallic isolate (MAT-1/2) (Faretra and Grindle 1992). There are no reports of the sexual state of *B. cinerea* occurring naturally on chickpea stubbles. However, it has been produced under laboratory conditions in India (Singh, 1997).

Characteristic Symptoms

All the aerial parts of chickpea are susceptible to the disease with growing tips and flowers being the most vulnerable (Bakr *et al.*, 1992; Grewal and Laha, 1982; Haware and McDonald, 1992; Haware, 1998; Bakr *et al.*, 1997). Drooping of the infected tender terminal branches is a common field symptom (Haware and McDonald, 1992; Pande *et al.*, 2005). According to Joshi and Singh (1969) initial symptoms appear on stem, leaves, inflorescence and pods as grey or dark brown lesions covered with erect hairy sporophores. Stem lesions are 10-30 mm long which later girdles the stem completely. Tender branches break off at the points where grey mould causes rotting. Affected leaves and flowers turn into a rotting mass. In the field, the disease first appears in isolated patches when the crop has achieved maximum canopy and the morning relative humidity is very high with low temperature. As the disease advances, patches of disease plant become more prominent, spreading slowly in the entire field. According to Laha and Grewal (1983) symptoms appeared on leaflets, petioles and growing tips as water soaked lesions. The lesions are brown and limited in size. However, under conditions of high humidity leaflets got blighted and bear abundant fungal fructifications.

On thick and hard stems, the grey mould growth are gradually transformed into a dirty, grey mass containing dark green to black sporodochia. The sclerotia are small, dark bodies and should not be confused with larger, black or dark brown sclerotia embedded in white mycelium of *Sclerotinia sclerotiorum* (Lib.) de Bary (Joshi and Singh, 1969).

Seedling Rot

The pathogen *Botrytis cinerea* is one of the many fungi associated with seedling disorders of chickpea (Cother 1977; Bretag and Mebalds 1987), creating a soft rot

(Burgess *et al.*, 1997). In most chickpea growing regions of the world, foliar infection is considered most important, whereas in Australia, soft rot of young seedlings resulting from seed- borne infection is also important and can result in total crop failure (Burgess *et al.*, 1997). Symptoms include poor emergence, yellowing, wilting, and death of seedlings and pale yellow to light tan discoloration of the tap root. Most plants that develop soft rot become flaccid and then die within a few days.

Epidemiology

Epiphytotics of botrytis grey mould from different parts of the world indicated the existence of definite and efficient mechanisms of survival of the pathogen from one season to another. The information regarding the survival and epidemiology is scanty so far as the botrytis grey mould of chickpea is concerned.

Survival

B. cinerea may survive in nature from one season to the next in several ways in the absence of chickpea crop. The survival may be through continued activity and growth, either parasitically on other hosts or saprophytically on available dead material or through entering upon an inactive phase of the life cycle or a dormant resting structure.

Crop Debris

Several workers have indicated the importance of infected debris as a source of survival of the pathogen and infection of the crop in next season (Mahmood and Sinha., 1990; Nene and Reddy, 1987; Grewal, 1988; Singh and Kaur, 1989; Singh and Tripathi, 1992). Coley-Smith (1980) found that conidia buried beneath the soil surface showed a rapid drop in viability and disappeared within 10-12 weeks, but those placed on soil surface survived for about 50 weeks. Being basically a saprophyte, it has been observed to survive saprophytically in soil and / or on infected crop debris (Mahmood and Sinha., 1990; Grewal, 1988; Singh and Kaur, 1989)

According to Meeta *et al.* (1986), *B. cinerea* survives in infected plant debris for 180 days at a soil depth of 6 cm but not at 2 or 4 cm. Grewal (1988) indicated that the fungus remained viable on infected plant debris present as admixture in seed lots. The fungus was found to be viable in infected seed and plant debris stored at 18C for 5 years (Grewal, 1988) while Singh and Kaur (1990) found that *B. cinerea* did not survive beyond 8 months in chickpea debris at a depth of 10cm in the soil and that the survival and recovery of the fungus was more at 5-10°C. Singh and Tripathi (1992, 1993) observed the survival of *B. cinerea* in infected chickpea debris at a depth of 25 cm for 8 months, *i.e.*, until the following season.

Survival in Seed

Cother (1977) was the first to demonstrate the seed-borne nature of *B. cinerea*. He showed that the majority of seeds inoculated with *B. cinerea*, failed to germinate in soil as they were colonized by the fungus and covered with sclerotia. It has been observed by several workers that in the event of severe infections, pods are also attacked resulting in no seeds or only small and shriveled infected seeds. In nature, thus, infected seed is thought to be the primary source of infection (Cother, 1977; Grewal and Laha, 1983; Haware *et al.*, 1997; Meeta *et al.*, 1988; Gurha *et al.*, 2003; Singh and Tripathi, 1993).

Fungus was found to be externally and internally seed-borne to the extent of 8.2 and 2.5 per cent, respectively Grewal (1988). In an another study, seeds collected from naturally infected plants from three cultivars G-543, H-208 and H-355 of chickpea, 8.0 to 18.5 per cent seeds were found infected externally and internally with *B. cinerea* (Grewal and Laha, 1983).

Haware *et al.* (1999) reported that *B. cinerea* was recorded up to 56 per cent in seed samples collected from Faisalabad (Pakistan). They found that naturally infected seed showed 7 to 17 per cent external and 1 to 1.5 per cent internal seed infection. Despite the fact that infected seeds carry the inoculums for a considerable period. The role of seed-borne inoculums, after getting exposed to soil environment, the epidemiology, is not yet been established (Tripathi and Rathi, 1992).

Sclerotia and Chlamydospores

The fungus *B. cinerea* is known to produce sclerotia on crop stubbles of many host species. The sclerotia are thought to be the main means of the fungal long-term survival (Coley Smith, 1980). In Europe, apothecia emerge from the fertilized sclerotia and wind dispersed ascospores are released mainly in the spring after chilling and periods of high rainfall on *Vicia* beans (Harrison, 1988). Sclerotia develop on the previous season's chickpea stubble in Australia after exposure to cold (>10°C) winter temperatures. As day time temperatures increases in spring, sclerotia germinate asexually, forming conidia on conidiophores. The sclerotia remain viable for the rest of the growing season but do not survive the following hot, dry summer conditions hence, sclerotia are not considered to be a means of long-term survival in Australia.

Alternative Hosts

Due to the wide host range of this pathogen, the role of alternative hosts is likely to play an important part in survival from one chickpea crop to another (Coley Smith, 1980; Haware, 1998; Knights and Siddique, 2002; Pande *et al.*, 2005). However, further studies are required to understand the host- specific pathogenecity of *Botrytis* isolates of chickpea.

Disease Development

There is a wealth of literature available on the temperature and relative humidity requirements of *B. cinerea* on many crops of importance. It should, however, be noted that the temperature and relative humidity requirements for *B. cinerea* appear to influenced by the host plant and even by the plant part being infected (Elad *et al.*, 1992). On chickpeas, the optimum temperature for sporulation and conidial germination is 25°C (Mahmood and Sinha 1990; Singh 1997) and 20°C (Rewal and Grewal 1989a), respectively, with 5°C and 30°C being the minimum and maximum extremes for conidial germination. However, different isolates were found to require differential light intensities and relative humidity for conidial germination (Rewal and Grewal 1989a).

BGM may develop rapidly over time and space, depending on the environmental conditions. Relative humidity, leaf wetness, and temperature are the most important factors (Tripathi and Rathi 1992; Butler 1993; Pande *et al.*, 2002). Bakr and Ahmed (1992) found that disease increased at temperatures of 17-28°C and 70-97 per cent relative humidity. In Bangladesh, maximum disease severity was recorded at a temperature range of 20-28°C (Bakr *et al.*, 1997) and 25-30C in India (Reddy *et al.*, 1990; Tripathi and Rathi, 1992). In the Indian sub-continent, BGM epidemics have occurred in years with high rainfall and a high number of rainy days (Bakr and Ahmed 1992; Joshi 1992; Tripathi and Rathi 1992; Davidson *et al.*, 2004). The duration of leaf wetness appears to have some influence on the development of BGM on chickpea. Disease severity increased with leaf wetness periods greater than 12h/day (Singh and Kapoor, 1984). The epidemics can spread rapidly at 95 per cent or above relative humidity and up to a maximum temperature of approximately 25°C in a dense crop canopy. Under such conditions the disease cycle can be completed in 7 days (Haware 1998).

Host Range

B. cinerea is a facultative parasite with a very wide host range in the temperate and sub- tropical regions (Singh, 1970). The inoculum is always more or less present in the environment waiting for the congenial weather to become active (Nene *et al.*, 1984). It causes grey mould diseases in a number of crop plants, such as strawberry, grapevine, apple, cabbage, carrot, cucumber, eggplant, lettuce, pepper, squash, tomato and several ornamentals such as chrysanthemum, dahlia, lily, roses, gladiolus and tulips etc. (Agrios, 1978). Rathi and Tripathi (1991) reported several hosts (cultivated and weeds) of *B. cinerea* in Nainaital *Tarai*.

Pathogenic Variability

The pathogen *B. cinerea* is reported to have extreme variability and adaptability to a wide range of environmental conditions. Joshi and Singh (1969) and Singh (1970) observed the formation of sclerotial and/or sporodochial bodies on *B. cinerea* infected chickpea plants in the *Tarai* region of Nainital, India, which were not found later from the same area (Pande, 1988). Singh and Bhan (1986) and Rewal and Grewal (1989b) identified 4 and 5 pathotypes respectively, among the *B. cinerea* isolates collected from Northern India. Kishore (2005) differentiated 8 chickpea isolates of *B. cinerea* collected from India and Nepal into distinct pathotypes based on their morpho-cultural characters and reaction on 39 differential lines and RAPD markers.

Molecular markers such as microsatellites are powerful tools for accurate detection of genetic diversity because they are highly polymorphic across numerous loci and are reproducible. In chickpea, microsatellites have revealed genetic variation among isolates of *Ascochyta rabiei*. A recent study that used microsatellite DNA markers developed specifically for the *B. cinerea* genome (Fournier *et al.*, 2002), revealed genetic variation in *B. cinerea* isolates of chickpea from 4 regions of Bangladesh, India, and Nepal (Isenegger *et al.*, 2005).

A UPGMA tree revealed that isolates from Bangladesh are quite diverse and several were closely related to isolates from India and Nepal (Isenegger *et al.*, 2005). Isolates from sub populations from Bangladesh showed potential for a highly adapted pathogenic group to chickpea, which can threaten (or break down) long term control with fungicides

Previously, molecular evidence revealed the role of genetic recombination in *B. cinerea* from grapevine in France (Giraud *et al.*, 1997). This is important, as genetic recombination can generate new genotypes, hence genetic diversity can spread quickly via asexual conidia. In other studies in *B. cinerea*, molecular markers have revealed high genetic diversity and high gene flow among populations from vegetable crops in Europe (Alfonso *et al.*, 2000; Moyano *et al.*, 2003).

Biochemical and Histopathological Basis of Host Plant Resistance

Unlike other major plant pathogen system of crop plants, detailed investigations have not been undertaken on the infection process of *B. cinerea* on chickpea plant and the biochemical basis of BGM resistance in chickpea has not been determined. Preliminary investigations on the infection process recorded that inoculated spores of the fungus germinate within 6-8 h and the germ tube proliferates saprophytically and forms a mycelial mat on the leaf surface. During the period of proliferation and formation of a mycelial mat, the hyphal tips, in direct contact with the host surface, swell to form appressoria and form the infection hyphae, which penetrate directly through the cuticle and form the sub cuticular and sub epidermal mycelium. In some cases the hyphae penetrate directly through the host surface, although penetration through stomata has also been observed (Pande, 1988). After penetration the infection hyphae grow and ramify in the leaf tissue sub cuticularly and sub epidermally. Mycelium grows within the mesophyll cells, which thickens and branches after penetration. The pathogen causes extensive damage to the leaf tissue by destroying epidermal and mesophyll cells, most probably by degrading the cell walls even in advance of invading hyphae.

It was observed that palisade and spongy parenchyma cells in the resistant genotype (ICC 10302) were more compact that in the moderately resistant (GG 588) and susceptible genotype (H 355). No significant anatomical alterations were observed in any of the genotypes up to 48h of germination. Degradation of mesophyll cells were quite evident in most parts of the susceptible cultivar 72h after inoculation, which became more pronounced after 96h resulting in complete necrosis of the leaf after 120h. In moderately resistant and resistant parents, the breakdown of mesophyll cells was first recorded 96h after inoculation. Consequently, yellowing was observed after 120h and complete degradation of mesophyll cells was quite pronounced after 120 or 144h (Pande, 1988). It was observed that under high humidity, even the field resistant genotypes were infected by *B. cinerea*. However, the infection and colonization in resistant cultivars were delayed by 24-48h as compared with the susceptible cultivars. Mohhamadi (1987) reported that leaf surface inhibitors, probably phenolic in nature, are important in resistance in chickpea under field conditions and a high correlation was observed between total phenolic content of the leaf washings and degree of resistance of the genotypes. In the presence of inhibitors, spore germination and germ tube growth were delayed for 6-8h and this time is sufficient for dessication of spores under tropical conditions. However, under humid conditions there was no dessication of the spores and germ tubes on the leaf surface, hence, even the field tolerant varieties became susceptible.

Total phenolic content, sugars, antifungal peptides and phytoalexins are observed to be associated with BGM resistance. *Botrytis* spp. is known to be 'high sugar' pathogen that usually attack plant tissues with more sugar content (Horsfall and Diamond, 1957). Mitter *et al.* (1977) observed that healthy chickpea plants of a BGM-resistant genotype ICC 1069 had significantly lower total soluble sugars and free amino acids and higher total phenol level that the susceptible var. BGM 408. The amount of sulfur containing amino acids, methionine and cystine, was almost double in genotype ICC 1069 compared with BGM 408. Further, shoot tips of both the cultivars had higher quantities of sugar and free amino acids and low content of phenols compared with the middle and lower leaves.

Two antifungal peptides with novel N-terminal sequences, designated cicerin and arietin with molecular weights of 8.2 and 5.6 kDa, respectively, were found in chickpea seeds. Arietin exhibited a higher translation- inhibiting activity in a rabbit reticulocyte lysate system and was found to be highly antifungal to *B. cinerea, Mycosphaerella arachidicola* and *F. oxysporum* (Ye *et al.*, 2002).

A pterocarbon phytoalexin. maackiain, was found associated with BGM resistance in *C. bijugum* Rech. F., a wild relative of chickpea. The concentration of maackiain in *C. bijugum* foliage was 200-300μg/g, compared with <70 μg/g in susceptible species. After inoculation with *B. cinerea,* maackiain concentration increase to >400 μg/g in *C. bijugum,* whereas no significant increase was recorded in the susceptible species.

Disease Management

Cultural Practices

Cultural management of BGM in recent years has been very well demonstrated in India, Bangladesh and Nepal. Early sowing, high seed rates, closer row spacing, and traditional use of bushy genotypes result in dense canopies, which favours development of BGM in chickpea. Thus, techniques which minimize canopy density may alleviate disease intensity. Haware and McDonald (1992) reported that delayed sowings reduced BGM incidence even in susceptible cultivars, but significantly reduced the grain yields. Singh (1997) also observed that the late sown crop (around 20 Nov.) in Punjab, India, showed significantly low incidence of BGM. Bakr *et al.* (1993) reported that wider spacings alone reduced BGM disease significantly in Bangladesh. Haware *et al.* (1997), Bakr *et al.* (1997) also reported that wider row spacing reduced BGM disease incidence and resulted in higher grain yields. Combination of wider row spacing, intercropping with linseed and two spray application of carbendazim @ 0.2 per cent significantly reduced BGM severity and increased grain yield of chickpea and linseed.

Chemical

Bakr *et al.* (1993) reported that seed treatment with bavistin + thiram (1:1), indofil M-45, thiabendazole, ronilan, rovral, bavistin @ 0.3 per cent controls seed borne inoculum of *B. cinerea.*

Foliar spray with ronilan, bavistin + thiram combination @ 0.1 per cent or bavistin alone @ 0.2 per cent provided complete protection to chickpea plants against aerial infection by *B. cinerea* (Grewal and Laha, 1983). Leaf spot of French bean caused by *B. cinerea* and *B. fabae* were successfully controlled by the use of Dithane M-45+ Triton in Egypt (Mansour, 1980). Singh and Kaur (1990) reported that seed treatment with triadimefon (0.1 per cent), Bavistin + Thiram (0.3 per cent), Dithane M-45 (0.3 per cent) or Baytan (0.1 per cent), together with one foliar spray of dithane M-45, hexacap, thiram, thiabendazole, baytan, or triademefon at 50 days after sowing or at the appearance of first symptoms completely controlled both primary and secondary infection of BGM. Haware and McDonald (1993) suggested the judicious of vinclozolin (0.2 per cent) in the integrated management of BGM. Singh *et al.* (1986) reported that foliar spray with indofil M-45, thiabendazole, baytan, bayleton, or thiram during crop growth stages in Feb-Mar controlled foliar infection.

Haware *et al.* (1997) reported that one spray with vinclozolin (0.2 per cent) at the time of flowering in the integrated management system reduced BGM incidence. However, Bakr *et al.* (1997) observed that two foliar sprays with vinclozolin (0.2 per cent) were required to control BGM in Bangladesh.

Biological Control

Although repeated fungicide application can alone achieve effective management of BGM in chickpea, biological control of *B. cinerea* using species of *Trichoderma* has been reported in some fruit and vegetable crops (Tronsmo, 1986; Nelson and Powelson, 1988; Elad, 1994). Mukherjee and Haware (1993) isolated species of *Trichoderma* from the rhizosphere of chickpea, tested them against *B. cinerea* in the laboratory, and identified the most effective isolate of *Trichoderma viride* in controlling BGM of chickpea. Mukherjee *et al.* (1995) in further tests, isolated fungicide- (vinclozolin) tolerant isolates for use along with vinclozolin in integrated BGM management system. Haware *et al.* (1997) reported that there were no difference in BGM incidence and grain yield of chickpea between three sprays of *T. viride* ($10^7 - 10^8$ spores / ml) and three sprays of vinclozolin (0.2 per cent). Burgees *et al.* (1997) reported that seed treatment with *Gliocladium roseum* suppressed the sporulation of *B. cinerea* on chickpea seed. They also found that sporulation of *B. cinerea* on chickpea seed naturally infected or inoculated with *B. cinerea* was suppressed by seed treatment with conidial suspension of *Gliocladium roseum* at 10^7 and 10^8 conidia / ml, respectively. There was no significant effect of *Rhizobium* on disease suppression by *G. roseum* and treatment with *G. roseum* at 10^8 did not reduce nodulation. Haware *et al.* (1999) reported biocontrol potential of *T. viride* isolate T-15 (isolated from chickpea rhizosphere) on *B. cinerea* in chickpea under controlled environmental conditions.

Host Plant Resistance

Extensive screening of the available chickpea material has been done at ICRISAT and some other Agricultural Universities to obtain resistance against Botrytis grey mould (Haware and Nene, 1982; Pandey *et al.*, 1982; Chaube *et al.*, 1983; Singh *et al.*, 1982. They have been screened more than 600 genotypes. Six lines GL-635, -699, -907, -926, -929, -930 were rated as resistant. At Pantnagar nine other lines *viz.*, ICC-466, -

478, -662, -755, -756, -799, -800, -1591 and NEC-138-2 were also found resistant in (rating 3) and eight lines *viz.*, GL-776, -777, -784, ICC-4000, -4950, -5033, P-1528-1-1; P-0.9 were moderately resistant (rating 5). Pandey *et al.* (1982) evaluated chickpea germplasms and varieties and classified GNG-3, C-235, BG-249 as resistant and GNG-16, H-77-4, BG-243, H-78-84, BG-406, DG-77-27, BC-250, BG-413, BG-407, BG-408, Type-3, DG-77-29, RSG-47, DG-77-40 and Radhey as tolerant for grey mould. Chaube *et al.* (1983), Rathi *et al.* (1983) screened chickpea for Botrytis grey mould resistance and reported two lines *viz.*, ICC-1069, ICC-7574 as resistant and one line, ICC-7574 as tolerant. Lines ICC-1069, ICC-7574 and NEC- 138-2 also have been found tolerant to disease in an isolation plant propagator at ICRISAT (Haware and Nene, 1982).

Tripathi and Rathi (1992) screened 10,000 chickpea (*Cicer arietinum* L.) accessions / germplasm lines against *Botrytis cinerea*. The following ICC lines were resistant: 466, 478, 662, 755, 756, 799, 800, 1069, 1591, 7574 and 10302. ICCL-87322 was also resistant. Lines GL-776, GL-777, GL-784, ICC-4000, ICC-4950, ICC-5033, P-1528-1-1 and P-0-9 were moderately resistant. Most of these resistant / moderately resistant accessions were *kabuli* type with erect habit.

Pande *et al.* (2006) screened a chickpea mini- core collection composed of 211 germplasm accessions representing the diversity of global chickpea germplasm collection of 16,991, maintained at the International Crop Research Institute for the semi arid tropics to identify sources of multiple disease resistance. They observed high level of resistance to fusarium wilt, where 21 accessions were asymptomatic and 25 resistant. In all, 355 and 6 accessions were moderately resistant to *Aschochyta* blight, BGM and dry root rot respectively.

Integrated Disease Management

An adequate level of genetic resistance to BGM is not available in the cultivated genotypes and fungicides become ineffective during conditions of high disease pressure. Hence, integrated disease management (IDM) using the available management options is essential to successfully manage the disease and mitigate yield losses. Chemical control of BGM combined with wider row spacing (Reddy *et al.*, 1993) or the use of *T. viride* (Agarwal *et al.*, 1999; Haware *et al.*, 1999) as a biocontrol agent has been attempted and use of tolerant genotype ICCL 87322 in combination with wider row spacing and spraying with bavistin was the best combination followed by the use of the tolerant genotype ICCL 87322 in combination with wider row spacing and intercropping with linseed. Judicious use of fungicides as a seed treatment and / or foliar spray in an IDM system could be economical and affordable to the resource poor farmer.

An IDM program involving cultivation of a BGM- tolerant Avarodhi, soil application of diammonium phosphate, wider row spacing (60 cm), seed treatment with carbendazim + thiram (3g/kg seed), and need based foliar application of carbendazim has been devised. This IDM program was evaluated in farmers' participatory research in 2 districts of Nepal during 1998-99 crop seasons as a collaborative research activity between ICRISAT, Nepal agricultural research council

(NARC), and natural resource institute (NRI), UK, and has resulted in a 400 per cent increase in grain yields and 300 per cent increase in net income.

Future Outlook

In chickpea, BGM is a devastating disease and extensive studies on the biology of the pathogen and screening programs to identify host-plant resistance have failed. Despite the extensive investigations in other hosts, the infection process of *B. cinerea* on chickpea has not been studied. Also, very little is known about the resistance mechanisms of chickpea against *B. cinerea*. Knowledge of the infection process and host defense mechanisms will help in devising management strategies for BGM. Hence, IDM programs suitable for adoption by resource-poor farmers should be emphasized. It is advised that BGM management in chickpea should be based on the location specific disease predictive models. Transgenic plant technology using PGIPs and other antifungal proteins could be the possible approach for imparting disease resistance to commonly adapted cultivars in the future.

References

Agarwal, A. and Tripathi, H.S. 1999. Biological and chemical control of botrytis grey mould of chickpea. *J.Mycol and Plant Pathol.* 29: 52-56.

Agarwal, A., Tripathi, H. S. and Rathi, Y. P. S. 1999. Integrated management of gray mould of chickpea. *J. Mycol and Plant Pathol.* 29: 116-117.

Agrios, G.N. (1978). Plant Pathology. Academic Press, London pp576.

Alfonso, C., Raposo, R. and Melgarejo, P. 2000. Genetic diversity in *Botrytis cinerea* populations on vegetable crops in greenhouses in south-eastern Spain. *Plant Pathology,* 49: 243- 251.

Bretag, T.W. and Mebalds, M.I. 1987. Pathogenicity of fungi isolated from *Cicer arietinum* L.grown in North-Western Victoria. *Australian Journal of Experimental Agriculture,* 27: 141-148.

Bakr, M.A., Rahman, M.M., Ahmed, F. and Kumar, J. 1993. Progress in the management of Botrytis gray mold of chickpea in Bangladesh. Haware,M.P.; Gowda,C.L.L.; McDonald,D. (eds.). International Crops Research Institute for the Semi-Arid Tropics (ICRISAT). Recent advances in research on Botrytis gray mold of chickpea: summary proceedings of the second Working Group Meeting to discuss collaborative research on Botrytis Gray Mold of Chickpea. Patancheru, A.P. (India). ICRISAT. p. 17-18.

Bakr, M.A., Hossain, M.S., and Ahmed, A.U. 1997. Research on Botrytis grey mould of chickpea in Bangladesh. Pages 15-18 *In* Recent advances in research on Botrytis grey mould of chickpea (Haware, M. P., Lenne, J. M., and Gowda, C. L. L., eds.). Patancheru 502324, Andhra Pradesh, India: ICRISAT.

Bakr, M.A., and Ahmed, F. 1992. Botrytis gray mold pf chickpea in Bangladesh. In 'Botrytis gray mold of chickpea. Summary Proceeding of the BARI/ICRISAT Working Group Metting'. (Eds M.P. Haware, D.G. Faris, C.L.L. Gowda) p. 10-12. (ICRISAT: Patancheru, AP, India).

Butler, D.R. 1993. How important is crop microclimate in chickpea botrytis gray mold? In "Recent advances in research on botrytis gray mold of chickpea'. (eds M.P. Haware, C.L.L. Gowda, D. McDonald) p. 7-9. (ICRISAT: Patancheru, AP, India)

Burgees, D.R., Bretag, T., and Keane, P.J. 1997. Biocontrol of seed borne *Botrytis cinerea* in chickpea with *Gliocladium roseum. Plant Pathology.* 46: 298-305.

Chaubey, H.S., Beniwal, S.P.S., Tripathi, H.S. and Nene, Y.L. 1983. Field screening of chickpea for resistance to Botrytis gray mold. *International Chickpea Newsletter,* 8: 20-21.

Cother, E.J. 1977. Isolation of important pathogenic fungi from seeds of chickpea. *Seed Science and Technology,* 5: 593-597.

Coley- Smith, J.R. 1980. Sclerotia and other structures in survival. In the biology of Botrytis. ed. J.R. Coley- Smith., K. Verhoeff and W.R. Jarvie. London, Academic Press. p. 85-114.

Davidson, J.A., Pande, S., Bretag, T.W., Lindbeck, K.D. and Kishore, G.K. 2004. Biology and management of *Botrytis* spp. In legume crops. In 'Botrytis: biology, pathology and control'. (eds Y. Eland, B. Williamson, P. Tudzynski, N. Delen) p. 295-318. Kluwer Academic Publishers: The Netherlands.

Elad, Y. 1994. Biological control of grape grey mould by *Trichoderma harzianum. Crop Protection.* 13: 35-38.

Elad, Y., Shtienberg, D., Yunis, H., and Mahrer, Y. 1992. Epidemiology of grey mould, caused by Botrytis. Research. Proceedings of the 10th International Botrytis Symposium'. Heraklion, Crete, Greece. (eds K. Verhoeff, N.E. Malthrakis, B. Williamson) p. 147-158. Pudoc Scientrific Publishers: Wageningen, The Netherlands.

Faretra, F. and Grindle, M. 1992. Genetic studies of *Botryotinia fuckellana* (*Botrytis cinerea*). In 'Recent Advances in Botrytis Symposium'. Heraklion, crete, Greece. (Eds K Verboeff, NE Malathrakis, B Williamson), p. 7-16. (Pudoc Scientific Publishers: Wageningen, The Nethrerlands).

Fournier, E., Giraud, T., Loiseau, A., Vautrin, D., Estoup, A., Solignac, M., Cornuet, J.M. and Brygoo, Y. 2002. Characterization of nine polymorphic microsatellite loci in the fungus *Botrytis cinerea* (Ascomycota). *Molecular Ecology Notes,* 2. 253-255.

F.A.O. 2005. 'FAO Bulletin of Statistics'. (Food and Agricultural Organizations of the United Nations, Rome).

Genoscope 2008. *Botrytis cinerea,* estimated losses for vineyards in France amount to 15-40 per cent of the harvest, depending on climatic conditions. In: Sequencing projects of *Botrytis cinerea* [online].

Giraud, T., Fortini, D., Levis, C., Leroux, P and Brygoo, Y. 1997.RFLP markers show genetic recombination in *Botrytinia fuckeliana* (*Botrytis cinerea*) and transposable elements reveal tow sympatric species. *Molecular Biology and Evolution,* 14: 1177-1185.

Grewal, J.S. and Laha, S.K. 1982. Chemical control of botrytis grey mould of chickpea. *Indian Phytopathology,* 36: 516-520.

Grewal, J.S. 1988. Diseases of pulse crops an overview. *Indian Phytopathology,* 41: 1-14.

Groves, J.W. and Loveland, C.A. 1985. The connection between *Botrytinia fukeliana* and *Botrytis cinerea. Mycologia,* 45: 415-425.

Guraha, S.N., Singh, G. and Sharma, Y.R. 2003. Diseases of chickpea and their management. In 'Chickpea research in India'. (eds Massod Ali, Shiv Kumar, N.B. Singh) p. 195-227. Indian Institute of Pulses Research: Kanpur, India.

Harrison, J.G. 1988. The biology of *Botrytis* spp. on *Vicia* beans and chocolate leaf spot disease-a review. *Plant Pathology* 37: 168-201.

Haware, M.P. and Nene, Y.L. 1982. Screening chickpea for resistance to botrytis grey mould. *International Chickpea Newsletter.* 6: 17-18.

Haware, M.P. and McDonald, D. 1992. Integrated management of botrytis gray mould of chickpea. In 'Botrytis gray mould of chickpea'. (eds M.P. Haware, D.G. Faris and C.L.L. Gowda) p. 3-6. (ICRISAT: Patancheru, AP, India)

Haware, M.P., Tripathi, H.S., Rathi, Y.P.S., Lenne, J.M. and Jayanthi, S. 1997. Integrated management of botrytis gray mould of chickpea: cultural, chemical, biological, and resistance options. Pages 9-12. *In* Recent advances in research on Botrytis grey mould of chickpea (eds Haware, M.P., Lenne, J.M., and Gowda, C.L.L.). Patancheru 502324, Andhra Pradesh, India: ICRISAT.

Haware, M.P. 1998. Diseases of chickpea. *In* 'The pathology of food and pasture legumes' (eds D.J. Allen, J.M. Lenne) p. 473-516. (ICARDA, CAB International: Wallingford, UK)

Haware, M.P., Mukherjee, P.K., Lenne, J.M., Jayanthi, S., Tripathi, H.S. and Rathi, Y.P.S. 1999. Integrated biological-chemical control of Botrytis gray mould of chickpea. *India Phytopath.* 52: 174-176.

Horsfall, J.G. and Diamond, A.E. 1957. Interaction of tissues sugar, growth substances and disease susceptibility. *A. Pflanzenkel. Pflanzenpathol. Pflanzenschutz. 64: 415-421.*

Isenegger, D.A., MacLeod, W.J., Ford, R., Pande, S., Abu Bakr M. and Taylor P.W.J. 2005. Genetic structure of *Botrytis cinerea* that causes Botrytis grey mold disease of chickpea in Bangladesh. Innovations for sustainable plant health-15th Australian Plant Pathological Society Handbook, p. 224.

Joshi, M.M. and Singh, R.S. 1969. A Botrytis grey mould of gram. *Indian Phytopathology,* 22: 125-126.

Joshi, S. 1992. Botrytis grey mould of chickpea in Nepal. In Botrytis gray mould of chickpea " (Eds. M.P. Haware, D.G. Faris and C.L.L. Gowada) p. 12-13. ICRISAT: Patancheru, A.P., India).

Knights, E.J. and Siddique, K.H.M. 2002. Manifestation of *Botrytis cinerea* on chickpeas in Australia. In: 'Workshop Proceedings Integrated Management of Botrytis Grey

Mould of Chickpea in Bangladesh and Australia'. p. 70-77. (Bangladesh Agricultural Research Institute: Joydebpur, Gazipur, Bangladesh).

Kishore, G.K. 2005. Cultural, morphological, pathogenic and genetic variation in *Botrytis cinerea,* causal agent of gray mold in chickpea. PhD thesis, Jawaharla Nehru Technological University, Hyderabad, Andhra Pradesh, India.

Laha, S.K. and Grewal, J.S. 1983. Botrytis blight of chickpea and its perpetuation through seed. *Indian Phytopatholgoy,* 36: 630-634.

Mansour, K. 1980. Chemical control of rust and leaf spots of field beans (*Vicia faba* L.). *Agricultural Research Review*. 58: 49-56.

Meeta, M., Bedi, P.S. and Kumar, K. 1986. Chemical control of gray mold of gram caused by *Botrytis cinerea* in Punjab. *Journal of Research,* 23: 435-438.

Meeta, M., Badi, P.S. and Jindal, K.K. 1988. Host range of *Botrytis cinerea* the incitant of grey mould of gram. *Plant Disease Research,* 3: 77-78.

Mahmood, M. and Sinha, B.K. 1990. Gray mould disease of Bengal gram in Bihar. Final Technical Bulletin, Tirhut College of Agriculture, Dholi, Bihar, India, p. 35.

Mitter, N., Grewal, J.S. and Pal, M. 1977. Biochemical changes in chickpea genotypes resistant and susceptible to grey mould. *Indian Phytopahtology,* 50: 490-498.

Mohhamadi, A.G. 1987. Phenols in relation to resistance against ascochyta blight and botrytis grey mould of Chickpeas. M.Sc.(Ag.) thesis, Govind Ballabh Pant University of Agriculture and Technology, Pantnagar, India.

Moyano, C., Alfonso, C., Gallego, J., Raposo, R. and Melgarejo, P. 2003. Comparison of RAPD and AFLP marker analysis as a means to study the genetic structure of *Botrytis cinerea* populations. *European Journal of Plant Pathology*. 109(5): 515-522.

Mukherjee, P.K., and Haware, M.P. 1993. Biological control of botrytis gray mold of chickpea. *International Chickpea Newsletter*. 28: 14-15.

Mukherjee, P.K. Haware, M.P. and Jayanthi, S. 1995. Preliminary investigations in integrated biocontrol of Botrytis gray mold of chickpea. *Indian-Phytopathol*. 48: 141-149.

Nelson, M.E. and Powelson, M.L. 1988. Biological control of gray mould of snap bean by *Trichoderma hamatum*. *Plant Disease*. 72: 727-729.

Nene, Y.L., Reddy, M.V. 1987. Chickpea diseases and their control. In 'The chickpea'. (eds M.C. Saxena and K.B. Singh) p. 233-270. (CAB International: Wallingford, UK).

Nene, Y.L. 1984. A proposed list of common names for diseases of chickpea. *International Chickpea Newsletter*. 4: 31-32.

Pandey, M.P., Beniwal, S.P.S. and Arora, P.P. 1982. Field reaction of chickpea varieties to chickpea grey mould. *International Chickpea Newsletter*. 7: 13.

Pandey, B.K. 1988. Studies on botrytis grey mould of chickpea. Ph.D. thesis, G. B. P. U. A and T. Pantnagar, Uttar Pradesh, India.

Pande, S., Singh, G., Rao, J.N., Bakr, M.A., Chaurasia, P.C.P., Joshi, S., Johansen, C., Singh, S.D., Kumar, J., Rahman, M.M. and Gowda, C.L.L. 2002. Integrated management of botrytis gray mold of chickpea. Information Bulletin No. 61, ICRISAT, Andhra Pradesh, India.

Pande, S., Neupane, R.K., Bakr, M.A., Stevenson, P.C., Rao, J.N., MacLeod, W.J., Siddique, K.H.M., Kishore, G.K., Chaudhary, R.N., Hoshi, S. and Johansen, C. 2005. Rehabilitation of chickpea in Nepal and Bangladesh through integrated management of Botrytis gray mold. *In* 'Paper presented in the 15th Australasian Plant Pathology Conference on Innovations for Sustainable Plant Health'. 26-29 September 2005, Deakin University Waterfront Campus, Geelong, Vic. p. 224. (Australian Plant Pathology Society: Canberra).

Pande, S., Stevenson, P.C., Rao, J.N., Neupane, R.K., Chaudhary, R.N., Grzywacz, D., Baurai, V.A. and Kishore, G.K. 2005. Reviving chickpea production in Nepal through integrated crop management, with emphasis on Botrytis gray mold. *Plant Disease*, 89: 1252-1262.

Pande, S., Sharma, M., Pathak, M. and Rao, J. N. 2006.Comparison of greenhouse and field screening techniques for Botrytis gray mold resistance. *International Chickpea and Pigeonpea Newsletter*, 13: 27-29.

Rathi, Y. P. S., Tripathi, H. S., Chaube, H. S., Beniwal, S. P. S. and Nene, Y. L. 1993. Screening chickpea for resistance to botrytis grey mould. *International Chickpea Newsletter*. 11: 31-33.

Rathi, Y. P. S. and Tripathi, H. S. 1991. Host range of *Botrytis cinerea*, the causal agent of grey mould of chickpea. *International Chickpea Newsletter*. 24: 37-38.

Reddy, M. V., Nene, Y. L.,Singh, G., Bashir, M. 1990. Strategies for management of foliar diseases of chickpea. Pages 117-127 in chickpea in the nineties: Proceedings of the second international workshop on chickpea improvement. 4-8 Dec 1989, ICRISAT, India. (Van Rheenen, H. A., Saxena, M. C., Walley, B. J., and Hall, S. D., eds.). Patancheru 502324, A. P. India.

Reddy, M. V., Ghanekar, A. M., Nene, Y. L., Haware, M. P., Tripathi, H. S., and Rathi, Y. P.S. 1993. Effect of vinclozolin spray, plant growth habit and inter-row spacing on botrytis gray mold and yield of chickpea. *Indian Journal of Plant Protection, 21: 112-113.*

Rewal, N., and Grewal, J.S. 1989a. Effect of temperature, light and relative humidity on conidial germination of three strains of *Botrytis cinerea* infecting chickpea. *Indian Phytopathology, 42: 265-268.*

Rewal, N., and Grewal, J.S. 1989b. Differential response of chickpea to grey mould. *Indian Phytopathology*. 42: 265-268.

Singh, G. 1997. Epidemiology of Botrytis grey mould of chickpea. p. 47-50. In: Recent advances in research on Botrytis grey mould of chickpea (Haware, M.P., Lenne, J.M., and Gowda, C.L.L., Eds.). Patancheru 502324, Andhra Pradesh,India: International Crop Research Institute for Semi Arid Tropics.

Singh, G. and Kaur, L. 1989. Genetic variability studies and scope for improvement in chickpea, Punjab, India. *International Chickpea Newsletter.*,20: 7.

Singh, G., Kapoor, S. and Singh, K. 1982. Screening of chickpea for grey mould resistance. *International Chickpea Newsletter*. 7: 13-14.

Singh, G. and Bhan, L.K. 1986. Chemical control of grey mould in chickpea. *International Chickpea Newsletter*. 15: 18-20.

Singh,G. and Kapor, S. 1984. Role of incubation and photoperiod on the intensity of botrytis grey mould of chickpea. *International Chickpea Newsletter*. 12: 23-24.

Singh, G. and Kaur, L. 1990. Chemical control of grey mould of chickpea. *Plant Disease Research*. 5: 132-137.

Singh, M.P. and Tripathi, H.S. 1992. Effect of temperature and depth of burial on survivability of *Botrytis cinerea* Pers. ex. Fr. causal agent of grey mould of chickpea (*Cicer arietinum* L.). *Indian J Mycol and Plant Pathol*. 22: 39-43.

Singh, M.P. and Tripathi, H.S. 1993. Effect of storage temperatures on survivability of *Botrytis cinerea* in chickpea seeds. *Indian J Mycol and Plant Pathol*. 23: 177-179.

Singh, R.B. 1970. Studies on the control of blight and wilt of gram. MSc. (Ag.) thesis, G.B.P.U.A and T, Pantnagar, UP, India

Tronsmo, A. 1986. *Trichoderma* used as a biocontrol agent against *Botrytis cinerea* rots on strawberry and apple. *Scientific Reports from the Agril University of Norwa*, 61: 1-22.

Tripathi, H.S. and Rathi, Y.P.S. 1992. Epidemiology of botrytis grey mould of chickpea. In 'Botrytis grey mould of chickpea'. (Eds MP Haware, DG Faris, CLL Gowda) p. 8-9. (ICRISAT: Patancheru, A.P. India).

Ye, X.Y. Ng TB, Rao, P.F. 2002. Cicerin and arietin, novel chickpea peptides with different antifungal potencies. *Peptides*. 23: 817-822.

2016, Diseases of Pulse Crops and their Sustainable Management 303–318
Editors: Samir Kumar Biswas, Santosh Kumar and Gireesh Chand
Published by: BIOTECH BOOKS, NEW DELHI

Chapter 17

Pigeonpea Wilt Complex and its Management with an Integrated Approach

P. Kishore Varma[1], K. Jyothirmai Madhavi[2] and Ch. Srilatha Vani[1]*

[1]*College of Agriculture, Acharya N.G. Ranga Agricultural University, Rajendranagar, Hyderabad*
[2]*Fruit Research Station, Dr. Y.S.R. Horticultural University, Sangareddy, Medak*

Introduction

Pigeonpea (*Cajanus cajan* (L) Millsp.) is a major grain legume crop in India. Pigeonpea restores soil fertility through atmospheric N fixation and hence is preferred by small and marginal farmers as one of the key crops in farming systems (Reddy *et al.*, 1990). The crop is grown in the world in 4.26 million hectares with the production of 3.05 million tonnes and average yield of about 716.5 kg/ha. In India, it accounts for 3.73 million hectares of area with a production of 2.90 million tonnes and yield of about 776 kg/ha (Anonymous, 2007).

Economic Importance

In India, Pigeonpea is known to be affected by economically important diseases like wilt, sterility mosaic, Phytophthora blight, Macrophomina stem canker and yellow

* Corresponding Author: E-mail: penumatsakishore@gmail.com

mosaic. Wilt disease complex in pigeonpea is a major production constraint in India (Hasan, 1984; Siddiqui and Mahmood, 1996 and 1999). The wilt incidence ranged from 1 to 55 per cent. The per cent wilt incidence in different states of India is presented in Table 17.1.

Table 17.1: Per cent Incidence of Pigeonpea Wilt in different States of India

Sl.No.	*State*	*Incidence (per cent)*	*Reference*
1.	Maharashtra (Marathwada region)	1 to 22 per cent	Pawar *et al.*, 2013
2.	Manipur (Motbung and Kanglatongbi)	55.22 per cent and 42.74 per cent	Chhetry and Ranjana Devi, 2014
3.	Andhrapradesh	5.3 per cent	Kannaiyan *et al.*, 1984
4.	Bihar	18.3 per cent	Kannaiyan *et al.*, 1984
5.	Karnataka	1.1 per cent	Kannaiyan *et al.*, 1984
6.	Maharashtra	22.6 per cent	Kannaiyan *et al.*, 1984
7.	Rajasthan	0.1 per cent	Kannaiyan *et al.*, 1984
8.	Uttar Pradesh	8.2 per cent	Kannaiyan *et al.*, 1984
9.	Tamil Nadu	1.4 per cent	Kannaiyan *et al.*, 1984
10.	West Bengal	6.1 per cent	Kannaiyan *et al.*, 1984

Yield losses due to Pigeonpea are estimated to be US $ 36 million in India under ambient conditions of disease development (Kannaiyan *et al.*, 1984). Saxena *et al.* (2002) estimated the production losses due to wilt as 97,000 tonnes per year. The losses due to wilt depends on the stage of the crop at which it wilts (Kannaiyan *et al.*, 1984). Disease progress is slow during the early phases of growth but accelerates during the flowering and podding stage. Kannaiyan and Nene (1981) reported that the loss in yield may reach 100 per cent when the crop wilts at pre-pod formation stage and partial losses may result if the plant wilts at pod filling stage or later.

Apart from the losses caused by the fungus alone, several nematodes also incite huge losses to pigeonpea production. For example, plant parasitic nematodes are estimated to cause losses in pigeonpea production to a tune of US $ 177 worldwide (Sharma and Nene, 1989). The cyst nematode, *Heterodera cajani*, is a major constraint in pigeonpea production causing yield loss of 30.1 per cent (Saxena and Reddy, 1987).

Disease Symptoms

Pigeonpea plants are susceptible to wilt pathogen at all stages of crop development. The reproductive stage is more susceptible to wilt compared to all other stages. The disease manifests itself as loss of turgidity in leaves followed by yellowing, withering and drying of leaves. Death of entire plant or some of its branches is seen under high disease severity. Isolated wilted plants may appear about 4-6 weeks after sowing and patches of dead plants in the field, usually at the flowering and podding stage, are the first indication of wilt.

Figure 17.1

The characteristic symptom of the disease in the adult plants is a purple band extending upwards from the base of the main stem. The band can clearly be seen in pigeonpea, when the green stems of healthy plants are compared to the diseased stems with coloured lines. Partial wilting is quite common and distinguishes the disease from termite damage, drought and Phytophthora blight, which also kill the plants. Partial wilting is due to infection of lateral roots, while total wilt is a result of tap root infection. If an infected plant is split open below the purple band, browning of the stem and brown to black discoloration of xylem vessels is visible. Sometimes lower branches show die-back with the purple band extending from the tip downward and intensive xylem blackening (Reddy *et al.*, 1998).

Pathogens Associated with Wilt Disease Complex

The disease is caused by a fungal pathogen, *Fusarium udum* Butler (Butler, 1906) causing mass mortality of the plants especially at flowering stage. The damage is further compounded when the wilt fungus is associated with the plant parasitic nematodes and the resistance of various wilt resistant cultivars is often broken by nematode association.

Several researchers reported the association of nematodes with wilt disease complex of pigeonpea. *Heterodera cajani* (Hasan, 1984), *Meloidogyne incognita* (Siddiqui and Mahmood, 1996) and *Rotylenchulus reniformis* (Sharma and Nene, 1990) are known to suppress the plant growth further in wilt susceptible pigeonpea genotypes like

ICP2376 and T21. The increased susceptibility of a genotype to wilt is found to be influenced by the cultivar reaction to wilt and the nematode associated. For example, *H. cajani* infection in pigeonpea increased pathogenicity of *F. udum* in the susceptible cultivar ICP 2376, but the resistant (ICP8863) and tolerant (BDN 1) cultivars are not influenced by nematode infection. (Sharma and Nene, 1989). Similarly, root knot nematode infection has broken the resistance of the cultivar, ICP 9145, but not in the resistant cultivar, ICP 8863 (Marley and Hillocks, 1996) suggesting different mechanism of resistance in different cultivars.

Biology of *Fusarium udum*

Pigeonpea wilt is caused by *Fusarium udum* Butler. The pathogen is both soil and seed borne, and survives on infected plant debris in the soil. The pathogen enters into the plant through primary and secondary roots and is carried through xylem vessels along with water transport system to stem and foliage. Histopathological studies of vascular bundles in resistant and susceptible pigeonpea lines has shown that wilt resistant and moderately resistant genotypes have narrow vascular bundles and xylem vessels as compared to susceptible genotypes. Since mycelium of *F. udum* travels from root to stem and other foliage parts through xylem vessels, the narrow xylem vessels in the resistant genotypes hinder growth and movement of pathogen reducing the chances of wilt development (Tiwari and Dhar, 2011). Moreover, the xylem vessels in the root and stem tissues are occluded and turn brown in susceptible genotypes (Pandey *et al.*, 1997).

The mycelium of *F. udum* is hyaline and produces 3 types of spores within the host tissue as well as in cultures namely microconidia, macroconidia and chlamydospores. Microconidia are produced on hyphal branches and are small, elliptical, unicellular or 1-2 septate. Macroconidia are produced in small cushions of stromatic mycelium on the surface of the host near ground level. The macroconidia are long, curved (fusaroid), with prominent apical hook, and notched at the base and septate (3-4 septa). The presence of the prominent apical hook distinguishes the species from *Fusarium oxysporum*. Considerable variation in size of micro- and macro-conidia is observed in different isolates. The size of macroconidia and microconidia ranged from 15.4-45.0 X 2.1-6.2 μm and 2.5-17.5 X 2.1-6.2 μm, respectively (Kumar and Upadhyay, 2014). Chlamydospores are also formed in the host as well as in old cultures. They develop from any cells of the hypha, often from cells of the macroconidium. The cells round off and become thick walled to form chlamydospores. These spores are oval to spherical, single or in chains, terminal or intercalary and persist in the soil for long time.

Fusarium species are capable of producing various primary and secondary metabolites of which enzymes, toxins and polysaccharides are known to be involved in wilt pathogenesis (Nema, 1992; Pandey *et al.*, 1995; Thomas, 1949). Enzymes help in the colonization of plant tissues by disintegration of middle lamella and matrix of cell wall. Cellulase, polygalactouronase and pectin methyl galactouronase are reported to play a role in pathogenesis. The production of these enzymes is known to be increased with incubation time (Shukla and Dwivedi, 2012). The production of polysaccharides in the xylem vessels also aids in hindering the flow of water and mineral nutrients from the root system to above ground parts, thus resulting in wilting of plants (Gothoskar *et al.*, 1955). Further, the production of the toxin, fusaric acid is known to be involved in virulence of the pathogen (Xu *et al.*, 1983). The enzyme and fusaric acid production varied with *F. udum* isolates (Kumar *et al.*, 2007).

Epidemiology

The population density of *F. udum* is found to be decreased with an increase in depth of the soil. The population of *F. udum* is found to be 236 cfu/g soil and 191cfu/g soil at a depth of 5-10 cm and 11-20cm, respectively. Incubation temperature of 25 to 30°C and soil water holding capacity of 75 per cent was found favourable for growth and multiplication of *F. udum*. It was also noticed that population density of the wilt pathogen decreased with increase in incubation period under *in vitro* conditions (Prasad and Saifulla, 2012). Soil moisture is also known to significant role on the population dynamics of *F.udum*. Increased population of *F. udum* was reported in irrigated fields compared to rainfed crop and *F. udum* population is further increased if pigeonpea is cultivated in the same field year after year (Chaudhari *et al.*, 2001).

Alfisols and vertisols are found to favour the disease incidence in susceptible cultivars. However, the alfisols supported slightly more pathogen population compared to vertisols (Naik *et al.*, 1997). The soils with 50 per cent or more soil particles with slightly acidic to alkaline pH also favoured disease incidence in susceptible cultivars (Upadhyay, 1979).

Identification of Pathogen and its Variability

Disease diagnosis and pathogen identification by traditional methods, which involve isolation of pathogen and characterizing it by inoculation and pathogenicity tests, are time consuming. The problems sometimes are further complicated by the occurrence of saprophytic strains of *Fusarium* on diseased plants which are morphologically similar to *F.udum*, but are non-pathogenic. To address this problem Geiser *et al.* (2004) have created FUSARIUM-ID database based on partial translation elongation factor 1-α (TEF) DNA sequences. Users can generate sequences using primers that are conserved across the genus, and use the sequence as a query to BLAST which can be accessed at http://fusarium.cbio.psu.edu. Correct identification of a known species often can be performed using this region alone. Apart from this, species specific primers for the detection of Fusarium species were developed by several researchers using conserved genes/regions such as Internal Transcribed Spacer (ITS), Cellobiohydrolase-C, Histone-3 and Topoisomerase-II (Moricca *et al.*, 1998; Hatsch *et al.*, 2002; Steenkamp *et al.*, 1999; Yadav *et al.*, 2011). The target genes and primer sequences used for identification of *Fusarium* sp. are presented in Table 17.2.

Several studies were undertaken to reveal the variability within the *F. udum* isolates. Fourty one isolates collected from different parts of India were grouped based on cultural characteristics on Potato Dextrose Agar (PDA). Variation with respect to mycelia colour (white and pink), pigmentation (light yellow to brown colour) and colony characters (fluffy, partially appressed and appressed growth) were observed (Mahesh *et al.*, 2010a). With the advent of molecular techniques, genetic

Table 17.2: Primer Sequences and the Target Genes for Identification of Fusarium Species

Primer Sequence	*Target Gene*	*Reference*
HFUSF (5'-ATCATCACTAACTTCATCACCAAT-3') HFUSR1 (3'-TGTCGAATGTTAGTAAGTGTTG-5')	Partial Histone-3 (H3) gene	Mesapogu *et al.*, 2011
ef1 (f5'-ATGGGTAAGGA(A/G)GACAAGAC-3') ef2 (5'-GGA(G/A)GTACCAGT(G/C)ATCATGTT-3')	TEF gene	O'Donnell *et al.*, 1998
Tof75 (5'CATCCTCGATGGGAGGTSGG-3')	Topoisomerase-II	Yadav *et al.*, 2011
98b (5'CGGCCATGATCATCAGATG3')	gene	
ITS1 (5'TCCGTAGGTGAACCTGCGG3') ITS4 (5'TCCTCCGCTTTATTGATATG3')	5.8s rRNA gene	Datta and Lal, 2013

diversity in *F. udum* was studied using RAPD and AFLP markers (Mesapogu *et al.*, 2012; Kiprop *et al.*, 2002; Kiprop *et al.*, 2005). The genetic diversity among *F. udum* isolates estimated using RAPD markers has produced a total of 126 loci of which 69 loci were polymorphic. Highest numbers of polymorphic bands were obtained with the OPB primer 17 revealing high degree of genetic diversity among *F. udum* isolates (Mesapogu *et al.*, 2012). Similarly, in Kenya, amplified fragment length polymorphism (AFLP) analysis of 56 isolates of *F. udum* has generated a total of 326 fragments of which 121 were polymorphic (Kiprop *et al.*, 2002). However, in most of the studies genetic diversity analysis had no relationship with cultural characteristics, aggressiveness and geographical origin of the isolates.

Management of Wilt Disease Complex

Use of Resistant Varieties

At present, no effective method of wilt management is available due to the soil borne nature of the pathogen and its ability to survive for longer period in the infested soil. Growing resistant varieties is the best option available for managing the disease. Resistance to wilt disease can be due to single dominant gene (Pandey *et al.*, 1996; Kotresh *et al.*, 2006) or due to a recessive gene (Jain and Reddy, 1995) in different crosses. Polygenic resistance is also reported against the disease (Shaw, 1936; Joshi, 1957). Though, reasonably good sources of resistance are available for wilt, pathogenic variability leads to instability of resistance. This has necessitated identifying donors for breeding of varieties with broad base multiple variant resistance. Mishra and Dhar (2010) identified three genotypes namely, ICP 8860, ICP 8863 and ICP 14722 as uniformly resistant to three variants of *F. udum* revealing their high and broad based resistance to Fusarium wilt at IIPR, Kanpur. Among the genotypes screened against *F. udum* variant-1 in Bangalore, the genotypes *viz.*, GRG 818, GRG 822, GRG 2009, MA 6, RVSA 073, BSMR 736, BSMR 853, KPL 43, KPL 44, BRG 3, IPA 8F, IPA 204, ICP 8858, ICPL 87119, ICP 8863 ICP 8863, ICP 20130, ICP 11298, ICP 20117, ICPL 20133, ICPL 20131, ICPL 20122, ICPL 99011, ICPL 20125, ICPL 20112, ICPL 99050, ICPL 99014, ICPL 99010, ICPL 99087, ICPL 20138, ICPL 20118, ICPL 99044, ICPL 20105, ICPL 20181, ICPL 20139, ICPL 20121, ICPL 20111, ICPL 99016, ICPL 20104, ICPL 99054, ICPL 14290, ICPL 99046, and ICPL 99088 showed resistant reaction (Asha *et al.*, 2012). Though different varieties with resistance to Fusarium wilt are available, the deployment of genotypes based on the existing race in a particular locality is of prime importance for effective management of wilt.

Dates of Sowing and Intercropping

Dates of sowing and intercropping are also known to influence the wilt incidence in pigeonpea. In early sown crop (1st July) wilt incidence was found to be more compared to late sown crop (16th September) and irrigated fields showed significantly increased pathogen population over rainfed crop. The effect of sowing dates on pathogen population dynamics was more pronounced under irrigated conditions than in rainfed cropping (Chaudhary, 2001). Intercropping with soybean and sunflower was also known to reduce wilt incidence (Bharathi and Chandrasekhar Rao, 2009).

Disease incidence in a wilt tolerant pigeonpea cultivar C 11 in sole and sorghum intercropped system at varied densities of *Fusarium udum* in Alfisol and Vertisol fields was studied at ICRISAT Asia Center, Patancheru, India during 1992 rainy season. A significant reduction in wilt incidence was observed in both the soils in sorghum intercropped pigeonpea compared to sole pigeonpea. (Naik *et al.*, 1997). Hence, intercropping of sorghum or soybean + sunflower in pigeonpea is a better option than sole cropping of pigeonpea in wilt prone areas.

Seed Treatment

The pigeonpea wilt pathogen is primarily a soil and seed inhabitant. Hence, protecting the seed with a chemical or biological barrier or a combination of both is of prime importance for the management of wilt. At present, seed treatment is one of the practical approaches to prevent the pathogen development in the early stage of crop development. Seed treatment with fungicides like carbendazim @ 2g/kg seed (Mahesh *et al.*, 2010b) or thiram + benomyl (1:1) @ 3g/kg seed (Gade *et al.*, 2007) is often recommended to reduce pigeonpea wilt. Apart from fungicidal treatment, seed treatment with *Trichoderma harzianum* (4g/kg seed) and *Pseudomonas fluorescens* (10g/kg seed) is also found to be effective at field level (Gade *et al.*, 2007; Mahesh *et al.*, 2010b).

Soil Amendments

Amendment of soil with Farm yard manure (FYM) and KNO_3 is known to reduce pigeonpea wilt. Soil application of FYM (20 kg N ha^{-1}) along with KNO_3 (20 kg N ha^{-1}) is shown to reduce the pathogen population and increase the antagonistic fungi in pigeonpea rhizosphere (Bharathi and Chandrasekhar Rao, 2009). The incorporation of cruciferous plant residues or crop rotation with *Brassica* sp. is known to suppress soil borne plant pathogens and is termed biofumigation (Kirkegaard *et al.*, 1993). Glucosinolates present in various cruciferous plant residues is known to be hydrolyzed by the enzyme myrosinase resulting in production of various sulphur containing toxic compounds like isothiocyanates (Sang *et al.*, 1984). Mustard oil cake amended soil showed decline in *Fusarium* population within 30 days of incorporation which was attributed to release of toxic volatiles like isothiocyanate and methyl sulphide (Gamliel and Stapleton, 1993).

Soil Solarisation

Soil solarization or slow soil pasteurization is the hydro/thermal soil heating accomplished by covering moist soil with polyethylene sheets as soil mulch during summer months for 4-6 weeks. The population of *F. udum* is reduced in solarised plots and the soil solarisation raised the soil temperature up to 54°C, 45.8°C and 39.8°C at 5, 10 and 15cm depths, respectively. The increase in soil temperature is known to reduce the *F. udum* population in the soil (Gade *et al.*, 2007).

Biological Control

Use of biocontrol agents in managing plant diseases is assuming significance in present day agriculture due to increased concerns on the use of chemical fungicides. Of different biocontrol agents, use of Plant Growth-Promoting Rhizobacteria (PGPR) is gaining importance owing to their potentiality in controlling plant diseases and in promoting growth and yields of plants (Glick, 1995). Disease management and yield enhancement with PGPR is by both direct and indirect mechanisms. Fluorescent Pseudomonads are one of the most widely used PGPR in managing several soil borne diseases of different crops (Thomashow and Weller, 1988; Lopez., 1988; Dileep Kumar *et al.*, 2001). Reduction in pathogen propagules in soil by these bacterial antagonists can be due to mechanisms such as antibiosis and hyperparasitism (Raaijmakers *et al.*, 1997). Disease suppressing ability of *P. fluorescens* strains is usually dependent on their ability to produce antifungal metabolites (Dowling and O'Gara, 1994; Keel *et al.*, 1992; Thomashow and Weller, 1996).

Extra-cellular metabolites produced by *P. fluorescens* isolates are highly antagonistic to several plant pathogens (Hass and Defago, 2005). For example, HCN; antibiotics such as 2, 4-diacetylphloroglucinol (2, 4-DAPG) and phenazine-1-carboxylic acid (PCA); enzymes and hormones of *P. fluorescens* play a vital role in suppressing soil borne diseases in various crops. Effective strains of *P. fluorescens* usually produce one or more diffusible antibiotics such as phenazine derivatives like pyoluteorin, pyrrolnitrin, oomycin A, viscosinamide and 2,4-diacetylphloroglucinol (2,4-DAPG) (Haas and Defago, 2005). Of them, 2, 4-DAPG is a broad spectrum antibiotic (phenolic metabolite) and its purified form has antifungal, antibacterial, antiviral, antihelminthic and phytotoxic properties (Bangera and Thomashow, 1999; Isnansetyo *et al.*, 2003; Keel *et al.*, 1992).

Besides disease controlling abilities, the potentiality of these *P. fluorescens* isolates in promoting plant growth is also well established (Nandakumar *et al.*, 2001; Karpagavalli *et al.*, 2002). These bacterial antagonists are known to produce indole acetic acid (IAA) and development of root system is evident (Pattern and Glick, 2002). Application of *P. fluorescens* strains to manage *Fusarium* wilts in different crops has earlier been reported (Lemanceau and Alabouvette, 1993). Wilt and root rot inhibition by various fluorescent Pseudomonads is attributed to activities such as production of siderophores under iron limiting conditions and through HCN production (Buyens *et al.*, 1996; Charest *et al.*, 2005). Besides, fluorescent Pseudomonads are known to induce systemic resistance in plants against various pests and diseases (Ramamoorthy *et al.*, 2001; Van Loon *et al.*, 1998; Zehnder *et al.*, 2001).

Siddiqui and Shakeel (2009) studied the efficacy of 21 isolates of fluorescent psudomonads isolated from pathogen suppressive soils against pigeonpea wilt complex. Some potent isolates of *P. fluorescens* (Pf 718, Pf 719 and Pf 736) and *P. aeruginosa* (Pa737) showed inhibition of *F. udum* and also reduced the hatching of *M. incognita* eggs. The use of these isolates along with *Rhizobium* (pigeonpea strain) further increased plant growth and reduced nematode multiplication and wilting index. In another study, co-inoculation of Root nodulating *Sinorhizobium fredii* (KCC5) and *P. fluorescens* (LPK2) enhanced the seedling growth and strongly inhibited the growth of *F. udum* (Harish Kumar *et al.*, 2010).

The *Bacillus* isolates, B615 and B603 were found to have antifungal activity against pigeonpea wilt pathogen and are also found to have inhibitory effect on hatching of nematode eggs and their penetration (Siddiqui and Shakeel, 2007). *Bacillus subtilis* and *Bacillus brevis* applied to soil were also found inhibitory to *F. udum* due to the production of the antibiotic bulbiformin (Vasudeva *et al.*, 1962; Bapat and Shah, 2000). The inhibition of pigeonpea wilt was correlated with the production of extracellular anatagonistic substances like bulbiformin that inhibited conidial germination. The bacterial antagonist, *Alcaligenes xylosoxydans* was shown to produce chitinase that inhibited pigeonpea wilt pathogen (Vaidya *et al.*, 2003).

Among different fungal biocontrol agents screened under *in-vitro* conditions against *F.udum*, high degree of antagonism was exhibited by *Aspergillus flavus, A. niger, Gliocladium virens, Penicillium citrinum, Trichoderma harzianum* and *Trichoderma viride* (Rajesh Singh *et al.*, 2002; Goudar and Srikant Kulkarni, 2000). When tested under green house and field conditions, of these bioagents, *G. virens* gave maximum reduction in *F. udum* populations and the disease (Rajesh Singh *et al.*, 2002; Ranjana Devi and Chhetry, 2012).

Integration of Management Practices

Use of resistant varieties, soil solarisation, seed treatment with fungicides and bioagents, amendment of soil with organic and inorganic substrates, crop rotation are some of the disease management options available for management of pigeonpea wilt complex. Despite the availability of several management options, with the exception of host resistance, it is likely that no one technique will result in the complete control of pigeonpea wilt disease complex. The integration of some of these approaches is considered as more practical approach for wilt disease management. For example, soil solarisation during summer months followe by seed treatment with thiram + benomyl was found effective in reducing pigeonpea wilt (Gade *et al.*, 2007). In another study, an integration of management practices like seed treatment with carbendazim (2kg/kg seed) and soil application of *Pseudomonas fluorescens* or *Trichoderma viride* (2.5kg/ha in FYM @ 50kg/ha) was found to reduce wilt incidence under field conditions (Mahesh *et al.*, 2010b).

Conclusions

Understanding the biology of plant pathogens and the disease epidemiology usually helps in devising effective management strategies. However, despite thorough investigations on pathogen biology, backdrop knowledge on aggravating factors, coupled with reasonable field resistance among cultivable germplasm, effective systemic fungicides and their conjunctive use with efficacious bioagents, wilt epidemics in pigeonpea are still common. This is attributed to high levels of variability among *Fusarium udum* propagules and their prevalence in pigeonpea soils at alarming levels. Coupled with, nemas populations are on rise in these pigeonpea cultivated soils due to various reasons. A plant protectionist should be able to develop a robust management strategy for this hitherto difficult to manage pigeonpea wilt with the available options in hand. For this, a thorough understanding on the aforementioned concepts such as pathogen biology, epidemiology, effective and specific bioagents with known antifungal activity against vegetative, reproductive and resting spores of pathogens is desirable. Besides, additional information on pathogen variability in field soils, population dynamics of *Fusarium* at various crop growth stages, their fluctuations as influenced by agronomic practices will be an asset to further devise location specific control measures through integration of various options.

Majority of the pigeonpea in Andhra pradesh is cultivated under rainfed conditions. Moreover, farmers from such areas are resource poor and thus cannot rely on chemical management that demands heavy budget and that too with little success. It is precisely at this juncture, the juxtapositioning of various compatible options under the gamut of IDM is required to bring down pigeonpea wilt below economic threshold levels. Besides, timely intervention measures and delivery methods and times of applications of various IDM components also have a greater impact in reducing the phenomenal losses due to this disease. Location specific IDM solutions through available management options with judicious chemical usage and inundative use of bioagents can be a long term solution to this pigeonpea wilt in a sustainable way.

References

Anonymous. 2007. Annual Report for 2007-08. All India Co-ordinated Research Project on Pigeonpea, IIPR, Kanpur, p. 239.

Asha, S.M., Ramappa, H.K. and Gowda, M.B. 2012. Screening of Pigeonpea Genotypes for Resistance to Fusarium Wilt. *Mysore Journal of Agricultural Science* **46 (3)**: 563-566.

Bangera, M. G. and Thomashow, L. S. 1999. Identification and characterization of a gene cluster for synthesis of the polyketide antibiotic 2, 4- diacetylphloroglucinol from *Pseudomonas fluorescens*. *Journal of Bacteriology* **2**: 87

Bapat, S. and Shah, A.K. 2000. Biological control of fusarial wilt of pigeonpea by *Bacillus brevis*. *Canadian Journal of Microbiology* **46(2)**: 125-132

Bharathi, V. and Chandrasekhar Rao, K. 2009. Effect of cropping systems and nitrogen on wilt incidence, Fusarium udum population and its antagonistic fungi in pigeonpea rhizosphere. *Indian Journal of Plant Protection* **2(1)**: 8-11.

Butler, E. J. 1906. The wilt disease of Pigeonpea and pepper. *Agriculture Journal of India*. **1**: 25-26.

Buyens, S., Heungens, K., Poppe, J., and Hofte, M. 1996. Involvement of pyochelin and pyoverdin in suppression of *Pythium*-induced damping-off of tomato by *Pseudomonas aerugionosa* 7NSK2. *Applied Environmental Microbiology*. **62**: 865-871.

Charest, M. H., Beauchamp, C. J., and Antoun, H. 2005. Effects of the humic substances of de-inking paper sludge on the antagonism between two compost bacteria and *Pythium ultimum*. *FEMS Microbiol. Ecol*. **52**: 219-227.

Chaudhary, R.G., Kumar, K. and Dhar, V. 2001. Influence of dates of sowing and moisture regimes in pigeonpea on population dynamics of *Fusarium udum* at different soil strata. *Indian Phytopathology* **54(1)**: 44-48

Chhetry, G.K.N. and Ranjana Devi, T. 2014. Wilt epidemiology of pigeonpea (*Cajanus cajan* (L.) Millsp.) in organic farming system. *IOSR Journal of Agriculture and Veterinary Science* **7 (1)**: 1-6.

Datta, J. and Lal, N. 2013. Genetic diversity of Fusarium wilt races of pigeonpea in major regions of India. *African Crop Science Journal* **21(3)**: 201-211.

Dileep Kumar, B. S., Berggren, I. And Martensson, A. M. 2001. Potential for improving pea production by co-inoculation with fluorescent *Pseudomonas* and *Rhizobium*. *Plant Soil*. **225**: 25-34.

Dowling, D.N. And O'Gara, F. 1994. Metabolites of *Pseudomonas* involved in the biocontrol of plant disease. *Trends Biotechnol*. **12**: 133-144.

Gade, R.M., Zote, K.K. and Mayee, C.D. 2007. Integrated management of pigeonpea wilt using fungicide and bioagent. *Indian phytopathology* **60(1)**: 24-30.

Gamliel, A. and Stapleton, J.J. 1993. Characterization of antifungal volatile compounds evolved from solarized soil amended with cabbage residue. *Phytopathology* **83:** 899-905.

Geiser, D.M., Gasco, M.M.J., Kang, S., Makalowska, I., *et al.*, 2004. FUSARIUM-ID v.1.0: A DNA sequence database for identifying Fusarium. *European Journal of Plant Pathology* **110**: 473-479.

Glick, B.R. 1995. Genotyping of antifungal compounds producing PGPR Pseudomonas. *Canadian Journal of Microbiology* **41**: 107-109.

Gothoskar, S.S., Scheffer. R.P., Walker, J.C. and Stehmann, M.A. (1955). The role of enzymes in development of *Fusarium* wilt of Tomato. *Phytopathology* **45**: 381-387.

Goudar, S.B. and Srikant Kulkarni. 2000. Bioassay of antagonists against *Fusarium udum*- the causal agent of pigeonpea wilt. *Karnataka Journal of Agricultural Sciences* **13(1)**: 64-67.

Harish Kumar, Bajpai, V.K., Dubey, R.C., Maheswari, D.K. and Kang, S.C. 2010. Wilt disease management and enhancement of growth and yield of *Cajanus cajan* (L) var. Manak by bacterial combinations amended with chemical fertilizer. *Crop Protection* **29**: 591-598

Hasan, A. 1984. Synergism between *Heterodera cajani* and *Fusarium udum* attacking *Cajanus cajan*. *Nematol. Medit*. 12: 159-162.

Hass, D. and Defago, G. 2005. Biological control of soil borne pathogens by fluorescent Pseudomonads. *Nature Rev. Microbiol.* **3**: 307-319.

Hatsch, D., Phalip, V. and Jeltsch, J.M. 2002. Development of a bipartite method for *Fusarium* identification based on cellobiohydrolase-C: CAPS and Western blot analysis. *FEMS Microbiol Lett.* **213**: 245–249.

Isnansetyo, A., Cui, L., HIramatsu, K. And Kamei, Y. 2003. Antibacterial activity of 2, 4-diacetylphloroglucinol produced by Pseudomonas spp. AMSN isolated from a marine algae against vancomycin-resistant *Staphylococcus aureus*. *Int J Antibmicrob Agents.* **22**: 545-547.

Jain, K. C., and Reddy, M. V. 1995. Inheritance of resistance to *Fusarium* wilt to Pigeonpea. *Indian Journal of Genetics and Plant Breeding* **55**: 434-437.

Joshi, A. R. 1957. Genetics of reistance to disease and pests. *Indian Journal of Genetics and Plant Breeding* **17**: 305-317.

Kannaiyan, J. and Nene, Y.L. 1981. Influence of wilt at different growth stages on yield loss in pigeonpea. *Tropical Pest Management* **27**: 141.

Kannaiyan, J., Nene, Y. L., Reddy, M. V., Rajan, J. G. and Raju, T. N. 1984. Prevalence of Pigeonpea diseases and associated crop losses in Asian, Africa and the Americas. *Tropical Pest Management.* **30**: 62-71.

Karpagavalli, S. Marimuthu, T., Jayaraj, J. And Ramabadram, R. 2002. An integrated approach to control rice blast through nutrients and biocontrol agent. *Res Crops.* **2**: 197-202.

Keel, C., Schnider, U., Maurhofer, M., Voisard, *et al.*, 1992. Suppression of root diseases by *Pseudomonas fluorescens* CHAO: importance of the bacterial secondary metabolite 2, 4- diacetylphloroglucinol. *Molecular Plant-Microbe Interactions.* **5**: 4-13.

Keel, C., Weller, D.M., Natsch, A., De'fago, *et al.*, 1996. Conservation of the 2, 4-diacetylphloroglucinol biosynthesis locus among fluorescent Pseudomonas strains from diverse geographic locations. *Appl. Environ Microbiol.* **62**: 552-562.

Kiprop, E.K., Baudoin, J.P., Mwang'ombe, A.W., Kimani, P.M., *et al.*, 2002. Characterization of Kenyan Isolates of *Fusarium udum* from Pigeonpea [*Cajanus cajan* (L.) Millsp.] by Cultural Characteristics, Aggressiveness and AFLP Analysis. *Journal of Phytopathology* **150 (10)**: 517-525

Kiprop, E.K., Mwang'ombe, A.W., Baudoin, J.P., Kimani, P.M. and Mergeai, G. 2005. Genetic variability among Fusarium udum isolates from pigeonpea. *African Crop Science Journal* **13(3)**: 163-172.

Kirkegaard, J.A., Gardner, P.A., Desmarcheleir, J.M. and Angus, J.F. 1993. Biofumigation- using *Brassica* species to control pest and diseases in horticulture and agriculture. In 9th Australian Research Assembly on Brassicas. (Eds. Wrotten, V. and Mailer. Agricultural Research Institute, Wagga. p. 77-82.

Kotresh, H., Fakruddin, B., Punnuri, S. M., Rajkumar, B. K. and Paramesh, H. 2006. Identificaiton of two RAPD marker genetically linked to recessive allele of a *Fusarium* wilt resistant gene in Pigeon pea. *Euphytica.* **149**: 113-120.

Kumar, S. and Upadhyay, J.P. 2014. Studies on cultural, morphological and pathogenic variability in isolates of *Fusarium udum* causing wilt of pigeonpea. *Indian Phytopathology* **67(1)**: 55-58

Kumar, V., Chauhan, V.B. and Srivastava, J.P. 2007. Pathogenic and biochemical variability in Fusarium udum causing pigeonpea wilt. *Indian Phytopathology* **60(3)**: 281-288

Lemanceau, P., and Alabouvette, C. 1993. Suppressionof *Fusarium* wilt by fluorescent *Pseudomonas*: Mechanisms and Applications. *Biocontrol Science and Technology*. **3**: 219-234.

Lopez, J. E. 1988. Role of fluorescent siderophore production in biological control of *Pythium ultimum* by a *Pseudomonas fluorescens* strain. *Phytopathology*. **78**: 166-172.

Mahesh, M., Muhammad Saifulla, Prasad, P.S. and Sreenivasa, S. 2010a. Studies on cultural variability of *Fusarium udum* isolates in india. *International Journal of Science and Nature* **1 (2)**: 219-225

Mahesh, M., Saifulla, M., Sreenivasa, S. and Shashidhar, K.R. 2010b. Integrated management of pigeonpea wilt caused by *Fusarium udum* Butler. *EJBS* **2(1)**: 1-7.

Marley, P.S. and Hillocks, R.J. 1996. Effect of root knot nematodes (*Meloidogyne* spp.) on Fusarium wilt in pigeonpea (*Cajanus cajan*). *Field Crops Research* **46**: 15-20.

Mesapogu, S., Bakshi, A., Babu, B K., Reddy, S.S., Saxena, S. and Arora, D. K. 2012 *Genetic diversity and pathogenic variability among Indian isolates of Fusarium udum infecting pigeonpea (Cajanus cajan (L.) millsp.). International Research Journal of Agricultural Science and Soil Science* **2 (1)**: 51-57

Mesapogu, S., Kishore Babu, B., Bakshi, A., Reddy, S.S., Saxena, S., Srinivasa, A.K. and Arora, D.K. 2011. Rapid detection and quantification of *Fusarium udum* in soil and plant samples using real-time PCR. *Journal of Plant Pathology and Microbiology* **2**: 107.

Mishra, S. and Dhar, V. 2010. Identification of broad base and stable sources of resistance to Fusarium wilt in pigeonpea. *Indian Phytopathology* **63(2)**: 165-167.

Moricca, S., Ragazzi, A., Kasuga, T. and Mitchelson, K.R., 1998. Detection of *Fusarium oxysporum* f. sp. *vasinfectum* in cotton tissue by polymerase chain reaction. *Plant Pathology* **47**: 486–494.

Naik, M.K., Reddy, M.V., Raju, T.N. and McDonald, D. 1997. Wilt incidence in sole and sorghum intercropped pigeonpea at different inoculum densities of *Fusarium udum*. *Indian Phytopathology* **50(3)**: 337-341.

Nandakumar, R., Babu, S., Viswanathan, R., Raguchander, T. and Samiyappan, R. 2001. Induction of systemic resistance in rice against sheath blight disease by *Pseudomonas fluorescens*. *Soil Biol Biochem*. **33**: 603-612.

Nema, A.G. 1992. Studies on Pectinolytic and cellulolytic enzymes produced by *Fusarium udum*, causing wilt of Pigeon pea- *Indian Jour. Forestry* **15**: 353-355.

O'Donnell, K., Kistler, H.C., Cigelnik, E. and Ploetz, R.C. 1998. *Multiple evolutionary origins of the fungus causing Panama disease of banana: Concordant evidence from nuclear and mitochondrial gene genealogies*. Proceedings of the National Academy of Sciences of the United States of America **95**: 2044–2049.

Pandey, R.N., Pawar, S.E. and Bhatia, C.R. 1996. Inheritance of wilt resistance in Pigeonpea. *Indian Journal of Genetics and Plant Breeding* **56**: 303-308.

Pandey, R.N., Pawar, S.E. and Bhatia, C.R. 1995. Effect of culture filtrate of *Fusarium udum* and fusaric acid on wilt susceptible and resistant pigeonpea cultivars. *Indian Phytopathology* **48**: 444-448.

Pandey, R.N., Pawar, S.E. and Bhatia, C.R. 1997. Interaction between *Fusarium udum* and wilt susceptible and resistant pigeonpea genotypes: Fungal growth and histopathology. *Indian Phytopathology* **50(1)**: 53-58

Pattern, C.L. and Glick, B.R. 2002. Role of *Pseudomonas putida* indoleacetic acid in development of host plant root system. *Applied Environmental Microbiology* **68**: 3795-3801.

Pawar, S.V., Deshpande, G.D., Dhutraj, D.N. and Utpal, D. 2013. Survey of pigeonpea wilt disease in marathwada region of Maharashtra state. *Bioinfolet* **10(1a)**: 175-176

Prasad, P.S. and Saifulla, M. 2012. Effect of soil moisture and temperature on population dynamics of Fusarium udum causing pigeonpea wilt. *Trends in Biosciences* **5(4)**: 303-305.

Raaijmakers, J., Weller, D.M. and Thomashow, L.S. 1997. Frequency of antibiotic producing *Pseudomonas* spp in natural environments. *Appl. Environ. Microbiol.* **63**: 881-887.

Rajesh Singh., Singh, B.K., Upadhyay, R. S., Bharat Rai., and Youn Su Lee. 2002. Biological control of *Fusarium* wilt disease of Pigeonpea. *Plant Pathol. J.* **18 (5)**: 279-283.

Ramamoorthy, V., Viswanathan, R., Raguchander, T., Prakasam, V., and Samiyappan, R. 2001. Induction of systemic resistance by plant growth-promoting rhizobacteria in crop plants against pests and diseases. *Crop Protection* **20**: 1-11.

Ranjana Devi, T. and Chhetry, G.K.N. 2012. Rhizosphere and non-rhizosphere microbial population dynamics and their effect on wilt causing pathogen of pigeonpea. *International Journal of Scientific and Research Publications* **2(5)**: 1-4.

Reddy, M.V., Raju, T. N. and Lenne, J.M. 1998. *Diseases of pigeonpea* In: D.J. Allen and J.M. Lenne (Eds.), The pathology of food and pasture legumes CAB International, NewYork, USA, p. 517-558.

Reddy, M.V., Sharma, S.B. and Mere Y.C. 1990. *Pigeonpea disease management*. In: Y.L.Mere, S.D. Hall and V.K. Sheila (Eds.), The Pigeonpea CAB. Int. Oxon, V.K. p. 303-37.

Sang, J.P., Minchinton, P.K. Johnstone, P.K. and Truscott, J.W. 1984. Glucosinolate profiles in the seed, root and leaf tissue of cabbage, mustard, rapeseed, radish and swede. *Can. J. Plant Sci.* **64**: 77-93.

Saxena, A. K., Lata, Annapurna, K. 2002. Training manual on plant growth promoting rhizobacteria. Division of Microbiology. Indian Agricultural Research Institute. New Delhi, India. p. 38.

Saxena, R. and Reddy, D.D.R. 1987. Crop losses in pigeonpea and mungbean by pigeonpea cyst nematode, Heterodera cajani. *Indian Journal of Nematology* **17**: 91-94.

Sharma, S.B. and Nene, Y.L. 1989. Interrelationship between *Heterodera cajani* and *Fusarium udum* in pigeonpea. *Nematropica* **19**: 12-28.

Sharma, S.B. and Nene, Y.L. 1990. Effect of Fusarium udum – alone and in combination with Rotylenchulus reniformis or Meloidogyne spp. on wilt incidence, growth of pigeonpea and multiplication of nematodes. *International Journal of Tropical Plant Diseases* **8**: 95-101.

Shaw, F.L.F. 1936. Inheritance of morphological characters and of wilt resistance in arhar. *Indian Journal of Agricultural Sciences* **6**: 439-487.

Shukla, A. and Dwivedi, S.K. 2012. Pathogenic action of Cx, PG and PMG enzymes of Fusarium udum and Fusarium *oxysporum* f.sp. *ciceri*. *International Journal of Current Research* **4(6)**: 111-113.

Siddiqui, Z. A. And Mahmood, I. 1996. Effects of *Heterodera cajani, Meloidogyne icongnita* and *Fusarium udum* on the wilt disease complex of Pigeonpea. *Indian Journal of Nematology* **26**: 102-104.

Siddiqui, Z.A. and Shakeel, U. 2007. Screening of Bacillus isolates for potential biocontrol of the wilt disease complex of pigeonpea (*Cajanus cajan*) under greenhouse and small-scale field conditions. *Journal of Plant Pathology* **89(2)**: 179-183.

Siddiqui, Z.A. and Shakeel, U. 2009. Biocontrol of wilt disease complex of pigeonpea (*Cajanus cajan* (L.) Millsp.) by isolates of *Pseudomonas* spp. *African Journal of Plant Science* 3(1): 1-12.

Steenkamp, E.T., Wingfield, B.D., Coutinho, T.A., Wingfield, M.J. and Marasas, W.F. 1999. Differentiation of *Fusarium subglutinans* f. sp. *pini* by Histone Gene Sequence Data. *Appl Environ Microbiol.* **65**: 3401–3406.

Thomas, C.A. 1949. A wilt inducing polysaccharide from *Fusarium solani f.sp. eumarnii.Phytopathology* 39: 572-579.

Thomashow, L.S. and Weller, D.M. 1988. Role of Phenazine antibiotic from Pseudomonas fluorescens biological control of *Gaeumannomyces graminis* var. *Tritici. J. Bacteriol.* **170**: 3499-3508.

Tiwari, S. and Dhar, V. 2011. Histopathological studies in pigeonpea genotypes, resistant and susceptible to Fusarium wilt. *Indian Phytopathology* **64(1)**: 89-90.

Upadhyay, R.S. 1979. Ecological studies on *Fusarium udum* Butler causing wilt disease of pigeonpea. *Ph.D. Thesis*, Banaras Hindu University.

Vaidya, R.J., Macmil, S.L, Vyas, P.R., Ghetiya, L.V., Thakor, K.J., Chhatpar, H.S. 2003. Biological control of Fusarium wilt of pigeonpea *Cajanus cajan* (L.) Millsp with chitinolytic *Alcaligenes xylosoxydans*. *Indian Journal of Experimental Biology* **41(12)**: 1469-72.

Van Loon, L.C., Bakker, P.A.H.M., and Pieterse, C.M.J. 1998. Systemic resistance induced by rhizosphere bacteria. *Annual Review of Phytopathology* **36**: 453-483.

Vasudeva, R.S., Singh, G.P. and Iyengar, M.R.S. 1962. Biological acivity of bulbiformin in soil. *Annals of applied biology* **50(1)**: 113-119.

Xu XH., Gu, G.F. and Shen, F.D. (1983). Studies on the relationship between fusaric acid and cotton fusarial wilt disease. *Alta. Agric. Univ. Pekinensis.* **9**: 29-35.

Yadav, M.K., Babu, B.K., Saxena, A.K., Singh, B.P., Singh, K., *et al.*, 2011. Real-time PCR assay based on Topoisomerase- II Gene for Detection of *Fusarium udum*. *Mycopathologia* **171**: 373-381.

Zehnder, G.W., Murphy, J.F., Sikora, E.J., and Kloepper, J.W. 2001. Application of rhizobacteria for induced resistance. *European Journal of Plant Pathology* **107**: 39-50.

2016, Diseases of Pulse Crops and their Sustainable Management *319–354*
Editors: **Samir Kumar Biswas, Santosh Kumar and Gireesh Chand**
Published by: **BIOTECH BOOKS, NEW DELHI**

Chapter 18

Uromyces fabae Inciting Pea Rust: An Overview

S.K. Singh[1]*, Lalan Sharma[2] and H.S. Tripathi[1]

[1]Centre of Advanced Studies in Plant Pathology, College of Agriculture, GBPUA&T, Pantnagar (U.S. Nagar) – 263 145, , Uttarakhand
[2]National Bureau of Agriculturally Important Microorganisms, Kusmaur, Maunath Bhanjan – 275 101, U.P.

Introduction

Rust of field pea caused by *Uromyces fabae* is very severe under warm and humid conditions in *Tarai* region. Rusts have plagued farmers around the globe throughout history. Cereals and legumes are two most important crop for humans, suffer from rust infections (Graham and Vance, 2003). Cereal rusts have been a recurring problem in Northern America, occasionally causing huge losses (Long, 2003). Legume rusts prevail in Africa, Asia, and Oceania. The rust fungi comprise more than 100 genera and around 7000 species (Maier *et al.*, 2003). *Puccinia* represents the largest genus with about 4000 species, followed by the genus *Uromyces* with about 600 species (Maier *et al.*, 2003). Maier *et al.* (2003) suggests that these two genera are polyphyletic. This implies that although morphologically similar but differences at the molecular level. Mainly five species of rust fungi have served as model organisms in the laboratory. *Melampsora lini* and its host *Linum usitatissimum*, for example, were used by Flor (1956) to demonstrate the gene-for-gene hypothesis. *Uromyces appendiculatus*, and *Puccinia graminis* have been used in a number of cytological and physiological

* Corresponding Author: E-mail: singh.sk30@gmail.com

studies (Zhou *et al.*, 1991; Leonard and Szabo, 2005). Today molecular analyses of rust fungi mainly focus on *P. triticina* (Thara *et al.*, 2003), *M. lini* (Catanzariti *et al.*, 2006), and *Uromyces fabae* (Jakupovic *et al.*, 2006).

Some of the most devastating plant pathogens are obligate biotrophic parasites (Brown and Hovmoller, 2002). This term characterizes a lifestyle in which the host suffers only minor damage over a longer period of time. The pathogen in turn depends on a living host to complete its life cycle (Staples, 2000). This form of parasitism stands in contrast to necrotrophic pathogens, which kill their hosts quickly and subsequently thrive on the dead plant material (Mendgen and Hahn, 2002). In order to mark off the true obligate biotrophic pathogens from hemibiotrophs or necrotrophs, the following criteria were suggested: (a) highly differentiated infection structures; (b) limited secretory activity; (c) a narrow contact zone separating fungal and plant plasma membranes; (d) long term suppression of host defense responses; (e) the formation of haustoria (Mendgen and Hahn, 2002). Accordingly the true obligate biotrophs comprise the downey mildews, the powdery mildews, and the rusts.

The pathogen has served as a model organism for almost half a century. It started in the 1960s and 1970s with physiological and cytological investigations (Thrower and Thrower, 1966; Abu Zinada *et al.*, 1975). These studies were continued during 1980s (Kapooria and Mendgen, 1985), before biochemical aspects became a new focus in the early 1990s (Deising *et al.*, 1991). Work on molecular aspects of biotrophy became possible with the advent of a procedure to isolates haustoria (Hahn and Mendgen, 1992) but presently it is very typical and required sophisticated laboratory. While much progress has been made during the last couple of years, two facts still impede research with obligate biotrophs: first, none of the true obligate biotrophs can be grown in culture and second there is currently still no stable transformation system available. Review of literature pertaining to different aspects of the present study is briefly presented under the following sub heading:

Historical Background

Two species of *Uromyces* have been reported to cause rust of pea. One of them *U. pisi* (Persoon) de Bary, has been reported from European countries (Mayer, 1947; Jorstad, 1948; Palter and Stetbiner, 1957). It is a heteroecious species having its aecial stage *Euphorbia cyparissias* and rarely occurs in India. In India, another species *U. fabae* (Pers.) de Bary has been found to cause pea rust (Butler, 1918; Prasada and Verma, 1948; Kapooria and Sinha, 1966). In addition to pea, *U. fabae* infects broad bean (*Vicia faba*) and other species of *Vicia* (Deutelmoser, 1926; Hiratsuka, 1933; Kapooria and Sinha, 1966), lentil (*Lens esculanta*) (Patil, 1933) and *Lathyrus* species (Patil, 1933; Anderson, 1952).

Biology of *Uromyces fabae*

Pathogen Description

Uromyces fabae (*Uromyces viciae-fabae*) causes rust of pea was first reported by DCH Persoon in 1801. Later de Bary (1862) changed the genus as *Uromyces fabae* (Pers.) de Bary. Thereafter, Kispatic (1949) described f. sp. *viciae-fabae* by including the host *Vicia fabae*. The pathogen *U. fabae* is described as autoecious rust with aeciospores, urediospores and teliospores found on the same host plant (Arhur and Cummins, 1962; Gaumann, 1998). Gaumann proposed that the fungus be classified into 9 forma speciales each with a host range limited to two or three species. Later it was observed that the isolates of *Uromyces viciae-fabae* share so many hosts in common that it is impossible to classify them into forma speciales (Conner and Bernier, 1982). Recently, based on the distinctive shape and dimensions of substomatal vesicle, *U. viciae-fabae* has been described as a species complex (Emeran *et al.*, 2005). It revealed that host specialized isolates of *U. viciae-fabae* were morphologically distinct, differing in both spore dimensions and infection structure morphology.

Taxonomy and Nomenclature

Rust of pea is caused by fungus *Uromyces fabae* (Pers.) de Bary. The systemic position of the fungus is as follows (Alexopolus *et al.*, 1996).

Kingdom	:	Fungi
Phylum	:	Basidiomycota
Class	:	Urediniomycetes
Order	:	Uredinales
Family	:	Pucciniaceae
Genus	:	*Uromyces*
Species	:	*fabae var. viciae fabae*

Uromyces fabae is an autoecious and heterothallic fungus forming all the four spore forms *viz.*, pycniospores/spermatiospores, aeciospores, urediospores and teliospores on pea only. Pycnia are small, flask shaped and produced as yellowish flecks on upper surface of leaves with a common nectar drop at mouth. As the time advanced and the haploid pustules remained unfertilized the formation of pycnia, with separate scanty nectar drops on the lower surface of the leaves was also observed after a week or two (Prasada and Singh, 1975).

As long as the pycnia remain unfertilized, the receptive hyphae of the monosporidial pustules increased gradually by formation of a large number of pycnia on both the surfaces of leaves with gradual increase in size of pustules. Pycnia formed on the lower side were also functional and not of the abortive type (Prasada and Singh, 1975).

Aecia are formed after diplodization by fertilization of pycnia through pycniospores or joining of pycnia of two opposite sexes when formed side by side. The peridium of aecium is short, whitish and cup shaped in which dikaryotic aeciospores are produced in chains. Aeciospores are round to angular or elliptical, yellow in colour with fine warts and 14.22 μ in diameter (Agrawal and Prasad, 1997; Singh, 2005).

Pathogenicity and Symptomatology

All the four stages develop on every green part of the host including the pods. The first symptom appeared with the development of aecia. The yellow aecia appear first on the undersurface of the leaves, stems and petioles. The formation of aecial stage is preceded by a slight yellowing which gradually turns brown. The uredopustules are powdery light brown in appearance. The teleutopustules occur in the same sorus as the uredia and develop from the same mycelium (Singh, 1973).

Host Range and Geographical Distribution

Uppal (1933) and Prasada and Verma (1948) found that several species of *Vicia, Lathyrus, Pisum* and Lentil are susceptible to *U. fabae* in India and abroad. In India species of *Vicia, Lathyrus,* and *Pisum* are described as host plant for *U. fabae* (Pers. De Bary) (Kapooria and Sinha, 1966). They observed natural infection on *Vicia sativa* L. and *V. hirsute* Gray, a common weed found in the field of lentil in India also. *Vicia faba* L., *V. biennes* L., *V. hirsuta* L., and *V. arborensis* L. were described as highly susceptible to *U. fabae* and *Vicia sativa* and *Lathyrus aphaca* were found to be disease free. Conner and Bernier (1982) reported a total of 52 species of *Vicia faba* and 22 species of *Lathyrus* to be infected by *Uromyces viciae-fabae.* They also found this pathogen on pea, lentil and faba bean. However, it is most commonly referred to as rust of faba bean where yield losses up to 50 per cent have been reported (Tissera and Ayres, 1986).

Butler (1918) reported that occurrence of rust pathogen *U. fabae* on pea and other leguminous crops from India. A random survey of pea growing region of three districts of Bihar (Lal *et al.*, 2007) and six districts of Himanchal Pradesh (Chauhan, 1988) of India. Now the disease has been reported from all over the world where climate is found temperate and sub-tropical (Sharma, 1998). The pathogen has very wide host range. It can survive on different cultivated and non-cultivated crops. Sydow and Butler (1912) reported this fungus from Maharastra.

There were reports of occurrence of *U. fabae* from most of the places in India. Prasada and Verma (1948) also reported the occurrence of *U. fabae* on lentil crop from Delhi. Roy (1949) from Bengal recorded the prevalence of *U. fabae* on the leaves of stems of *Pisum sativum.* Mitter and Tondon (1930); Patel (1934); Pavgi and Upadhyay (1966) and Kapooria and Sinha (1966) reported the distribution of this pathogen in the regions of Uttar Pradesh, respectively. Bilgrami *et al.* (1979) reported the occurrence of this pathogen on various host species of pea, lentil and *Lathyrus.* Baruah *et al.* (1980) reported that rust infection on the pea plants is caused by both *U. fabae* and *U. pisi.* of which *U. pisi* is of rare occurrence in India. Occurrence of *U. fabae* have been reported from Canada, Europe, Ethiopia, Australia and Iran in mild to severe forms on pea, lentil, alfalfa, broad bean and faba bean are also available (Conner and Bernier, 1992; Xue and Warkentin, 2002; Sadravi, 2007) (Figure 18.1).

Life Cycle

Uromyces fabae is a macrocyclic fungus, it exhibits all five spore forms known for the Uredinales. It is also autoecious, as all spore forms are produced on a single host (Mendgen, 1997) (Figure 18.2). After overwintering on residual plant material, diploid teliospores germinate in the spring with a metabasidium. After meiosis, the latter produces four haploid basidiospores with two different mating types. These are ejected from the metabasidium and after landing on a leaf of a host germinate and produce infection structures. Pycnia are produced which contain pycniospores and

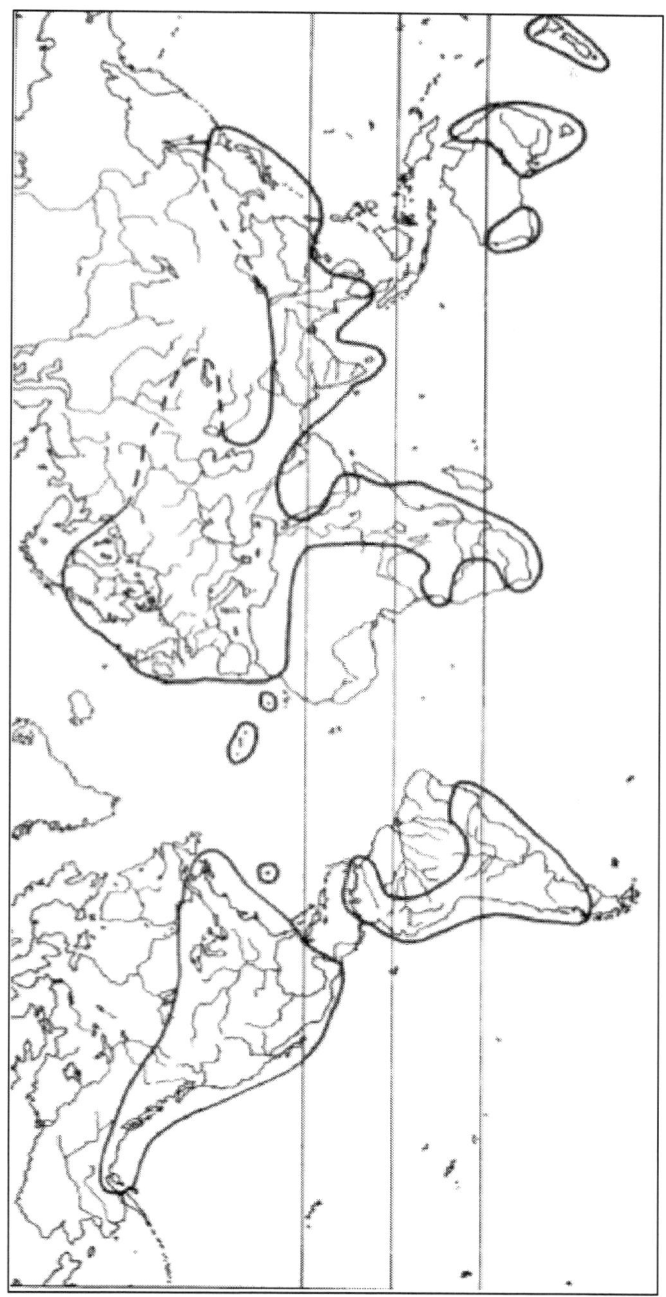

Figure 18.1: Geographical Distribution of *Uromyces fabae* and Occurrence of Pea Rust (CMI, Distribution maps of plant diseases).

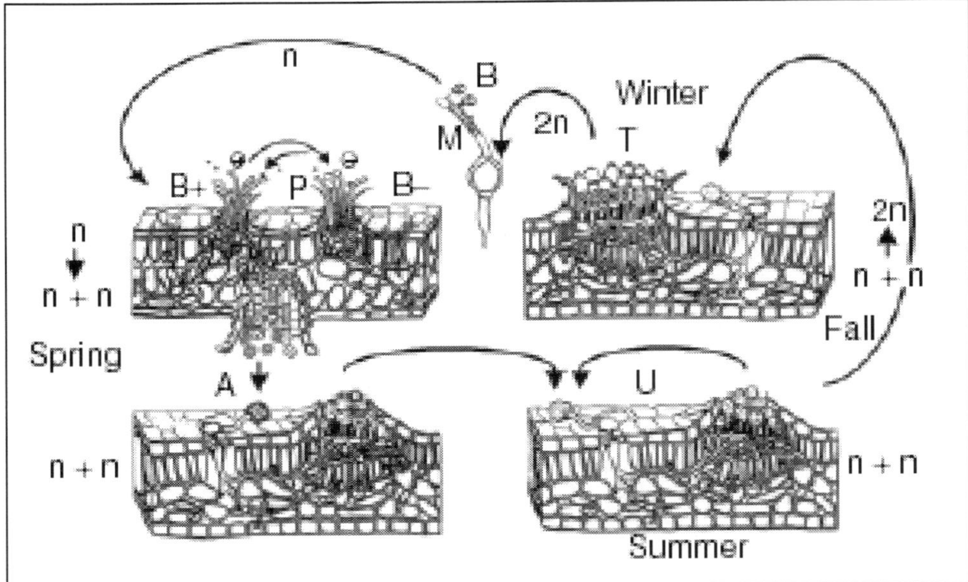

Figure 18.2: Life Cycle of *Uromyces fabae*. Overwintering diploid (2n) teliospores (T) germinate in the spring with a metabasidium (M) from which four haploid (n) basidiospores (B) of two mating types (+, -) are formed. Haploid pycniospores (P) are exchanged between pycnia of different mating types on the upper surface of a leaf. After spermatization dikaryotic (n+n) aeciospores (A) are formed in aecia at the lower surface of the leaf. Infecting aeciospores produce uredia from which dikaryotic urediospores (U) are formed. At the end of summer uredia differentiate into telia from which teliospores are formed and the cycle closes (Mendgen, 1997).

receptive hyphae. Pycniospores are exchanged between pycnia of different mating types and after spermatization, dikaryotization occurs in aecial primodia. An aecium differentiates and dikaryotic aeciospores are produced. These aeciospores germinate and form infection structures from which uredia which produce urediospores are formed. Urediospores are the major asexual spore form of rust fungi produced in massive amounts through repeated infection of host plants during the summer. Urediospores are aerially dispersed and can travel thousands of kilometers carried by the wind (Brown and Hovmoller, 2002). In the fall, uredia differentiate into telia, the nuclei fuse during sporogenesis and single-celled diploid teliospores develop for the winter.

Variability in the Pathogen

Pathogenic variability has been reported in field collection of *Uromyces fabae* (Singh and Sokhi, 1980; Conner and Bernier, 1982; Xue and Warkentin, 2002). The urediospores of *U. fabae* were the only infective spores and are used in various resistance-screening programme in pea (Xue and Warkentin, 2002), faba bean (Sillero *et al.*, 2000), lentil (Chauhan *et al.*, 1996) and sweet pea (Sokhi, 1984).

Morphological Characteristics

The peridium of aecium in *U. fabae* is short, whitish and cup shaped. The aeciospores are round to angular or elliptical, yellow in colour with fine warts. They measure 14-22μ in diameter. The servediospores are round to ovate light brown spinny with 3-4 germpores and measure 20-30 × 18.26μ. The teliospores are subglobose to ovate, thick walled, with flattened apex, smooth, single cell, pedecellate and measure 25-38 × 18.27μ (Arthur and Cummins 1962).

The height of stomatal lips was reported to induce the differentiation of the appressorium on the stomata in most of the rust pathogens (Allen *et al.*, 1991) and by *Uromyces appendiculatus* in bean rust (Wynn, 1976). They reported that a height of 0.5-0.75 μm to induce appressorium by urediospore of *Uromyces striatus* in alfalfa (Kemen *et al.*, 2005) for subsequent infection of the host leaves. The differences in epidermal features and structures may lead to avoidance of the pathogen for infection (Sillero and Rubiales, 2002). The role of haustoria in nutrient uptake has been extensively studied in *U. fabae* (Voegele and Mendgen, 2003) and specific genes expressed during differentiation of infection structure in urediospores have also been characterized (Xuei *et al.*, 1992).

Host Pathogen Interaction

Information is not available in the literature for histopathological and ultrastructural studies of pea rust disease. Although, we are mentioned here the literature on other rust to the related aspect. Urediospores are single-celled, hydrophobic, and carry spines on their surface (Woods and Beckett, 1987). An important morphological feature used to distinguish different rust species is the number and position of germ pores. Three to four germ pores in an equatorial or near equatorial position are typical for *Uromyces fabae* (Emeran *et al.*, 2005). Fully developed urediospores are almost completely dehydrated which gives them an irregular shape (Clement *et al.*, 1998). Only upon hydration do spores adopt an ellipsoid form. Although dry urediospores hydrate rapidly, their surface is nonwettable (Clement *et al.*, 1994). This hydrophobicity is responsible for the initial adhesion to the host surface (Clement *et al.*, 1993b). The initial contact is quickly followed by production of an extracellular matrix consisting of low-molecular-weight carbohydrates and glycosylated polypeptides (Clement *et al.*, 1993a). The next step is the formation of an adhesion pad. Cutinases and esterases are involved in the adhesion process because while spores treated with esterase inhibitors do form an adhesion pad, they fail to adhere (Deising *et al.*, 1992). Besides hydration (Clement *et al.*, 1997), two other factors have been found to have an influence on germination. One of these factors is light. A period of at least 40 min of darkness is required to induce germination (Joseph and Hering, 1997). The other parameter is temperature. Urediospores will germinate in a range between 5 and 26 °C with the optimal germination temperature being 20°C (Joseph and Hering, 1997). The cytoplasm moves into the germ tube as the developing germ tube meanders across the surface attached to it via matrix like material (Clement *et al.*, 1994). In order to produce infection structures downstream of the germ tube further signals are required. It has been shown that a topographical signal is needed for the differentiation of an appressorium (Allen *et al.*, 1991). In *U. appendiculatus* a mechanosensitive channel has been identified which might be involved in the transduction of the topographic signal into a differentiation response (Zhou *et al.*, 1991). The cytoplasm transfers to the appressorium and the vacuolated germ tube is separated by a septum. Differentiation of the appressorium coincides with the detection of a number of lytic enzymes (Deising *et al.*, 1995b). Acidic cellulases (Heiler *et al.*, 1993), extracellular proteases (Rauscher *et al.*, 1995), and chitin deacetylase (Deising and Siegrist, 1995) were found. At the base of the appressorium a penetration hypha is formed (Terhune *et al.*, 1993). For *U. appendiculatus* a turgor pressure of 0.35MPa has been reported. This pressure is high enough to distort artificial surfaces or stomatal guard cell lips (Terhune *et al.*, 1993). Within the stomatal cavity a substomatal vesicle is formed which is separated from the penetration hypha by a septum. The vesicle is a stretched cylindrical structure which narrows into an infection hypha (Kapooria and Mendgen, 1985). More enzymes can be detected which may have a role in the local breakdown of the host cell wall (Deising *et al.*, 1995b). Pectin esterases (Frittrang *et al.*, 1992), pectin methylesterases (Deising *et al.*, 1995a), and neutral cellulases (Heiler *et al.*, 1993) can be found. A haustorial mother cell is differentiated upon contact with a mesophyll cell, and it is separated from the infection hypha by a septum. The cytoplasm moves into the haustorial mother cell and earlier structures are vacuolated. Formation of the haustorial mother cell coincides with the onset of polygalacturonate lyase activity (Deising *et al.*, 1995a). Up to the haustorial mother cell, infection structures of this 'penetration phase' can be generated *in vitro* by germinating spores on colloidon membranes (Kapooria and Mendgen, 1985) or on structured polyethylene sheets (Deising *et al.*, 1991). Haustoria, and structures of the 'parasitic phase' and the 'sporulation phase' are only formed *in planta*. Haustorial mother cells have a thick, multilayered wall that attaches firmly to the host wall and forms a penetration hypha to invade the host cell (Heath, 1997). One or more signals of the host are needed to complete the differentiation of the haustorium (Heath, 1990). The haustorium develops from the haustorial mother cell with a slender neck and a haustorial body (Heath and Skalamera, 1997). During the formation of the haustorium, the cell wall of the host cell is breeched. The expanding haustorium invaginates the host plasma membrane. With development of the haustorium, a zone of separation is formed between the plasma membranes of parasite and host. It is composed of the fungal cell wall and the extrahaustorial matrix (Hahn *et al.*, 1997a). It seems likely that this zone plays an important role in maintaining the biotrophic life style. Undoubtedly the extrahaustorial matrix represents a formidable trading place for the exchange of nutrients and information (Heath and Skalamera, 1997). There is some evidence that the cytoplasmic membrane of the host enclosing the haustorial body, the so-called extrahaustorial membrane. The neck region of the haustorium is characterized by electron-dense material joining the plasma membranes of host and parasite (Harder and Chong, 1984). Thatcher (1939) studied the effect of *U. fabae* on pea. He pointed out that fungus increased the permeability of the host cell by secreting some metabolites, which ultimately proves fatal and recorded by Ehrlich and Ehrlich (1963) for *Puccinia graminis* var. *tritici*.

Weintraub and Ragetli (1966) evaluated the infection of *Vicia faba* leaves infected with bean mosaic virus causes ultrastructural changes in the chloroplasts. These are generally similar to the changes observed in chloroplasts of other plants infected with rusts or viruses, treated with various chemicals or during the process of leaf senescence. It appears that one of the mechanisms by which the rust may attack the leaf tissues of *Vicia faba* is by exudation of enzymes involved in the degradation of starch and synthesis of glycogen.

Transmission electron microscopy was used to examine details of the host–pathogen interface in daylily leaf cells infected by the rust fungus *Puccinia hemerocallidis*. The extrahaustorial membrane that separated each dikaryotic haustorium from the cytoplasm of its host cell was especially well preserved and appeared almost completely smooth in profile. Large aggregations of tubular cytoplasmic elements were present near haustoria in infected host cells. Many of these tubular elements were found to be continuous with the extrahaustorial membrane and conspicuous electron-dense deposits present in the extrahaustorial matrix extended into these elements. The use of gold-conjugated wheat germ agglutinin for labeling of chitin revealed that these deposits were not part of the haustorial wall. Portions of many of the tubular elements associated with haustoria were conspicuously beaded in appearance. Some tubular elements were found to be continuous with flattened cisternae that in turn bore short beaded chains. Distinctive tubular-vesicular complexes previously reported only in cryofixed rust haustoria also were found in the haustoria of *P. hemerocallidis* (Mim *et al.*, 2002).

Cymadothea trifolii, a biotrophic leaf pathogen, forms a unique structure within its own hyphae, presumably for nutrient uptake from its host. This structure, called an interaction apparatus, consists of long, thin, often net-like cisternae surrounded by a membrane continuous with the fungal plasma membrane. The plant plasmalemma opposite the interaction apparatus invaginates to produce a host bubble. The interaction apparatus and host bubble are apoplastic and are linked by a tube with an electron dense sheath that may channel nutrients from the host to the pathogen. Within the tube, the cell walls of host and parasite appear altered. The interaction apparatus and host bubble may be analogous to haustoria in other obligately biotrophic fungi while the electron dense sheath of the tube may be equivalent to the haustorial neckband (Simon *et al.*, 2004).

Disease Development

Effect of Temperature on Germination

The differences in the spore germination and its subsequent differentiation on the resistance and susceptible genotypes are well documented. It was observed that in both resistant and susceptible genotypes only 20 per cent of the spores usually germinate on the host surface. Significant number of urediospores formed appressorium on the stomata in the susceptible genotypes compared to resistant germplasm lines (Sillero and Rubiales, 2002). The effect of wetness duration and temperature on the germination and subsequent infection of the host surface was studied in rust of alfalfa (Webb and Nutter, 1997). According to them the time of appearance of rust in alfalfa ranged from 7 days at 30°C to 21 days at 15°C and appearance of pustule is greatly effected by post inoculation temperature.

Prasad and Verma (1948) working with *Uromyces fabae* from lentil found that infection with aeciospores at lower temperatures (17-26°C) results in the formation of secondary aecia, while at 25°C the infection causes development of uredia. No infection by aeciospores occurs at 30°C. Optimum temperature for germination of uredospores is 16-22°C, while no germination occurs at 28-29°C. The teleutosores of lentil rust can germinate at 12-22°C. The fungus completes its life cycle on peas and is further endowed with survival potential in the telial stage (Singh, 1973).

El-Fiki *et al.* (1999) suggested that uredospores of isolate A grown on modified MS medium alone showed higher germination than those produced on the same medium supplemented with broad bean leaf extract, while an opposite trend was observed for isolate B. Percentage of uredospore germination was generally higher in isolate B than isolate A. The highest germination percentage was obtained when 2 per cent sucrose or glucose was used as a germination substrate. Polygalacturonase (PG) and cellulolytic (Cx) enzymes induced in the growth medium were higher in isolate A than isolate B.

Mode of Survival of the Pathogen

The aeciospores and urediospores of *Uromyces fabae* did not survive at a temperature more than 30°C for one week. Therefore, they were not supposed to survive the high temperature of the intervening crop season. Teliospores survive the intervening season and germinate to produce basidiospore-producing pycnia, which subsequently cause infection in pea (Singh, 1999). The *U. fabae* has a wide host range. Conner and Bernier (1982) have suggested the role of collateral host for their survival. They reported that *Vicia species* and *Lathyrus species* served as a collateral host to *U. fabae* and helped in its survival during the absence of the main crop.

Effect of Temperatures on Survival of Urediospore

Very limited information is available in the literature regarding the effect of temperature on the perpetuation of fungus (*Uromyces viciae-fabae*) in nature.
Prasada and Verma (1948) reported that at relatively low temperatures of 17-22°C results in formation of secondary aecia while at 25°C the infection causes development of uredia. No infection by aeciospores occurs at 30°C. These spores remain viable for 8, 6, 4, 3 and 2 weeks at temperatures of 3-8°C, 10-12°C, 17-18°C and 30°C, respectively. No viability in retained after 6 weeks showing that the aeciospores do not survive during the off season for the crop.

Singh (1998) reported that optimum temperature for germination of urediospores in 16-25°C. No germination occurs at 28-29°C. In the plain of North India, where warm season sets towards the end of March, these spores from lentil were found to remain viable for 16-17 week, when stored at 3-8°C and only for 2 weeks at 36-37°C. Thus these spores do not survive in the hot summer interning two successive crop of lentil. The teliospore of the lentil rust fungus have been found to have no dormancy and can germinate at 12-22°C soon after their formation.

Batra and Stavely (1994) working on *Uromyces appendiculatus* (Pers.) Fr. reported that the replanting spore (urediospore) germinate best at 15 to 24°C, while the teliospores in crop debris germinate at 10°C to 15°C under favourable conditions the spore complete the infection cycle with in 5 day and a new crop of urediospores is produced within next 5-10 days under favourable conditions.

Gross *et al.* (2001) reported that, the source of initial inoculum is not clearly understood. This study determined the potential for urediniospore survival overwinter. Uredia-bearing bean leaves from artificially inoculated greenhouse grown plants were kept outside near a field from November to May from 1990 to 1996. Based on bioassay urediniospores survived overwinter but viability declined over time. Overwinter survival indicates urediniospores may function as initial inoculum in state (USA).

In rust of chickpea caused by *Uromyces ciceris-arietini* (Gregnon), the urediospore lose viability in open within 2-4 weeks but can survive for longer period if kept in sealed vials at 6°C (Singh, 1998). They germinate at any temperature between 5° and 30°C but the optimum between 11-20°C. No germination occurs at 35°C and above. The incubation period depends on prevailing temperature, being 15 days at 11-12°C, 11 days at 20-25°C and 19 days at 25-30°C. The spores were killed when exposed to 45°C for 72 h to 40°C for 96 h or to 35°C for 8 days, obviously these spores of the fungus are not likely to survive the summer heat in the plains of North India. Attempts to germinate teliospores in laboratory have been unsuccessful. Their formation also does not take place at temperature below 20-25°C.

Chanda *et al.* (2006) reported that the dominance of aeciospores at all growth stages of pea in Northern East Plain Zone. Urediospore production was erratic and was only observed in a few samples of stems and tendrils (5-10 per cent). Inoculation of pea plants either by aeciospores or urediospores resulted in the production of aeciospores. Production of aeciospores was observed at a temperature range of 10-25°C, with a maximum at 25+2°C. Among the different growth stages of pea, the pod formation stage was highly susceptible and produced the maximum number (744) of aecidia/leaf at 20-25°C. Urediospore production mainly coincided with the senescence of the pea plants. Maximum germination (2 per cent) of aeciospores was observed at 25°C, whereas maximum urediospore germination (3.5 per cent) was at 15°C. Temperatures >15°C decreased urediospore germination. A relative humidity (RH) of 100 per cent was favourable for aeciospore germination while 98 per cent RH favoured urediospore germination.

Role Environmental Factor affecting Disease Development

In the literature information pertaining to severity of infection by *Uromyces viciae-fabae* and pustules per plant increased progressively with an increase in the duration of leaf wetness upto 24h, but did not increase further significantly. Both were high at 20°C under greenhouse and laboratory conditions. It was observed that relationship between severity of pea rust and duration of leaf wetness at above 20°C temperature may be useful in predicting disease outbreak if initial inoculum is present (Chauhan and Singh, 1994).

Beckett *et al.* (1990) concluded that dry urediospores of *U. viciae-fabae* will not germinate but will adhere to substrata (faba bean and pea leaves and synthetic surfaces). High humidity or free water initiates germination. Spores imbibe free water, swell and germ pores are lysed. Extracellular matrix is produced and possibly released through the germ pore. Emerging germ tubes are enclosed in a sheath of matrix and adhere tightly to the substratum. Several factors, either singly or in combination, may induce the formation of appressorium like structures.

Periodically rust infection reaches severe epidemic proportions in North Dakaka. Under normal weather conditions, the disease develops too late in the growing season (about early pod striping) to cause serious damage. Rust attacks when growing conditions are cool (60-75°F, 15-24°C) and moist with frequent, dew or rains. Rust is most serious on late planted beans, heavily fertilized plants, bean delayed in maturity by weather damage such as hail or on plants planted or adjacent to old plant ground.

Sharma (1998) observed that free moisture (rain, dew) or high relative humidity (96 per cent plus) are essential for infection of *Uromyces appendiculatus* when moisture and temperature are favourable for more than eight hours and spores are present. Sumartini (1998) reported that rust is a major disease of Phaseolus caused by *Uromyces phaseoli* (*Uromyces appendiculatus*). Bean planted either in the dry or wet season could be infected and the wind, insect, farm machinery and human being play an important role in disease spread. Urdbean rust intensity was recorded on 1-9 scale during the crop season and compared with data on weather parameter such as rainfall, relative humidity, rainy days and minimum temperature. The results showed that rainfall and relative humidity during December were highly critical in determining rust disease intensity on urdbean (Srinivasulu *et al.*, 1999).

Similarly the occurrence of rust disease (*Uromyces appendiculatus*) in French bean in mid-hills of Uttar Pradesh is erratic. Analysis of meteorological data of eight years (1986-1993) showed that excessive precipitation is deleterious to disease development, which explains the occurrence of severe infection during periods of low rainfall (Sharma, 1998). Atmospheric temperatures around 20°C maximum and 5°C minimum with high RH (60-70 per cent mean weekly) and light shower or drizzle favour *Uromyces viciae-fabae* development and spread. Higher temperature 25°C maximum and 7-8°C minimum and less or more rains disfavour rust spread (Mittal, 1997). Khare and Agrawal (1978) reported that high humidity, cloudy or drizzling weather with temperature of 20-22°C favours disease and that plants are more susceptible at flowering in lentil for *Uromyces viciae-fabae.* Similar conditions were reported for rust development and spread in chickpea (Grewal, 1988).

Silva *et al.* (2001) conducted experiments to evaluate the influence of genetic resistance and wetness period on the infection and lesion development of rust disease in common bean (*Phaseolus vulgaris*). Susceptible, moderately resistant and resistant cultivars were subjected to 4, 8, 12, 16, 20 and 24 h of wetness periods after inoculation with *Uromyces appendiculatus.* There was significant effect of genetic resistance and wetness period on the pathogen infection and lesion development. Resistance cultivars did not exhibit lesions at any wetness period. In these cultivars, the number of lesion increased upto 24 h of wetness. Moderately resistant cultivars also exhibited wetness. There was no significant effect of 16, 20 and 24 h wetness duration in relation to infected leaf area or lesion per cm^2. A wetness period of 16 h with a temperature of 22°C was sufficient for rust level resistance evaluation in common bean.

Schwartz *et al.* (2001) reported that rust of bean (*Uromyces appendiculatusa*) development is favoured by cool to moderate temperatures (65 to 85°F) with moist conditions that resulted in prolonged period of free water on the leaf surface for more than 10 hours. Weather conditions during late July and August usually are the most favourable because of cooler night. Repeating disease cycle may occur at 10-14 day intervals under favourable conditions. Negussie *et al.* (2005) reported that at 20°C, dew period of at least 3 h was required for minimum infection of lentil rust, whereas maximum infection occurred with a dew period of 24 h. Infection efficiency increased linearly as the duration of dew period increased from 0 to 24 h. Decision to apply one or more fungicide treatment will depend on the risk of rust epidemic in a particular year. Bean rust prefer cool, moist weather and is further enhanced by varieties susceptible to the pathogen. To assist growers in determining rust potential, a plant rust forecast model with a worksheet was developed (Steadman, 2000).

Disease Management

Cultural Practices

In the literature information in respect to effect of cultural practices on fieldpea rust and grain yield of fieldpea is scanty. However, information is available on other crops. Cultural practices mostly affect the environmental conditions favourable for growth and buildup of inoculum. These can be utilized alone or in combination as a means of plant disease control.

Effect of Sowing Date

Delay in sowing after 5 October, increased the incidence of *Uromyces viciae-fabae* and decreased grain yield. Cultivar Khaparkheda gave the highest seed yield (1.54 t/ha) and had the lowest incidence (Sangar and Singh, 1994). Bhardwaj and Sharma (1996) reported that plants from 15 October sowing were taller, produced the highest number of marketable pods and highest green pod yield (4.74 t/ha). Among the genotype, Arkel and VL-7 were earlier to flower and recorded the higher green pod yield (4.34 and 4.22 t/ha, respectively). Percentage disease index of rust (*Uromyces viciae-fabae*) was lowest in 15 October sowing followed by 30 September. VL-7 and Arkel recorded the lowest disease index of rust.

Singh *et al.* (1996) reported that three pea cultivars were sown on 4 dates between 5 October and 4 December in a field trail. Mean seed yield was 2.23, 1.80, 1.29 and 0.61 with sowing on 5 October, 25 October, 14 November and 4 December, respectively. Incidence of rust (*Uromyces viciae-fabae*) increased as sowing was delayed. Disease incidence was lower in cv. Khaperkheda than Rachna or JP 789, but seed yields were not significantly different between cultivars.

Bhardwaj and Sharma (1996) tested fourteen sowing dates of pea cv. Lincoin, crop sown on 18 October, powdery mildew (*Erysiphe polygoni*) appeared late by 21 and 18 days during 1989 and 1990, respectively, whereas rust (*Uromyces viciae-fabae*) infection was delayed by 8 and 10 days, respectively. Powdery mildew density was maximum in mid October sown crop in comparison to September and November sowing, being highest in the 23 October sown (78.2 average per cent disease index) followed by 18, 28 and 30 October sowing. Per cent disease index of rust was maximum (30.6) in early (25 September) sown crop in comparison to late sowings. However, sowing of pea on 13, 18 and 23 October gave maximum pod yield of 90.6, 84.2 and 82.3 kg/ha, where yield losses were estimated to the tune of 17.81, 23.56 and 26.03 per cent respectively. Mittal (1997) recorded the effect of sowing dates on rust disease occurrence and crop yield in lentil cv. VL-Massor-1, sown on four dates (9, 19 and 29 October and 9 November) during five *rabi* seasons (1989 to 1993). He observed that incidence of disease declined from the first to last sowing date. Sowing on 19 October was most effective in reducing the disease and increasing yield of lentil.

Host Resistance

For evolvement of rust resistant varieties there is need to screening of existing lines / germplasms / cultivars against the disease.

Kumar *et al.* (1994) tested thirty tall genotypes of fieldpea were screened against rust severity. Variety Pant peas had lowest peak rust cover. Under disease progress curve (AUDPC) value, growth rate (c) and apparent infection rate (r). However, in general KFP 106, DMR 11, HUP 8603, type 163, KPMR 22 showed high level of slow resistance, being conditioned by a number of genes with small effects is more desirable.

Xue and Warkentin (2002) observed that 93 fieldpea varieties to three isolates of *U. viciae-fabae* with symptoms (LAS) under control condition. Significant difference ($P<0.5$) was observed from pea varieties and rust isolates, and variety × isolate interaction. The varieties, Tara and Century were the most resistant to both UF-1 and UF-3 isolates. Victoria and Topper were the most resistant to UF-2 only.

The 648 accessions of *Vicia faba* was screened for resistance to faba bean rust (*Uromyces viciae-fabae*). Two distinct types of resistance severity (DS) and area under the disease progress curve (AUDPC), but differing in the expression of hypersensitivity. These two types of resistance were characterized by three macroscopic components of resistance, increased latent period, decreased colony size and relatively reduced infection frequency, both on seedlings and on adult plant (Sillero *et al.*, 2000). The genetics of resistance to *Erysiphae polygoni* and *U. viciae-fabae* in peas in Varanasi, India (Tyagi, 1999). Mishra (2009) evaluated the out of 107 genotypes tested, genotypes P 9-77, P 2432, P 2572 and P 2930 were found resistant, whereas 27 exhibited moderate reaction.

Screening of Germplasm Lines

Screening for rust severity indicated wide range of variations for rust resistance in the germplasm lines of pea and none of the genotypes tested were found to be free from infection (Xue and Warkentin, 2002; Kumar *et al.*, 1994; Chand *et al.*, 2004a; Gupta, 1990; Singh and Srivastava, 1985, Narshinghani *et al.*, 1980). Rust is described as one of the major diseases of the field pea and is responsible for substantial losses in grain yield (Kumar *et al.*, 1994). They used area under disease progress curve (AUDPC) to depict the overall disease stress that the plants were subjected to and described the pea varieties Pant P 8, KEP 106, DMR 11, HUP 8063, Type 163, KPMR 22 to possess good levels of partial resistance that could be effectively used in the breeding porgramme. Xue and Warkentin (2002) reported significant interaction between variety and isolate. These observations indicated that screening of pea germplasm lines cannot be based on any single isolate of *U. fabae*. Screening technique was standardized using aeciospore under polyhouse conitions. Successful epiphytotics were created using this technique repeatedly both under field and polyhouse conditions (Chand *et al.*, 2004a).

Chemical Control

Several inorganic sulphur preparations are reported to give effective control of *U. fabae* (El-Helaly, 1939; Jacks, 1954; Accantino, 1964). El-Helaly (1939) reported that fortnightly sulphur sprays at 5 per cent concentration with 0.2 per cent soap brought broad bean rust infection to 25 per cent from 100 per cent in unsprayed plots.

Jacks (1954), found that broad bean leaves sprayed with water gave a mean number of 295.8 rust lesions per leaf as against no lesions on leaves sprayed with either 0.75 per cent lime sulphur, 0.5 per cent time sulphur plus 0.2 per cent colsul 40 (40 per cent colloidal sulphur), 1 per cent colsul 40 or 0.5 per cent cosan (75 per cent wettable sulphur). It was reported by El-Helaly (1939) that 40 per cent sulphur in suspension (sulsol) at 0.3 per cent concentration with 0.2 per cent soap, when sprayed on broad bean plants, reduced the rust severity from 100 per cent in unsprayed plots to 35 per cent in sprayed plots.

Organic sulphur fungicides like ferbam, ziram, thiram and zineb have been reported to give good control of *U. fabae* (Jacks, 1954; Jacks and Webb, 1956; Accantino, 1964). Jacks (1954) reported that when broad bean plants were sprayed with a fungicide suspension and inoculated 24 hours later with a spore suspension of *U. fabae*, it was observed that in unsprayed plants, the mean number of rust lesions were 295.8 per leaf against none in plants sprayed with 0.45 per cent dithane Z-78 (65 per cent zineb), 0.5 per cent ferm spray (70 per cent ferbam) and 0.5 per cent manzate (70 per cent maneb). On the other hand the plants sprayed with 0.4 per cent thiospray (70 per cent zineb) gave 0.3 or less number of lesion per leaf.

Under field tests with dithane Z-78 (zineb @ 0.14 per cent) and fudasin ultra (ziram @ 0.14 per cent) sprayed at 3 week interval, the number of lesions per leaf was reduced to 1.6 and 5.8, respectively, as compared to 67.8 in case of water sprayed plants (Jacks and Webbs, 1956). Manzate (maneb @ 0.03 per cent) sprayed at 14 days interval and Thiospray (thirma @ 0.1 per cent) sprayed at 7 days interval were also equally effective against *U. fabae* (Jacks and Webbs, 1956).

Different copper fungicides have been used against *U. fabae*. These included Bordeaux mixture (Deutelmoser, 1926; El-Helaly, 1939; Kispactic, 1949), Copper oxychloride (El-Helaly, 1939) and Cuprous oxide (Reichert and Patil, 1946). Dose of bordeaux mixture varying from 0.25 to 2.0 per cent has been reported to give good control of the disease on bean (Deutelmoser, 1926; El-Helaly, 1959; Kispatic, 1949). El-Helaly (1939) reported that unsprayed plots of broad bean showed 100 per cent infection while the pots sprayed with 1 per cent bordeaux mixture had a disease severity of 15 per cent only. He also reported the effectiveness of a 25 per cent copper oxychloride preparation (boaisol) @ 0.2 per cent concentration. Cuprous oxide (Perenox) at 0.3 per cent concentration sprayed @ 10-14 day interval controlled the disease and increased the yield by 50 per cent (Reichert and Palti, 1946).

Very few systemic fungicides have been tested against *Uromyces,* plantvax (oxycarboxin), vitavax (carboxin), pyra carbolid, indar (RH-124; 4-n butyl 1,2,4 triazole and triforine have been reported to be effective against *U. appendiculatus* causing bean rust. Okioga and Jaffer (1972) reported that bean rust best controlled by Plantavax (2.5 kg a.i./ha), triforine (0.1 per cent solution) and RH-124 (675 g/ha).

Hiremath and Pavgi (1971) obtained complete inhibition of aeciospore germination of *U. fabae* with aureofungin to 20μg/ml and recommended early application of higher aureofungin concentrations to control rust disease. Chauhan (1988) have reported that G 696, tecto, benlate, bavistin and topsin M were effective in suppressing aeciospore germination and germ tube elongation, the best being G 696 and Benlate. Sugha *et al.* (1994) reported sensitivities of aeciospores and urediospores to benzimidazole and triazole fugicides. They concluded that benomyl, carbendazim, thiabendazole and thiophanate methyl have very good potential for suppressing the early establishment of pea rust due to aeciospores, whereas benomyl, flutriafol and myclobutanil should be effective in suppressing the late infections due to urediospores.

Efficacy of plantvax and vitavax against *U. appendiculatus* infecting cowpea plats and found both these to the highly effective as protective and curative sprays (Parambaramani, 1971). Pyracarbolid, plantvax and triforine to be effective against bean rust. The treatments improved seed quality but not necessarily yield (Okioga and Jaffer, 1972).

Hedge *et al.* (1974) observed that pea crop sprayed twice with benlate (0.05 per cent), calixin (0.1 per cent), plantvax (0.1 per cent), a combinations of benlate + plantvax and calixin + plantvax. The results indicated that calixin and combination of calixin + plantvax effectively controlled pea rust. Phygon XL (0.25 per cent) and spergon (0.5 per cent) under glasshouse conditions reduced the number of lesions per leaf to nil and 0.3 or less, respectively, against 295.8 lesions per leaf in plants sprayed with water (Jacks, 1954). The effectiveness of spergon was also observed under field conditions (Jacks and Webbs, 1956), when 0.1 per cent spergon was sprayed at 10 day interval on broad bean plants, the disease incidence was considerably reduced. Jacks (1954) reported that broad bean plants sprayed with tilt 406 at 0.2 per cent and reported that mean number of rust lesions per leaf was 295.8 on unsprayed plants as against 0.3 or less on plants sprayed with tlit 406 thereby giving an effective control of *U. fabae* on beans.

Chemical like sulpha preparations and antibiotics have also been reported effective against *U. fabae* (Tuffery, 1954). Tuffery (1954) reported that broad bean plants sprayed with 30 ppm solutions of streptomycin was free from rust, which was present in neighboring unsprayed plants. Jones (1956) found that effectiveness of sulpha preparations like sulphanilamide containing 100 mg per ml sulphanamide translocated rapidly into leaves and consequently was very effective against *U. fabae,* whereas sulphaghanidin and sulphadiazine tended to accumulate in the roots and hence were less effective.

Accatino (1964) reported from Chile, the control of lentil rust by nickel nitrate. Lentil plants spayed with 0.5 per cent nickel nitrate solutions at flowering time remained free from the rust. Upadhyay and Gupta (1994) studied the effect of bayleton (triadimefon) calixin (tridemorph), calixin-M (maneb + tridemorph), dithane M-45 (mancozeb), karathane (dinocap) and sulfex (wettable sulfur) were tested against rust (*Uromyces viciae-fabae*) on pea variety T-163 and PV-3 susceptible to rust. Triadimefon, maneb + tridemorph and tridemorph were effective against rust disease field conditions. Diaz Franco and Perez Garcia (1995) reported the effect of propiconazole, triademefon and triforine to control rust of chickpea (*Uromyces ciceris-arietini*). Observed that propiconazole decreased the infection, but yield was unaffected. Fuzi (1995) the result of field trails with control fungicides to compare their efficacy with that of systemic sterol biosynthesis inhibiting fungicides against rust (*Uromyces pisi*) and powdery mildew (*Erysiphe pisi*) of peas.

Khaled *et al.* (1995) reported chemical control of pea rust using fungicides (benomyl, carboxin, metalaxyl, oxycarboxin, thiram, triadimefon and triforine) alone or with dithane M-45 (Macozeb) gave good protection against rust. Pande *et al.* (1995) reported the control of lentil rust with fungicides. The best control was recorded by tridemorph (as calixin) followed by metiram (as compogram) and benomyl (as Benlate).

Ayub *et al.* (1996) evaluated six fungicides antracol (0.2 per cent) [propineb], calixin (0.1 per cent) [tridemorph], folicur (0.1 per cent) [tebuconazole], plantvax (0.2 per cent) (oxycarboxin), knowin (0.2 per cent) (carbendazim) and tilt (0.05 per cent) [propiconazole], were assessed for their ability to control *Uromyces viciae-fabae*, the cause of lentil rust. Tilt gave the best control, reducing rust intensity and increased pod yield. Folicur and calixin were also effective against the disease.

Gonzalez and Garcia (1996) studied the effectiveness of bitertanol, dichlobutrazole, diniconazole, hexaconazole, iprodione, metiram, oxycarboxin, penconazole, puracarbolid, triadimefon, triadimenal and dtridemorph + maneb, compared with sulfur, used as spray with an untreated control plot, against *U. appendiculatus* on *Phaseolus vulgaris* and their effect on yield. The best result were obtained with bitertanol 30 EC, hexaconazole 5 EC and oxycarboxin 75 WP sprayed at 0.5 kg/ha at 14 days interval. No significant differences were reported among the treatments, which statistically excelled sulfex 80 WP sprayed weekly at 3 kg/ha and raised yield significantly compared with sulphur.

Kale and Anahosur (1996) reported that triadimefon and mancozeb were effective in controlling *Uromyces phaseoli* var. *vignae* (*U. appendiculatus*) on cowpea. Diclobutrazol (0.25 per cent) reduced disease after one spraying.

Istran (1996) attempted formulation of contact fungicides against *Uromyces viciae-fabae* on pea along with other fungicides against *Uromyces viciae-fabae* and *Erysiphe pisi*. All the treatments reduced infection, the best results was recorded with treatment opus (epoxyconazole). The most treatments reduced infection by both pathogens, leading to increase in yield. Opus was again among the most effective in increasing yield, while several formulation combining polyram DF (metiram) with Altro combi (cyproconazole + carbendazim) or Kumulus-S (sulfur).

Singh *et al.* (1997) studied the efficacy of 9 fungicides topsin-M [thiophanate-methyl], calixin (tridemorph), benlate (benomyl), bavistin (carbendazim), indofil-M 45 (mancozeb and thiophanate-methyl), indofil Z-78, ridomil (metalaxyl) and PP-450 (fletriafol) for the control of pea rust caused by *Uromyces viciae-fabae*. The highest susceptible pea cultivar Rachna was sown and first spray as fungicides was applied just after the appearance of symptoms, followed by 2 more sprays at 10 days interval. All the fungicides treatments significantly reduced the disease severity and increased the grain yield of pea. PP-450 was the most effective fungicides given 74.7 per cent disease control. Form the pooled data from 2 seasons, it was concluded that 3 sprays at 10 day interval of PP-450 at 0.1 per cent, ridomil at 0.2 per cent or calixin at 0.06 per cent may be recommended for the management of pea rust.

Huge and Nahar (1997), tested four fungicides, propiconazole, tridemorph (calixin 75 EC) tebuconzole (folicur 25 EC) and sulfur (as thiouit) were tested under field conditions for their ability to control rust (*Uromyces viciae-fabae*) in peas (*Pisum sativum*). Four sprays with tebuconazole (0.05 per cent) was highly effective in controlling the disease and was the economical in term of crop yield (605.87 kg/ha) and net profit, with a benefit cost ratio 5.69. Sharma (1998) reported that mancozeb proved to be the most effective fungicides in controlling rust of French bean caused by *U. appendiculatus*. The extent of avoidable losses due to this disease was 60.8 per cent during the epidemic year.

Fontem and Bounda (1998) reported that rust control and ethylene-bis-dithiocarbamate (EBDC) residues in green beans (*Phaseulus valgaris* cv. *Morgn*) were evaluated in 2 seasons of 1993 in Dschang, Cameroon. Four weekly application of mancozeb (2.8 kg ha^{-1}), sulfur (2.24 kg ha^{-1}) or mancozeb + sulfur (1.4 + 1.12 kg ha^{-1}) were initiated from appearance of rust symptoms (causal agent *Uromyces appendiculatus*). In both season, treatments with mancozeb alone resulted in a low area under pustule count progress curve (AUPPC) and a significant increase in leaf area index, and pod and green fodder yields. Pod yields were negatively correlated with AUPC. Pod yield loss was estimated at 61 per cent and 32 per cent in the early and late seasons, respectively.

Gupta and Shyam (1998) observed that, the efficacy of triademefon, hexaconazole, difenaconazole, flusilazole, fenarimol, penconazole, mancozeb and chloro-thalonil were tested for the rust control. Among these hexaconazole (0.10 per cent) and difenoconazole (0.01 per cent) were best against rust and increased yield.

Sumartini (1998) studied on rust of french bean (*Phaseolus vulgaris*) caused by *Uromyces phaseoli* (*Uromyces appendiculatus*), observed that Triadimefon fungicides were effective. The proper application of triadimefon (Based on rust disease intensity) increased its effectiveness and efficiency – Triadimefon sprayed 3-5 times on one-third of the lower-canopy when the disease intensity is 5 per cent and then at 10 days interval afterwards decreased rust intensity by 22 per cent or 55 per cent in the dry or wet seasons, respectively. The fungicide treatment reduced the fresh pod and grain yield losses by 32 per cent and 62 per cent, respectively.

Srivastava (1999) tried 7 fungicides [bavistin (carbendazim), Topsin-M (thiophanate-methyle), captafal (captan), wettable sulfur, cumin-L (ziram), dithane M-45 (mancozeb) and kavach (chlorothalonill), against rice bean (*Vigna umbellate* cv RBL-1] rust caused by *Uromyces appendiculatus* under field. The maximum disease control and highest increase in yield were achieved by topsin-M followed by bavistin and wettable sulfur.

Efficacy of mancozeb, benomyl, captan and thiophanale-methyl as seed dressing and foliar sprays, carboxin as a seed dressing and oxycarboxin as a foliar spray were evaluated for the control of lentil rust caused by *U. viciae-fabae* (Mohy *et al.*, 1999). Mancozeb seed treatment was the most effective followed by carboxin and benomyl. Mancozeb when applied as foliar spray was also the most effective fungicides followed by oxycarboxin and benomyl. Mancozeb when applied as foliar spray was also the most effective fungicides followed by oxycarboxin and benomyl. Spraying twice after the appearance of disease symptoms was the most effective fungicides followed by oxycarboxin and benomyl. A single application before the appearance of rust symptoms was the least effective in controlling the disease.

Gupta and Shyam (2000) observed that potted three-week-old pea cv. Lincoln plant were treated with 7 ergosteral biosynthesis inhibiting fungicides *viz.*, Fenarimal, 0.04 per cent, defenoconzole, 0.015 per cent, triadimefon, 0.05 per cent, cyper conazole, 0.03 per cent, Flusilazole, 0.04 per cent, penconazole, 0.05 per cent and hexaconazole, 0.05 per cent and 2 non-systemic (mancozeb, 0.25 per cent and chlorothalnil, 0.25 per cent) fungicides after being inoculated with *Uromyces viciae fabae* aeciospores for 72 h. cyperconazole, flusilezole, penconazole and hexaconazole completely inhibited rust incidence and rust severity on leaves. The inhibition exhibited by fenarimol, defenoconazole and triadimefon was lesser, although it was significantly higher than that of two non-systemic fungicides. Mancozeb and chlorthalonil resulted in a rust severity of 22.91 and 38.75 per cent, respectively and a rust incidence of 100 per cent.

Alam (2007) evaluated the eight different fungicides were tested and all fungicides resulted significantly better performance over the control. Based on the percent disease index (PDI), pod yield and yield contributing characters (plant height, pods per plant, length of pod, breadth of pod and seeds per pod), sedozole 5 EC performed better than the other fungicides and control treatment, which was closely followed by conza.

Sugha (2008) evaluate the efficacy of 22 fungicides against pea rust during crop (rabi) at farmers field. Three fortnightly foliar sprays, starting with the appearance of disease, individually of bayleton, score, tebuconazole (folicur and tebuconazole) and hexaconazole (contaf and sitara) among systemic and at 10 day intervals of antracol, microsul and share among non-systemic fungicides proved effective for combating the disease and in ameliorating the crop yield significantly.

During 1995-96, seed treatment with bavistin (35.3 per cent) or thiram (37.0 per cent) was significantly better in reducing leaf spot and rust with concomitant increase in yield than control. Bavistin (PDI = 24.5 per cent) had the least disease incidence fallowed by mancozeb (PDI = 28.21 per cent) and topsin-M (PDI = 33.00). Crop yield was also highest in bavistin spray treatment (20.18 kg/ha) followed by dunet (19.4 kg/ha). Effect of seed treatment versus spray of chemicals was significant for disease reduction but not for yield during all the years. Seed treatments either with bavistin or thiram resulted in significant disease reduction and increase in grain yield over the untreated control while seed treatment followed by spraying the crop twice with mancozeb was recorded the best.

Effect of Disease on Yield of Pea

The protection efficiency of slow rusting had been described in faba bean, where in some of the slow rusting lines the yield losses varied from 1-2 per cent whereas in other lines the losses varied from 6-43 per cent. These findings indicated that some of the lines were tolerant to rust in faba bean (Rashid and Bernier, 1991). Effect of disease levels on the yield components of pea had been described in case of *Mycospharella pinodes* (Tivoli *et al.*, 1996). It indicated a reduction of 40 per cent in yield under diseased conditions. Similarly, effect of rust disease on the yield of pea had been demonstrated (Chand *et al.*, 2004b). The protection provided by slow rusting components was conclusively demonstrated in germplasm lines of pea by sowing two sets of experiment. One set was protected with fungicidal spray and the other set was inoculated with aeciospore suspension. The differences in 100-seed weight of these two sets were high for susceptible genotype of low for resistant genotypes in pea. Therefore, the difference in the 100-seed weight could be an indicator of resistance. In case of powdery mildew infection at an early reproductive stage reduced the number of seeds per pod as well as the seed weight. But only seed weight was reduced in plants infected at a late reproductive stage (Faloon *et al.*, 1993). This indicates the effect of time of incidence of the disease on the yield of crop.

Pea rust is one of the most important diseases in India and is becoming a limiting factor in pea cultivation and production. Heavy yield reductions were reported due to this disease. The disease completely destroyed the crop in sub-mountainous region of North India (Singh, 1999). Yield losses in pea due to rust was also reported by Upadhyay and Singh (1994). Singh (1999) reported the losses in pea yield due to this disease from *Tarai* regions. An average of about 56.81 per cent grain yield loss was reported in the year 1986-1988. They further reported that about 22.21 per cent of loss occurs in the seed weight of the pea.

Singh *et al.* (2004) investigated that the effect of time of mancozeb (Indofil M-45) sprays on pea rust. Foliar applications of mancozeb started just after the initiation of the disease (70 and 75 days after sowing) and repeated thrice, twice and once at an interval of 10 days. Similarly, delayed sprays began 10 days after the disease appearance (80 and 85 days after sowing) and repeated twice and once at 10 days interval as per schedule. The 4 sprays at 10-day intervals commencing with the initiation of the disease were the most effective in reducing rust severity from 69.7 to 10.2 per cent and increasing seed yield from 1108.3 to 1824.9 kg/ha. This was followed by 3 timely sprays with 24.1 per cent disease severity and 1683.3 kg/ha seed yield. However, under 2 timely sprays applied at inception of the disease and subsequently 10 days later, the net profit was Rs 4720/ha with benefit:cost ratio of 6.05 and appeared feasible for resource-poor farmers. One timely spray just after the disease appearance gave benefit:cost ratio of 4.77 with net profit of Rs 1860/ha. Spray schedule started immediately after the rust appearance was more effective and beneficial than sprays applied 10 days after the disease appearance.

Botanical Pesticides

Several informations are available on the control of *U. fabae* by botanical pesticides. Tripathi and Rathi (2003) have reported foliar spray with neem oil was effective for management of field pea and lentil rust. *Ghewande et al.* (1993) reported that *Azadirachta indica* (neem) and *Lawsonia inermis* (henna) extracts were effective in controlling *Puccinia arachidis* causing rust of groundnut. Similar observations have been reported by Ganapthy and Narayanasamy (1990).

A total of 206 plant species have been tested against the actively of 26 species of phytopathogenic fungi, including assays on spore germinations, mycelial growth, sporulation and in some cases experiments in green house and field. Plant products in the form of aqueous and hexanic extracts, powder, essential oils and secondary plant metabolites have been tried. Significant field results have been obtained with aqueous extracts against bean rust (*Uromyces appendiculatus*), squash powdery mildew (*Erysiphe cichoracearum*) and squash downy mildew (*Pseudopernospora cubensis*), where plant products increased from 38 to more than 200 per cent. The control of *U. appendiculatus* was obtained with essential oil (Montes *et al.*, 2000).

Sajid *et al.* (1995) studied comparative efficacy of neem products against rust of wheat and neem oil was reported to inhibit the germination of urediospores completely. Rizvi *et al.* (1988) have reported that the botanical pesticides containing azadirachtin as active ingredients were highly effective in reducing the spore germination of *U. striatus* causing lucerne rust.

Biological Control

Information in respect of biological control of rust severity and influence on grain yield of fieldpea is lacking. Bilogical control is the reduction of amount of inoculum or disease producing activity of a pathogen accomplished by or through one or more living organism other than mass (Cook and Baker, 1983). Biological control is use of living agents to control plant pathogen (Mukhopadhyay, 1987). It can be achieved by either promoting the native antagonists to reach a density sufficient to suppress a pathogen(s) or by introducing alien antagonists by using organic amendments or other cultural practices, the recent trend is to isolate, multiply and introduce the antagonists to soil or specific court of infection to achieve a successful biological suppression of a disease.

Antagonist and Antagonism

The basis of biological control is the exploitation of the antagonistic potential of biocontrol agents. Bioagents are known as antagonists. An antagonist is a micro-organism that adversely affect another organism (*e.g.* a faster pathogen) growing in association with it (Baker and Cook, 1974). In biological control of plant pathogens, antagonist are biological agents, with the potential to interfere in the life process of plant pathogen (Cook and Baker, 1983).

Antagonism (Wood and Tveit, 1955) includes

(a) Antibiosis, which results from the liberation of an antibiotic or other chemical by one microorganism and harmful to the pathogen.

(b) Completion for some nutrients or other conditions in limited supply but needed by the pathogen.

(c) Predation, hyperparasitism, mycoparasitism or other forms of direct exploitation of a pathogen by other organism.

Direct exploitation is the only form of antagonism where the beneficial organism may exhibit a host dependent population trend. The antagonists act by :

1. Reducing the population of pathogen
2. Preventing the pathogen from infecting the plant and
3. Limiting disease development after infection

(Mukherjee *et al.*, 1995)

Example of Fungal Antagonists

Gliocladium virens Miller, Giddlens and Foster

Petch (1939) studied the genus *Gliocladium corda* and characterization of *G. virens* was done from forest and cultivated soil of Georgia by Miller *et al.* (1957).

Taxonomy and Morphology

The systemic position of *Gliocladium* is given below :

Sub division	:	Deuteromycotina
Form class	:	Deuteromycetes
Form sub class	:	Hyphomycetidae
Form family	:	Moniliaceae
Form sub family	:	Gliosporae
Genus	:	*Gliocladium*
Species	:	*virens*

(Alexopoulus *et al.*, 1996)

Colonies on PDA broadly defused with dark green in colour. Conidiophores more or less erect, hyaline, terminally branched in penicillate fashion, phialides divergent or appressed, bottle shaped phialospores hyaline or pigmented, non-separate, slimy (forming globes or loose columnar heads), common in soil (Cook and Baker, 1983).

Gliocladium virens (Miller, Giddens and Foster) side branches or conidiophores closely approach the supporting conidiophore and closely apprised apical flask shaped phialides bearing a large drop of slimy conidia. Phialospores smooth, green, elliptical 4-6 × 3-4 μm chlamydospores globose, thin walled 6-8 μm in diameter (Miller *et al.*, 1957). The perfect state of *G. virens* is *Hypocera gelatinosa* (Tode : Fr). *G. virens* differs from *Trichoderma* spp. in penicillium like sterigmata, closely apprised rather than divergent.

Trichoderma harzianum Rifai

Taxonomy and Morphology

Sub division	:	Deuteromycotina
Form class	:	Deuteromycetes
Form sub class	:	Hyphomycetidae
Form order	:	Moniliales
Form family	:	Moniliaceae
Form sub family	:	Gliosporae
Genus	:	*Tridhoderma*
Species	:	*harzianum*

(Alexopoulus *et al.*, 1996)

Conidiophores erect or straggling, highly ramified more or less conical, verticillate blending divergent flask shaped phialides singly or in clusters, from which sub-globose to ellipsoidal, slimy no-separate phialospores are borne, often gathered in bale at the opening of the phialides. Chlamydospores are commonly formed intercalary or terminal, globose to ellipsoidal, hyaline and smooth walled. Phialospores smooth, green, sub-globose to short abovoid, 2.4-3.2 × 2.2-2.8 μm. colonies in culture usually grow rapidly, floccose, treated and white to green.

Pseudomonas fluorescens

Pseudononas fluorescens make up a diverse group of bacteria that can generally be distinguished from other *Pseudomonas* by their ability to produce a water soluble pigments. They are typically gram-negative chemoheterotrophic, motile rods with polar flagella and are grouped in r-RNA homology group I. This classification method divides all *Pseudomonas* spp. into five groups based on the relatedness of their r-RNA genes, which undergo fewer changes in the course of evolution. *P. fluorescens* have simple nutritional requirements, and this is reflected by the related abundance of these organisms in nature.

However, Yuen *et al.* (2001) control the bean rust (*Uromyces appendiculatus*) by bacterial antagonists (*Pantoea agglomerans* B_1 and *Stenotrephomonas maltophilia* C_3) in USA and Carrion *et al.* (1999) tried to control bean rust (*Uromyces appendiculatus*) by (*Verticillium lecanii*) as biocontrol agent.

Essential Oils

Meager information is available on the control of *U. fabae* by essential oils however, Oxenham *et al.* (2005) evaluated the essential oil on mycelial growth of the plant pathogenic fungus *Botrytis fabae* and *Uromyces fabae* was reduced significantly by both the methyl chavicol chemotype oil and the linalol chemotype oil, and the major individual components of the oils all reduced fungal growth, with methyl chavicol, linalol, eugenol and eucalyptol reducing growth significantly. Combining the pure oil components in the same proportions as found in the whole oil led to very similar reductions in fungal growth, suggesting that the antifungal effects of the whole oils were due primarily to the major components. When the fungus was exposed to the oils in liquid culture, growth was reduced by concentrations considerably smaller than those used in the Petri dish studies. *Botrytis fabae* and the rust fungus *Uromyces fabae* were also controlled in vivo, with the whole oils of both chemotypes, as well as pure methyl chavicol and linalol, reducing infection of broad bean leaves significantly. Most effective control of fungal infection was achieved if the treatments were applied 3 h post-inoculation.

Induced Resistance

Phaseolin and other phytoalexins also accumulated rapidly when bean cells expressed resistance to avirulent strain of bean rust fungus, *U. appendiculatus.* Tyihak *et al.* (1989) found that localized induction of resistance in bean against the rust fungus *Uromyces appendiculatus* by an amino acid derivatives. Aqueous solution of N-trimethyl-L-lysine (TML) at various concentration were sprayed onto the under surface of primary leaves of bean seedlings and leaves were challenged 1, 6 or 8 days later by inoculation with a suspension of urediospore (6×10^4 spore/ml) of *U. appendiculatus.* Pustule densities were assessed ten days after later and results varied enormously depending on the concentration of TML and the interval between induction and challenge. Maximum protection was observed as 36.5 per cent of the infection density on control plants when 10^{-9} mol/l TML was applied 6 days before the challenge inoculation with *U. appendiculatus.*

There was no protection effect by the TML treatment when the interval between induction and challenge was only 1 day. Application of TML was shown not to inhibit germination of urediospore or formation of appressoria over stomata, suggesting that it is acted through physiology of the bean plant.

The most recent development on systemic induced resistance in the bean system comes from Ciba-Gigy's AG of 2, 6-dichloro-isonicotinic acid (CGA-41396). Application of its formulation with 20 μg ml^{-1} a.i. as a spray of first leaves not only protected the second leaves against *C. lindemuthianum,* but also against *Pseudomonas syringae* pv. *phaseolicola,* the cause of halo blight disease and *U. appendiculatus,* the cause of bean rust disease.

Walters and Murray (1992) observed that inoculation of the lowest two leaves of broad bean (*Vicia faba*) with urediospore of the rust fungus *Uromyces viciae-fabae,* when plants also had two other fully developed leaves and two young developing leaves, caused these upper leaves to become resistance to challenge inoculation with the same rust fungus 1, 3, 6 and 9 days later. The resistance was seen as diminished infected areas on the leaves and as fewer uredio per standard area for upto 29 days from challenge. The resistance was very high when 1 day separated the two inoculations but had disappeared when 12 days separated the two.

In further experiments with plants at the same stage and using the same isolates of the rust fungus, Walters and Murray (1992) found that resistance in the upper leaves of broad bean (*Viciae faba* L.) plants to infection by the rust fungus. *Uromyces viciae-fabae* was increased following treatment of the lower leaves with 10 mM potassium phosphate or 5 mM EDTA increasing the interval between treatment of the lower leaves and inoculation of the upper leaves had little effect on rust infection. Thus, rust infection was reduced by 50 and 34 per cent if the upper leaves were inoculated 1 day, after treatments of the lower leaves with potassium phosphate or EDTA, while there is 77 per cent reduction in infection if the interval between treatment and inoculation was increased to 12 days. Application of calcium nitrate (10 mM) after the phosphate or EDTA treatments prevented the induction of systemic resistance.

The failure of microorganisms to colonize plants has often been attributed to the presence of inhibitory compounds within challenged tissues. Antimicrobial compounds isolated from plants generally fall into two categories (i) constitutive compound, which are present in healthy plants, and (ii) induced compound, synthesized from remote precursor following infection (Deverall, 1977).

Induced resistance exists into different forms : localized and systemic. As the term implies localized resistance can be detected only in the area immediately adjacent to the site of attempted penetration by the pathogen (elicitation) while systemic resistance refers to resistance that occurs at sites in the host distant from the point of interaction with a potential pathogen. Systemic resistance sometimes follows localized resistance (Sequeira, 1983). The terms acquired resistance is used as synonymous with induced resistance, localized and systemic types share several characteristics.

1. They are time dependent, that is, resistance in established only after certain metabolic change occur in the host during a specific interval following the inoculation.
2. They are temperature dependent, systemic resistance, being for more sensitive to this factor then localized resistance.
3. They are affected by light to different degree, but light is essential only in certain case of systemic resistance.
4. In both cases, resistance is persistent once the machinery for induction of the plant response has been turned on and
5. Most important, both types are nonspecific in terms of the organism that induces the response and the range of pathogen against which the plant is protected.

Mandryk (1962) reported that induced resistance in an energy demanding process, thus, it places demands on the plant that can not be satisfied if the plant is growing under nutritional conditions that are already limiting. For this reason, supplemented nutrient must be supplied to plants during the inoculation period to maintain the yield at a normal level. But contrary to this, preliminary experiments by Madamachi and Kuc (1991) suggested that high nutrients level make induction of systemic resistance in tobacco is also effective.

The biological spectrum of phosphate and oxalate-induced resistance was reported by Madamachi and Kuc (1991). They used both chemicals to induce resistance to the fungal disease *viz.*, anthracnose, scab, rust, gummy stem blight and powdery mildew. Good resistance was also induced against the bacterium *Pseudomonas syringae* pv *lachrymans* and tobacco necrosis virus (TNV). Systemic acquired resistance (SAR), has become a subject of increasing inquiry (Kessmann *et al.*, 1994). SAR is a broad physiological immunity that can be triggered biologically with a necrogenic pathogen on its cell component or chemically with certain natural or synthetic chemical compounds. Kessmann *et al.* (1994) reported that a second potential principle chemically induced SAR are based on compounds including various inorganic salts, silicon, oxalate, phosphate, 2-thiouracid, polyacrytic acid, benzoic acid, nucleic acid, unsaturated fatty acid, N-trimethyl-L-lysin etc. Fosetyl-Al; Metalaxyl (Ward, 1984) and more recently, triazoles (Hauthal, 1993) are claimed to have resistance inducing activity. He observed that a chemical will be considered as activator of the SAR response of the chemical induces resistance to the some spectrum of pathogens and induces expression of the same biochemical markers as in the biological model. Furthermore, the chemical should have no direct antimicrobial activity. In the case of antimicrobial activity, an SAR-inducing activity could be established by demonstrating to pathogen isolates that are genetically resistant to the antibiotic effects of the chemical.

Induced resistance is effective under field conditions in protecting crop plants (Tuzun *et al.*, 1986). Growth and yield of induced plant in either unaffected or enhanced in the absence of pathogen (Tuzun and Kuc, 1985; Tuzun *et al.*, 1986). Induced systemic resistance activates several natural resistance mechanism of the host and therefore may have long lasting broad spectrum durable and stable, unlike current systemic fungicides on a single site for action (Kassmann *et al.*, 1994). It is worthwhile to mention herewith, that induced systemic resistance lasts for the life of annual plants and since the mechanism for resistance activated are inherent in the plants, appear at safe as resistance developed by breeding (Madamachi and Kuc, 1991).

The ability to induce systemic resistance in even very susceptible plants may help to utilize current high yielding, high quality crops. The signals for immunization may permit foliar, soil or seed application of immunizing agents with utilization of existing equipments for application of pesticides. Recent development indicates that common, inexpensive and nontoxic chemical may release immunity signals and be useful for crop protection (Doubrava *et al.*, 1988). Unfortunately, the concepts that plants can be effectively protected by inducing resistance are generally not recognized outside the scientific community. Consequently, farmers have been unaware of this additional of disease control in plants and thereby unable to utilize and help improve the technology (Madamachi and Kuc, 1991; Kassmann *et al.*, 1994).

References

Abu Zinada, A.A.H., Cobb, A. and Boulter, D. 1975. An electron microscopic study of the effects of parasite interaction between *Vicia faba* L. and *Uromyces fabae. Physiol. Plant. Pathol.*, 5: 113–118.

Accantino, P. 1964. Chemical control on lentil rust. *Agric. Tech.*, 23-24: 7-14.

Agrawal, S.C. and Prasad, K.V.V. 1997. Diseases of lentil. Oxford and IBH Publishing Co. Pvt. Ltd., New Delhi, p. 155.

Alam, M.M., Sadat, M.A., Hoque, M.Z. and Rashid, M.H. 2007. Management of powdery mildew and rust diseases of garden pea using fungicides. *Inter. J. Sust. Crop Prod.*, 2(3): 56-60.

Alexopolus, C.J., Mims, C.J. and Blackwell, M. 1996. Introductory Mycology. John Wiley and Sons. Inc., p. 869.

Allen, E.A., Hazen, B.E. and Hoch, H.C. 1991. Appressorium formation in response to topographical signals by 27 rust species. *Phytopathology,* 81: 323–331.

Anderson, J.P. 1952. The uredinales of Alaska and adjacent parts of Canada. *Iowa St. Call J. Sci.,* 26: 507-526.

Arthur, J.C. and Cummins, G.B. 1962. Manual of Rusts in United States and Canada. Hafner Publishing Co., p. 438.

Ayub, A., Rahaman, M.Z., Ali, S. and Khatun, A. 1996. Fungicidal spray to control leaf rust of lentil. *Bangladesh J. Plant Pathol.,* 12-: 61-62.

Baruah, H.K. 1980. Text book of plant pathology. Oxford and IBH, New Delhi.

Batra, L.R. and Stavely, R. 1994. Attraction of two spotted spider mites to bean rust uredinia. *Plant Dis.,* 78: 282-284.

Beckett, A. Tatnell, J.A. and Taylor, N. 1990. Adhesion and pre-invasion behaviour of urediospores of *Uromyces viciae-fabae* during germination on host and synthetic surfaces. *Mycological Research,* 69: 865-875.

Bhardwaj, M.L. and Sharma, J.M. 1996. Performance of early pea genotypes under different sowing times in the hills of Himachal Pradesh. *J. Hill Res.,* 9: 62-64.

Bilgrami, K. Jamaluddin, S. and Rizvi, M.A. 1979. List of fungi. Part-I. List of References. Today and Tommorrow's Printers and Publishers New Delhi, p. 264.

Brown, J.K. and Hovmoller, M.S. 2002. Aerial dispersal of pathogens on the global and continental scales and its impact on plant disease. *Science,* 297: 537–541.

Butler, E.J. 1918. Fungi and diseases in plants. Thatcher, Spink Co. Calcutta, p. 547.

Carrion, G., Romero, A. and Rico-Vray, V. 1999. Use of *Verticillicum lecarini* as biocontrol agent against bean rust (*Uromyces appendiculatus*). *Fitopathologia,* 34(4): 214-219.

Catanzariti, A.M., Dodds, P.N., Lawrence, G.J., Ayliffe, M.A. and Ellis, J.G. 2006. Haustorially expressed secreted proteins from flax rust are highly enriched for avirulence elicitors. *Plant Cell,* 18: 243–256.

Chand, R., Srivastava, C.P. and Kushwaha, C. 2004a. Screeing technique for pea (*Pisum sativum* L.) genotypes against rust disease (*Uromyces fabae* (Pers.) de Bary. *Indian J. Agril. Sci.,* 74: 166-167.

Chand, R., Srivastava, C.P., Singh, B.D. and Sarode, S.B. 2004b. Identification and characterization of slow rusting components in pea (*Pisum sativum* L.). *Genetic Resources and Crop Evaluation.* 34: 48-52.

Chanda, Chand, R. and Srivastava, C.P. 2006. Role of aeciospores in outbreaks of pea (*Pisum sativum*) rust (*Uromyces fabae*). *European J. Plant Pathl.,* 115: 323-330.

Chauhan, M.P., Singh, I.S. and Singh, R.S. 1996. Genetics of rust resistance in lentil. *Phytopathology,* 49: 387-388.

Chauhan, R.S. 1988. Epidemiology and control of pea rust caused by *Uromyces viciae fabae* (Pers.) de Bary. M.Sc. thesis, Deptt. of Plant Pathology, HPKV, Plampur, India, p. 109.

Chauhan, R.S. and Singh, B.M. 1994. Effect of different duration of leaf wetness on pea rust development. *Pl. Dis.Res.*, 9: 200-201.

Clement, J.A., Butt, T.M. and Beckett, A. 1993a. Characterization of the extracellular matrix produced in vitro by urediniospores and sporelings of *Uromyces viciae-fabae*. *Mycol. Res.*, 97: 594–602.

Clement, J.A., Martin, S.G., Porter, R., Butt, T.M. and Beckett, A. 1993b. Germination and the role of extracellular matrix in adhesion of urediniospores of *Uromyces viciae-fabae* to synthetic surfaces. *Mycol. Res.*, 97: 585–593.

Clement, J.A., Porter, R., Butt, T.M. and Beckett, A. 1994. The role of hydrophobicity in attachment of urediniospores and sporelings of *Uromyces viciae-fabae*. *Mycol. Res.*, 98: 1217–1228.

Clement, J.A., Porter, R., Butt, T.M. and Beckett, A. 1997. Characteristics of adhesion pads formed during imbibition and germination of urediniospores of *Uromyces viciae-fabae*. *Mycol. Res.*, 101: 1445–1458.

Conner, R.L. and Bernier, C.C. 1982. Host range of *Uromyces viciae-fabae*. *Phytopathology*, 72: 687-689.

Cook, R.J. and Baker, K.F. 1983. The nature and practices of biological control of plant pathogen. APS Books, St. Paul, MN USA, p. 539.

de Bary, A. 1962. Morphologie and Physiologie-der plize Flechten and Myxomyceter.

Deising, H., Frittrang, A.K., Kunz, S. and Mendgen, K. 1995a. Regulation of pectin methylesterase and polygalacturonate lyase activity during differentiation of infection structures in *Uromyces viciae-fabae*. *Microbiology*, 141: 561–571.

Deising, H., Jungblut, P.R. and Mendgen, K. 1991. Differentiation related proteins of the broad bean rust fungus *Uromyces viciae fabae*, as revealed by high resolution two-dimensional polyacrylamide gel electrophoresis. *Arch. Microbiol.*, 155: 191–198.

Deising, H., Nicholson, R.L., Haug, M., Howard, R.J. and Mendgen, K. 1992. Adhesion pad formation and the involvement of cutinase and esterases in the attachment of uredospores to the host cuticle. *Plant Cell*, 4: 1101–1111.

Deising, H., Rauscher, M., Haug, M. and Heiler, S. 1995b. Differentiation and cell wall degrading enzymes in the obligately biotrophic rust fungus *Uromyces viciae-fabae*. *Can. J. Bot.*, 73: S624–S631.

Deutelmoser, E. 1926. Plant protection measures in vegetable culture. I. Measures against fungus pests. Obst.-U. Gemusebu I. XXII, 19: 291-292. (*Rev. Appl. Mycol.*, 1927: 137).

Deverall, B.J. 1977. Defence mechanisms of plants. Cambridge monographs in experimental biology. No. 19 Cambridge University Press, Cambridge.

Diaz Franco, A. and Perez Garcia, P. 1995. Chemical control of rust and 'rabia' of chickpea and its influence on yield. *Revista mexicana de Fitopathologia,* 13: 123-125.

Doubrava, N.S., Dean, R.S. and Kue, J. 1988. Induction of systemic resistance to anthracnose caused by Colletotriclum lagenarium in cucumber by oxalate and extracts from spinach and rhubarb leaves. *Physiol. Mol. Plant Pathol.,* 33: 69-80.

Ehrlich, H.G. and Ehrlich, M.A. 1963. Electron microscropy of the host-parasite relationships in stem rust of wheat. *American J. Bot.,* 50: 123-130.

El-Fiki, A.I.I., El-Habaa, G.M.D., Esmail, I.A. and Eid, K.E. 1999. Induction of better growth and sporulation of *Uromyces fabae* (Pers.) de Bary, grown in axenic culture. *Annals of Agricultural Science,* 37: 1187-1200

El-Helaly, A.F. 1939. Preliminary studies on the control of bean rust. *Bull. Minst. Agric. Egypt,* 201: 19.

Emeran, A.A., Sillero, J.C., Niks, R.E. and Rubiales, D. 2005. Infection structures of host-specialized isolates of *Uromyces viciae-fabae* and of other species of *Uromyces* infecting leguminous crops. *Plant. Dis.,* 89: 17–22.

Faloon, S.L.H., Bresford, R.M. and Wallace, A.R. 1993. Powdery mildew reduces yield of pea plants. Proceedings, 6th International Congress o Plant Patholgy, 26 July-6August, Montreal, Canada.

Flor, H.H. 1956. The complementary genetic systems in flax and flax rust. *Adv. Gen.,* 8: 29–54.

Fontem, D.A. and Bounda, H. 1998. Rust control and EBDC residues in gree beans sprayed with Mancozeb and Sulphur. *Inter, J. Pest Manag.,* 44: 211-214.

Frittrang, A.K., Deising, H. and Mendgen, K. 1992. Characterization and partial purification of pectinesterase, a differentiationspecific enzyme of *Uromyces viciae-fabae. J. Gen. Microbiol.,* 138: 2213–2218.

Fuzi, I. 1995. Fungicides against diseases of pea, *Pisum sativum* L., In Hungary. *Pesticides Science,* 45: 292-295.

Ganapathy, T. and Narayansamy, P. 1990. Effect of plant products on the incidence of major diseases of groundnut. *Int. Arachis Newslett.,* 7: 20-21.

Gaumann, E.A. 1998. Comparative morphology of fungi. Translated by Caroll William Dodge, Biotech Books, Delhi. pp: 563.

Ghewande, M.P., Desai, S., Narayan, P. and Ingle, A.P. 1993. Integrated management of foliar diseases of groundnut (*Arachis hypogea* L.) in India. *Int. J. Pest Management,* 39: 375-378.

Gonzalez, M. and Garcia, E. 1996. Evaluation of fungicides for the control of bean rust (*Uromyces appendiculatus*). *Agronomica Mesoamericana,* 7: 86-89.

Graham, P.H. and Vance, C.P. 2003. Legumes: importance and constraints to greater use. *Plant. Physiol.,* 131: 872–877.

Grewal, J.S. 1988. Diseases of pulse crops: An overview. *Indian Phytopathology*, 41: 1-14.

Gross, P.L. and Venette, J. 2001. Over winter survival of bean rust urediospores in North Dakota. *Plant Disease*, 85: 226-227.

Gupta, R.P. 1990. Evaluation of pea germplasm for their reaction to powdery mildew and rust. *Indian Journal of Pulses Reseach*, 3: 186-188.

Gupta, S.K. and Shyam, K.R. 1998. Control of powdery mildew and rust of pea by fungicide. *Indian Phytopathology*, 51: 184-186.

Gupta, S.K. and Shyam, K.R. 2000. Post infection activity of ergosterol biosysthesis inhibiting fungicides against pea rust. *J. Mycol. Pl. Pathol.*, 30: 414-415.

Hahn, M. and Mendgen, K. 1992. Isolation of ConA binding haustoria from different rust fungi and comparison of their surface qualities. *Protoplasma*, 170: 95–103.

Hahn, M., Deising, H., Struck, C. and Mendgen, K. 1997a. Fungal morphogenesis and enzyme secretion during pathogenesis. Resistance of Crop Plants against Fungi (Hartleb H, Heitefuss R and Hoppe H-H, eds), p. 33–57.

Harder, D.E. and Chong, J. 1984. Structure and physiology of haustoria. The Cereal Rusts Origins, Specificity, Structure, and Physiology, Vol. I. (Bushnell WR and Roelfs AP, eds), p. 431–476. Academic Press Inc., Orlando.

Hauthal, H.G. 1993. New wag interlligenter chemic: *Beispiel Fllanzenschutz aus Chemic. Tech. Lab.*, 41: 566-570.

Heath, M.C. 1990. Influence of carbohydrates on the induction of haustoria of the cowpea rust fungus in vitro. *Exp. Mycol.*, 14: 84–88.

Heath, M.C. 1997. Signalling between pathogenic rust fungi and resistant or susceptible host plants. *Ann. Bot.*, 80: 713–720.

Heath, M.C. and Skalamera, D. 1997. Cellular interactions between plants and biotrophic fungal parasites. *Adv. Bot. Res.*, 24: 195–225.

Hedge, R.K. 1974. Chemical control of powdery mildew and rust in peas. *Pesticides*, 8: 21-22.

Heiler, S., Mendgen, K. and Deising, H. 1993. Cellulolytic enzymes of the obligated biotrophic rust fungus *Uromyces viciae-fabae* are regulated differentiation-specifically. *Mycol. Res.*, 97: 77–85.

Hiratsuka, N. 1933. Studies on *Uromyces fabae* its related species. *Jap. J. Bot.* 6.: 329-379.

Hiremath, R.V. and Pavgi, M.S. 1971. *In vitro* assay of aureofungin against some rust fungi, *Hind. Anti. Bull.* 13: 83-86.

Huge, H.I. and Nahar, M.S. 1997. Efficacy and economics of different fungicides in controlling rust and powdery mildew of garden pea, Bangladesh. *J. Sci. Indust. Res.*, 32(4): 533-536.

Istran. F. 1996. Control fungicides against fungal disease of peas. *Novenyvedelem*, 32(3): 144-146.

Jacks, H. 1954. Screening test with fungicides for control of broad bean rust. *N.Z.J. Sci. Tech. Sect. A.*, 36: 274-279.

Jacks, H. and Webb. A.J. 1956. Field tests for control of broad bean rust. *M.Z.J. Sci. Tech. Sect. A.*, 38: 157-159.

Jakupovic, M., Heintz, M., Reichmann, P., Mendgen, K. and Hahn, M. 2006. Microarray analysis of expressed sequence tags from haustoria of the rust fungus *Uromyces fabae*. *Fungal Genet. Biol.* 43: 8–19.

Jorstad, I. 1948. Erterrust, *Uromyces pisi* (D.C.) Fuck., 1. Norge (Pea rust: *Uromyces pisi* (D.C.) Fuck., in Narway). *Nord, Jordbr-Forskn.*, 7-8: 198-207. (*Rev. Appl. Mycol.*, 1950: 242-243).

Joseph, M.E. and Hering, T.F. 1997. Effects of environment on spore germination and infection by broad bean rust (*Uromyces viciae-fabae*). *J. Agric. Sci.*, 128: 73–78.

Kale, J.K. and Anahousur, K.H. 1996. Chemical control of cowpea rust. *Karnataka, J. Agric. Sci.*, 9: 179-181.

Kapooria, R.G. and Mendgen, K. 1985. Infection structures and their surface changes during differentiation in *Uromyces fabae*. *J. Phytopathol.*, 113: 317–323.

Kapooria, R.G. and Sinha, S. 1966. Studies on the host range of *Uromyces fabae* (Pers.) de Bary. *Indian Phytopathol.*, 19: 229-230.

Kemen, E., Hahn, M., Mendgen, K. and Struck, C. 2005. Different mechanisms of *Medicago truncataula* ecotypes against the rust fungus *Uromyces striatus*. *Phytopathology*, 95: 153-157.

Kessmann, H., Staub, T., Hofmann, C., Maetzke, T. and Herzog, T. 1994. Induction of systemic acquired disease resistance in plants by chemicals. *Ann. Rev. Phytopathol.*, 32: 439-459.

Khaled, A.A., El-Moity, SMHA, and Omar, SAM, 1995. Chemical control of some faba bean diseases with fungicides. *Egyptian J. Agri. Res.*, 73: 45-56.

Khare, M.N. and Agrawal, S.C. 1978. Lentil rust severity survey in Madhya Pradesh. Proceedings of All India Pulse Workshop, Baroda, p. 3.

Kispatic, J. 1949. Prilong proznounju biologije in suzbijanaja bobve rdij *Uromyces fabae* (Pers.) de Bary *f. sp. viciae-fabae* de Bary. *Annals of Transaction Agriculture Society*. 1: 61.

Kumar, T.B.A., Rangaswamy, K.T. and Ravi, K. 1994. Assessment of tall field pea genotypes for slow rusting resistance. *Legume Research*, 17: 79-82.

Lal, H.C., Upadhyay, J.P., Jha, A.K. and Kumar, Atul 2007. Survey and surveillance of lentil rust and its cross infectivity on different host. *J. Res.* (BAU), 19: 111-113.

Leonard, K.J. and Szabo, L.J. 2005. Stem rust of small grains and grasses caused by *Puccinia graminis*. *Mol. Plant Pathol.*, 6: 99–111.

Long, D.L. 2003. Cereal Rust Bulletin: Final Report August 7, 2003. Cereal Disease Laboratory, U.S. Department of Agriculture.

Madamachi, N.R. and Kuc, J. 1991. Induced systemic resistance in plants. *In:* The Fungal Spore and Disease Initiation in Plants and Animals Cole, G.T. and Hoch, H.C. (eds.). New York, Plenum, p. 347-362.

Maier, W., Begerow, D., Weib, M. and Oberwinkler, F. 2003. Phylogeny of the rust fungi: an approach using nuclear large subunit ribosomal DNA sequences. *Can. J. Bot.*, 81: 12–23.

Mandryk, M. 1962. The relationship between acquired resistance to *Peronospora tobacina* in *Nicotiana tabacum* and soil nitrogen levels. *Aust. J. Agric. Res.*, 13: 10-16.

Mayer, E. 1947. Mycological notes-XII. *Bull. Soc. Neuchatel, Sci. Nat.* IXX: 33-60 (*Rev. Appl. Mycol.*, 1948: 158).

Mendgen, K. 1997. The Uredinales. The Mycota Vol. V Plant Relationships Part B (Esser K and Lemke PA, eds), p. 79–94. Springer-Verlag, Berlin. 57: 267–276.

Mendgen, K. and Hahn, M. 2002. Plant infection and the establishment of fungal biotrophy. *Trends Plant Sci.*, 7: 352–356.

Mims, C.W., Rodriguez-Lother, C. and Richardson, E.A. 2002. Ultrastructure of the host pathogen interface in daylily leaves infected by the rust *Puccinia hemerocallisdis*. *Protoplasma*, 219: 221-226.

Mishra, R.K., Pandey, K.K. and Pandey, P.K. 2009. Screening of pea (Pisum sativum) genotypes against rust caused by Uromyces fabae. *Indian Journal of Agricultural Sciences*, 79: 402-403

Mittal, R.K. 1997. Effect of sowing dates and disease development in lentil as sole and mixed crop with wheat. *J. Mycol. and Plant Pathol.*, 27: 203-209.

Mitter, J.H. and Tondon, R.N. 1930. Fungi flora of Allahabad, India. Journal of India Botanical Society, 9: 190-196.

Mohy, U.D., Din, G., Kahn, M.A. and Khan, S.M.C. 1999. Evaluation of seed and foliar applied fungicides to control lentil rust. *Pakistan J. Phytopathol.*, 11: 77-80.

Montes-Belmont, R., Cruz-Cruz, V., Martinez-Martinez. G. and Sandoval-Garcia, G. 2000. Antifungal properties in higher plants. *Retrosp. Analy. Invest.*, 18: 125-131.

Mukherjee, P.K., Mukhopadhyay, A.N., Sarmah, D.K. and Shrestha, S.M. 1995. Comparative antagonistic properties of *Gliocladium virens* and *Trichoderma Harzianum* on *Sclerotium rolfsii* and *Rhizoctonia solani* its relevance to understanding the mechanism of biocontrol. *J. Phytopathol.*, 143: 275-279.

Mukhopadhyay, A.N 1987. Recent innovations in plant disease control by ecofriendly biopesticides. In 83rd Annual Meeting of Indian Science Congress, Patiala, Jan. 1-8, 1996.

Narsinghani, V.G., Singh, S.P. and Pal B.S. 1980. Note on rust resistance pea varieties. *Indian J. Agril. Sci.*, 50: 453.

Negussie, T., Pretorius, Z.A. and Bender, C.M. 2005. Effect of some environmental factors on *in vitro* germination of urediniospores and infection of lentils by rust. *J. Phytopathol.*, 153: 43-47.

Okioga, D.M. and Jaffer, A.A. 1972. Studiesa on efficacy of various systemic fungicides against bean rust [(*U. appendiculatus* (Pers.) Lev.] Mics. *Rep. Trop. Pestic. Res. Inst.*, 805: 13.

Oxenham, S.K., Svoboda, K.P. and Walters, D.R. 2005. Antifungal activity of the essential oil of basil (*Ocimum basilicum*). *J. Phytopathol.*, 153: 174-180.

Palter, J. and Stettiner, M. 1957. The principal pests and diseases affecting legume crops in spring. *Adv. Leadl. Agric. Consult.*, 20: 15 (*Rev. Appl. Mycol.*, 1957: 530).

Pande, S.K., Srivastava, R.K. and Shahi, H.N. 1995. Fungicidal evaluation against rust of lentil. *Ann. Plant Prot. Sci.*, 3: 164.

Parambaramani, C., Vidhyasekaran, P. and Kandaswamy, T.K. 1971. Preliminary studies on use of new oxathin fungicides for the control of some rusts. *Madras Agric. J.*, 58: 705-706.

Patel, M.K. 1934. *Indian Bulletin of Plant Protection* (8): M1999-200.

Patil, M.K. 1933. India: disease in the Bombay presidency. *Int. Bull. Pl. Prot.*, 11: 246.

Pavgi, M.S. and Upadhyay, H.P. 1966. Parasitic fungi from North India IV Mycopathology et. *Mycological Applications*. 30: 527-260.

Persoon, D.C.H. 1801. *Synopsis methodica fungorum.* 1: 224.

Petch, T. 1939. Gliocladium. *Trans. Br. Mycol. Soc.*, 257-263.

Prasada, R. and Singh, S.P. 1975. Sexual behaviour of *Uromyces fabae. Indian J. Mycol. and Plant Pathol.*, 5: 139-144.

Prasada, R. and Verma, U.N. 1948. Studies on lentil rust, *Uromyces fabae. Indian Phytopathol.*, 1: 142-146.

Rashid, K.Y. and Bernier, C.C. 1991. The effect of rust on yield of feba bean cultivars and slow rusting populations. *Canadian J. Pl. Sci.*, 71: 967-972.

Reichert, I. and Patil, J. 1946. Rust and choclate sports of broad beans in Palestine. *Palest. J. Bot. Res. Ser.*, 2: 202-213.

Rizvi, G., Saxena, P., Faruqui, S.A. and Chandra, N. 1998. The effect of botanical pesticides on the germination of rust sproes of Lucerne. (Abs). ICPPMSA, 11-13 Dec., 1998, CSAUA and T, Kanpur, p. 141.

Roy, T.C. 1949. Fungi of Bengal: Directorate of Agriculture. Govt. of West Bengal.

Sadravi, M., Ono, Y., Pei, M. and Rahnama, K. 2007. Fourteen rusts from Northeast Iran. *Journal of Plant Pathology*, 89: 191-202.

Sajid, M.N., Ihsan, J. Nasir, M.A. and Shakir, A.S. 1995. Comparative efficacy of neem products and Baytan against leaf rust of wheat. *Pakistan J. Phytopatho.*, 7: 71-75.

Sangar, R.B. and Singh V.K. 1994. Effect of sowing dates and pea varieties on the severity of rust, powdery mildew and yield. *Indian J. Pul. Res.*, 7: 88-89.

Schwartz, H. F., Steadman, J.R. and Lindgren, D.T. 2001. Rust of dry beans. Cooperative Extension, Colorado State University, USA.

Sequeira, L. 1983. Mechanisms of induced resistance in plants. *Annu. Rev. Microbiol.*, 37: 51-79.

Sharma, A.K. 1998. Epidemiology and management of rust disease of French bean. *Vegetable Science*, 25: 85-88.

Sillero, J.C. and Rubiales, D. 2002. Histological characterization of resistance to *Uromyces viciae fabae* in feba bean. *Phytopathology*, 92: 294-299.

Sillero, J.C., Moreno, M.T. and Rubiales, D. 2000. Characterization of new sources of resistance to *Uromyces viciae-fabae* in a germplasm collection of *Vicia faba*. *Plant Pathology*, 49: 389-395.

Silva, S.R., Rios, G.P., Silva, S.C. 2001. Influence of genetic resistance and wetness period on infection and lesion development of common bean rust. *Fitopathologia Brasileria*, 26: 726-731.

Simon, U.K., Robert, B. and Franz, O. 2004. The unique cellular interaction between the leaf pathogen *Cymadothea trifolii* and *Trifolium repens*. *Mycologia*, 96: 1209-1217.

Singh, N.I. 1998. The fungal air-spora of Imphal, its seasonal fluctuations and relationship with occurrence and intensity of major crop diseases. *International Aerobiology Newsletter*, 28: 15-17

Singh, R.A., De, R.K. and Chaudhary, R.G. 2004. Influence of spray time of mancozeb on pea rust caused by *Uromyces viciae-fabae*. *Indian J. of Agril. Sci.*, 74: 502-504

Singh, R.M. and Srivastava, C.P. 1985. Evaluation classification and usefulness of pea germplasm lines for quantitative characters. *Legume Research*, 8: 68-73.

Singh, R.R., Mohit, Singh and Singh, M. 1997. Chemical control of pea rust. *Ann. Pl. Prot. Sci.*, 5(1): 118119.

Singh, R.S. 1973. Plant diseases. Oxford and IBH, New Delhi, p. 512.

Singh, R.S. 1999. Plant Diseases. Oxford and IBH, New Delhi, p. 686.

Singh, R.S. 2005. Plant Diseases. Oxford and IBH Publishing Co. Pvt. Ltd., New Delhi, p. 396.

Singh, S.J. and Sokhi, S.S. 1980. Pathogenic variability in *Uromyces viciae-fabae*. *Plant Disease*, 64: 671-672.

Singh, V.K., Sangar, R.B.S. and Singh, R.N. 1996. Effect of varieties and sowing dates on disease incidence and productivity of fieldpea (*Pisum sativum*). *Indian J. Agron.*, 4: 451-453.

Sokhi, S.S. 1984. A distinct physiologic race of *Uromyces viciae fabae* on sweet pea. *Indian Phytopathology*, 37: 123-125.

Srinivasulu, B., Rao, M.P.P., Varaprasad, Y., Satyanarayana, A., Reddy, M.V. and Reddy, D.R. 1999. Influence of rainfall and relative humidity on urdbean rust in rice fallows. *Indian J. Pul. Res.*, 12: 277-278.

Srivastava, L.S. 1999. Efficacy of fungitozicants in managing rice bean rust in Sikkim. *J. Hill Res.*, 12: 86-87.

Staples, R.C. 2000. Research on the rust fungi during the twentieth century. *Annu. Rev. Phytopathol.*, 38: 49–69.

Steadman, J.R. and Harveson, R.M. 2000. Use of fungicides for managing bean rust. University of Nebraska. USA (online data-www.un.edu.).

Sugha, S.K., Banyal, D.K. and Rana, S.K. 2008. Management of pea (Pisum sativum) rust (Uromyces fabae) with fungicides. *Indian Journal of Agricultural Sciences*, 78: 269-271.

Sugha, S.K., Chauhan, R.S. and Singh, B.M. 1994. Sensitivity of aeciospores and uredospores of the pea rust pathogen to selected systemic fungicides. *Tropical Agriculture*, 71: 27-30

Sumartini. 1998. Rust disease in French bean and its control. *Journal Peneletian and Pengembangan Pertanian*, 17: 149-153.

Sydow, H. and Butler, E.J. 1912. Fungi Indiae Orientalis pars IV. *Annals of Mycology*, 9: 372-421.

Thara, V.K., Fellers, J.P. and Zhou, J.M. 2003. *In planta* induced genes of Puccinia triticina. *Mol. Plant Pathol.*, 4: 51–56.

Thatcher, P.S. 1939. Osmatic and permeability relations in the nutrition of fungus parasite. *Amer. J. Bot.*, 26: 499-458.

Thrower, L.B. and Thrower, S.L. 1966. The effect of infection with Uromyces fabae on translocation in Broad Bean. *J. Phytopathol.*, 57: 267–276.

Tissera, P. and Ayres, P.G. (1986) Transpiration and the water relations of faba bean (Vicia faba) infected by rust (*Uromyces viciae fabae*). *New Phytol.*, 102: 385–395.

Tivoli, B., Beasse, C., Lemarchand, E. and Masson, E. 1996. Effect of Aschochyta blight (*Mycosphaerella pinodes*) on yield components of single pea (*Pisum sativum*) plants under field conditions. *Annals of Applied Biology*, 129: 207-216.

Tripathi, H.S. and Rathi, Y.P.S. 2003. Studies on Epidemiology and Management of Rust of Field pea. Final Technical Report, CAS in Plant Patholgoy, G.B.P.U.A. and T., Pantnagar, Uttaranchal, p. 125.

Tuffery, L. 1954. Antibiotics and disease, Experiments with plants. N. 2. *Odnr.*, 10: 340-341.

Tuzun, S. and Kuc, J. 1985. A modified technique for inducing systemic resistance to blue mold and increasing growth of tobacco. *Phytopathology*, 75: 1127-1129.

Tuzun, S., Nesmith, W., Ferriss, R.S. and Kuc, J. 1986. Effect of stem injections with *Peronospora tobacum* on growth of tobacco and protection against blue mod in the field. *Phytopathology*, 76: 938-941.

Tyagi, M.K. 1999. General mean analysis of adult plant resistance to powdery mildew and rust disease in fieldpea. *Plant Dis. Res.*, 14: 171-174.

Tyihak, E., Steiner, U. and Schonbeck, F. 1989. Induction of disease resistance by N-trimethyl-L-lysine in bean plants against *Uromyces phaseoli*. *J. Phytopatholgoy*, 126: 253-256.

Upadhyay, A.L. and Gupta, R.P. 1994. Fungicidal evaluation against powdery mildew and rust of pea (*Pisum sativum* L.). *Ann. Agric. Res.*, 15: 114-116.

Uppal, B.N. 1993. Host range of *U. viciae fabae. Int. Bull. Prot.* 7: M 103, and M746.

Vieira, R.F., Paula, T.J. De. (Jr.) and Benger, P.G. 1998. Effect of fungicides and the insecticide cartap on the control of foliar bean disease during the winter spraying season. *Summa Phytopathologica.*, 2: 17-22.

Voegele, R.T. and Mendgen, K. 2003. Rust haustoria: nutrient uptake and beyond. *New. Phytol.* 159: 93–100.

Walters, D.R. and Murray, D.C. 1992. Induction of systemic resistance to rust in *Viciae faba* by phosphate and EDTA: effects of calcium. *Plant Pathology*, 41: 444-448.

Ward, E.W.B. 1984. Suppression of metalaxyl activity by glybhosbhate: evidence that host defense mechanisms contribute to metalaxyl inhibition of *Phytopathora megasperma* f. sp. glycines in soybeans. *Physiol. Plant Pathol.*, 25: 381-386.

Webb, D.H. and Nutter, F.W. 1997. Effect of leaf wetness duration and temperature on infection efficiency, latent period, and rate of pustule appearance of rust in alfalfa. *Phytopathology*, 87: 946-950.

Weintraub, M. and Ragetli, H.W.J. 1966. Fine structure of inclusion and organelles in *Vicia faba* infected bean mosaic virus. *Virology*, 28: 290-302.

Wood, R.K.S. and Treit, M. 1995. Control of plant diseases by use of antagonistic organisms. *Bot. Rev.*, 21: 441-492.

Woods, A.M. and Beckett, A. 1987. Wall structure and ornamentation of the urediniospores of *Uromyces viciae-fabae*. *Can. J. Bot.*, 65: 2007–2016.

Wynn, W.K. 1976. Appressorium formation over stomata byt the bean rust fungus response to a surface contact stimulus. *Phytopathology*, 66: 136-146.

Xue, A.G. and Warkentin, T.D. 2002. Reactions of field pea varieties to three isolates of *Uromyces fabae*. *Cana. J. Pl. Sci.*, 82: 253-255.

Xuei, X., Bhairi, S., Staples, R.C. and Yonder, O.C. 1992. Characterization of INF56, a gene expressed during infection structure development of *Uromyces appendiculatus*. *Gene*, 110: 49-55.

Yuen, G.Y., Steadman, J.R., Lindren, D.T., Schaff, D. and Jochum, C. 2001. Bean rust biological control using bacterial agents. *Crop Prot.*, 20: 395-402.

Zhou, X.L., Stumpf, M.A., Hoch, H.C. and Kung C. 1991. A mechanosensitive channel in whole cells and in membrane patches of the fungus *Uromyces*. *Science*, 253: 1415–1417.

2016, Diseases of Pulse Crops and their Sustainable Management 355–369
Editors: Samir Kumar Biswas, Santosh Kumar and Gireesh Chand
Published by: BIOTECH BOOKS, NEW DELHI

Chapter 19

Fusarium Wilt of Chickpea: An Overview

Vivek Singh[1]*, Santosh Kumar[2], Mehi Lal[3] and B.K. Gupta[4]

[1]Department of Plant Pathology,
[4]Department of Agriculture Extension,
Banda University of Agriculture and Technology, Banda – 210 001, U.P.
[2]Department of Plant Pathology, Bihar Agricultural University,
Sabour, Bhagalpur – 813 210, Bihar
[3]Central Potato Research Institute Campus,
Modipuram, Meerut – 250 110, U.P.

Introduction

Chickpea (*Cicer arietinum* L.) is the world's third most important pulse crop after bean and pea and India accounting for approximately 75 per cent of world chickpea production. It occupies an area of 8.00 mha and its production is 7.1 mt with an average productivity of 885 kg/ha (Economic survey, 2009-10). Globally, chickpea is cultivated in 12.14 million hectare of land with 11.3 million tonnes being produced (FAOSTAT 2012). India ranks first in terms of production and productivity, followed by Pakistan, Turkey, Iran, Myanmar, Ethiopia, Mexico, Australia, Mexico, Canada and the United States. Chickpea seeds are highly nutritious, comprising 18–24 per cent protein, 4–10 per cent fat, 52–71 per cent carbohydrate, and 10–23 per cent fibre,

* Corresponding Author: E-mail: vsinghiitk@gmail.com

minerals and vitamins (Jukanti *et al.*, 2012). In addition to its importance as a food crop, it is valued for its beneficial effects in improving soil fertility. It is a crop of both tropical and temperate regions. Chickpea is a diploid with 2n = 2x = 16 having a genome size of approximately 931 Mbp (Arumuganathan and Earle, 1991). It is a highly self-pollinated crop with an out crossing rate of less than 1 per cent. Two main types of chickpea cultivars are grown globally– *kabuli* and *desi*, representing two diverse gene pools. *Kabuli* type is grown in temperate regions while the *desi* type is grown in the semi-arid tropics.

Low yield of chickpea attributed to its susceptibility to several fungal, bacterial and viral diseases. The yield losses by individual insects and diseases range from 5 to 10 per cent in temperate regions and 50 to 100 per cent in tropical regions (Van Emden *et al.*, 1988). Among the diseases affecting chickpea, wilt caused by *Fusarium oxysporum* f.sp. *ciceri* (Padwick) Matuo and K. Sato is one of the most severe disease of chickpea throughout the world (Gupta *et al.*, 2009). The disease is wide spread in nature and reported from at least 33 countries (Nene *et al.*, 1996). In India, it is prevalent in all chickpea growing states and causes an annual loss of 10 per cent (Singh and Dahiya, 1973). However, it was observed that early wilting causes 77-94 per cent losses, while late wilting causes 24-65 per cent loss (Haware and Nene, 1980). Recently an extensive survey has been made in the major chickpea growing states of India and it was observed that the incidence of chickpea wilt varied from 14.1 to 32.0 per cent, with the highest in the state of Rajasthan, followed by Jharkhand and the lowest in the states of Punjab. Disease samples collected from different parts of India, observed that about 45.5 per cent samples were infected with *Fusarium oxysporum* f.sp. *ciceri* (*Foc*) and remaining were other pathogens (Dubey *et al.*, 2010). The pathogen appears to be highly variable. The pathogen is both seed and soil borne in nature. Drenching with fungicides is very expensive and impractical. *F. oxysporum* survive as mycelium and chlamydospores in seed and soil, and also on infected crop residues, roots and stem tissue buried in the soil, for up to 6 years. Therefore, management of the disable is very difficult.

Symptoms

The disease appears at any stage of crop growth. Early wilt symptoms can be observed within 4 weeks of sowing on susceptible cultivars. The infected seedlings showed sudden drooping of leaves and petioles. These seedlings retain their dull green colour. They usually show uneven shrinking of the stem above and below the collar region. Whole seedling collapse and lie on the ground. These plants are not showing any external rotting of roots. Wilt pathogen has also been reported to produce yellowing symptoms resulted in a progressive foliar yellowing followed by necrosis, 30-40 days after inoculation from southern Spain (Trapero-Cases and Jimenez-Diaz, 1985). Late wilt also occurred during flowering stage. Disease/wilted plants did not show external root-rot symptoms but exhibited internal discoloration of the xylem and pith, if the collar region is cut transversely or diseased stem split vertically. Partial wilting also occurs occasionally.

The Pathogen

Chickpea wilt was first reported by (Butler, 1918) from India. However, Rhind (1926) expressed the doubt about fungal nature of the disease and in 1928 Robertson also failed to isolate a specific organism from wilted chickpea plant. It was McRae (1929) who confirmed the association of *Fusarium* species with wilted chickpea plants. Earlier it was believed that two different causes were associated with chickpea wilt, one was the species of *Rhizoctonia* resembling to *R. bataticola* while others was of physiological disorder (Dastur, 1935). Finally the causal agent was confirmed a *Fusarium oxysporum* f. sp. *ciceri* (Chattopadhyay and Sen Gupta, 1967).

A new fungus *F. solani* was also found associated with wilted chickpea plants (Grewal *et al.*, 1974a and1974b). They collected large number of wilted chickpea plants of different cultivars from IARI (New Delhi), Hisar (Haryana) and Ludhiana (Punjab) and F. *o.* f. sp. *ciceri* (15.5 per cent), *Rhizoctonia solani* (8.5 per cent), *R. bataticola* (8.3 per cent), *Operculella padwickii* (6.2 per cent) and *F. moniliforme* (1.6 per cent) were found to be associated with these samples. *F. solani, F. o.* f. sp. *ciceri* and *R. solani* proved potential pathogens and caused 82-100 per cent mortality.

Considering the complex nature of the wilt disease, a symposium was organized by Dr. H.K. Jain, Director of IARI, New Delhi in 1973 on *Problems of wilt and breeding for wilt resistance in Bengal gram*. During the symposium it was concluded that gram wilt comprises of two components, one being principally affected by environmental factors, like soil moisture, soil and atmospheric temperature, soil aeration, nutrient status of soil, plant density, date of sowing etc. and second the most important being contributed by the pathogen where in *F. o.* f. sp. *ciceri and F. solani* were involved. In nature, however, there is no demarcation between these two components and the different causal factors interact freely and hence the role of each factor remains to be determined (Jain and Behl, 1974). Subsequently, Fusarial component was recognized as one of the most important part of chickpea wilt (Pal, 1998). Later-on, the diseases involved in wilt complex were characterized and clearly distinguished on the basis of symptoms caused by them and were designated as black root rot (*F. solani*), dry root rot (*R. bataticola*), wet root rot (*Rhizoctonia solani*), collar rot (*Sclerotium rolfsii*), wilt (*F. o.* f. sp. *ciceri*) and foot rot (*Operculella padwickii*) (Nene *et al.*, 1978).

The mycelium of *F. oxysporum* f. sp. *ciceris* is delicate, white, cottony, becoming felted and wrinkled with age. Hyphae are septate and profusely branched. Microconidia are oval to cylindrical, straight to curved, 2.5 – 3.5 x 5-11 μm on simple short conidiophores. Macroconidia develop on the same conidiophores; thin walled, 3-5 septate, fusioid, pointed at both ends and measure 3.5-4.5 X 25-65 μm in size. Chlamydospores are smooth or rough walled, terminal or intercalary and may develop singly, in pairs or in a chains. Recently, Dubey *et al.* (2010) observed high level of variability in morphological characters of 112 isolates of *F.o.* f. sp. *ciceri*.

The isolates were highly variable in their colony growth pattern, size of colony and pigmentation. The size of micro-conidia varied from 5.1-12.8×2.5-5.0 μm, whereas macro-conidia arranged from 16.5-37.9×4.0-5.9 μm with1-5 septations. The isolates were grouped into 12 categories on the basis of their radial growth, size of macro-conidia and growth pattern.

Pathogenic Variability

The pathogen appears to be highly variable. Fusarium wilt is one of the most important disease, limiting chickpea production worldwide, has been the target of breeding for resistance (Kumar *et al.*, 1985). Cultivation of resistant cultivars is one of the most practical and cost efficient strategies for managing this disease. However, the efficiency of resistant cultivars in disease management is limited by high pathogenic variability in *F. oxysporum* f. sp. *ciceris*. The pathogen has extreme genotypic and phenotypic variability and can adapt a wide range of environmental conditions.

The isolates of the pathogen were variable in their pathogenic behaviour and 300 chickpea isolates of *Fusarium* were categorized into three groups as non-pathogenic, pathogenic and those causing seed-rot (Padwick, 1939). Internationally at present 8 physiological races of the pathogen (0, 1A, 1B/C, 2, 3, 4, 5, and 6) have been identified by reaction on a set of differential chickpea cultivars (Haware and Nene, 1982; Jimenez-Diaz *et al.*, 1993). The distribution pattern of these races in different part of the world indicates regional specificity for there occurrence. Race 2, 3 and 4 has been reported only from India (Haware and Nene, 1982), whereas, 0, 1 B/C, 5 and 6 are found mainly in the Mediterranean region and United States (Jimenez-Gasco *et al.*, 2001). Unlike the other races, race 1A is more widespread and has been reported in India, California (USA) and the Mediterranean region (Haware and Nene, 1982; Jimenez-Gasco *et al.*, 2001). Pathotypes also exhibit differences in their disease symptoms. Races 0 and 1 B/C cause yellowing whereas, 1A, 2, 3, 4, 5 and 6 showed wilting syndrome (Trapero-cases and Jimenez-Diaz, 1985). The races causing yellowing symptoms are not prevalent in India. The yellowing pathotypes are less aggressive than the wilting pathotypes. Forty-seven pathogenic isolates of Foc from California, India, Tunisia and Spain and 25 non-pathogenic *F. oxysporum* isolates from chickpea roots in Algeria, Italy, Moracco, Pakistan and Spain were analyzed for vegetative compatibility. Only one vegetative compatibility group (VCG) found among the 47 isolates of *Foc*, however 25 non-pathogenic isolates produces three multi-member VCG, each group containing two to five isolate, most of the non-pathogenic isolates were compatible with a yellowing-type low virulent isolate and suggested the possibility of the existence of transition between pathogenic and non-pathogenic populations (Nogales-Moncada *et al.*, 2009).

Honnareddy and Dubey (2006) established 3 new races of the pathogen in India based on the reactions on set of known differential cultivars. The reactions noticed on the cultivars were not similar to the earlier report even with the isolates of reported races taken from culture collections (Haware and Nene, 1982). Thus, indicated that the present set of differential cultivars should be re-standardized with a new set of cultivars to get clear cut virulence differentiation of present pathogenic population. Recent, study of Dubey *et al.*(2010) on the virulence analysis of 64 isolates collected from major chickpea growing states of India on 14 varieties including 10 international differentials revealed that more than one races are prevalent in one state. Majority of the isolates were not matched with the race specific reactions, therefore, it was suggested that the cultivars GPF-2, DCP 92-3 and KWR 108 should be included in the set of differentials to get clear cut differential reactions. In another study Dubey *et al.* (2012) found that seventy isolates of *Fusarium oxysporum* f. sp. *ciceris* (*Foc*) causing chickpea wilt representing 13 states and four crop cultivation zone of India were analyzed for their virulence and genetic diversity. The isolates of the pathogen and showed high variability in causing wilt incidence on a new set of differential cultivars of chickpea, namely C104, JG74, CPS1, BG212, WR315, KWR108, GPF2, DCP92-3, Cahffa and JG62. New differential cultivars for each race were identified, and based on differential responses, the isolates were characterized into eight races of the pathogen.

Genetic Variability

The identification of pathogenic races generally based on the differential reactions to selected host genotypes. After development of PCR, several biotechnological approaches like Random Amplified Polymorphic DNA (RAPD), Amplified Fragment Length Polymorphism (AFLP), Simple Sequence Repeat (SSR) and Inter-Simple Sequence Repeat (ISSR) have been increasingly used to study variability in pathogenic population. Barve *et al.* (2001) distinguished four races prevalent in India based on oligonucleotide probes. Sivaramakrishanan *et al.* (2002) studied genetic variability among 43 Indian isolates of chickpea wilt pathogen by using RAPD and AFLP and reported that molecular marker gave three different clusters. Of these, two clusters represented race-1 and race-2 and third cluster consisted of race-3 and race-4. They also observed that the high levels of DNA polymorphism with the molecular markers suggest the rapid evolution of new recombinants of the pathogen in the chickpea growing fields. A phylogenetic relationship among the eight known pathogenic races of chickpea wilt pathogen showed that the inferred intraspecific phylogeny observed in each of those races forms a monophyletic lineage. Moreover, virulence of races to resistant chickpea cultivars has been acquired in a simple stepwise pattern, with few parallel gains or losses (Jimenez-Gasco *et al.*, 2004). Sequence tagged microsatellite site (STMS) markers linked to six genes governing resistance to six races (0, 1A, 2, 3, 4 and 5) of the pathogen were also identified (Sharma and Muehlbauer, 2007).

Differences in the stages of plant-pathogen interaction account for the difference in aggressiveness between *F. oxysporum* f.sp. *ciceris* races (Jimenez-Diaz *et al.*, 1989). The faster and more extensive colonization of xylem vessels of chickpea cultivar P-2245 in roots and stem was observed by race 5 than by race 0. Genetics of resistance in chickpea cultivar WR-315 against races 1A, 2, 3, 4, and 5 of the pathogen concluded that the resistance was governed by single genes to each of the 5 races (Sharma *et al.*, 2005). Inheritance of wilt resistance in chickpea using allele-specific associated primer (ASAP) marker (CS-27_{700}) indicated that among the 100 F_5 progenies, 28 showed early wilting (Less than 25 days for complete wilting), 43 showed late wilting (25-55 days) while 29 showed no wilting and segregating in the ratio of 1:2:1 in the F_5 generation, indicated that more than one gene was involved in resistance (Brindha and Ravikumar, 2005).

Random Amplified Polymorphic DNA (RAPD), ISSR and SSR markers were also used to assess the genetic diversity of 64 isolates of the pathogen collected from 5 major chickpea growing states of the country (Dubey and Singh, 2008). The cluster generated by RAPD, grouped all isolates into 2 major groups at 30 per cent genetic similarity, whereas, ISSR and SSR analysis also grouped all isolates into 2 major clusters. Majority of Punjab state isolates and few from Rajasthan distinguished separately by the molecular marker evaluated, whereas, second group contains isolates from all other states suggest the existence of diverse genetic population of the pathogen at same location. The RAPD (OPM6, OPI 9, OPN 4, OPF 1, P 17, P 21 and SC 1), ISSR (ISSR 7, ISSR 11 and ISSR 12) and SSR (MB 17) markers clearly distinguished area specific isolates. One hundred eight isolates from 13 different provinces of Turkey was analyzed using RAPD and ISSR. UPGMA (Unweighted Pair-Group Method with Arithmetic mean) average cluster analysis provided substantially similar discrimination among Turkish isolates and divided into three major groups. Molecular variance confirmed that the genetic variability resulted from the differences among isolates within region. They also indicated that the low-genetic differentiation and high gene flow among populations had a significant effect on the emergence and evolutionary development of *Foc* (Bayraktar *et al.*, 2008).

Forty eight isolates of *Foc* collected from different chickpea growing region of India were evaluated using AFLP. UPGMA cluster analysis and principle coordinate analysis distinctly classified 48 isolates into two major pathogenic and non-pathogenic groups. The pathogenic isolates further clusterd into six major groups at 0.77 genetic similarties. Region specific grouping was observed within few isolates (Sharma *et al.*, 2009).

Variability of pathotypes of *F. oxysporum* f. sp. *ciceris* and breakdown of natural resistance are the main hindrances to developing resistant plants by applying resistant breeding strategies and lack of information of potential resistant genes limits gene-transfer technology. Understanding of *Fusarium* spp.-chickpea interaction at a cellular and molecular level is essential for isolation of potential genes involved in counteracting disease progression, cDNA amplified fragment length polymorphism followed by homology search helped in differentiating and analyzing the up and down regulated gene fragments. Several detected DNA fragments appeared to have relevance with pathogen-mediated defense. Some of the important transcript-derived fragments were homologous to genes for sucrose synthase, isoflavonoid biosynthesis, drought stress response, serine threonine kinases, cystatins, arginase, and so on. Reverse-transcriptase polymerase chain reaction (RT-PCR) performed with samples collected at 48 and 96 h postinfection confirmed a similar type of differential expression pattern. Based on this finding, interacting pathways of cellular processes were generated. This study has an implication towards functional identification of genes involved in wilt resistance (Gupta *et al.*, 2009).

Nogales-Moncada *et al.* (2009) studied 47 pathogenic and 25 non-pathogenic isolates of *Foc* using RAPD-PCR and reported that total DNA of the heterokaryon showed the presence of the two genomes in it. The existence of a single vegetative compatibility group in the population supports the hypothesis of its monophyletic origin. Bayraktar *et al.* (2008) also indicated that the low-genetic differentiation and high gene flow among populations of *Foc* collected from 13 different provinces of Turkey had significant effect on the emergence and evolutionary development of chickpea wilt pathogen caused by *F. oxysporum* f.sp. *ciceris*. Recently Dubey *et al.* (2014) analyzed the genetic diversity of 70 isolates of *F. oxysporum* f.sp. *ciceris* originated from various states of India representing eight races causing wilt in chickpea using translation elongation factor -1α (TEF-1α), β-tubulin and internal transcribed spacer (ITS) gene-regions. TEF-1α, β-tubulin and ITS gene-specific markers produced –720, -500, and 550 bp amplicons respectively, in all the isolates of the pathogen. A phylogenetic tree constructed from the sequences generated in the present study along with the sequences generated in the present study along with the sequences of foreign isolates of *Fusarium* species available in NCBI database sharing more than 90 per cent nucleotide sequence similarity grouped the isolates into two major clusters. Most of the isolates showed more or less similar grouping pattern in case of the three gene sequences. Each group had the isolates representing different races as well as place of origin indicating low level of diversity among the isolates in respect of these genes sequences. Except TEF-1α, the groups generated by β-tubulin and ITS gene-sequences did not correspond to the state of origin and races of the pathogen.

Management of Disease

Management practices directed toward pathogen for checking the progression of the disease occurrence could be exclusion and eradication of the pathogen and to reduce its inoculum. By the varied nature of pathogen involved, evolving resistant varieties has so far proved to be the best bet, although other conventional chemical, cultural methods and biological control have also yielded good results. Since this crop is grown principally in rainfed areas, many of the known conventional chemical methods have not found wide adoption (Hari Chand and Khirbat, 2009).

Cultural Control

Cultural practices are invaluable in reducing disease losses of chickpea. A disease control program is enhanced whenever one can utilize as many methods of control as possible. By proper adjustment of cultural practices, crop diseases can be avoided or at least adverse effect to be minimized. Suitable adjustment in cultural practices can modify the environment in such a manner that it becomes unfavorable for the pathogen and disease development (Khoury and Makkouk, 2010).

The Fusarium wilt is soil and internally seed borne disease. The chlamydospore like structure was detected in the helium region of the seed (Haware *et al.*, 1978). The disease development is faster at 24-27°C soil and air temperature. Below 17°C, infection remains restricted in the root without any wilt symptoms. The importance of the soil temperature has also been substantiated by the observation that late sowing of the crop reduces the incidence of the disease. Deep ploughing during summer and removal of host debris from the field reduces inoculum levels. A PCR-based method was developed for the detection of pathogen in natural and artificial soils (Pedrajas *et al.*, 1999).

Host Resistance

The use of resistant or tolerant cultivars is easy, cheap, environmentally sound and effective (Dodds and Rathjen, 2010), unless pathogens overcome the resistance. Host plant resistance is an important tool to control diseases of major food crops in developing countries, especially wheat, rice, potato, cassava, chickpea, peanut and cowpea. The use of resistant varieties is very much welcomed by resource poor farmers because it does not require additional cost and it is environment-friendly.

Use of resistant cultivars and adjustment of sowing dates are important measures for management of Fusarium wilt in chickpea. Landa *et al.* (2006) examined the effect of temperature on resistance of chickpea cultivars to wilt caused by various races of *Foc*. The chickpea cultivar Ayala was moderately resistant to *Foc* when inoculated plants were maintained at a day/night temperature regime of 24/21°C but was highly susceptible to the pathogen at 27/25°C. Field experiments over three consecutive years indicated that the high level of resistance of Ayala to wilt in mid- to late January sown crop differed from a moderately susceptible reaction under warmer temperatures in delayed sown crop (late February or early March). Experiments in growth chambers showed that a temperature increase from 24 to 27°C was sufficient for the resistance reaction of cultivars Ayala and PV-1 to race 1A of the pathogen to shift from moderately or highly resistant at constant 24°C to highly susceptible at 27°C. A similar but less pronounced effect was found when Ayala plants were inoculated with *Foc* race 6. Conversely, the reaction of cultivar JG-62 to races 1A and 6 was not influenced by temperature, but less disease developed on JG-62 plants inoculated with a variant of race 5 at 27°C compared with plants inoculated at 24°C. These finding indicated the importance of appropriate adjustment of temperature for characterization of resistance of chickpea cultivars as well as the races of *Foc*. The sowing time may be considered as an important component for expression of level of resistance in chickpea cultivars. Therefore, proper sowing time may be recommended for the management of the disease. The influence of *Meloidogyne artiella* on the reactions of chickpea genotypes with *Foc* races 0,1A and 2 was determined. The plants co-infected with race-0 and *M. artiella* decreases root colonization by the pathogen in genotypes CA 336.14.3.0 and PV 61 but not in ICC 14216 K 1 and UC 27. Whereas plants co-infected with race-1A and *M. artiella* significantly increased root colonization in all genotypes, except in cultivar BG 212. The cultivars UC 27 and ICC 14216 K showed resistant against races 1A and 2 and it was not modified by co-infection with the pathogen and *M. artiella* (Nauas Cortes *et al.*, 2008).

The identification and use of host plant resistance has the great potential in the long-term management of wilt. The genotypes H99-9, Pusa 212, JG 315, JG 322, PCS 1(Sel.ICCV-11), PCS 2 (Sel.KPG 142-1), PCS 5 (Sel.BGD-112), and PCS 6 (Sel. Pusa-1073) showed resistant (<10 per cent wilt incidence) reaction during 3 years (2001-2004) were deposited to Gene Bank, NBPGR, New Delhi with ACC No. 405202-405209 (Dubey and Singh, 2004). Recently, two genotypes namely H01-36 and PCS 8 (Sel.H82-2) showed < 10 per cent wilt were graded as resistant during three consecutive years *i.e.* 2004-07 of testing under wilt sick field and deposited to Gene Bank, NBPGR, New Delhi with ACC No. 553468 and 553469 (Dubey and Singh, 2008). In addition to these a large number of wilt resistant cultivars namely Avrodhi, Haryana Channa 1, BGD 72, BGM 547, GNG 469 (Smrat), GNG 663 (Vardan), RSG 693 (Aadhar), KPG-59 (Uday), K 3256 (Pragati), Phule G-87207 (Vishal), Phule G 9425-9, JG 322, GPF 2, PBG 1, Pusa 372 and Pusa 1053 (Chamtkar) were identified. Out of 117 *desi* chickpea genotype evaluated for resistance to wilt, three genotypes ICCV 05526, ICCV 05530 and ICCV 05533 showed asymptomatic, 11 resistant and 4 moderately resistant until harvest time (Pande *et al.*, 2007). Two *kabuli* accessions ICC 14194 and ICC 17109 from Mexico showed complete resistance to wilt (Gaur *et al.*,

2006). Three high yielding varieties of chickpea (GJG 0809 for NHZ, CSJ 515 and GLK 28127 for NWPZ) were identified along with three other state released varieties *viz.*, IPC 2004-98 (*desi* large seeded), IPC 2004- 1 (medium large seeded) and IPC 2005-62 for UP (IIPR, 2013-14).

The resistance to wilt for race 2 in cultivar WR 315 is controlled by a single recessive gene (Sharma *et al.*, 2005). Tullu *et al.* (1998) reported single recessive gene for resistance to race 4 and identified a RAPD marker linked with resistance. In order to identify DNA marker linked to H_2 locus of wilt resistance, the recombinant inbred lines derived from the cross K 850 (late wilting) x WR-315 (resistant), segregating for only H_2 locus were utilized. The recombinant inbred lines (RIL) showed 1:1 segregation for late wilting and resistance. The linkage analysis indicated that the $A07C_{417}$ marker is linked to H_2 locus and susceptibility and were separated by 21.7 centi Morgan (cM). The RILs of another cross JG-62 x WR-315 segregate for both H_1 and H_2 loci; DNA markers linked to H_1 and H_2 also showed independent segregation in the RILs of a cross JG-62 x WR-315. The $A07C_{417}$ marker was also found linked to H_2 locus of wilt susceptibility in different genotypes tested. The DNA marker $A07C_{417}$ showed linkage with H_2 locus across genetic backgrounds. The identification of DNA markers linked to both H_1 and H_2 of wilt resistance will facilitate marker-assisted selection of resistance genes to susceptible varieties. (Soregaon *et al.*, 2007).

Biological Control

Presently, there is worldwide swing to the use of eco-friendly methods for protecting the crops from pests and diseases. The use of potential harmful chemical sprays is viewed with dis-satisfaction in many countries. As such in the present context, biological control of wilt with bioagents offers a great promise. The most obvious and apparent environment friendly alternative to pesticides is to use naturally occurring biological approaches, to manage agriculturally important pest and diseases. Biological control may be effective either upon introduction by application or through strengthening their natural occurrence. A biological control agent colonizes the rhizosphere, the site requiring protection and leaves no toxic residues as opposed to chemicals. The first requirement of biological control is the identification and deployment of highly effective strains. The filamentous fungi, *Trichoderma* have attracted the attention because of their strong action against various plant pathogens (Harman *et al.*, 2004). The species of *Trichoderma* have been evaluated against the wilt pathogen and have exhibited greater potential in managing chickpea wilt under glasshouse and field conditions (Kaur and Mukhopadhayay, 1992). Earlier, Padwick (1941) also observed that a species of *Trichoderma* was highly antagonistic to wilt pathogen of chickpea. Mane (1995) isolated and evaluated 88 different fungal and bacterial agents against *F. oxysporum* f. sp. *ciceris* and four isolates of *Trichoderma* spp., three isolates of fluorescent Pseudomonas, one isolate of *Acrophilophora* sp. and three isolates of *Gliocladium* spp. showed antagonistic activity *in vitro*. Under field conditions, maximum wilt reduction (28.3 per cent) in cultivar Pusa 256 was observed when *T. viride* was applied as seed coating with talc as carrier with gum.

The soil inoculum can be reduced by addition of 15-20 tonnes of farmyard manure with *Trichoderma* sp. at 4-5 kg/ha before sowing (Singh and Dubey, 2007). The inhibitory activity of the *Trichoderma* against the pathogen was tested with four biocontrol agents. *Trichoderma viride* showed highest 77.2 per cent and 80.2 per cent growth inhibition at 8 and 16 days after inoculation, respectively. The per cent growth inhibition of fungal pathogen was positively correlated with activities of peroxidase at 8 days after inoculation (DAI). Mycoparasitism process was clearly observed during 8 to 16 DAI, peroxidase and protease correlated negatively, however, chitinase and ß-1,3 glucanase correlated positively for growth of the pathogen (Gajera *et al.*, 2009)

Recently, 10 isolates belonging to three species of *Trichoderma i.e. T. viride, T. harzianum and T. virens* were evaluated against four isolates of the pathogen representing 4 different races, Dharwad (race 1), Kanpur (race 2), Ludhiana (race 3) and Delhi (race 4) commonly prevalent in India. *T. viride* (IARI P-1) followed by *T. harzianum* (IARI P-4) and *T. viride* (IARI P-19) inhibited maximum mycelial growth of the pathogen. These bioagents also enhanced seed germination, root and shoot length and decreased wilt incidence under green house condition. The isolates proved potential *in vitro* tests were evaluated individually and in combination with carboxin under field condition. The efficacy of *Trichoderma* species was enhanced in combination with carboxin. The integration of *T. harzianum* (10^6 spores/ml/10g seed) and carboxin (2.0 g/kg seed) enhanced seed germination by 12.0-14.0 per cent and grain yield by 42.6-72.9 per cent and reduced wilt incidence (44.1-60.3 per cent) during 3 years (2002-05) of field experimentations under wilt sick field (Dubey *et al.*, 2007). The isolates of *Trichoderma* could be distinguished among the species by using oligonucleotide primers. Primer OPA 13 efficiently differentiated all the isolates of *Trichoderma* in to their species except T 2 of *T. viride*. Some of the primer as primer OPA 10 also showed species specific banding pattern (Dubey and Suresh, 2006). Biocontrol activity and plant growth promotion of bacterial strains were evaluated under greenhouse conditions, in which *P. aeuroginosa* (P10 and P12), *B. subtilis* (B1, B6, B28 and B99) and *P. aeuroginosa* (P12 and B28) provided better control than untreated control (15.8-44.8 per cent) in seed treatment and soil-inoculation, respectively (Karimi *et al.*, 2012).

Chemical Control

Seed dressing with benlate T (Benomyl + thiram) eradicated seed borne inoculum (Haware *et al.*, 1978). The disease can be managed by seed treatment with various seed dressing fungicides. Two years field data clearly indicates that seed treatment with Bavistin + thiram (1:1) at 2.5 g/kg seed before sowing decreased seedling mortality 7 per cent and increased seed germination 11.2 per cent and grain yield 22.30 per cent (Pal and Singh, 1993).

Application of fungicides, either though seed treatment or foliar application, is a very effective control measure. But intensive use of fungicides, especially multiple applications in a single growing season, causes concerns for increased selection pressure for fungicide resistance, increased production costs and negative impact on the environment. Furthermore, the majority of chickpea production is in developing countries and on subsistence farms. Limited resources are available for production inputs. Integrated management strategies and practices are required to minimize crop losses due to diseases and pests. Combining cultural practices and chemical methods increases production efficiency and reduces disease losses.

Conclusion

The pathogen is both seed and soil borne. Drenching with fungicides is very expensive and impractical. *F. oxysporum* survive as mycelium and chlamydospores in seed and soil, and also on infected crop residues, roots and stem tissue buried in the soil for up to 6 years and yield losses of up to 60 per cent may occur under favourable conditions. Therefore, integrated management strategies are the only solution to maintain plant health. Integrated disease management (IDM), which combines biological, cultural, physical and chemical control strategies in a holistic way rather than using a single component strategy proved to be more effective and sustainable (Khoury, 2010). In this approach, promotion of ecological balance, environmental safety and sustenance of productivity are given due attention. IDM includes encouragement of beneficial biological agents to reduce pathogen inoculum, modification of cultural practices, use of resistant varieties and minimum use of chemicals for checking the pathogen population.

References

Annonymous 2010. *Economic survey 2009-10.* Oxford University press, YMCA Library Building, Jai Singh Road, New Delhi p294.

Annual Report, 2013-14. Outreach Project on *Phytophthora, Fusarium* and *Ralstonia* Diseases of Horticultural and Field Crops. Indian Institute of Pulses Research, Kanpur.

Arumunganathan, K. and Earle, E.D. 1991. Nuclear DNA content of some important plant species. *Pl. Mol. Biol. Rep.* **9**: 208-218.

Barve, M.P., Haware, M.P. Sainani, M.N. Ranjekar, P.K. and Gupta, V.S. 2001. Potential of microsatellites to distinguish four races of *F. oxysporum*. f. sp. *ciceri* prevalent in India. *Theoretl. Appl. Genet.* **102:** 138-147.

Bayraktar, H., Dolar, F. S. and Madan, S. 2008. Use of RAPD and ISSR markers in detection of genetic variation and population structure among *F. oxysporum* f sp. *ciceris* isolates on chickpea in Turkey. *Phytopathology* **156:** 146-154.

Brindha, S. and Ravikmar, R.L. 2005. Inheritance of wilt resistance in chickpea – a molecular marker analysis. *Curr. Sci.* **88:** 701-702.

Butler, E.J. 1918. *Fungi and Plant Disease.* Bishen Singh Mahendra Pal Singh and Periodical Experts, Dehradun and Delhi. p547.

Chattopadhyay, S.B. and Sen Gupta, P.K. 1967. Studies on wilt diseases of pulses, I. variation and taxonomy of *Fusarium* spp. associated with wilt diseases of pulses. *Ind. J. Mycol. Res.* **5:** 45.

Dastur, J.F. 1935. Gram wilt in the central provinces- *Agric.liv-stk,* India. **5**: 615-627.

Dubey, S.C. and Singh, B. 2004. Reaction of chickpea genotypes against *Fusarium oxysporum* f. sp. *ciceris* causing vascular wilt. *Indian Phytopathol.* **57**: 233.

Dubey, S.C. and Suresh, M. 2006. Randomly amplified polymorphic DNA markers for *Trichoderma* species and antagonism against *Fusarium oxysporum* f. sp. *ciceris* causing chickpea wilt. *J. Phytopathol.* **154**: 663-669.

Dubey, S.C. and Singh S.R. 2008. Virulence analysis and oligonucleotide fingerprinting to detect diversity among Indian isolates of *Fusarium oxysporum* f. sp. *ciceris* causing chickpea wilt. *Mycopathologia* **165**: 389-406.

Dubey, S.C., Singh, B. and Bahadur, P. 2007. Diseases of pulse crops and their ecofriendly management In: *Ecofriendly Management of Plant Diseases*. (eds: S. Ahamad and U. Narain) Daya Publishing House, Tri Nagar, Delhi. pp 16–44.

Dubey, S.C., Kumari P., Singh, V. and Singh, V. 2012. Race profiling and molecular diversity analysis of *Fusarium oxysporum* f. sp. *ciceris* causing chickpea wilt in India. *Journal of Phytopathology*. doi: 10.1111/j.1439-0434.

Dubey, S.C., Suresh, M. and Singh, B. 2007. Evaluation of *Trichoderma* species against *Fusarium oxysporum* f. sp. *ciceris* for integrated management of chickpea wilt. *Biological Control* **40**: 118–127.

Dubey, S.C., Singh, S.R. and Singh, B. 2010. Morphological and pathogenic variability of Indian isolates of *Fusarium oxysporum* f. sp. *ciceris* causing chickpea wilt. *Archives of Phytopathology and Plant Protection* **43:** 174-189.

FAOSTAT 2012. Final 2012 data now available. (Food and Agriculture Organisation of the United Nations: Rome) Available online at: http: //faostat.fao.org/site/567/DextopDEfault.aspx?PageID=567#ancor [Verified 9 June 2014].

Gajera, H.P., Bambharolia, R.M., Patel, S.V., Mandavia, M.K. and Golakiya, B.A. 2009. Significance of lytic enzymes from *Trichoderma* in the *in vitro* biocontrol of fungal plant pathogen *Fusarium oxysporum* f. sp. *ciceris. Ind. J. Agricul.Biochem.* **22:** 31-37.

Gaur, P.M., Pande, S. Upadhyaya, H.D. and Rao, B.V. 2006. Extra large *kabuli* chickpea with high resistance to Fusarium wilt. *Intern. Chickpea and Pigeonpea Newsl.* **13**: 5-7.

Grewal, J.S., Pal, M. and Kulshrestha D.D. 1974b. Fungi associated with gram wilt. *Ind. J. Genet. Pl. Breed.* **34**: 242-248.

Grewal, J.S., Pal, M. and Kulshrestha, D.D. 1974a. A new record of wilt of gram caused by *Fusarium solani. Curr. Sci.* **43**: 767.

Gupta, S., Chakraborti, D., Rangi, R.K., Basu, D. and Das, S. 2009. A molecular insight into the early events of chickpea and *Fusarium oxysporum* f. sp. *ciceris* (race 1) interaction through cDNA-AFLP analysis. *Phytopathology* **99**: 1245-1257.

Gurjar, G., Barve, M. Giri A. and Gupta, V. 2009. Identification of Indian pathogenic races of *Fusarium oxysporum* f. sp. *ciceris* with gene specific ITS and Random markers. *Mycologia* **101:** 484-495.

Harman, G.E., Howell, C.R., Viterbo, A., Chet I. and Lorito, M. 2004. *Trichoderma* species – opportunistic, avirulent plant symbionts. *Nature Reviews* **2**: 43-56.

Haware, M.P. 1998. Disease of chickpeas. Pages 473-506. *In*: *The Pathology of Food and Pasture Legumes.* D.J. Allen and J.M. Lenne, eds. CAB International, Wallingford, U.K.

Haware, M.P. and Nene, Y.L. 1980. Influence of wilt at different growth stages on yield loss of chickpea. *Trop. Grain Legume Bull.* **19**: 38-40.

Haware, M.P. and Nene, Y.L. 1982. Races of *Fusarium oxysporum* f. sp. *ciceri. Pl. Dis.* **66**: 809-810.

Haware, M.P., Nene, Y.L. and Rajeshwari, R. 1978. Eradication of *Fusarium oxysporum* f. sp. *ciceri* transmitted in chickpea seeds. *Phytopathology* **68**: 1364-1367.

Honnareddy, N. and Dubey, S.C. 2006. Pathogenic and Molecular characterization of Indian isolates of *Fusarium oxysporum* f. sp. *ciceris* causing chickpea wilt. *Curr. Sci.* **91**: 661-666.

Jain, H.K. and Bahl, P.N. 1974. Recommendations of symposium on gram wilt. *Ind. J. Genet. Pl. Breed.* **34**: 236-238.

Jimenez-Diaz, R.M., Trapero-Casas, A. and Cabrea de la Colina, J. 1989. Races of *Fusarium oxysporum* f. sp. *ciceri* infecting chickpeas in southern Spain. *In: Vascular wilt diseases of plants. NATO ASI Ser.* (E.C. Tjamos and C.H. Beckman, eds.) Springer-Verlag, Berlin H **28**: 515-520.

Jimenez-Diaz, R.M., Alcala-Jimenez, A.R. Hervas, A. and Trapero-Casas, J.L. 1993. Pathogenic variability and host resistance in the *Fusarium oxysporum* f. sp. *ciceris* / *Cicer arietinum* pathosystem. In: Proc. 3rd Eur. Semin. Fusarium Mycotoxins, Taxonomy Pathogenicity and Host Resistence. Plant breeding and Acclamatization Institute, Poland. p87-94

Jimenez-Gasco, M.M., Perez-Artes, E. and Jimenez-Diaz, R.M. 2001. Identification of pathogenic races 0, 1B / C, 5 and 6 of *Fusarium oxysporum* f. sp. *ciceri* with Random Amplified Polymorphic DNA (RAPD), *Euro. J. Pl. Pathol.* **107**: 237- 248.

Jimenez-Gasco, M.M., Navas-Cortes, J.A. and Jimenez-Diaz, R.M. 2004. The *Fusarium oxysporum f. sp. ciceris / Cicer arietinum* pathostystem: a case study of the evolution of plant-pathogenic fungi into races and pathotypes. *Int. Microbiol.* **7**: 95-104.

Jukanti, A.K., Gaur, P.M., Gowda, C.L.L., Chibbar, R.N. 2012. Chickpea: nutritional properties and its benefits. *The British Journal of Nutrition* **108**, S11–S26. doi: 10.1017/S0007114512000797.

Kaur, N.P. and Mukhopadhayay A.N. 1992. Integrated control of chickpea wilt complex by *Trichoderma* spp. and chemical methods in India. *Trop. Pest Management.* **38**: 372-375.

Kumar, J., Haware, M.P. and Smithson, J.B. 1985. Registration of four short duration *Fusarium* wilt resistant kabuli chickpea germplasm. *Crop Sci.* **25**: 576-577.

Landa, B.B., Navas-Cortes, J.A., del, M., Jimenez- Gasco, M., Katan, J., Retig, B. and Jimenez-Diaz, R.M. 2006. Temperature response of chickpea cultivars to races of *Fusarium oxysporum* f. sp. *ciceris*, causal agent of Fusarium wilt. *Pl. Dis.* **90:** 365-374.

Mane, S.S. 1995. *Studies on Fusarium oxysporum f. sp. ciceri causing chickpea wilt with special reference to its management by bio-agents.* Ph.D. thesis, IARI, New Delhi. p.72.

Mc Rae, W. 1929. India; new disease reported during the year 1928. *Int. bull. Pl. Prot.* III: 21-22.

Navas- Cortes, J.A., Landa, B.B. Lopez, J.R. Jimenez- Diaz, R.M. and Castillo, P. 2008. Infection by *Meloidogyne artiellia* does not break down resistance to races 0,1A and 2 of *Fusarium oxysporum* f. sp. *ciceris* in chickpea genotypes. *Phytopathology* **98:** 709-718.

Nene, Y.L., Haware, M.P. and Reddy, M.V. 1978. *Diagnosis of some wilt like disorders of chickpea,* ICRISAT Information Bulletin-**3**, p.44.

Nogales- Moncada, A.M., Jimenez Diaz, R.M. and Perez Artes, E. 2009. Vegetative compatibility groups in *Fusarium oxysporum* f. sp. *ciceris* and *Fusarium oxysporum* non- pathogenic to chickpea. *J. Phytopathology* **157:** 729-735.

Padwick, G.W. 1939. *Report of the Imperial Mycologist*. Scient. Rept. Agric. Res. Inst., New Delhi, 1937-38; 105-112.

Padwick, G.W. 1941. *Report of the Imperial Mycologist*. Scient. Rept. Agric. Res. Inst., New Delhi, 1937-38; 94-101.

Pal, M. 1998. Diseases of pulse crops, their relative importance and management. *J. Mycol. Pl. Pathol.* **28**: 114-122.

Pal, M. and Singh, B. 1993. Channe koo Uktha Rog Se Bachayen. *Kheti* **47**(6)**:** 24-25.

Pande, S., Gaur, P.M., Sharma, M. Rao, J.N., Rao, B.V. and Kishore, G.K. 2007. Identification of single and multiple disease resistance in *desi* chickpea genotypes to Ascochyta blight, Botrytis gray mold and Fusarium wilt. *J. SAT Agric. Res.* **3:** 1-3.

Pedrajas, M.D.G., Bainbridge, B.W. Heale, J.B. Artes, E.P. and Diaz, R.M.J. 1999. A simple PCR-based method for the detection of the chickpea wilt pathogen *F. oxysporum* f. sp. *ciceris* in artificial and natural soils. *Euro. J. Pl. Pathol.* **105:** 251-259.

Rhind, D. 1926. Annual report of the Mycologist of Burma, for the year ended the 30th June 1925. Rangoon. Supdt. Govt. Printing and Stationary, Burma. p5.

Robertson, H.F. 1928. Mycology-Report on the operation of the Dept. of Burma for the year ended 30th June, 1927. p11-12.

Sharma, K.D. and Muehlbauer, F.J. 2007. Fusarium wilt of chickpea: physiological specialization, genetics of resistance and resistance gene tagging. *Euphytica* **157:** 1-14.

Sharma, K.D., Chen, W.D. and Muehlbauer, F.J. 2005. Genetics of chickpea resistance to five races of Fusarium wilt and a concise set of race differentials for *Fusarium oxysporum* f. sp. *ciceris*. *Pl. Dis.* **89:** 385-390.

Sharma, M., Varshney, R.K. Rao, J.N., Kannan, S., Hoisington D. and Pande, S. 2009. Genetic diversity in Indian isolates of *Fusarium oxysporum* f. sp. *ciceris* chickpea wilt pathogen. *African J. Biotech.* **8:** 1016-1023.

Singh, B. and Dubey, S.C. 2007. Channe Kaa Mallyni Rog - Bachaw Ke Uppaya. *Kheti* **60**(8)**:** 18-20.

Singh, K.B. and Dahiya, B.S. 1973. Breeding for wilt resistance in Chickpea. *Symposium on problem and breeding for wilt resistance in Bengal gram*. Sept. 1973 at IARI, New Delhi, p. 13-14.

Sivaramakrishnan, S., Kannan, S. and Singh, S.D. 2002. Genetic variability of Fusarium wilt pathogen isolates of chickpea (*Cicer arietinum* L.) assessed by molecular markers. *Mycopathologia* **155**: 171- 178.

Soregaon, C.D., Ravikumar, R.L. and Thippeswamy, S. 2007. Identification of DNA markers linked to H_2 locus of Fusarium wilt resistance in chickpea. *Ind. J. Genet. Pl. Breed.* **67:** 323-328.

Trapero-Casas, A. and Jimenez-Diaz, R.M. 1985. Fungal wilt and root rot diseases of chickpea in southern Spain. *Phytopathology* **75**: 1146-1151.

Tullu, A., Muehlbauer, F.J., Simon, C.J., Mayer, M.S., Kumar, J., Kaiser, W.J. and Kraft, J.M. 1998. Inheritance and linkage of a gene for resistance to race 4 of Fusarium wilt and RAPD markers in chickpea. *Euphytica* **102**: 227-232.

Van Emden, H.F., Ball, S.L. and Rao, R. 1988. Pest diseases and weed problems in pea lentil and faba bean and chickpea. In: (ed. R.J. Summerfield) *World Crops: Cool season Food Legumes*. ISBN 90-247-3641-2. Kluwer Academic Publishers. Dordrecht, The Netherlands. p519-534.

Khoury, W. El. and Makkouk 2010. Integrated Plant disease management in developing countries. *Journal of Plant Pathology*, 92(S4): 35- 42.

Karimi, K. Amini, J., Harighi, B. and Bahramnejad, B. 2012. Evaluation of biocontrol potential of Pseudomonas and Bacillus spp. against Fusarium wilt of chickpea. *Australian Journal of Crop Science*, 6(4): 695-703.

Sunil, C. Dubey, Kumari P. and Singh, V. 2014. Phylogenetic relationship between different race representative populations of *Fusarium oxysporum* f. sp. *ciceris* in respect of translation elongation factor-1α, β-tubulin, and internal transcribed spacer region genes. *Archives of Microbiology*, 196: 445–452.

2016, Diseases of Pulse Crops and their Sustainable Management 371–384
Editors: Samir Kumar Biswas, Santosh Kumar and Gireesh Chand
Published by: BIOTECH BOOKS, NEW DELHI

Chapter 20

Fungal Diseases of Lentil and their Integrated Management

Santosh Kumar[1]*, Prabhat Kumar[2], Mehi Lal[3] and Gireesh Chand[1]

[1]Department of Plant Pathology, Bihar Agricultural University, Sabour, Bhagalpur – 813 210, Bihar
[2]Betel Vine Research Centre, Islampur, Nalanda (Bihar Agricultural University, Sabour)
[3]Central Potato Research Institute, Campus Modipuram, Meerut, U.P.

Introduction

Lentil (*Lens culinaris* Medikus) commonly known as "Poor man's meat", as it is one of affordable protein rich legume. It contains, high amount of fibre, Vitamin A, calcium, starch, iron, phosphorus, copper and manganese and approximately 22 per cent protein (Bhatty, 1988). Additionally, because of its high lysine and tryptophan contents, its consumption with wheat or rice provides a balance in essential amino acids in human nutrition. It is widely used in India, Southwest Asia, and Mediterranean areas in the form of split lentil (dhal) and it is still an important source of dietary protein in these areas. While the lentil seed is used mainly as food, the straw can also be used as a high quality animal feed or as a source of organic material

* Corresponding Author: E-mail: santosh35433@gmail.com

for soil improvement through nitrogen fixation and green manuring (Erskine *et al.*, 1990). The area under lentil in India is around 1.59 million hectare with a production of 0.94 million tonnes and productivity 697 Kg/ha (Anonymous, 2011).

Lentil plants encounter numerous diseases that are caused by fungi, viruses and nematodes. Under favourable environmental conditions, the disease spread very fast and may appeared as epiphytotic condition. Diseases not only affect the plant growth and reduction in yield, but also reduce quality and grading of grain which affect market price. The seed also transmits diseases if the infected grain seeds are used as planting material. Among the important fungal diseases including, Fusarium wilt, rust, Ascochyta blight and Botrytis gray mold (BGM) are the major threat in the cultivation of lentil crop. These diseases may cause heavy losses ranging from 25 to 72 per cent depending on the cultivars and pathogen type (Sepulveda, 1985; Singh *et al.*, 1986). The concept of integrated disease management is to maintain increase production, to ensure food security with future dimension while preserving the underlying resources base. The changed agriculture scenario in India has to focus our attention to manage plant diseases, cut crop losses, avoid wide fluctuations in production and sustain the high levels of productivity.

1. Fusarium Wilt

Lentil wilt caused by *Fusarium* spp. was first reported by Prissyajnyuk from Russia in 1931. Padwick (1941) observed wilt of lentil in Delhi and Karnal and showed association of *Fusarium* spp. with the wilt. Vasudeva and Srinivasan (1952) were the first to study the fungus causing lentil in details. Now wilt disease is widely distributed in all the lentil growing states as Uttar Pradesh, Himanchal Pradesh, Bihar, West Bengal, Assam, Orissa, Rajasthan, Haryana, Uttarakhand and Punjab (Agarwal *et al.*, 1991).

Economic Importance

The losses caused by the disease ranges from 20-24 per cent annually (Ali, 2007). The extent of the damage to the crop due to the disease ranges from 20-24 per cent annually (Ali, 2007). However, the extent of losses by wilt depends upon the susceptibility of the cultivars sown, prevalent atmospheric conditions and soil type (Ahmad, 2010). Vasudeva and Srinivasan (1952) have reported wilt disease of lentil caused by *Fusarium* spp. is one of the serious diseases and it causes huge loss of the standing crop throughout the world. The pathogen is responsible for severe grain losses under conducive environment condition (Stoilova and Chavdarov, 2006).

The Pathogen

The genus *Fusarium* is one of the most economically important genera since it includes many pathogenic species which cause a wide range of plant diseases (Nelson *et al.*, 1981). Three types of asexual spores are produced by the pathogen. Microconidia: single celled, hyaline, ovoid, cylindrical, oblong or slightly curved. Macroconidia: nearly straight to fusiform, falcate, slender, thin-walled with indistinct septa, mostly single celled, rarely 1-5 septate. Conidia form freely on mycelium at the end of free conidiophores. Base of conidia papillate or tending to foot-cell formation. Chlamydospores are terminal or intercalary, borne on mycelium, rarely on conidia, spherical to pyriform, smooth, hyaline, rich in protoplasmic contents, one celled, occasionally 2-celled, abundantly formed on certain media but rare to none on others (Vasudeva and Srinivasan, 1952).

Symptoms

The disease appears either in the early stages of crop growth (seedling) or during the reproductive stages (adult stage) (Stoilova and Chavdarov, 2006; Khare, 1981). seedling wilt is characterized by sudden drooping followed by drying of leaves and the whole seedling and apparently healthy looking roots exhibited reduced proliferation. The roots do not show any external rotting but look apparently healthy. Such roots, when split vertically from the collar region to downward, show a brown discolouration of the vascular tissues. During the adult stage the growth of infected plant is checked, leaves shrink on get curled starting from the lower part of the plants and extending upwards become yellow and the whole plants gets dried. Root and stem of the wilted plant when split vertically, show internal discolouration of the pith and xylem tissues.

Vasudeva and Srinivasan (1952) reported that in a broadcast crop, wilt occurs in isolated patches, more or less circular in outline, which enlarge as the disease advances when the crop is sown in rows, the disease appear to progress along the lines (Figure 20.1).

Figure 20.1: Lentil Plants Showing Typical Wilt Sympton.

Disease Cycle

Fusarium oxysporum is an abundant and active saprophyte in soil and organic matter, with some specific forms that are plant pathogenic (Smith *et al.*, 1988). Its saprophytic ability enables it to survive in the soil between crop cycles in infected plant debris either as mycelium, or as any of its three different spore types. Healthy plants can become infected by *F. oxysporum* if the soil in which they are growing is contaminated with the fungus. The fungus can invade a plant either with its sporangial germ tube or mycelium by invading the plant's roots. The roots can be infected directly through the root tips, through wounds in the roots, or at the formation point of lateral roots (Agrios, 1997). Once the fungus enter inside the plant, the mycelium grows through the root cortex intercellulary. When the mycelium reaches the xylem, it invades the vessels through the xylem's pits. At this point, the mycelium remains in the vessels, where it usually advances upwards toward the stem and crown of the plant. As the fungus grows the mycelium branches and produces microconidia, which are carried upward within the vessel by way of the plants sap stream. When the microconidia germinate, the mycelium can penetrate the upper wall of the xylem vessel, enabling more microconidia to be produced in the next vessel. The fungus can also advance laterally as the mycelium penetrates the adjacent xylem vessels through the xylem pits (Agrios, 1997).

Due to the growth of the fungus within the plant's vascular tissue, the plant's water supply is greatly affected. Thus lack of water induces the leaves' stomata to close, the leaves wilt and the plant eventually dies. It is at this point that the fungus invades the plant's parenchymatous tissue, until it finally reaches the surface of the dead tissue, where it sporulates abundantly (Agrios, 1997). The resulting spores can then be used as new inoculum for further spread of the fungus.

Epidemiology

The fungus is soil borne, which can survive in the soil and plant debris in the absence of its host for a period of 3-4 years. The disease is favoured by low soil temperature, 30 per cent soil water holding capacity and increasing plant maturity. Kaushal and Sharma (1998) reported highest wilt incidence (75.1 per cent) at temperature range of 24-27°C and the lowest (13.0 per cent) in the temperature range of 13-16°C.

Host Pathogen Interaction

The pathogen *F. oxysporum* f.sp. *lentis* enters the host through natural openings, such as emergence of secondary roots, or through wounds, which expose xylem (Khare, 1980) and ultimately it colonizes the metaxylem and stops the supply of water and mineral ions, resulting in loss in turgidity and finally wilting (Saxena *et al.*, 1992). The vascular tissues of lentil roots infected by *F. oxysporum* f. sp. *lentis* showed the presence of inter- as well as intra-cellular mycelium. The xylem vessels are plugged with the mycelial growth and formation of tyloses. Chlamydospores-like bodies has also been detected in the vessels (Agarwal and Prasad, 1997). Mehrotra and Claudius (1973) reported production of hydrolytic enzyme by *F. oxysporum* f.sp. *lentis* which caused same disease of wilting in lentil as caused by the pathogen. The enzymes induced loss of turgor in foliage leading to drooping of top before finally withering. The production of certain toxin metabolites by *F. oxysporum* f.sp. *lentis* also responsible for the reduced vigour of seedling of lentil, which formed very small darkgreen leaves with loss of turgidity and finally the leaves turned yellow (Agarwal and Khare 1975).

Integrated Management

- ✰ Use of tolerant varieties such as Pant L-4, Pant L-6, Pant L-8 and Noori etc.
- ✰ Ploughing of the field during summer. Planting of seed at proper depth (2-4 cm). Burning of straw and stubbles after harvest.
- ✰ Crop rotation with resistant crops and soil amendments with organic matter enhances antagonism with other soil microflora can reduce the chances of infection. The application of organic amendments, manures and composts that are rich in nitrogen, may reduce soil-borne diseases by releasing allelochemicals during microbial decomposition.
- ✰ Treat the seeds with talc based formulation of *Trichoderma harzianum* @ 4 g/kg of seed or with fungicide such as benornyl (0.3 per cent) or thiram + benomyl (1:1, 0.3 per cent) reduces wilt incidence and increases grain yield. De and Chaudhary, (1999) reported that *Gliocladium virens, T. harzianum* or *T. viride* reduced the wilt incidence of lentil caused by *F.oxysporum* f.sp. *lentis* by 79 per cent and increased yield by 140, 224 and 241 per cent, respectively.

2. Rust

Rust caused by fungus *Uromyces viciae-fabae* (Pers.) Schroet is considered as a major limiting factor in successful lentil cultivation especially in North-Eastern and Northern plain of India. This disease causes heavy damage in Uttar Pradesh, Bihar, Punjab and Madhya Pradesh. The earliest recorded of the disease come from India when it was recorded by Butler in 1918 in the Gangetic plains of India. The disease has been reported from several countries including Cyprus (Nattrass, 1932), Morocoo (Melencon, 1936), Sicily (Canonaco, 1937) Palestine (Royss, 1937), Portugal (De Sousa *et al.*, 1939), Chile (Accantino, 1964) Hungary (Podhradscky 1968), Nepal (Karki, 1991) and Iran (Abbasi and Pooralibaba, 2002).

Rust disease is a potential threat to lentil cultivation and causes substantial yield losses ranging from 60-69 per cent (Sepulveda, 1985). In 1978 severe outbreak of lentil rust was recorded in the Narmada Valley of Madhya Pradesh (Khare and Agrawal, 1978). Rust has been a major constraint affecting yield adversely. *Tarai* region of Uttarakhand and its surrounding areas, earlier disease has appeared in almost epiphytotic form in this area (Khare and Agarwal, 1978).

The Pathogen

Uromyces viciae fabae is an autoecious and heterothallic fungus forming all the four spore forms *viz.* pycniospores/spermatiospores, aeciospores, urediospores and teliospores on lentil only as Pycnia. It is small, flask shaped and produced as yellowish flecks on upper surface of leaves with a common neactor drop at mouth. As the time advanced and the haploid pustules remained unfertilized, the formation of pycnia with separate scanty neater drops on the lower surface of the leaves was observed after a week or two week (Prasad and Singh 1975).

The uredial stage is followed by the formation of teliospore in the same sorus and on the same mycelium. In nature, the teliospore are generally formed when the host reaches maturity and environmental conditions are unfavorable for the propagation of rust in the uredial stage. The teliospore are subglobuse to ovate or elliptical, single celled, pedicillate, thick walled with flattened apex and 25-38 × 18-27 μ in size. The teleutospores can germinate at temperature ranging from 12-22°C to provide a four celled basidium on which four, single celled, hyaline basidiospores are produced (Prasad and Verma, 1948; Agrwal and Prasad, 1997; Singh, 2005).

Symptoms

Rust pustules can be seen on leaf blade, petiole and stem. Rust starts with the formation of yellowish-white pycnidia and aecial cups on the lower surface of leaflets and on pods, singly or in small groups in a circular form. Later, brown uredial pustules emerge on either surface of leaflets, stem and pods. Pustules are oval to circular and up to 1 mm in diameter. They may coalesce to form larger pustules. In severe infections, leaves are shed and plants dry prematurely, the affected plant dries without forming any seeds in pods or with small shrivelled seeds.

Disease Cycle

The dissemination of lentil rust pathogen takes place by aeciospores which form secondary aecia after infection of leaves. The secondary aecia are formed at a temperature of 17-22° C but at 25° C, the infection causes development of uredia. The aeciospores and probably the uredospores do not survive during off season. The teliospores can withstand the summer heat and hence the lentil rust perpetuates in its telial stage in the left over diseased plants trash or on seed as external contaminant and infects the new crop in the next season. Aeciospores from lentil have also been found to infect pea and *Vicia faba* and Lathyrus. However, another species of *Uromyces, V. pisi* (Pers) Wint is a heteroceous rust and commonly occur on cultivated pea. The aecial stage of this pathogen develops on *Euphorbia cyparissias* L. This rust is not common in India.

Epidemiology

The disease generally starts from low-lying patches in the paddock and radiates towards the border. Rust is an autoecious fungus, completing its life cycle on lentil. Rust usually appears at the later stage of crop growth, *i.e.* flowering or early podding stage. High humidity, cloudy or drizzly weather with temperature range 20 to 22°C favour the disease development. Age of plant had no direct relationship with rust appearance in lentil, while 24 hours leaf wetness after inoculation was found to be optimum for rust development (Joshi and Tripathi, 2012). Under cool humid weather, it may result in 70-100 per cent yield losses (Sepulveda, 1885). The optimum germination of aeciospores, urediospores and teliospores were recorded at 20°C (Joshi and Tripathi, 2012).

Integrated Management

- ✰ Lentil varieties such as Pant L-639, Pant L-406, Pant L-6, pant L-7 and Pant L-8 are rust resistant can be the most effective way to check the disease on more economical and sustainable basis.
- ✰ Eradication of volunteer groundnut and alternate host plants is important in reducing the primary source of inoculums.
- ✰ Field management should include a time gap of at least a month, between successive lentil crops.
- ✰ Foliar spray of fungicides such as Mancozeb (0.2 per cent a.i.), Bayleton (0.05 per cent a.i) and Calixin (0.2 per cent a.i.) are found effective against the pathogen. Singh *et al.* (2004) reported that four spray of mancozeb at 10 days interval commencing with the initiation of the disease were most effective in reducing rust severity from 69.7 to 10.2 per cent and increasing yield.

3. Ascochyta Blight

Ascochyta blight, caused by *Ascochyta fabae* f. sp. *lentis*, is a widespread disease of lentil. It causes severe yield loss and stains the seed which can cause downgrading or rejection (Bedi and Morrall, 1990). Under favourable conditions yield losses can be as high as 50 per cent in susceptible varieties. Additional economic losses occur due to poor grading of seed.

The Pathogen

Ascochyta blight of lentil, caused by *Ascochyta fabae* f. sp. *lentis*. The pathogen is highly host specific and teleomorph stage is *Didymella fabae* Jellis and Punith. The mycelium is hyline to brownish and septate. Conidia are formed from hylie, ampliform phialides. These conidia are hyline, oval to oblong, straight or slightly curved, septate or aseptate, rounded at each end and measure 10-16 x 3.5 micrometer in size. The conidia germinate by long germ tubes. Within the pycnidia, the conidia can remain viable up to a year or more.

Symptoms

Lesions begin as small brown spots on leaflets, petioles, stems and pods initially which later enlarge and fade to a tan colour with a darker margin with numerous tiny black fruiting bodies (pycnidia), which are the fruiting bodies of the *Ascochyta* fungus. In wet weather, the lesions may coalesce, resulting in premature leaf drop and defoliation; the tips of severely diseased stems wilt, turn brown, and die (Figure 20.2). Later in the season, lesions may form on the pods, resulting in seed infection. Infected seed becomes partially or entirely discoloured brownish-purple. Severely infected seed may be shrivelled. Tiny black fruiting bodies may be present on the surface of infected seeds (Figure 20.3).

Disease Cycle

Emerging plants become infected from infected seed or from spores growing on debris near the plants, especially when seed is planted into cold (8° C), wet soil. Lesions

Figure 20.2: Leaves showing Ascochyta Blight Symptoms.

Figure 20.3: Infected Grains.

appear on the plants within 7 to 9 days of infection. Within 10 to 14 days, fruiting bodies in the lesions start to produce spores, which are spread by rain splash to new infection sites on the plant or to adjacent plants. The disease develops and spreads most rapidly during cool (15° C), wet weather. Seed infection can increase rapidly when lentils are left in the swath during wet or humid weather conditions (Seid and Beniwal, 1991).

Seed borne Ascochyta is an important source of the pathogen. The Ascochyta fungus also is spread in the spring by wind borne spores formed in fruiting bodies on infected crop debris. Later in the summer, it is primarily spread by water-splashed spores. Wet weather in late summer can result in extensive pod and seed infection. When infected seed is planted, small proportions of the seedlings emerge diseased. The fungus then spreads from plant to plant by rain-splashed spores.

Epidemiology

Ascochyta fabae f.sp. *lentis* is a seed-borne pathogen; the disease is transmitted from infected seeds to seedlings. Disease is favoured by prolonged wet, cool conditions, often before flowering. Frequent showers can result in an epidemic. However, it is most damaging if rain occur during pod formation stage results in pod infection and subsequent downgrading of lentil seed quality due to seed staining. An extended period of leaf wetness is required for disease development, with maximum disease developing occurring after 24 to 48 hours of leaf wetness. Temperatures between 50F and 68F are highly favourable for disease development, and maximum disease development occurs at approximately 59F.

Integrated Management

- ✰ The use of certified, disease-free seed will help to minimize the disease, although seed with up to 5 per cent *Ascochyta lentis* infection will not significantly affect yield as long as soil and weather conditions favor quick germination and good plant vigor.
- ✰ The most economical and sustainable strategies to control ascochyta blight are through resistance breeding along with cultural practices like removal of plant debris from the field, crop rotation, deep sowing, mixed cropping with wheat and deep ploughing during summer.
- ✰ Follow 4 to 5 year crop rotation between lentil crops on the same field and do not plant lentils near fields that were infected during the previous season.
- ✰ Bury crop debris after harvest.
- ✰ Early sowing to escape moist weather at harvest can minimize disease.
- ✰ Fungicides as benomyl, carbendazim, carbathiin, iprodione and thiobendazole @ 0.1 per cent are effective.

4. Botrytis Grey Mould (BGM)

Botrytis grey mould (BGM) of lentil is caused by *Botrytis fabea* and *Botrytis cinerea*. It is a sporadic disease. *Botrytis cinerea* has also been isolated from lentil seed in India. The disease is widely distributed on Australia, Argentina, Nepal, Myanmar, Bangladesh and Pakistan and causing 70-80 per cent yield losses under favourable conditions (Haware and McDonald, 1992). In Indian the disease is particularly improvement in North Indian.

The Pathogen

BGM of lentil is caused by *Botrytis fabea* and *Botrytis cinerea*, a necrotrophic fungus is dominant on lentil crop. *B. cinerea* grown on potato dextrose agar (PDA) has a white cottony appearance, which turns light grey with age. The mycelium is septate, brown and 8-10μm wide. Young hyphae are thin and hyaline. Conidia and conidiophores are not in pycnidia or acervuli. Conidiophores lighter brown than hyphae, with hyaline tip, septate, 8-24μm wide. Tips of conidiophores or their branches are slightly enlarged and bear small pointed sterigmata. Conidia are hyaline, one- celled oval or globose or short cylindrical and borne in clusters at the tips of conidiophores branches.

Symptoms

All aboveground plant parts of lentil can be affected by botrytis grey mould. Symptoms may initially appear either on flowers and pods, or lower in the crop canopy, depending on the location of the crop. The most damaging symptoms become apparent after the crop has reached canopy closure, and a humid microclimate is produced under the crop canopy. The disease appears on the lower foliage as discrete lesions on leaves which are initially dark green, but turn greyish-brown, then cream as they age, that enlarge and coalesce to infect whole leaflets. Severely infected leaves senesce and fall to the ground. Lesions girdle the stem and cover it with a furry layer of grey mould, eventually causing stem and whole plant death. Pods which become infected will be covered in a grey mouldy growth, will roted and turn brown when dried out. Seeds within these pods fail to fill properly and are often discoloured and shrivelled.

Disease Cycle

Both the pathogens *Botrytis cinerea* and *Botrytis fabae* that cause botrytis grey mould can survive in infected seed, sclerotia in the soil or in infected trash and on alternate host plants. Sowing botrytis infected seed can give rise to infected seedlings reducing seedling survival and establishment of crop. Old infected trash is an important source of fungal inoculums. Spores are produced on old trash and are carried by the wind into new crops. Under favourable conditions of high humidity and moderate temperatures, the disease can spread rapidly; producing spores on newly infected tissue, and further spreading the disease within crops. The development of botrytis grey mould epidemics is largely determined by the prevailing environmental conditions, especially the presence of moisture. Environmental conditions and canopy density are primary factors that influence the development of botrytis grey mould epidemics in lentil crops. The formation of a favourable microclimate under the crop canopy, especially following canopy closure and humid conditions after rain, favour the development and further dispersal of the botrytis grey mould pathogens.

Epidemiology

There are several main sources of inoculums of botrytis grey mould, these include; seed-borne inoculums, sclerotia, mycelium in old infected trash, and alternate host plants. The disease develops quickly in high humid condition (RH over 80 per cent) for 4 – 5 days and moderate temperatures (15-25°C) with high moisture particularly at flowering stage. The disease is capable of causing serious yield losses in years when spring rainfall is high and / or when there are prolonged wet periods.

Integrated Management

- ☆ Avoid some field for continuation of crop. Only use seed with less than 5 per cent botrytis infection or preferably use healthy seed.
- ☆ Avoid early sowing of crop and reduce seed rate early sowing and high sowing rates can cause rank crop growth, lodging and increased risk of grey mould.
- ☆ Control volunteer lentils plants, faba beans, vetch, lathyrus and chickpeas early to limit the build-up of disease inoculum.
- ☆ Seed treatments with fungicides such as benomyl, carboxin, chlorothalonil or thiabendazole (0.1 per cent) can reduce seed-borne inoculum levels.
- ☆ Lentil varieties- Pant L-639 and Pant L-406 are resistant.

5. Anthracnose

Anthracnose of lentil is caused by *Colletotrichum truncatum*, is a relatively new disease of lentil. The disease is causing yield losses in excess of 50 per cent, which is much more destructive than ascochyta blight.

Symptoms

Anthracnose kills lower leaves and may cause premature death of affected plants. Affected plants may exhibit severe defoliation. Sunken, small, brownish with black margins develop on the stem that enlarge and may girdle the stem. Stem girdling causes plants to wilt and die (unlike ascochyta, which does not cause wilt or plant death). Small black bristle-like fruiting structures develop on the lesions of affected stems and leaves. Typical field symptoms are lodged plants with abnormally dark brown stems. Signs of the pathogen, black fruiting bodies are usually present on the lower stem are diagnostic symptom of disease.

Disease Cycle

The fungus that causes lentil anthracnose (*Colletotrichum truncatum*) is both soil-borne, seed-borne in nature pathogen survives in soil and crop residue from year to year. Emerging plants generally become infected from spores growing on debris in the soil. Spread within the field is by means of spores that are dispersed by rain splash and wind.

Epidemiology

The disease is favored by warm (20°C to 24°C) weather and infection will only take place if the leaves remain wet for 18 to 24 hours.

Integrated Management

- ✰ Crop rotation with non-host plant, eradication of weed hosts and cultivation in well drained soils are helpful in minimizing/preventing the disease.
- ✰ Seed free from anthracnose infection should be used. Before sowing seeds should be treated with ferbam, Ziram or thiram @ 0.5 g/100 g seeds.
- ✰ Follow the recommended crop rotation of at least 3 years. Avoid peas and faba beans in the rotation since there is evidence that they are susceptible to at least some of the same strains of the anthracnose fungus that attack lentils. Do not plant seed in the fields where lentils were previously grown.
- ✰ Application of many protectent or systemic fungicides such as maneb and Zineb @ 3.5 g/L, benomyl @ 0.55 g/L, captafol @ 3.5 kg/ha, carbendazim @ 0.5 kg/ha have been effective to control anthracnose.
- ✰ The cultivation of disease resistant varieties that have shown some resistance to anthracnose

Conclusion

In this chapter, sustainable management of major fungal diseases of lentil have been discussed. Earlier research was mainly focused on resistant sources and chemical control of diseases. Now the major emphasis is on identifying, evaluating and integrating location specific components of IDM. The management of these diseases in lentil, integrated approaches have been recommended by different authors from time to time who added crop rotation, field sanitation, use of disease free seeds and seed treatment with different bioagents or fungicides, hot water, dry heat treatment; and foliar spray of fungicides (Russell *et al.*, 1987; Seid and Beniwal, 1991).

References

Abbasi, M. and Pooralibaba, H.R. 2002. First report of lentil rust caused by *Uromyces viciae-fabae* in Iran. *Rostaniha,* 3(1/4): En48-En49, Pe109-Pe110.

Accantino, P. 1964. Chemical control experiment on the lentil rust. Agric. Tech. 23(24): 7-14.

Agarwal, S.C., Singh, K. and Lal, S.S. 1991. Plant protection of lentil in India. Paper presented in ICAR-ICARDA seminar 'Lentil in South Asia', New Delhi, 11-15 March 1991, p37.

Agrawal, S.C and Prasad K.V.V. 1997. Diseases of lentil oxford and IBH publishing Co. Pvt. Ltd. New Delhi. p155.

Agrios, G.N. 1997. Plant Pathology (IV ed.), Academic Press, New York, London, p.365.

Ahmad M.A. 2010. Variability in *Fusarium oxysporum* f.sp. *ciceris* for Chickpea Wilt Resistance in Pakistan, Ph.D. Thesis.

Ali, M. 2007. Augmentation of Pulses Production. Mission2007: IIPR Kanpur-24: 16-21.

Anonymous 2011-12. Project Coordinator's Report (Rabi crops). All India Coordinated Research Project on MULLaRP. Indian Institute of Pulses Research, Kanpur.

Bedi, S. and Morrall, R.A.A. 1990. Ascochyta blight in lentil crops and seed samples in Saskatchewan in 1988. *Canadian Plant Disease Survey* 70 (2): 123-125.

Bhatty, R.S. 1988. Composition and quality of lentil (*Lens culinaris* Medik.): a review. *Canadian Institute of Food Science and Technology* 21 (2), 144-160.

Butler, E.J. 1918. Fungi and Disease in Plants. Thacker Spink and Co. Calcutta, India. p547.

Canonaco, A. 1937. A heavy attack by rust or lentil. *Rev. Path. Veg.* XXVII 9(10): 281-285.

De Sousa, D.O. Camara, E. De Oloiveira, A.L.B. and De Luz G.G. 1939. Some rust of Portygal. *I Agron Lufit.* 1: 410-434.

De, R.K. and Chaudhary, R.G. 1999. Biological and chemical seed treatment against lentil wilt. *Lens Newsletter,* 26(1-2): 28-31.

Erskine, W., Rihawe, S. and Capper, B.S. 1990. Variation in lentil straw quality. *Annals of Feed Science Technology* 28: 61–69.

Haware, M. P. and Donald, D. 1992. Integrated Management of Botrytis Gray Mold. Summary Proceedings of the BARI/ICRISAT Working Group Meeting to Discuss Collaborntive Research on Botrytis Gray Mold of Chickpea. International Crops Research Institute for the Semi-Arid Tropics, Patancheru, Andhra Pradesh, India.

Joshi, A. and Tripathi, H. S. 2012. Studies on epidemiology of lentil rust (*Uromyces viciae fabae*). *Indian Phytopath.* 65(1): 392-393.

Karki, P.B. 1991. Present status and future prospects of plant protection research in Nepal paper presented in ICAR–ICARDA Seminar 'Lentil in South Asia' New Delhi, 11-15. March 1991. p13.

Kaushal, R.P. and Sharma, C. 1998. Effect of temperature, soil pH and phosphorus level on lentil wilt development. *LENS Newsletter,* 245(1 and 2): 65-67.

Khare, M.N. 1980. Wilt of Lentis. *Jawaharlal Nehru Krishi Vishwa Vidyalaya,* Jabalpur, M.P. India. p155.

Khare, M.N. 1981. In: Diseases of Lentils, (Eds.): C. Webb and G. Hawtin. Farnham Royal. UK: ICARDA/CAB, p163-172.

Khare, M.N. and Agrawal, S.C. 1978. Lentil rust survey in MP. Paper presented in All India Pulse Workshop held during September 1978 at Gujarat Agricultural University, Baroda (Gujarat).

Malencon, G. 1936. Moroccon Mycological Notes. *Rev. Mycol. N.S.* 1: 257-275.

Mehrotra, R.S. and Claudius,G.R. 1973. Role of metabolites and enzymes in the root rot and wilt diseases of *Lens culinaris. Indian Journal of Mycology and Plant Pathology* 3: 8-16.

Nattrass, R.M. 1932. Annual report of the mycologist for 1931. Annual report department of agriculture. *Cyprus.* p. 56-64.

Nelson, P.E., Toussoun, T.A. and Cook, R.J. 1981. *Fusarium, Diseases, Biology, and Taxonomy.* Univ. Park, London, U.S.A. The Penn. State Univ. Press. p457.

Padwick, G.W. 1941. Report of the imperial mycologist. Sci. Rep. Agric. Res. Inst., New Delhi, 1939-40, p94-101.

Podhradszky, J. 1968. Diseases of lentil. Novenyvedelmi Enciklopedia II (ed.: G. Ubrizsky) *Akudemi Kiado, Budepest,* p. 122.

Prasad, R. and Singh S.P. 1975. Sexual behavious of *Uromyces fabae. Indian J. Mycol. and Plant Pathol.* 5: 139-144.

Prasad, R. and Verma, U.N. 1948. Studies on lentil rust, *Uromyces fabae* (Pers.) de Bary. *Indian Phytopathol.,* 1: 142-146.

Royss, T. 1937. Contribution to the knowledge of the Uredinae of Palsetine Homage to Prof. E.C. Teodoresco, Buchorest. p. 13.

Russell, A.C., M.G. Cromey and W.A. Jermyn. 1987. Effect of seed treatment on seed-borne ascochyta of lentil. *Proceedings Agronomy Society New Zealand* 17: 15-18.

Saxena, D.R.; Mohy, S.; Saxena, R.R. and Khare, M.N. 1992. Root morphology and anatomy in relation to wilt incidence in lentil. *Lens Newsltt.,* 2: 46-49.

Seid, A. and Beniwal, S.P.S. 1991. Ascochyta blight of lentil and its control in Ethiopia. *Tropical Pest Management* 37: 368-373.

Sepulveda, R. P. 1885. Effect of rust caused by *Uromyces viciae-fabae* Pers, de Bary on yield of lentil. *Agriculture technical* 45: 335-339.

Singh, K., J.S. Jhooty and Cheema, H.S. 1986. Assessment of losses in lentil yield due to rust caused by *Uromyces fabae*. *LENS Newsletter* 13: 28.

Singh, R.A., De, R.K. and Chaudhary, R.G. 2004. Influence of spray time of mancozeb of pea rust caused by *Uromyces viciae-fabae*. *Indian J. Agric. Sci.*, 74(9): 502-504.

Singh, R.S. 2005. Plant Disease. Oxford and IBH, New Delhi. p396.

Smith, I.M., Dunez J., Phillips, D.H., Lelliott, R.A. and S.A. Archer, eds. 1988. European handbook of plant diseases. Blackwell Scientific Publications: Oxford. p583.

Stoilova, S. and Chavdarov, P. 2006. Evaluation of Lentil Germplasm for Disease Resistance to *Fusarium* wilt (*Fusarium oxysporum* f. sp. *lentis*). *Cent. Eur. Agr.* **7**: 121-126.

Vasudeva, R.S. and Srinivasan, K.V. 1952. Studies on the wilt disease of lentil (*Lens esculenta* Moench.). *Indian Phytopath.* 5: 23-32.

2016, Diseases of Pulse Crops and their Sustainable Management 385–410
Editors: **Samir Kumar Biswas, Santosh Kumar and Gireesh Chand**
Published by: **BIOTECH BOOKS, NEW DELHI**

Chapter 21

Integrated Disease Management of Greengram (*Vigna radiata* (L.) Wilczek) and Blackgram (*Vigna mungo* (L.) Hepper.

K. Jyothirmai Madhavi[1] and P. Kishore Varma[2]*

[1]*Fruit Research Station,*
Sangareddy, Dr. Y.S.R. Horticultural University
[2]*Department of Plant Pathology, College of Agriculture,*
Acharya N.G. Ranga Agricultural University, Rajendranagar, Hyderabad

Introduction

India is the largest producer and consumer of the pulses in the world, accounting for 30 per cent of global production, 27 per cent of consumption, and 34 percent of food use (FAO, 2000). It is also top importer, with an 11 per cent share of world imports during 1995–2001. Pulses are very important crops in India, with annual production of about 12-15 million tons from 22-24 million hectares (Moe *et al.*, 2008). Pulses play an important role in meeting the dietary requirements of proteins in India and contribute sustainability to the production systems by enriching the soil through biological nitrogen fixation. Among pulses, blackgram (urdbean - *Vigna mungo* L. Hepper) and greengram (mungbean - *Vigna radiata* L. Wilczek) are major grain legumes

* Corresponding Author: E-mail: kjm.agri@gmail.com

and are cultivated in both upland and rice fallows in India. They come under warm season legumes. Both are important short duration grain legume crops with wide adaptability, low input requirement. They also have the ability to improve soil fertility by fixing atmospheric nitrogen. Mungbean is grown on about 3.70 million hectares with annual production of 1.57 million tons. Similarly Urdbean is grown on about 3.24 million hectares with annual production of 1.52 million tons. These two crops together contribute 80 per cent to the total pulse production. Major Blackgram states are Madhya Pradesh, Maharashtra, Andhra Pradesh, Tamil Nadu and Uttar Pradesh while Orissa, Maharashtra, Andhra Pradesh, Rajasthan, Madhya Pradesh, Bihar, Karnataka and Uttar Pradesh are major greengram growing states.

Blackgram and greengram have been found being infected with lot of pathogens, which include fungi, bacteria and viruses. Only a few of them cause economic losses. Sufficient knowledge regarding the symptoms of damage and timely management of these diseases will help in increasing the productivity.

Integrated Disease Management (IDM) in Greengram and Blackgram

Both mungbean and urdbean are infested by similar fungal pathogens and viruses. Some of these diseases are causing severe losses in yield in the epidemic conditions. Majority of diseases are common across the agro ecological zones. Greengram and blackgram suffer from many diseases caused by fungi, bacteria, viruses, nematodes and also abiotic stresses. Four diseases (yellow mosaic, anthracnose, Cercospora leaf spot and powdery mildew) that attack both the pulses are considered economically important. Yellow mosaic caused by mungbean yellow mosaic virus (MYMV) is the most serious limiting factor in mungbean and urdbean cultivation.

A successful integrated disease management (IDM) strategy is one under which grain legumes have been protected from the yield-reducing effects of the pathogen rendering the later to economic insignificance. The IDM involves a total system approach to the suppression of pathogen populations to a level where higher yields can be obtained and enables the farmers to achieve maximum economic return. In the IDM system, the individual component of disease management such as host plant resistance (HPR), agronomic practices, judicious use of fungicides, pesticides for vector control, biopesticides for pathogen control etc., need to be compatible or complimentary.

The main emphasis in research and development to combat food legume diseases is on host resistance and chemical control where ever applicable, and quite often these components of disease management were practiced in isolation to each other. Recently a shift in scientific thinking and practice in the management of grain legume diseases has been seen and greater emphasis was on identifying, evaluating, and integrating location specific components of integrated disease management (IDM). In general IDM has followed the principles of IPM (Jeger, 2000). The location specific IDM of food legumes is primarily based on host plant resistance (HPR) or genetic resistance; additionally other components of diseases management. In some environments, IDM may require a single component used alone (usually HPR) or in combination with one other component (such as fungicide seed treatment) to adequately combat diseases of food legumes. The components of IDM employed in the production of food legumes are Host plant resistance (HPR), Disease modeling (prediction) for avoidance of high risk or disease pressure, Chemical sprays (fungicides, pesticides), Biological control, and Cultural (agronomic) practices (sowing dates, plant population etc.). In this review, disease management practices in integrated manner to control economically important viral and foliar diseases of greengram and blackgram have been discussed.

Viral Diseases

The development of an effective disease management package for virus diseases is dependent on the availability of basic information required to design an appropriate combination of interventions which can slow down virus disease development. The most important of these are identity of the causal agent, mode of transmission, ecology of the virus disease including that of its vector, extent and value of crop losses, availability of genetic resistance, and available crop-protection methodologies and their applicability to specific farming systems and socio-economic situations. In many locations, however, such complete basic information is not yet available. Control is optimized through IDM approaches, which combine all possible measures that operate in different ways such that they complement each other and applicable in farmers' fields. Thus, control measures can be classified as those that control the virus, those that are directed towards avoidance of vectors or reducing their incidence, and those that integrate more than one method. Cultural practices such as healthy seed, rouging, alteration in sowing dates and use of early maturing cultivars are effective in minimizing virus disease incidence. Almost 50 per cent of viruses affecting leguminous crops are seed-borne (Bos *et al.*, 1988). Seed-borne infections permit the introduction of primary virus inoculum into the field which facilitates secondary spread to reach a serious level in locations where the environmental conditions permit high vector activity. A close relationship between sowing date and the extent of subsequent virus spread is well documented for many crops.

Blackgram and greengram have been subjected to the attack of several biotic stresses such as fungi, bacteria and viruses, affecting the productivity and among them, viral diseases became great menace and are the great yield reducers. Blackgram and greengram are attacked by *Mungbean yellow mosaic virus* (MYMV), *Urdbean leaf crinkle virus* (ULCV), Leaf curl/necrosis causing viruses, *Peanut bud necrosis virus* (PBNV) (= *Groundnut bud necrosis virus* – GBNV) and *Tobacco streak virus* (TSV) which are the major threats of production.

Yellow Mosaic (Mungbean Yellow Mosaic Virus)

This disease is caused by the *Mungbean yellow mosaic virus* (MYMV) belonging to Gemini group of viruses, which is transmitted by the whitefly (*Bemisia tabaci*). This viral disease is found on several alternate and collateral hosts which act as primary sources of inoculum. Nariani (1960) was the first person ever to report the MYMV disease from the fields of IARI, New Delhi. The disease is characterized by the presence of bright yellow patches on leaves interspersed with green areas, complete yellowing and stunting of the plants and he reported 20-30 per cent incidence at institute areas. Nene (1972) conducted a survey in different districts of Uttar Pradesh and he reported that the yellow mosaic incidence in mungbean in different districts of Uttar Pradesh ranged from 5-100 per cent and depending upon the stage at which the plants were infected, the yield loss varied from 10-100 per cent. From his results, he also concluded that the virus causing urd bean mosaic was the same as the one causing mungbean yellow mosaic. Nariani (1960) has reported the disease and studied the symptomatology of MYMV for the first time. In mungbean first symptoms of the disease appear on the young leaves in the form of mild scattered yellow specks or spots. The next trifoliate leaf emerging from the growing apex shows irregular yellow and green patches alternating with each other. The leaf size is generally not much affected but sometimes the green areas are slightly raised and the leaves show slight puckering and reduction in size. The size of yellow areas goes on increasing in the new growth and ultimately some of the apical leaves turn completely yellow. The diseased plants usually mature late and bear very few flowers and pods. The size of the pod is reduced and more frequently immature and small sized seeds are obtained from the pods of diseased plants. Nariani (1960) found that mungbean seed does not transmit mungbean yellow mosaic virus. Similar observations have been reported by Ahmed and Harwood (1973), Nair and Nene (1973), Dhingra and Chenulu (1985), Grewal (1988) and Gautam (1990). Nariani (1960) first to report the occurrence of mung yellow mosaic and its transmission by the whitefly *Bemisia tabaci* (Genn.) predominantly and it has been reported to be the vector of similar diseases on *Phaseolus lunatus* L. by Capoor and Varma (1948) and on *Dolichos lablab* by Capoor and Varma (1950).

Nair and Nene (1973) reported that, the whitefly, *Bemisia tabaci* transmits MYMV in a circulatory manner. Fifteen minutes starvation before acquisition increases transmission from 15.5 to 39 per cent and starvation before inoculation increases transmission from 12.5 to 50 per cent. Chenulu *et al.* (1979) pre-acquisition and pre-inoculation starvation have no effect on efficiency of transmission of the virus and also reported that the virus, MYMV is not transmitted transovarially by the vector.The pathogen is transmitted by the white fly Bemisia tabaci Genn. Transmission of the virus is discussed in detail by Singh *et al.* (1998). Nath (1994) studied the effect of weather parameters on whitefly population and incidence of yellow mosaic virus on green gram during 1990-91 in Assam. He reported a simple positive significant correlation between MYMV disease incidence and whitefly population, temperature, relative humidity, rainfall and number of rainy days and yield of green gram had a negative correlation with disease incidence.

Management

Host Plant Resistance (HPR)

Chenulu *et al.* (1979) taken up varietal screening for resistance against MYMV at IARI, New Delhi and reported that Jalgaon-781, T-2, Khargaon and Mung local showed cent per cent infection, however, Pusa baisakhi showed least infection. Singh *et al.* (1980) reported that mungbean, Hyb-4-3 A and Hyb-12-4, remained free from disease at Ludhiana. Mungbean selections 15229, L 24-2 in Punjab, ML-220, PLS 274 in Tamil Nadu and L 80, ML 326, PDM 54 and PDM 62 in UP were also reported to be resistant to yellow mosaic. Kuldip Singh *et al.* (1996) screened 126 mungbean germplasm lines for resistance against MYMV. They found ML 267, ML 337, ML 393, ML 395, ML 409, ML 443, ML 591, ML 593, ML 605, MUG 225, PDM 84-143, PDM 219 and Pusa 8731 germplasms were resistant against MYMV. Asthana (1998) from Indian Institute of Pulses Research, Kanpur reported PDM-11, PDM-54, PDM-84-139, PDM-84-143 varieties as yellow mosaic resistant and can be utilized in breeding programmes. Siddiqui *et al.* (1999) evaluated some indigenous soybean lines for resistance against mungbean yellow mosaic bigemini virus, transmitted by *Bemisia tabaci* at IARI, New Delhi and found PK-1189, PK-1180, SL-443 and SL-444 were consistent and most promising sources of resistance. Basandrai *et al.* (1999) evaluated one hundred diverse stocks of blackgram (*Phaseolus mungo* L.) for resistance against five different diseases widely prevalent in Himachal Pradesh. They found HPBU 38, HPBU 153, LBG 626 and UG 367 were resistant against mung bean yellow mosaic and web blight. WVG 108 was found resistant against *Cercospora* leaf spot and MYMV and UG 407 was resistant against *Cercospora* leaf spot, MYMV and powdery mildew.

Raje and Rao (2002) found that the genotypes, PLM 19, PLM 25, PLM 32, PLM 42, PLM 113, PLM 122, PLM 618, IC-1396-3, IC-2153, IC-43591, EL-3902-A-EC-5551 and J-45 were resistant to yellow mosaic virus, *Cercospora* leaf spot and powdery mildew under field conditions. Marappa *et al.* (2003) evaluated mungbean genotypes for resistance against powdery mildew, yellow mosaic and bacterial blight at Bangalore, and they found that the genotypes AKM-9911, AKM-8803, Co-4, KM-2194, KM-188-3, MIVT-842, MIVT-854, MIVT-867, ML-173, ML-1380, DBGG-11, PMB-43, PS-16, SML-151, UPM-99-3, V-2964 and LM-56 were free from MYMV incidence. Ganapathy *et al.* (2003) in view of identifying resistance against mungbean yellow mosaic virus, urdbean leaf crinkle virus and leaf curl virus in urdbean, evaluated 71 entries at NPRC, Vamban, Tamil Nadu. They found that RU 2229, VBG 86, 2KU 54, VBG 89, SU16 were highly resistant to MYMV. Pathak and Jhamaria (2004) evaluated fourteen mungbean varieties for resistance against yellow mosaic virus at ARS Navgaon. They found ML-5 and MUM-2 were resistant with only 2.22 and 3.12 per cent infection as against cent per cent infection in K-851, a check cultivar. Peerajade *et al.* (2004) tested 85 genotypes against MYMV at MARS, Dharwad. Among them, GG 41 and GG 42 were found resistant and GG 52 showed moderate resistance.

Cultural Practices

Cultivation of resistant varieties, manipulation in sowing dates, inter/mixed cropping of mungbean and urdbean with non-host crops like sorghum, pearl millet and maize and application of systemic insecticides such as aldicarb, disyston and foliar application of metasystox has been found effective in controlling the disease by reducing vector control (Vishwa Dhar *et al.*, 2004). The effectiveness of sticky traps in controlling whiteflies has been reported by Raghupathi and Sabitha (1994) in soybean, Uthamasamy (1989) in cotton, Cohen and Marco (1973) for the management of cucumber mosaic virus and potato viruses. Maize as a barrier crop was found ineffective by recording 6.20 whiteflies per plant. The effectiveness of barrier crop depends on many factors like vigour, thickness and height, environmental factors like wind velocity and direction and growth of the crop (Singh, 1985 and Raghupathi and Sabitha, 1994).

Using Botanicals

Phadke *et al.* (1988) studied the effect of Neemark formulations on the incidence of whiteflies and yield of cotton and they reported that Neemark @ 0.4 per cent was on par with endosulphan @ 0.1 per cent in controlling whiteflies and bollworm incidence. Neemark @ 0.5 per cent was superior to fenvalerate @ 0.0125 per cent in controlling whitefly eggs. As far as yield is concerned, Neemark @ 0.5 per cent recorded the best yield. Verma and Verma (1993) reported two fold increase in nodulation and grain yield with 50 per cent reduction in mungbean yellow mosaic incidence when dry leaf powder of *Clerodendron aculeatum* was applied as soil amendment in addition to six times foliar spray at weekly interval. Patel *et al.* (1994) tested the efficacy of Neemark (azadirachtin) as ovipositional deterrent against pests of cotton. They reported that neemark 0.5 per cent recorded only 30.39 per cent egg laying as against significantly high percentage 69.61 per cent of egg laying in untreated check by *Bemisia tabaci* Genn. Somashekara *et al.* (1997) reported that among neem products tested, NSKE (4 per cent) induced high whitefly mortality while neem leaf extract and neem cake powder were ineffective against whiteflies. Further, hundred per cent mortality of *B. tabaci* was observed up to eight days in triazophos (0.15 per cent) mixed with either neemark/RD9/Replin/Aphidin. Chandrasekharan and Balasubramanian (2002) evaluated the efficacy of botanicals and insecticides against sucking pests, *viz.*, aphid, *Aphis craccivora* Koch. and whitefly, *Bemisia tabaci* Genn. on greengram. They reported that among the treatments, acephate 75 SP @ 0.075 per cent and TNAU neem oil (C) 60 EC at 3.0 per cent were found significantly superior by recording higher percentage of reduction in aphid population and yellow mosaic virus (YMV) incidence due to whitefly and also with grain yield recording 8.5 and 7.4 q/ha, respectively.

Chemical Control

The effectiveness of insecticides was attributed to greater residual activity, high level of protection, quick knock down effect of insecticides on viruliferous vectors compared to botanicals, plant products and cultural practices that act indirectly by enhancing growth of plant, delaying disease appearance as reported by Baranwal and Ahmed (1997), by inducing resistance as reported by Verma and Varsha (1995) or by changing feeding behavior, by deterring settling activity of vector or becoming toxic to vector before inoculating virus as reported by Somashekara *et al.* (1997) or by acting as ovipositional deterrents (Patel *et al.*, 1994) which are not sharp and accurate enough to restrict the activity of viruliferous vectors. Sastry and Singh (1973) evaluated different insecticides for control of whiteflies under field conditions and reported that ekatox (0.02 per cent), metasystox (0.02 per cent) and rogor (0.05 per cent) were found superior over other treatments by registering lowest whitefly population and yellow vein mosaic incidence on okra. Borah (1995), Sastry and Singh (1973) and Mote (1976) reported that the foliar spray of dimethoate @ 0.03 per cent was highly effective in reducing whitefly incidence in greengram fields. Borah and Nath (1995) reported the reduced incidence of whitefly transmitted yellow vein mosaic disease due to dimethoate sprays at 15 and 30 days after germination. Sastry and Singh (1974) conducted experiments to restrict the spread of MYMV by controlling its vector population. They reported that four sprays, each of parathion (0.02 per cent), oxydematon methyl (0.02 per cent) and dimethoate (0.05 per cent) at 10 day intervals starting from the germination of okra seeds or only one application of phorate 10G (15 kg/ha) at the time of sowing the seeds not only reduced the vector, whitefly (*Bemisia tabaci*) but also restricted the spread of virus to a greater extent. Mote (1976) reported that among different systemic insecticides evaluated, three applications of dimethoate or monocrotophos at 0.05 per cent at fortnightly intervals starting from 2 weeks after transplanting were found promising in controlling whitefly and also increased the yield to 307.48 and 265.19 per cent respectively. Borah (1995) reported that the foliar application of cypermethrin (0.01, 0.015 per cent), deltamethrin (0.0028, 0.0042 per cent) and dimethoate (0.03, 0.04 – 5 per cent) were effective in reducing whitefly incidence in green gram.

Borah and Nath (1995) reported that 2 sprays of dimethoate (0.03 per cent) at 15 and 30 days after germination were found highly effective in reducing the incidence of whitefly, transmitting bhendi yellow vein mosaic virus disease. Ahmed (2001) reported that soil application of carbofuran 3G (15 kg/ha) twice (once at sowing and 20 days after sowing) + 5 sprays of metasystox (0.2 per cent) at 15 days interval recorded least (29.33 per cent) incidence of bhendi yellow vein mosaic as against 91.26 per cent incidence in control plots. treatment with lowest tomato mosaic disease incidence (8.22 per cent) as well as least aphid population (2.71 aphids/leaf) as against 12.97 per cent disease and 11.12 aphids/leaf in control. Mote *et al.* (1993) evaluated different insecticides in managing sucking pests of cotton *viz.*, aphids, jassids, thrips, mites and whiteflies. They reported that imidacloprid seed treatment (@ 10 g/kg) effectively checked aphid and jassid population up to 60 days but for thrips and whiteflies a higher dose of 15 g/kg was found effective. Plant growth characters like plant height, number of leaves, leaf area and yield were quite superior in imidacloprid treated plots.

Blackgram Leaf Crinkle [Urdbean Leaf Crinkle Virus (ULCV)]

The pathogen involved is leaf crinkle virus which is seed transmitted as well as by vectors like whitefly and aphid. The notable symptoms are enlargement of leaves followed by crinkled surface of leaf lamina, which is more pronounced in younger leaves. Due to the infection the flowering is delayed, inflorescence turns bushy appearance, poor pod setting and reduction in yield. This disease is caused by urdbean leaf crinkle virus (ULCV) belonging to Tospovirus. The virus is transmitted by aphids, whitefly and leaf hoppers and through sap. Disease symptoms include crinkling, curling, and puckering of leaves often coupled with stunting and malformation of floral organs. Enlargement in size followed by crinkle surface of laminae are the haracteristics symptoms on affected trifoliate leaves. Pollen production, fertility and subsequent pod formation is severely reduced with affect on seed weight and size of seeds in infected plants leading to decrease in yield.

Among the viral diseases, Urdbean leaf crinkle virus (ULCV) is considered to be the most serious depending on the season and variety cultivated (Reddy *et al.*, 2005). ULCV is known to spread through seeds and insects (Nene, 1972; Kadian, 1980). According to Ahmad *et al.* (1997), ULCV is transmitted through seed at the rate of 2.7 to 46 per cent. Leaves feeding beetele (*Henoewpilachna dodecastigma* Wied), whitefly and two aphid species have been reported as vector of ULCV (Narayanasamy and Jaganathan, 1973; Dhingra, 1976; Beniwal and Bharathan, 1980). The initiation of primary infection was considered to be originated through seed borne inoculums of the virus (Negi and Vishunavat, 2004). Field environment was highly conducive for natural spread of the disease due to high vector population and the buildup of inoculums potential of virus from the very beginning. It also indicates that there are minimum chances that any disease escape mechanism could become operative except the changes in environmental factors.

Management

Use of disease free seeds, rogueing of affected plants in the field, hot water treatment of seed at 55 0 C for 30 minutes, timely management of insect vectors will effectively control the disease. Seeds from diseased crops should not be used. Treating the seeds with imidacloprid 70 WS@ 5ml/kg and rogueing the infected plants to avoid contact between healthy and diseased plants during intercultural operations would also be effective. Varieties like, D-3-9, K 12, ML 26, RI 59, T44 RII (Mungbean); HUP 27, 102, 164, 315 (Urdbean) are resistant to ULCV. One foliar spray of insecticide (dimethoate 30 EC @ 1.7ml/ha) on 30 days after sowing will discourage the development of the disease.

Ravinder Reddy *et al.* (2005) have studied the susceptible stage and symptomatology of urd bean leaf crinkle virus (ULCV) on urd bean. They reported that the incubation period was short and symptoms developed on second trifoliate leaf stage onwards when the plants were inoculated at younger stage as compared to older plants. In the infected plants the increase in leaf size was evident from third trifoliate leaf stage onwards. Reduction in rachis length of terminal leaflet of infected trifoliate and thickening of stem and petiole were evident in infected plant. The size of stipules increased prior to the symptom development in lamina in all the infected plants. These symptoms are of great use in eliminating infected plants in the early stages for effective management of the disease.

The plant extracts effect has been attributed to inhibitory principles present in the leaf extracts. Consequently the number of virus particle entering the plants might have been reduced. According to Chowdhury and Shah (1985) the ULCV can be inhibited to 30 per cent over control but Reddy *et al.* (2006) reported 40 per cent reduction over control. In our study it was observed that neem extract reduced 50 per cent of disease incidence over control. As the disease incidence and whitefly population both reduced due to plant extracts spray so the mechanism of disease reduction was quite clear, because the whitefly was the major vector of ULCV. The results obtained in the present studies have indicated a great potential of plant extracts for the control of plant disease. Thus there is a need to identify and purify the active ingredients and their formulation will lead to a practical method for viral disease management.

Environmental factors showed positive interaction in case of maximum and minimum temperature but in case of relative humidity the interaction was negative. With an increase in temperature the infection rate increased and with the increase in relative humidity the infection rate decreased. The neem extract gave better results for reducing vector population as well as disease incidence (Binyamin *et al.*, 2011).

Leaf Curl/Necrosis Causing Viruses

Of several viral diseases attacking greengram and blackgram leafcurl disease caused by *Peanut bud necrosis virus* (PBNV) (= *Groundnut bud necrosis virus* – GBNV) (Amin *et al.*, 1985) transmitted by *Thrips palmi* (Karny) in a propagative manner (Sreekanth *et al.*, 2002) was considered to be a major threat, causing 40 per cent yield loss (Nene, 1972). Recently, *Tobacco streak virus* (TSV) has also been reported to be a cause of leaf curl symptoms on blackgram (Prasada Rao *et al.*, 2003d; Ladhalakshmi *et al.*, 2005) and greengram (Bhat *et al.*, 2002; Prasada Rao *et al.*, 2003c) paving confusion in field diagnosis to assess the disease incidence. Although both the viruses cause necrotic symptoms and are transmitted by thrips, the method of transmission and the virus vector relationship vary and hence need different approaches of management practices. It is necessary to identify and differentiate necrosis-causing viruses and their incidence on blackgram and greengram to follow the appropriate management practices. The most economical way to manage these necrosis-causing viruses is to grow resistant varieties.

Nene (1972) observed leafcurl symptoms for the first time on greengram and considered to be a major threat, causing 40 per cent yield loss. Ghanekar and Beniwal (1975) used cowpea cv. C-20 as a local lesion host for mungbean leafcurl virus. Amin *et al.* (1985) reported TSWV as the causal virus of leafcurl of mungbean and urdbean based on host range, thrip transmission and serological relationship of leaf curl to bud necrosis disease of groundnut reported to be caused by TSWV (Ghanekar *et al.*, 1979). However, it has been subsequently reported upon identification that bud necrosis disease of groundnut is caused by a distinct tospovirus, PBNV (Reddy *et al.*, 1992). Bhat *et al.* (2001) conducted a survey for the presence of tospovirus infections on mungbean and urdbean fields during 1999 where the samples were reacted positive to GBNV antiserum and the presence of PBNV was also proved by nucleic acid hybridization technique. Prasada Rao *et al.* (2003a) conducted ELISA and infectivity based surveys for the detection of PBNV in mungbean and urdbean in Andhra Pradesh during 2000-2002 and reported the presence of PBNV in 337 samples out of 372 by DAC-ELISA. The leaf curl disease of greengram (Sreekanth *et al.*, 2002a) and blackgram (Rajkumar, 2004) caused by PBNV was conclusively proved to be transmitted by *Thrips palmi* (Karny).

Natural infection of TSV in mungbean from Tamil Nadu was detected by RT-PCR and the sequence and phylogenetic analyses revealed that the CP gene of mungbean TSV isolate was highly conserved (99-100 per cent) (Bhat *et al.*, 2002b). At NBPGR, during routine testing of mungbean and urdbean samples by ELISA, three urdbean and two mungbean samples were found positive to TSV and rest all to PBNV (Prasada Rao, 2003). Ladhalakshmi *et al.* (2006) reported urdbean stem necrosis disease, caused by TSV in Tamil Nadu, on the basis of host range, serological relationship, electron microscopy, and sequence analysis of the CP region. TSV infection on mungbean was reported from Tamil Nadu and polyclonal antiserum was developed for TSV, which could detect the TSV isolates of urdbean, sunflower and soybean (Vinod *et al.*, 2007).

Nene (1972) described that the leaf curl symptoms in blackgram appear throughout the season, starting within 20 DAS till the plants are able to throw last new leaves. The earliest symptoms, on youngest leaves, are appearance of chlorosis around some lateral view and its branches near the margin of a leaf. The leaves show curling of margins downwards although rolling up of a few affected young leaves is not uncommon, some of the leaves show twisting. These leaves become brittle and whole trifoliate including the stalk, may fall down from the plant on giving a jerk. The veins show reddish brown discoloration on the under surface, which also extends to the petiole. Majority of the plants that show symptoms within five weeks after sowing the seed die due to top necrosis within a week or two. A few plants which escape the death remain stunted and fail to flower. Thus, like other viral diseases recorded on blackgram and greengram, leaf curl also appears to be a serious problem. Plants infected late in the season produce reduced number of pods containing smaller and lighter grains. Some pods do not contain grains at all. The grains produced by diseased plants are unfit both for seed as well as for consumption purposes.

Symptoms of the necrosis disease of blackgram consisted of brown necrotic lesions on young leaves, with brown streaks on petioles and stems and in severe cases infected plants were found dead (Ladhalakshmi *et al.*, 2006). TSV infection on mungbean under field condition was characterized by brown necrotic areas in the leaves, petiole, necrosis of stem and drying of the plants from the tip, whereas mixed infection of TSV and PBNV resulted in reduction in internodal length and axillary shoots with severe stunting, vein necrosis along with chlorotic spots on young leaves and necrotic streaks, reduced lamina area, vein necrosis, yellowing on the leaf margin and chlorotic spots on matured trifoliate leaf (Vinod *et al.*, 2007).

Evidence is overwhelming that tospovirus is not seed transmitted either in groundnut (Reddy *et al.*, 1983) or in other legumes (Reddy and Wightman, 1988). Several workers have reported that there is no seed transmission of PBNV in mungbean and urdbean (Nene, 1972; Prasada Rao *et al.*, 2003c; Rajkumar, 2004: Jyothirmai Madhavi, 2009). In India, to date TSV has not been reported to be seed-transmitted in any plants belonging to Asteraceae (previously compositae). Studies conducted on field infected and as mechanically inoculated plants of groundnut, sunflower, okra, soybean, urdbean, mungbean, marigold and parthenium failed to show seed transmission of TSV (Prasada Rao *et al.*, 2003b, 2009; Ladhalkshmi *et al.*, 2006; Reddy *et al.*, 2007).

Sreekanth (2002) and Jyothirmai Madhavi *et al.* (2009) reported the greengram and blackgram leaf curl disease was transmitted by *T. palmi* but not *S. dorsalis* or *F. schultzei.* All the four thrips species *viz., Frankliniella schultzei, Scirtothrips dorsalis, Megalurothrips usitatus* and *Thrips palmi,* transmitted TSV isolates of both blackgram and greengram in presence of TSV-infected pollen (Jyothirmai Madhavi *et al.*, 2011).

Management

Controlling thrip vectors is difficult because physical, chemical and biological methods of control are not very effective. They can reduce thrips population to economically viable direct damage level, but not enough to prevent the spread of disease. Consequently, the development of genetic resistance has become the best management strategy for these diseases in both medium and the long term, especially in case of TSWV (Rosello *et al.*, 1996). Nene (1972) screened blackgram genotypes against blackgram leaf curl virus under field conditions at Pantnagar and identified the lines N-212 and Khargoan to be free from the disease. Sreenivasulu (1994) studied the reaction of 38 genotypes against the blackgram leaf curl disease, among them seven genotypes, *viz.*, PLV-807, PLV-1149, PANT-U-30, PANT-U-426, NP-3, LBG-648 and LBG-667 were moderately susceptible, 16 genotypes were susceptible and rest 15 genotypes were highly susceptible. Identification of resistant genotypes by field screening and adopting the proven resistant or tolerant lines would be most economic and efficient for solving the problems involved with blackgram leaf curl disease incidence and consequent thrips population. Jain *et al.* (2000) screened 30 entries of blackgram varieties with two different dates of sowing under field conditions and among them six genotypes had shown susceptibility reaction, rest of them had shown highly susceptible reaction and none showed resistance.

Out of 38 greengram genotypes screened for mungbean leaf curl disease and *Thrips palmi* in both *rabi* and *kharif* seasons of 2000, four genotypes *viz.*, LGG-460, LGG-480, LGG-491 and LGG-582 consistently exhibited resistant reaction registering negligible number of thrips and leaf curl disease incidence both in *kharif* and *rabi* seasons. Whereas, 18 genotypes were rated as moderately susceptible, 11 were rated as highly susceptible and five were rated as very highly susceptible (Sreekanth *et al.*, 2002b). 83-6 were identified as resistant to thrips (Chhabra and Kooner, 1988; Chhabra *et al.*, 1993). In the All India Coordinated Pulses Improvement Project (AICPIP), preliminary screening of summer mungbean cultivars showed that PMS 2, PMS 3, 12-333 and CO 3, ML 5 and ML 337 (All India Coordinated Pulses Improvement Project, 1981-82 and 1991-92, respectively) were resistant to thrips. The most economical and convenient way to manage TSV is to grow resistant varieties. However, resistance to TSV was not found in cultivated species in India.

High seed rates (30 and 25 Kg/ha) led to reduced incidence of leaf curl disease and thrips populations, and more yields as compared to low and recommended (10 and 15 kg/ha) seed rates in both seasons studied. Incidence of leaf curl disease and vector population was significantly influenced by different seed rates in both the seasons (Rajakumar *et al.*, 2007).

Planting date has been reported as an important factor in epidemics of peanut bud necrosis in India. Singh and Srivastava (1995) reported higher incidences of bud necrosis in peanuts planted two weeks prior to the normal planting time of 1 July, and lower incidences of bud necrosis in peanuts planted in mid- or late July. In areas where there are strong prevailing directional winds, sequentially planting peanuts in fields that are down-wind from earlier planted peanuts may have increased risks of damage by spotted wilt (Black, 1990). Explanations for the differing effects of planting date have been based on circumstantial evidence and remain speculative. A common explanation has been that thrips populations are often variable from year to year as well as across planting dates (Mitchell, 1996; Mitchell and Smith, 1991). The optimum planting dates for minimizing spotted wilt may vary among years and locations, and should be determined as closely as possible for the specific location or region in which peanuts are planted. Furthermore, use of even the optimum planting date for suppression of spotted wilt still may not be adequate as a sole practice to prevent significant losses to the disease. Infection of an individual peanut plant with TSWV is of greater probability among sparse plant populations than among dense populations. Establishing higher plant populations does not appear to reduce the number of infections in a particular field, but likely reduces the percentage of plants that are infected. Similar observations have been reported with PBNV. Reddy *et al.* (1983) reported that increases in plant population density resulted in corresponding decreases in incidence of bud necrosis but did not affect the number of infected plants per unit area.

Insecticide applications have been indicated for reduction of peanut bud necrosis (Singh and Srivastava, 1995), but levels of reduction in incidence often have not been substantial. Application of imidacloprid as a seed treatment or as an in-furrow treatment has resulted insubstantial increases in incidence of spotted wilt relative to plots receiving other insecticide treatments or no insecticide (Todd *et al.*, 1994). This effect may be due to modified thrips feeding behavior caused by the insecticide. Because of the mobility of thrips vectors, the potential for interplot interference from nontreated plots to confound results from insecticide trials have been recognized (Black *et al.*, 1993, Todd *et al.*, 1996). Large field tests were conducted in which winter or spring applications of carbofuran were made in efforts to kill overwintering thrips in fallow fields or on volunteer peanut plants (Todd *et al.*, 1996).

The plant defense activator acibenzolar-S-methyl provides significant suppression of spotted wilt in tobacco (Pappu *et al.*, 2000). Results with this material on peanut have not been as consistent. Wells *et al.* (2002) found fewer feeding scars by *F. fusca* on excised peanut leaves treated with acibenzolar-S-methyl than on nontreated leaves. In field studies, Wells *et al.* (2002a) reported that acibenzolar-S-methyl applied in-furrow or in combination with in-furrow and foliar applications had incidence of spotted wilt similar to those treated with phorate and less than in nontreated plots in two of five tests. Neem extracts have been reported to provide suppression of peanut bud necrosis in India (Singh and Srivastava, 1995).

The planting of twin rows spaced 18–24 cm apart at the same seeding rate per ha as single rows has become increasingly popular in Georgia (Baldwin *et al.*, 1997). Changing from single to twin rows requires considerable effort and expense. Twin rows may require adjustments in cultural practices, such as tillage, and in the digger-inverter at harvest. Furthermore, cultivars that lack a prominent main stem and produce excessive vine growth may be extremely difficult to manage and harvest when planted in a twin-row pattern. Many growers have found that their initial investment in these adjustments is justified by lower incidence of spotted wilt and resultant increased yields. Biological control of thrips by the entomoparasitic nematode *Thripinema fuscum* has been reported to correlate with lower incidence of spotted wilt in peanut (Funderburk *et al.*, 2002). However the potential for utility of this agent for management of thrips or spotted wilt has not been demonstrated.

Combination of seed treatment with imidacloprid, removal of weeds, border crop with maize, intercrop with red gram and dimethoate spray @ 2 ml/l recorded zero per cent disease incidence followed by seed treatment with imidacloprid @ 1 ml/kg of seed+ Removal of TSV susceptible weeds+ Spray with dimethoate @ 2 ml/l 30 days after sowing recorded 1.8 per cent disease incidence (Nagamani *et al.*, 2013).

A combined effort of biotechnology and traditional breeding may further enhance opportunities for development of disease-resistant cultivars that meet the market requirements However, integrated management systems, using moderately resistant cultivars and suppressive chemical and cultural practices, have been developed and successfully deployed for minimizing losses to this disease. Although bud necrosis(PBNV) and stem necrosis(TSV) have seldom been controlled completely, integrated management practices have had a huge impact on production.

Fungal Diseases

Anthracnose (*Colletotrichum lindemuthianum* and *C. capsici*)

In recent years, greengram anthracnose caused by *Colletotrichum truncatum* (Schw.) Andrus and Moore has become one of the major diseases which is known to occur in many countries *viz.* India, Nigeria, Thailand, Philippines, Upper volta, Zambia, Palmira, Columbia, *etc.* (Agarwal, 1991). It occurs in all the parts of the world, wherever greengram is cultivated. In India, the greengram anthracnose was first reported from Jorhat of Assam state in 1951 (Majid, 1953). The disease has been reported from all major mungbean growing regions of India in mild to severe form and in tropical and subtropical areas it causes considerable damage by reducing seed quality and yield. The disease causes qualitative as well as quantitative losses (Sharma *et al.*, 1971). Losses in yield due to anthracnose have been estimated to be in the range of 24 to 67 per cent (Deeksha and Tripathi, 2002a) and 18.2 to 86.57 per cent disease index of anthracnose have been reported in northern Karnataka (Laxman, 2006).

The disease occurs on several legume crops including mungbean and urdbean. The fungus *Colletotrichum* spp. is the causal organism of the anthracnose affecting aerial plant parts, however, the leaves and pods are more vulnerable. The characteristic symptoms of this disease are circular brown sunken spots with dark centers and bright red orange margins on leaves and pods. In severe infection, affected part withers off. Infection just after germination causes seedling blight. Five species of *Colletotrichum* are known to attack mungbean and urdbean. The pathogen survives from one crop season to the next on infected seeds and crop residue. Intermittent rains at frequent intervals favor the epidemic development of the disease. The optimum temperature and relative humidity for disease development is 17-24°C and 100 per cent, respectively.

Management

Hot water treatment of seeds at 55–60°C for 15 minutes is effective in seed-borne infection. Seed treatment with captan or thiram @ 2 g/kg seed and spraying chemicals like thiram @ 2 ml or mancozeb @ 3g/litre of water are effective in preventing/ controlling the disease. Hot water seed treatment at 58°C for 15 minutes has been found effective in checking the seed-borne infection and increasing proportion of seed germination. Seed treatment with thiram 80 per cent WP @ 2 g/l or captan 75 WP @ 2.5 g/l helps in eliminating the seed borne infection.

Foliar spray of propiconazole and foliar spray of hexaconazole followed by benomyl seed treatment+ foliar spray of benomyl and seed treatment with carbendazim + foliar spray of carbendazim are effective in minimizing the per cent disease index and getting higher grain and stalk yields. Deeksha and Tripathi (2002) and Laxman (2006) reported propiconazole, hexaconazole and carbendazim were effective fungicides against anthracnose of blackgram and greengram, respectively.

The biorationals are less effective in controlling the disease when compared to chemical and do not enhance the yield. Chandrasekaran *et al.* (2000), Varaprasad (2000) and Laxman (2006), reported the ineffectiveness of plant products and bioagents over the fungicides in controlling anthracnose of soybean, chickpea and greengram, respectively. The low effectiveness of the bioagents might be attributed to the low level of relative humidity prevailing in the field (Belanger *et al.*, 1994). From the practical point of view, the chemical, which gives the maximum returns is more important rather than the control of the disease. So calculation of benefit:cost ratio gives an information on whether the technology could be adopted in the farmers fields or not. Hence, benefit:cost ratio is an important parameter for recommendation of any treatment for successful control of plant disease. Kulkarni (2009) reported that though foliar spray of propiconazole gave significant control of anthracnose, maximum cost:benefit ratio of 23.10 was realized with foliar spray of hexaconazole followed by carbendazim seed treatment + foliar spray of carbendazim (22.62), foliar spray of propiconazole (14.16) and benomyl seed treatment + foliar spray of benomyl (12.73). This clearly indicates that one foliar spray hexaconazole (0.1 per cent) was more useful not only in reducing the cost of protection but also gave higher benefits as compared to other treatments and can be recommended for the management of greengram anthracnose. Similar types of findings were observed by many workers (Bharadwaj and Thakur, 1991; Shirshikar, 1995 and Laxman, 2006). Hence, spraying of hexaconazole (0.1 per cent) could be considered as an effective management practice to manage anthracnose of greengram. Integration of moderately resistant genotypes (BGS-9, TM-98-50 and TM-97-55) coupled with hexaconazole spray will be effective in reducing the disease pressure and enhancing the yields of greengram.

Powdery Mildew (*Erysiphe polygoni*)

Powdery mildew caused by *Erysiphe polygoni* DC, is a problem in cool dry weather. Pathogen is obligate parasite and has wide host range. Limited information is available on the etiology and biology of *E. polygoni* on blackgram and greengram. Available information on etiology and biology of the fungus is described in detail by Singh *et al.* (1998). Powdery mildew is a major problem for urdbean and mungbean cultivation and causes severe yield loss. The fungus *E. polygonii* produces conidiophores carrying chains of white conidia. The disease appears on all the part of plants above soil surface. Disease initiates as faint dark spots, which develop into small white powdery spots, coalesceing to form white powdery coating on leaves, stems and pods. At the advance stages, the color of the powdery mass turns dirty white. The disease induces forced maturity of the infected plant causing heavy yield losses and its intensity increases in stress condition. The pathogen overwinters and survives on the host tissue in the form of cleistothecia. Unlike most fungal plant pathogens, the powdery mildews do not require the surface of the host plant to be wet for infection, favored by high humidity. It is favored by cooler conditions and is widespread in the late planted crop.

Management

Many resistant sources are available against powdery mildew (Vishwa Dhar *et al.*, 2004). Growing tolerant variety like Krishnaya and two sprayings with carbendazim (1g) or thiophanate methyl (1ml) or tridemorph (1ml) in one litre of water, one immediately after the disease appearance and the second after 15 days found effective against the disease. Adopt clean cultivation by destroying diseased plant refuge. Delayed sowing of mungbean and urdbean with wider spacings considerably reduce the disease severity. Opt for resistant varieties as per recommendation of local agricultural authorities (Mungbean: LM 223, LM 24, P115, ML 131, MI 322, ML 337, ML 395 SS1, JRUM 1, TARM 1 and AVRDC 1381; Urdbean: COBG10, LBG 648, 17, Prabha, IPU 02-43, AKU 15 and UG 301). Spray with NSKE @ 50 g/l or neem oil 3000 ppm @ 20 ml/l twice at 10 days interval from initial disease appearance. Spray with eucalyptus leaf extract 10 per cent at initiation of the disease and 10 days later also if necessary. Spray with water soluble sulphur 80 wp @ 4 kg/l or carbendazin 50 WP @ 1 g/l (0.05 per cent), benlate (0.05 per cent) and topsin-M (0.15 per cent). Rotate chemicals with different modes of action. Also its incidence can be reduced by adjusting the date of sowing with wider spacing. Chemical control with fungicides karathane, calixin, bavistin, benlate, topsin M, sulfur dust etc. has been found effective to control the disease under field conditions.

Cercospora Leaf Spot (*Cercospora canescens*)

Cercospora leaf spot (CLS) is caused by several species dominated by *Cercospora canesens* and may cause severe losses of yield under humid weather conditions. Leaf spots with brown to greyish centre and reddish brown border are its characteristic symptoms. The petioles, stems and pods also get affected by the pathogen. During favourable condition the spots increase in size and at the time of flowering and pod formation lead to defoliation. Five species of Cercospora infect mungbean and urdbean with slight variation in their symptoms. The fungus survives on the infected seeds and crop debris. Spots produced are small, numerous in numbers with pale brown centre and reddish brown margin. Similar spots also occur on branches and pods. Under favourable environmental conditions, severe leaf spotting and defoliation occurs at the time of flowering and pod formation. The fungus is seed-borne and also survives on plant debris in the soil. High humidity favours disease development.

Management

Following the integrated pattern of management practices like growing tolerant variety like LBG 167, use of clean seed and removal of crop debris, spraying with 1 per cent Bordeux mixture or chlorothalonil or mancozeb @ 2 g/litre of water, field sanitation, crop rotation, destruction of infected crop debris and avoiding the collateral hosts in the vicinity of the crop would greatly help in reducing the incidence of the disease. Growing resistant varieties like LM 113, LM 168, LM 170, JM 171(Mungbean); Naveen, Jawahar Urd-3, Gujarat Urd-1 and Barkha (Urdbean) and treating the seeds with thiram or captan @ 2.5g/kg of seed would be effective in managing the disease in the early stages of the crop. On appearance of the symptoms spraying with carbendazim 50 WP @1.0 g/l or mancozeb 45 WP @ 2.0 g/l and subsequent spray after 10 to 15 days would be effective in later stages of the crop. Spraying with copper oxychloride @ 3 to 4 g/liter water has also been found effective in management of the disease. Singh *et al.* (2001) reported yield losses to the tune of 50 per cent in severely diseased field. Since there is low level of resistance to *cercospora* leaf spot, the cultural practices and chemical control play an important role in its management. Cultural practices such as field sanitation, crop rotation, destruction of infected crop debris, and avoiding collateral hosts in the vicinity of the crop may help in reducing the incidence. Mancozeb, carbendazim, copper oxychloride and benomyl are reported to reduce disease incidence considerably.

Macrophomina Leaf Blight (*Macrophomina phaseolina*)

Macrophomina phaseolina (Tassi) Goid is one of the most virulent and destructive pathogen which incite diseases in wide range of hosts, while the symptoms produced were seedling rot, collar rot, leaf blight and pod rot in mothbean (Sandhu and Singh, 1998). Root rot incited by *M. phaseolina* has been rated as most devasting disease of mungbean. The pathogen attacks on all parts of plant *i.e.* root, stem, branches, petiols, leaves, pods and seeds. Moreover, seed infection of *Rhizoctonia bataticola* (*M. phaseolina*) ranges from 2.2-15.7 per cent which causes 10.8 per cent in grain yield and 12.3 per cent in protein content of seed in mungbean (Kaushik *et al.*, 1987). The infected seeds act as an important source of primary inoculum for new areas (Sandhu and Singh, 1998). Soil and seed borne nature of the disease possesses problems for an effective disease management. Therefore, it has become a serious problem in hampering the production of the mungbean in all growing areas of India.

In pre-emergence stage, the fungus causes seed rot and rotting of germinating seedlings. In post-emergence stage, seedlings get blighted due to soil or seed borne infection. Prominent symptoms are decay of secondary roots and shredding of the cortex region of the tap root. Small, circular, brown spots appear on the cotyledons or on young leaves. At podding stage, some of the veins in the leaf develop copper colour. As the severity increases, drooping of leaves occurs due to weakening and breakage of the veins. Such leaves droop, dry and shed. The pathogen can survive through seed, soil, diseased plant parts and host plants. The severity of the disease increases with the increase in temperatures. Fungus survives in upper layers of the soil and enters plant through stem.

Management

Practicing clean cultivation practices and crop rotation with non pulse crop found helpful in avoiding the disease. Destroying the diseased plant debris by burning or burying in the soil, seed treatment with carbendazim + Thiram (1:2) @ 3 g/kg of seed, basal application of zinc sulphate @ 25kg/ha or neem cake @ 150 kg/ha or soil application and *P. fluorescens* (1 x 1010cfu/g) or *T. viride* (1 x 108cfu/g) @ 2.5 kg/ha +50 kg of well decomposed FYM at the time of sowing helps in prevention of the disease. The diseased plants should be uprooted and destroyed so that the sclerotia do not form or survive. Seeds treated with *Trichoderma* (1 x 108cfu/g) 5 – 10 g/kg of seed or captan 75WP @ 2.5 g/l and thiram 80 per cent WP @ 2 g/l before sowing provides significant protection. Spraying with carbendazim 50 WP @ 1.0 g/l at an interval of 15 days with the appearance of the symptoms would also be effective.

Latha and Rajappan (2001) reported that root rot incidence was significantly reduced by FYM in blackgram caused by *M. phaseolina*. Mathur *et al.* (2003) observed that vermicompost was significantly superior in reducing incidence of *F. oxysporum* causing wilt of fenugreek in field conditions. Root rot of mungbean caused by *M. phaseolina* and considered a soil borne disease. Upmanu *et al.* (2002) found that soil amendment + seed treatment with bavisitn was highly effective in reducing pre and post emergence mortality in french bean. Besides, observed that *T. harizianum* was sensitive to bavistin. Gupta and Sharma (2007) reported that bavistin (0.1 or 0.05 per cent) in combination with different antagonistic *viz., T. viride* and *T. harizianum* were found effective against white root rot of apple. *T. harzianum* was found the most effective against the fungus under *in vitro* and in pots conditions followed by *T. viride and T. polysporum. P. fluorescens* was the least effective in reducing root rot incidence. The relative efficacy of herbal oils and plant products under pots house condition exhibited palmarosa oil to be the most effective as seed dresser in reducing root rot incidence. Bavistin was considered to be the most effective to inhibit mycelial growth of pathogen as well as reducing root rot incidence, followed by captan or thiram, indofil M-45 and vitavax or raxil, while copper sulphate was the least effective treatment in both conditions. In the case of organic manures, vermicompost was the most effective in reducing the root rot incidence under pots conditions. FYM and goat manure was found moderately effective in controlling root rot incidence. Integrated management approach showed that vermicompost and bavistin in combination was more effective in reducing the root rot incidence in pots conditions(Kumari *et al.*, 2012).

Multiple resistant genotypes against different diseases of mungbean were found by Jameel Akhtar *et al.* (2014). They have conducted field screening during *Kharif*, 2008 and 2009 which revealed that out of 31 genotypes of greengram, only 1 genotype ML 1299 and out of 14 genotypes of blackgram, only 3 genotypes *viz*, BS 2-3, IPU 02-43 and B 3-8-8 showed resistant or highly resistant response against multiple diseases including Cercospora leaf spot, web blight and powdery mildew. In another trial conducted for management of foliar diseases of greengram and blackgram through chemical revealed that two foliar sprays of propiconazole @ 0.1 per cent at an interval of 8-10 days was most effective as it reduced severity of Cercospra leaf spot (85.77 and 80.10 per cent), web blight (77.87 and 85.29 per cent) and powdery mildew (100.00 per cent each) with average yield of 9.08 and 8.83q/ha of greengram and blackgram, respectively.

In this review, developments on management of important viral and fungal diseases of both Greengram and blackgram have been discussed. Major emphasis should be given on identifying, evaluating and integrating location specific components of IDM. Despite the development of various IDM modules to tackle economically important diseases of legumes, gap persists between scientists and farmers, particularly in the developing countries. This is due to dynamic agro-eco systems and socio-economic conditions, and due to dynamic nature of host-pathogen interactions / co-evolution that necessitates refinement of IDM strategies. Therefore, persistent need exists for refinement, validation, transfer and adoption of IDM modules. Experience over the last few decades has shown that plant viruses cannot be controlled by preventive measures only. The complicated ecology of many of the viruses which affect legume crops, and for this matter crops in general, calls for a strategy which is long term, sustainable, economically acceptable to the resource-poor farmers, and friendly to the environment. The most sound approach is to exploit all available measures and use them in an integrated fashion. Effective virus-disease control has always been and will continue to be by integrated management, that is by crop management or ecosystem management.

Future IDM strategies will also account effects of climate variability-on pathogen dynamics that is contributing to emergence and resurgence of new strains / pathogens, necessitating the appropriate refinement for disease control. HPR, fungicides, natural plant products, biofungicides, botanicals and agronomic practices will remain the potentially viable options for IDM. In addition, the future modules will incorporate the latest advancements being made in the identification of biocontrol agents and products from biotechnological approaches such as marker-assisted selection, genetic engineering, and wide hybridization to develop cultivars with resistance to diseases. Emphasis will be on using information technology for disease modeling, developing decision support systems, and utilization of remote sensing to refine up-scale and disseminate IDM technologies.

References

Abtahi, F.S. and Habibi, M.K., 2008. Host range and some characterization of *Tobacco streak virus* isolated from lettuce in Iran. *African Journal of Biotechnology*, **7(23)**: 4260-4264.

Agarwal, S.C., 1991. *Diseases of Greengram and Blackgram*, International Book Distributors, Dehradun, p. 321.

Ahmad, Z., Bashir, M. and Mtsueda, T., 1997. Evaluation of legume germplasm for seed borne viruses in Harmonizing Agricultural Productivity and Conservation of Biodiversity, Breeding and Ecology Proc. 8th SABRAO J. Cong. Annu. Meeting Korean Breeding Soc. Seoul. Korea, p. 117–120

Ahmed, M. and Harwood, R.F., 1973. Studies on whitefly transmitted yellow mosaic of urd bean (*Phaseolus mungo*). *Plant Disease Reporter*, **57**: 800-802.

Amin, P.W., 1985. Studies on arthropod vectors of groundnut viruses, their ecology and control. Progress Report for 1978-1983, ICRISAT, p. 166.

Amin, P.W., Ghanekar, A.M., Rajeshwari, R. and Reddy, D.V.R., 1985. *Tomato spotted wilt virus* as the causal pathogen of leaf curl of mungbean, *Vigna radiata* (L.) Wilczek and urdbean, *Vigna mungo* (L.) Hepper in A P, India. *Indian Journal of Plant Protection*, **13**: 9-13.

Baldwin, J.A., Beasley, J.P. Jr., Culbreath, A.K. and Brown, S.L., 1997. Twin versus single row patterns for peanut production. *Proc. Am. Peanut Res. Ed. Soc.*, **29**: 20 (Abstr.)

Baranwal, V.K. and Ahmed, N., 1997. Effect of *Clerodendrum aculeatum* leaf extract on tomato leaf curl virus. *Indian Phytopathology*, **50(2)**: 297-299.

Basandrai, A.K., Gartan, S.L., Basandrai, D. and Kalia, V., 1999. Blackgram (*Phseolus mungo*) germplasm evaluation against different diseases. *Indian Journal of Agricultural Sciences*, **69(7)**: 506-508.

Belanger, R.R., Labbe, C. and Jarwis, W.R., 1994. Commercial scale control of rose powdery mildew with fungal antagonist. *Plant Disease*, **78:** 420-424.

Beniwal, S.P.S. and Bharathan, N., 1980. Beetle transmission of urdbean leaf crinkle virus. *Indian Phytopathol.*, **33**: 600–601

Bharadwaj, C.L. and Thakur, D.R., 199. Efficacy and economics of fungicide spray schedules for control of leaf spots and pod blights in urdbean. *Indian Phytopathol.*, **44(4)**: 470-475.

Bhat, A.I., Jain R.K., Chaudhary, V., Krishnareddy, M., Ramiah, M., Chattannavar, S.N. and Varma, A., 2002b. Sequence conservation in the coat protein gene of *Tobacco streak virus* isolates causing necrosis disease in cotton, mung bean, sunflower and sunnhemp in India. *Indian Journal of Biotechnology*, **1**: 350-356.

Bhat, A.I., Jain, R.K., Varma, A., Naresh Chandra and Lal, S.K., 2001a. Tospovirus(es) infecting grain legumes in Delhi-their identification by serology and nucleic acid hybridization. *Indian Phytopathology*, **54**: 112-116.

Binyamin, R., Khan, M.A., Ahmad, N. and Ali, S., 2011. Relationship of epidemiological factors ith urdbean leaf crinkle virus disease and its management using plant extracts. *Int. J. Agric. Biol.*, **13**: 411–414

Black, M.C., Andrews, T.D. and Smith, D.H., 1993. Interplot interference in field experiments with spotted wilt disease of peanut. *Proc. Am. Peanut Res. Ed. Soc.*, **25**: 65 (Abstr.)

Black, M.C., 1990. Predicting spotted wilt in south Texas peanuts. *Proc. Am. Peanut Res. Ed. Soc.*, **22**: 83 (Abstr.)

Borah, R.K. and Nath, P.D., 1995. Evaluation of an insecticide schedule on the incidence of whitefly, *Bemisia tabaci* (Genn.) and yellow vein mosaic in okra. *Indian Journal of Virology*, **11(2)**: 65-67.

Borah, R.K., 1995. Effect of synthetic pyrethroids and organophosphorus insecticides on the incidence of whitefly, *Bemisia tabaci* (Genn.) and yellow mosaic virus in greengram, *Vigna radiata* (L.) Wilczek. *Indian Journal of Virology*, **11(1)**: 75-76.

Bos, L., Hampton, R.O. and Makkouk, K.M., 1988. Viruses and virus diseases of pea, lentil, faba bean and chickpea. In: *World Crops: Cool Season Food Legumes* (eds. Summerfield R.J.). Kluwer Academic Publishers, Dodrecht, The Netherlands, p. 591-615.

Capoor, S.P. and Varma, P.M., 1948. Yellow mosaic of *Phaseolus lunatus* L. *Current Science*, **17(5)**: 152-153.

Capoor, S.P. and Varma, P.M., 1950. New virus disease of *Dolichos lablab*. *Current Science*, **19**: 248-249.

Chandrasekaran, A. and Rajappan, K., 2002. Effect of plant extracts, antagonists and chemicals individual and combined on foliar anthracnose and pod blight of soybean. *J. Mycol. Pl. Path.*, **32(1)**: 25-27.

Chandrasekharan, M. and Balasubramanian, G., 2002. Evaluation of plant products and insecticides against sucking pests of greengram. *Pestology*, **26(1)**: 48-50.

Chenulu, V.V., Venkateswarlu, V. and Rangaraju, R., 1979. Studies on yellow mosaic disease of mungbean. *Indian Phytopathology*, **32**: 230-235.

Chhabra, K.S. and Kooner, B.S., 1988. Varietal resistance in mungbean (*Vigna radiate* (L.) Wilczek) to thrips (*Megalurothrips distalis* Karny). *The Tropical Vegetables Information Service (TVIS)*, **3**: 10-11.

Chhabra, K.S., Kooner, B.S., Saxena, A.K. and Sharma, A.K., 1993. Role of allomones in varietal resistance in summer mungbean against thrips, *Megalurothrips distalis* (Karny). In: *Proceedings of the International Symposium on Pulses Research*, December 4-8, 1993, Kanpur, India.

Chowdhury, A.K. and Shah, N.K., 1985. Inhibition of urdbean leaf crinkle virus by different plant extracts. *Indian Phytopathol.*, **38**: 566–568.

Cohen, S. and Marco, S., 1973. Reducing the spread of aphid-transmitted viruses in peppers by trapping the aphids on sticky yellow polyethylene sheets. *Phytopathology*, **63**: 1207-1209.

Deeksha, J. and Tripathi, H.S., 2002a. Cultural, biological and chemical control of nthracnose of urdbean. *J. Mycol. Pl. Path.*, **32(1)**: 52-55.

Deeksha, J. and Tripathi, H.S., 2002b. Effect of Indofil M-45 in disease severity of anthracnose in urdbean. *J. Mycol. Pl. Path.*, **32(1)**: 86-87.

Dhingra, K.L. and Chenulu, V.V., 1985, Effect of yellow mosaic on yield and nodulation of soybean. *Indian Phytopathology*, **38**: 248-251.

Dhingra, K.L., 1976. Transmission of urdbean leaf crinkle virus by two aphid species. *Indian Phytopathol.*, **28**: 80–82

Food and Agricultural Organization (FAO) 2000. Agricultural Marketing in Myanmar. Ministry of Agriculture and Irrigation, Yangon.

Funderburk, J., Stavisky, J., Tipping, C., Gorbet, D. and Momol, T., 2002. Infection of *Frankliniella fusca* (Thysanoptera: Thripidae) in peanut by the parasitic nematode *Thripinema fuscum* (Tylenchidae: Allantonematidae). *Environ. Entomol.*, **31**: 558–563.

Ganapathy, T., Kuruppiah, R. and Gunasekaran, K., 2003. Identifying the source of resistance for mungbean yellow mosaic virus (MYMV), urd bean leaf crinkle virus and leaf curl virus disease in urd bean (*Vigna mungo* (L.) Hepper). In *Annual Meeting and Symposium on recent Developments in the diagnosis and Management of Plant Diseases for Meeting Global Challenges*, December 18-20, 2003, University of Agricultural Sciences, Dharwad, p. 30.

Gautam, H.C., 1990. Identification of soybean diseases and their integrated control. *Farmer and Parliament*, p. 15-22.

Ghanekar, A.M. and Beniwal, S.P.S., 1975. Cowpea a local lesion host for mungbean leaf curl virus. *Indian Phytopathology*, **28**: 527-528.

Ghanekar, A.M., Reddy, D.V.R., Lizuka, N., Amin, P.W. and Gibbons, R.W., 1979. Bud necrosis of groundnut (*Arachis hypogaea*) in India caused by *Tomato spotted wilt virus*. *Annals of Applied Biology*, **93**: 173-179.

Grewal, J.S., 1988. Diseases of pulse crops – an overview. *Indian Phytopathology*, **41(1)**: 1-14.

Gupta, V.K. and Sharma, K., 2007. Integration of chemicals and biocontrol agents for managing while root rot of apple. *Acta Horticulturae*, p. 635.

Jain, K.L., Gupta, A.K. and Trivedi, Amit, 2000. Evaluation of mung bean germplasm for resistance to powdery mildew and virus disease. *Indian Phytopathology*, Golden Jubilee – Proceedings, p. 547-548.

Jameel, Akhtar, Lal, H.C., Kumar, Yogesh, Singh, P.K., Ghosh, Jyotirmoy, Khan, Zakauallah and Gautam, N.K., 2014. Multiple Disease Resistance in Greengram and Blackgram Germplasm and Management through Chemicals under Rain-fed Conditions. *Legume Research*, **37(1)**: 101-109.

Jeger, M.J., 2000. Bottlenecks in IPM. *Crop Protection*, **19:** 787-792.

Jyothirmai Madhavi, K., 2009. Identification and characterization of *Peanut bud necrosis virus* (PBNV) and *Tobacco streak virus* (TSV) on blackgram (*Vigna mungo* L. Hepper) and greengram (*Vigna radiata* L. Wilczek). *Ph.D.Thesis*, Acharya N.G. Ranga Agricultural University, Hyderabad, Andhra Pradesh.

Jyothirmai Madhavi, K., Prasada Rao, R.D.V.J. and Subbarao, M., 2011. Pollen associated thrip transmission of Tobacco streak virus in blackgram and greengram. *Indian Journal of Plant Protection*, **39(1 and 2)**: 92-96.

Kadian, O.P., 1980. Studies on leaf crinkle disease of urdbean [*Vigna mungo* (L.) Hepper] mungbean [*V. radiata* (L.) Wilczek] and its control. *Ph.D. Thesis*, Department of Plant Pathology Haryana Agricultural University, Hisar, India, p. 177.

Kaushik, C.D., Chand, J.N. and Saryavir, 1987. Seedborne nature of *Rhizoctonia bataticola* causing leaf blight of mung bean. *Indian J. Mycol. Plant Pathol.*, 17: 154-157.

Kulkarni, S. and Ramakrishnan, K., 1977. Epidemiology and control of brown leaf spot of rice.

Kumar, R., 1982, Aerospora in a pine forest in India.

Kumari, R., Shekhawat, K.S., Gupta, R. and Khokhar, M.K., 2012. Integrated Management against Root-rot of Mungbean [*Vigna radiata* (L.) Wilczek] incited by *Macrophomina phaseolina*. *J. Plant Pathol. Microb.*, **3**: 136.

Ladhalakshmi, D., Ramiah, M., Ganapathy, T., Krishnareddy, M., Khabbaz, S.E., Merin Babu and Kamalakannan, A., 2006. First report of the natural occurrence of *Tobacco streak virus* on blackgram (*Vigna mungo*). *Plant Pathology*, **55**: 1395.

Latha, T.K.S. and Rajappan, K., 2001. Effect of organic amendments on fungi nematodes disease complex in blackgram. *Madras Agriculture Journal*, **87**: 707-709.

Laxman, R., 2006. Studies on leaf spot of greengram caused by *Colletotrichum truncatum* (Schw.) Andrus and Moore. *M.Sc. (Agri.) Thesis*, Univ. Agric. Sci., Dharwad, Karnataka, India.

Majid, S., 1953. Annals Report of department of agriculture, Assam for year ending 31st March 1950. II. *The Grow More Food Campaign*, **11:** 107.

Marappa, N., Savithramma, D.L., Nagaraju, Prameela, H.A. and Krishnamurthy, R.A., 2003. Evaluation of mungbean genotypes against powdery mildew, yellow mosaic virus and bacterial blight diseases at Bangalore. In: *Annual Meeting and Symposium on Recent Developments in the Diagnosis and Management of Plant Diseases for Meeting Global Challenges*, December 18.

Mathur, K., Bansal, R.K. and Gurjar, R.B.S., 2003. Organic management of wilt of fenugreek A seed spice. *Journal of Mycology and Plant Pathology*, **33**: 491.

Mitchell, F.L. and Smith, J.W. Jr., 1991. Epidemiology of *Tomato spotted wilt virus* relative to thrips populations. *See Ref.*, **75**: 46–52

Mitchell, F.L., 1996. Implementation of the IPM planting window for management of *Tomato spotted wilt virus* and avoidance of peanut yellowing death. *Final Compliance Report, Texas Pest Management Association, Biologically Intensive Integrated Pest Management Grant Program.*

Moe, A.K., Yutaka, T., Fukuda, S. and Kai, S., 2008. Impact of Market Liberalization on International Pulses Trade of Myanmar and India. *Journal of Faculty Agriculture,* Kyushu University, **53(2)**: 553–561.

Nagamani, P., Viswanath, K., Jyosthna, M.K., Reddy, C. Srinivasa, 2013. Management of Peanut Stem Necrosis Disease of Groundnut. *Indian Journal of Plant Protection,* **41(4)**: 349-352.

Nair, N.G. and Nene, Y.L., 1973. Studies on yellow mosaic of urd bean (*Phaseolus mungo* L.) caused by mungbean yellow mosaic virus I. transmission studies. *Indian Journal of Farm Science,* **1**: 109-110.

Narayanasamy, P. and Jaganathan, T., 1973. Vector transmission of black gram leaf crinkle virus. *Madras Agric. J.*, **60**: 651–652

Nariani, T.K., 1960. Yellow mosaic of mung (*Phaseolus aureus*). *Indian Phytopathology,* **13**: 24-29.

Nath, P.D., 1994. Effect of sowing time on the incidence of yellow mosaic virus disease and whitefly population on greengram. *Annals of Agricultural Research,* **15(2)**: 174-177.

Negi, H. and Vishunavat, K., 2004. Role of seed-borne inocula of leaf crinkle virus in the development and yield of urdbean. *Annl. Plant Prot. Sci.,***12**: 452–453

Nene, Y.L., 1972. In: *A Survey of Viral Diseasesof Pulse Crops in Uttar Pradesh,* GB Pant University Agricultural Technical Research Bulletin, **4**: 19.

Pappu, H.R., Csinos, C.S., McPherson, R.M., Jones, D.C. and Stephenson, M.G., 2000. Effect of acibenzolar-S-methyl and imidacloprid on suppression of tomato spotted wilt Tospovirus in flue- flue-cured tobacco. *Crop Prot.,* **19**: 349–354.

Patel, M.S., Yadav, D.N. and Rai, A.B., 1994. Neemark (azadirachtin) as ovipositional deterrent against cotton pests. *Pestology,* **18(8)**: 17-19.

Pathak, A.K. and Jhamaria, S.L., 2004. Evaluation of mungbean (*Vigna radiata* L.) varieties to yellow mosaic virus. *Journal of Mycology and Plant Pathology,* **34(1)**: 64-65.

Peerajade, D.A., Ravikumar, R.L. and Rao, M.S.L., 2004. Screening of local mungbean collections for powdery mildew and yellow mosaic virus resistance. *Indian Journal of Pulses Research,* **17(2)**: 190-191.

Phadke, A.D., Khandal, V.S. And Rahalkar, S.R., 1988. Use of a neem product in insecticide resistance management (IRM) in cotton. *Pesticides,* **22(4)**: 36-37.

Prasada Rao, R.D.V.J., 2003c. Integrated Management of Viral Disease Problems of Mungbean (*Vigna radiata*) and Urdbean (*Vigna mungo*). Final Report of NATP-PSR Project RPPS-03.

Prasada Rao, R.D.V.J., Jyothirmai Madhavi, K., Vara Prasad, K.S. and Khetarpal, R., 2005. Present status of stem necrosis disease of groundnut caused by *Tobacco streak virus*. *Indian Journal of Virology,* **16**: 66.

Prasada Rao, R.D.V.J., Sarath Babu, B., Sreekanth, M. and Manoj Kumar, V., 2003a. ELISA and infectivity assay based survey for the detection of *Peanut bud necrosis virus* in mungbean and urdbean in Andhra Pradesh. *Journal of Plant Protection,* **31**: 26-28.

Prasada Rao, R.D.V.J., Jyothirmai Madhavi, K., Reddy, A.S., Varaprasad, K.S., Nigam S.S., Sharma, K.K., Kumar, P. Lava and Waliyar, F., 2009. Non-transmission of *Tobacco streak virus* isolate occurring in India through the seeds of some crop and weed hosts. *Indian Journal of Plant Protection,* **37(1 and 2)**: 92-96.

Prasada Rao, R.D.V.J., Reddy, D.V.R., Nigam, S.N., Reddy, A.S., Waliyar, F., Yellamanda Reddy, T., Subramniam, K., John Sudheer, M., Naik, K.S.S., Bandopadhyay, A., Desai, S., Ghewande, M.P., Basu, M.S. and Somasekhar, 2003c. Peanut Stem Necrosis: A New Disease of Groundnut in India. Information Bulletin no. 67. Patancheru 502 324, Andhra Pradesh, India: International Crops Research Institute for the Semi-Arid Tropics, p. 16.

Raghupathi, N. and Sabitha, D., 1994. Effect of barrier crops, mulching and yellow sticky traps in the management of yellow mosaic virus disease of soybean. In

Crop Diseases: Innovative Techniques and Management. Eds. Sivaprakasam, K. and Seetharaman, K., Kalyani Publishers, New Delhi, p. 419-420.

Raja Kumar, N., Subba Reddy, C., Krishnamurthy, K.V.M. and Reddy, M.V., 2007. Influence of Seed Rate on Leaf Curl Disease and Thrips Vector Population in Blackgram. *Indian Journal of Plant Protection*, **35(1)**: 90-92.

Raje, R.S. and Rao, S.K., 2002. Screening of mungbean (*Vigna radiata* L. Wilczek) germplasm for yellow mosaic virus, cercospora leaf spot and powdery mildew. *Legume Research*, **25(2)**: 99-104.

Rajkumar, N., 2004. Management of *Peanut bud necrosis virus* (PBNV) in blackgram (*Vigna mungo* L. Hepper). *Ph.D. Thesis*, Acharya N.G. Ranga Agricultural University, Hyderabad, Andhra Pradesh.

Ravindrababu, R., 1987. Studies on mung bean yellow mosaic virus disease. *Ph.D. Thesis*, Tamil Nadu Agricultural University, Coimbatore.

Reddy, D.V.R., Ratna, A.S., Sudarshana, M.R., Poul, F. and Kiran Kumar, I., 1992. Serological relationship and purification of bud necrosis virus, a tospovirus occurring in peanut (*Arachis hypogaea* L.) in India. *Annals of Applied Biology*, **120**: 279-286.

Reddy, D.V.R., Amin, P.W., McDonald, D. and Ghanekar, A.M., 1983. Epidemiology and control of groundnut bud necrosis and other diseases of legume crops in India caused by tomato spotted wilt virus. In: *Plant Virus Epidemiology*, ed. RT Plumb, JM Thrash, p. 93–102. Oxford: Blackwell Sci., p. 377.

Reddy, C., Tonapi, V.A., Navi, S.S. and Jayarajan, R., 2005. Influence of plant age on infection and symptomological studies on urdbean leaf crinkle viruse in urdbean (*Vigna mungo*). *Int. J. Agric. Sci.*, **1**: 1–6.

Reddy, Ch. Ravandar, Tonapi, Vilas A., Varanavasiappn, S., Navi, S.S. and Jayarajan, R., 2006. Management ofurdbean leaf crinkle virus in urdbean (*Vigna mungo* L. Hepper). *Int. J. Agric. Sci.*, **2**: 8–10

Rosello, S., Soler, S., Diez, M.J., Rambla, J.L. and Nuez, F., 1999. New sources for high resistance of tomato to the *Tomato spotted wilt virus* from *Lycopersicon peruvianum Plant Breeding*, **118(5)**: 231-235.

Sandhu, A. and Singh, R.D., 1998. Role of seed borne inoculum in development of charcoal rot of cowpea. *J. Mycol. Plant Pathol.*, **28**: 193-195.

Sastry, K.S.M. and Singh, S.J., 1973. Field evaluation of insecticides for the control of whitefly (*Bemisia tabaci*) in relation to the incidence of yellow vein mosaic of okra (*Abelmoschus esculentus*). *Indian Phytopathology*, **26(1)**: 129-138.

Sharma, H.C., Khare, M.N., Joshi, L.K. and Kumar, S.M., 1971. Efficacy of fungicides in the control of diseases of *kharif* pulses mung and urid. *All India Workshop on Kharif Pulses*, p. 2.

Shirshikar, S.P., Singh, R.A., De, R.K., Gurha, S.N. and Ghosh, A. 2001. Management of cercospora leaf spot of mungbean. In: National Symposium on pulses for

Sustainable Agriculture and National Security., Nov 17-19, 2001, New Delhi, ISPRD, IIPR, Kanpur, India, p. 146.

Siddiqui, K.H., Trimohan, Rana, V.K.S. and Lal, S.K., 1999. Location of sources of resistance amongst soybean genotypes to yellow mosaic virus disease due to whitefly, *Bemisia tabaci* Genn. under natural conditions. *Shashpa*, **6(1)**: 37-40.

Singh, A.B. and Srivastava, A.K., 1995. Status and control strategy of peanut bud necrosis disease in Uttar Pradesh. *See Ref.*, **34**: 65–69.

Singh, G., Singh, K., Gill, A.S. and Chhabra, K.S., 1980. *National Seminar on Disease Resistance in Crop Plants*. Tamil Nadu Agricultural University, Coimbatore, 22 and 23 December, 1980.

Singh, Kuldip, Singh, Sarvjeet and Gumber, R.K., 1996. Resistance to mungbean yellow mosaic virus in mungbean. *Indian Journal of Pulses Research*, **9(1)**: 90.

Singh, R.A., Gurha, S.N. and Ghosh A. 1998. Diseases of mungbean and urdbean and their management. In: *Diseases of Field Crops and their Management* (ed. Thind T.S.), National Agricultural Technology Information Centre, Ludhiana, India, p. 179-204.

Singh, S.J., 1985. Management of virus diseases of vegetable crops. In: *Integrated Pest and Disease Management*. Eds. Jayaraj, S., Proceedings National Seminar, Tamil Nadu Agricultural University, Coimbatore, p. 193-201.

Somashekara, Y.M., Nateshan, H.M. and Muniyappa, V., 1997. Evaluation of neem products and insecticides against whitefly (*Bemisia tabaci*), a vector of tomato leaf curl geminivirus disease. *Indian Journal of Plant Protection*, **25(1)**: 56-59.

Sreekanth, 2002. Bioecology and management of thrips vector(s) *Peanut budnecrosis virus* (PBNV) in mungbean (*Vigna radiata* L. Wilezek). *Ph.D. Thesis*, Acharya N.G. Ranga Agricultural University, Hyderabad, Andhra Pradesh.

Sreekanth, M., Sreeramulu, M., Prasada Rao, R.D.V.J., Sarath Babu, B. and Ramesh Babu, T., 2002a. Effect of sowing date on *Thrips palmi*. Karny population *and Peanut budnecrosis virus* incidence in greengram (*Vigna radiata* L. Wikzek). *Indian Journal of Plant Protection*, **30**: 16-21.

Sreekanth, M., Sreeramulu, M., Prasada Rao, R.D.V.J., Sarath Babu, B. and Ramesh Babu, T., 2002b. Evaluation of greengram genotypes (*Vigna radiata* L. Wilezek) for resistance to *Thrips palmi* Karny and *Peanut bud necrosis virus*. *Indian Journal of Plant Protection*, **30**: 109-114.

Sreenivasulu, A., 1994. Effect of certain management practices on the occurrence of thrips and leaf curl virus on blackgram (*Vigna mungo* L. Hepper). *M.Sc. Thesis*, Acharya N.G. Ranga Agricultural University, Rajendranagar, Hyderabad.

Todd, J.W., Culbreath, A.K. and Brown, S.L., 1996. Dynamics of vector populations and progress of spotted wilt disease relative to insecticide use in peanuts. *Acta Hortic.*, **431**: 483–490.

Todd, J.W., Culbreath, A.K., Rogers, D. and Demski, J.W., 1994. Contraindications of insecticide use relative to vector control and spotted wilt disease progress in peanut. *Proc. Am. Peanut Res. Ed. Soc.* **26**: 42 (Abstr.)

Upmanyu, S., Gupta, S.K. and Shyam, K.R., 2002. Innovative approaches for the management of root rot and web blight (*Rhizoctonia solani*) of french bean. *Journal of Mycology and Plant Pathology*, **32**: 317-331.

Uthamasamy, S., 1989. Whitefly - the emerging pest. In: *State Level Plant Protection Workers Project*, May 2-3, CPPS, Tamil Nadu Agricultural University, Coimbatore, p. 62-63.

Varaprasad, C. H., 2000. Studies on blight disease of chickpea caused by *Colletotrichum dematium* (Pers. Ex. Fr.) Grove. *M. Sc. (Agri.) Thesis*, Univ. Agric. Sci., Dharwad, Karnataka, India.

Verma, A. and Verma, H.N., 1993. Management of viral diseases of mungbean by *Clerodendrum* leaf extract. *Indian Journal of Plant Pathology*, **11(1 and 2)**: 63-65.

Verma, H.N. and Varsha, 1995. Induction of systemic resistance by leaf extract of *Clerodendrum aculeatum* in sunnhemp against sunnhemp rosette virus. *Indian Phytopathology*, **48(2)**: 218-221.

Vinod, J., Ganapathy, T., Rabindran, R. and Bharathi, M., 2007. Sero-diagnosis of mixed infection of Ilar and Tospo viruses in mungbean in Tamil Nadu, India. In abstracts of 10th International Plant Virus Epidemiology Symposium, Controlling Epidemics of Emerging and Established Plant Virus Diseases – The Way Forward. 172 p.

Vishwa, D., Singh, R.A. and Gurha, S.N., 2004. Integrated disease management in pulse crops. In: *Pulses in New Perspective* (eds. Masood Ali., Singh B.B., Shiv Kumar and Vishwa Dhar). Indian Society of Pulses research and development, IIPR, Kanpur, p. 324-344.

Wells, M.L., Culbreath, A.K., Csinos, A.S. and Todd, J.W., 2002. Effects of a plant activator and insecticides on tobacco thrips (Thysanoptera: Thripidae) feeding, behavior, and survival. *J. Agric. Urban Entomol.*, **19**: 117–120.

Wells, M.L., Culbreath, A.K. and Todd, J.W., 2002a. The effect of applications of acibenzolar-S-methyl on *Tomato spotted wilt virus* and thrips in peanut. *Peanut Sci.*, **29**: 136–141.

2016, Diseases of Pulse Crops and their Sustainable Management 411–427
Editors: Samir Kumar Biswas, Santosh Kumar and Gireesh Chand
Published by: BIOTECH BOOKS, NEW DELHI

Chapter 22

Major Virus Problems in Pulse Production in India and its Management

Amarendra Kumar[1]*, Santosh Kumar[1], Prabhat Kumar[2] and Tamoghna Saha[3]

[1]Department of Plant Pathology,
[3]Department of Entomology,
Bihar Agricultural University, Sabour, Bhagalpur – 813 210, Bihar
[2]Betelvine Research Centre, Islampur, Nalanda-801303
(Bihar Agricultural University, Sabour, Bhagalpur, Bihar)

Introduction

India is the largest producer, consumer, importer and processor of pulses in the world, with 24 per cent share in the global production. The important pulse crops are chickpea (48 per cent), pigeonpea (15 per cent), mungbean (7 per cent), urdbean (7 per cent), lentil (5 per cent) and fieldpea (5 per cent). The major pulse-producing states are Madhya Pradesh, Maharashtra, Rajasthan, Uttar Pradesh, Karnataka and Andhra Pradesh, which together account for about 80 per cent of the total production (Ali and Gupta, 2012). Among these, viruses are the second largest and most important group affecting all parts of the plant at all stages of growth of pulse crops. Viruses which cause major economic importance on pulses belong to the Luteoviruses, Nanoviruses,

* Corresponding Author: E-mail: kumaramar05@gmail.com

Potyviruses, Carlaviruses and Furoviruses. Yield losses vary from year to year and from location to location and can be considerable in some circumstances. Seed quality can also be impaired. Viral symptoms in plant foliage are often confused with nutritional deficiencies, herbicide damage or water-logging, leading to underestimation of the extent of damage caused. Many of the viruses are seed borne in their pulse hosts; some are sufficient to have enabled worldwide distribution.

A successful integrated disease management (IDM) strategy is one under which pulses have been protected from the yield-reducing effects of the viruses rendering the later to economic insignificance. The IDM involves the individual component of disease management such as host plant resistance (HPR), agronomic practices, judicious use of fungicides, pesticides for vector control, biopesticides for viral control, risk forecasting that operate on different aspects of the viral disease epidemiology, such that they complement each other and can be applied together in farmers' fields collectively to provide farmers with maximum economic return

Major Viral Diseases of Pulses in India

List of ajor viral diseases of pulses in India in showing in Table 22.1

Types of Virus

Viruses infecting pluses are either transmitted persistently or non-persistently by aphids. The persistently transmitted viruses are often called *Luteoviruses*. These are not seed-borne and require a continuous 'bridge' of live plant material to survive from one growing season to the next. Non-persistently transmitted viruses are seed-borne to varying extents, sometimes infecting commercial seed stocks of pulses. Persistent transmission means that once the insect becomes infectious, it remains so for the rest of its life. After an insect vector feeds on an infected plant, the virus has to pass through its body and lodge in the salivary glands before it can be transmitted to healthy plants. Not all aphid species are vectors of this kind of virus in pulses so the identification of aphid species is very important. Persistently transmitted viruses typically start with a random distribution of infected plants in autumn and increases during the season as vectors colonise the crop. Transmission rates can dramatically increase with large aphid flights that will often coincide with aphid activity and build up prior to sowing.

Non-persistently transmitted viruses can be seed-borne (depending on the virus – crop combination), but require aphid vectors to spread during the season. Non-persistent transmission or stylet-borne viruses, insects with sucking mouthparts can inoculate the virus into plants for only a few minutes after acquisition and the insect loses the virus after short feeding periods of a few seconds to a few minutes and upon moulting. Non-persistent viruses are specifically confined on the upper most tissues *i.e.*, epidermal tissues of the host. These non-persistently seed borne viruses, AMV and BYMV can sometimes cause heavy yield losses.

Symptoms Caused by Viruses

Viruses exhibit different kinds of symptoms in pulses. These varies in severity from death of the plant at one extreme to symptomless or unapparent infection at the

Table 22.1: Major Viral Diseases of Pulses in India

Name of Disease	*Name of Virus*	*Host Crop*	*Losses*	*Referances*
Sterility mosaic	*Pigeonpea Sterility Mosaic Virus* (PSMV)	Pigionpea	Up to 90 per cent	Kulkarni *et al.*, 2002
Stunt virus	*Bean leaf roll luteo virus* (BLRV)	Chickpea	58.7-89.1 per cent	Saxsena *et al.*, 1991
Mosaic	*Alfalfa Mosaic Virus*	Chickpea	–	–
Narrow leaf	*Bean Yellow Mosaic Virus*	Chickpea	–	–
Proliferation	*Cucumber Mosaic Virus*	Chickpea	–	–
Yellow mosaic	*Bean Yellow MosaicVirus*	Mungbean, Urdbean, Soybean and cowpea	10-100 per cent	Nene 1972; Singh 1980; Rathi 2002
Mosaic mottle	*Bean Common Mosaic Virus*	Mungbean, Urdbean, cowpea	100 per cent	Nene, 1972
Leaf crinkle	*Urdbean Leaf Crinkle Virus*	Mungbean, Urdbean, cowpea	35 to 81 per cent	Bashir *et al.*, 1991
Cowpea golden mosaic	*Cowpea Golden Mosaic Virus*	Cowpea	60-100 per cent	Mali, 1996
Cowpea aphid-bornemosaic	*Cowpea Aphid-borne MosaicVirus*	Cowpea	13 - 87 per cent	Bashir *et al.*, 1996
Stunt	*Red Clover Vein Mosaic Virus* (RCVMV)	Pea	87.8-100 per cent	Khan and Singh, 1997
Pea seed-borne mosaic	*Pea Seed-borne Mosaic Virus* (PSbMV)	Pea	11-36 per cent	Kraft and Hampton, 1980
Pea enation mosaic (PEMD)	*Pea Enation Mosaic Virus* (PEMV)	Pea	10-100 per cent	Hull, 1981
Pea streak mosaic	*Pea Streak Mosaic Virus*	Pea	–	–

other. The intensity and kinds of symptoms depend on virus and pulse species and to a lesser extent on virus strain, pulse variety, climatic conditions and plant stage at infection. Plants infected via seed at an early stage may be stunted and all their leaves usually may show more uniform discoloration. In contrast, with current-season infection older leaves may not show any symptoms. Foliage symptoms are often clearly visible on young leaves and can include yellowing (sometimes reddening), vein clearing, leaf mottle, leaf distortion, curling of leaves, reduced size, chlorotic or necrotic spotting, or more widespread necrosis. Shoot symptoms may be seen as bunching of young leaves, growth of auxiliary shoots, bending over of the growing point, tip or apical necrosis, streaking of stems, stunting and wilting or plant death. Symptoms such as leaf yellowing, veining, mottling, and wilting can often be confused with nutrient deficiencies, herbicide damage or water stress unless sufficiently distinct. It is also difficult to tell which virus is present without resorting to laboratory tests on plant samples.

Epidemics

High summer, autumn rainfall and close planting are the three most important conditions that predispose pulse crops to severe virus infection. Summer and autumn rainfall will stimulate early growth of pastures, weeds and crop volunteers upon which aphids build up before the growing season starts. This result in early aphid flights to newly emerged crops and early infection.The infected plants can then act as a reservoir for further spread of infection within the crop so the final virus incidence is high.

Management of Viral Diseases of Pulses

The development of an effective management package for virus diseases is based on the availability of basic information required to design an appropriate combination of interventions which can slow down virus disease development. The most important of these are: (a) diagnosis and identity of the causal agent, (b) mode of transmission, (c) ecology of the virus disease including that of its vector, (d) availability of genetic resistance, (e) extent and value of crop losses and (f) available crop-protection methodologies and their applicability to specific farming systems and socio-economic situations.

Control is optimized through IDM approaches, which combine all possible measures that operate in different ways such that they complement each other and applicable in farmers' fields. Thus, control measures can be classified as (i) those that control the virus, (ii) those that are directed towards avoidance of vectors or reducing their incidence, and (iii) those that integrate more than one method. Cultural practices such as healthy seed, rouging, alteration in sowing dates and use of early maturing cultivars are effective in minimizing virus disease incidence. 45-50 per cent of viruses affecting leguminous crops are seed-borne in nature (Bos *et al.*, 1988). Seed-borne infections permit the introduction of primary virus inoculum into the field which facilitates secondary spread to reach a serious level in locations where the environmental conditions permit high vector activity. A close relationship between sowing date and the extent of subsequent virus spread is well documented for many crops. Host resistance is the most acceptable component in virus control because it is environment-friendly, practical and economically acceptable to farmers. There are many crop cultivars with adequate levels of virus resistance; in lentil against Pea seed-borne mosaic virus (PSbMV), Bean yellow mosaic virus (BYMV), Alfalfa mosaic virus (AMV), Pea enation mosaic virus (PEMV), Bean leaf roll virus (BLRV), FBNYV and Soybean dwarf virus (SbDV), in chickpea against CMV, BYMV and PSbMV, in faba bean against BYMV, CMV, AMV, BLRV and PEMV, and in pea against BYMV, PEMV, PSbMV and BLRV. Another area of host resistance which is not well exploited is resistance to vector(s). Success in reducing virus spread by chemical control of vectors is likely with persistently rather than with non-persistently transmitted viruses. Each of the control measures mentioned provides only partial control, but combining genetic resistance, cultural practices, and chemical sprays is expected to lead to improvements. The use of host resistance, and one or two well-timed sprays coupled with optimal planting date and early rouging of virus-infected plants could offer reasonable and economic control and stabilize production. Each strategy needs to be affordable by farmers and fulfil the requirements of being environmentally and socially responsible. It must also be compatible with control measures already in use against other pests and pathogens (Pande *et al.*, 2009).

Pigeonpea (*Cajanus cajan* (L.) Millsp.

Sterility Mosaic

Sterility mosaic disease (SMD) was first described in 1931 from Pusa Bihar (Mitra, 1931) and subsequently from rest of India and other countries. This disease is of a major concern in UP, Bihar, Gujarat, Tamil Nadu and Karnataka. Patches of bushy pale and green plants with smaller leaves, without flowers or pods are the common field symptoms and is a serious disease in India and Nepal (Reddy and Vishwa Dhar, 2000). Etiology of sterility mosaic is unknown despite of numerous attempts during the past 20 years. Lava Kumar *et al.* (2000) reported a tenui virus of asymmetric morphology as the cause of Sterility Mosaic disease and retained the name of virus as Pigeonpea sterility mosaic virus (PPSMV). The PPSMV is flexous, branched filaments measuring 3-8nm in diameter. The virus is not transmitted through seed or sap. It is transmitted by the vector eriophyid mite (*Aceria cajani*), which are very small, pink in colour, spindle shaped and normally feed on the underside of leaflets. The vector survives on perennial pigeonpea during summer months. It lays milky white eggs on the vegetative terminals. Many of the nymphs are found on young folded leaflets. The wind dispersal of infective mites may cause plant to plant infestation. Sterility mosaic virus also causes green chlorotic patches on the leaflets. If plants get infected in early vegetative stage, it results the complete sterility of the plants. Infection at an early stage (<45-day-old plants) results in a 95 to 100 per cent loss in yield while losses from late infection (>45-day-old plants) depend on the level of infection (*i.e.*, number of affected branches per plant) and range from 26 to 97 per cent (Kannaiyan *et al.*, 1984), Late infection may result into partial sterility that means only some parts of the plants exhibit the disease symptoms. Reduction in leaf size with light and dark green mosaic pattern, stunting of plants and profuse branching etc. are some of the characteristic symptoms of the disease. The seeds produced from partially affected plants are discolored and shriveled, with about 20 per cent reduction in dry weight.

Maximum virus incidence and yield losses occur in ratooned and perennial pigeonpea. The severely affected plants are completely sterile. SMD infection often predisposes plants to powdery mildew (*Oidiopsis taurica*) (Reddy *et al.*, 1984) infection and infestation by spider mites (*Schizotetranychus cajani*) (Sithananthamet *al.*, 1989), compounding the damage. The mite vector predominantly survives in the vegetative buds and growing points. Weather parameters comprising temperature range of 10-25°C and 25-35°C with relative humidity of 60 per cent help in built up of the vector (Sharma *et al.*, 2010).

Information on physiological specialization of SMD is limited and is based on symptoms observed on the inoculated plants using the mite vector (Reddy *et al.*, 1990). Recently, PCR based techniques are being developed for distinction of different species of eriophyid mites (Vishwa Dhar *et al.*, 2005). Host Plant Resistance is the most reliable and sustainable method for the management of pigeonpea sterility mosaic disease. Considerable progress has been made in identifying resistance sources and developing resistant cultivars to the disease. Some attention has also been paid to cultural and chemical control of sterility mosaic. Cultural practices such as destroy ratooned pigeonpea, uproot and destroy infected plants at the initial stage of disease development, crop rotation to reduce inoculum levels and vector populations, chemical control as seed treatment with 25 per cent carbofuran or 10 per cent aldicarb (3g kg-1 seed) and spraying acaricides or insecticides like karathane, metasystox to control the mite vector in the early stages of plant growth (Reddy *et al.*, 1990).

Other minor viral diseases of pigeonpea are yellow mosaic, mung bean yellow mosaic virus, mild mosaic and tobacco mosaic virus

Chickpea (*Cicer arietinum* L.)

Fifteen viruses are reported from chickpea but only few of them are of economic importance. The virus diseases in order of economic importance are stunt, mosaic, proliferation and narrow leaf.

Stunt

Stunt is the most important viral disease of chickpea prevalent in most of chickpea growing countries. The disease is more serious in north India, Pakistan, Ethiopia and Tunisia and cause losses upto 100 per cent. The disease is characterized by stunting of plant, shortening of internodes and phloem discolouration at the collar region (Nene *et al.*, 1978). Phloem browning at collar region is the characteristic symptom of the disease. Stems and leaves of the affected plants are thick and bristle. A transverse cut of root shows a brown ring. Affected plants can be easily spotted in the field by their yellow, orange or brown discolouration and stunted growth (Nene *et al.*, 1978). Some time stunt and fusarium wilt infection occurs together. In such case, xylem discolouration, which is typical of fusarium wilt, is also seen. The wilting is results of combine infection (Nene *et al.*, 1991).

Chickpea stunt is caused by luteovirus, probably bean leaf roll virus (BLRV). Brunt *et al.* (1990) refers the chickpea virus both as BLRV and chickpea stunt luteovirus (CpLV). CpLV is a strain of BLRV but no detailed study is reported. In contrast to previous studies, it has shown that a Gemini virus, The host range of BLRV includes common legume crop such as bean (*Phaseolus vulgaris* L.) faba bean (*Vicia faba* L.), lentils (*Lens cularis* Medik), peas (*Pisum sativum* L.), lucerne (*Medicago sativa* L.) white clover (*Trifolium repens* L.), and alfalfa plant are symptomless reservoir of the virus. Through leafhoppers the virus could be successfully inoculated to species of leguminosea, solanaceae and chenopodiaceae (Horn *et al.*, 1993). BLRV as well as CpLV is transmitted by grafting and number of aphid vectors (*Aphis crassivora* and *Myzus persiceae*) in a persistent manner (Kaiser and Danesh, 1971; Brunt *et al.*, 1990), more detailed worked required to identify the key natural vectors of the luteovirus associated with stunt of chickpea. Horn *et al.* (1996) suggested that the spread and sources of infection are limited. They reported that incidence of stunt was greater in sparsely planted chickpea field where the crop was grown monoculture than mixed cropping systems. Nene *et al.* (1991) also reported that early sowing (September), and wide spacing favour stunt incidence.

Close spacing of sowing should be done when aphid vector actively is low. Most effective managing stunt is by growing resistant varieties.

Mosaic

This is a minor disease reported from Algeria, India, Iran, Morocco, New Zealand and USA (Nene *et al.*, 1991). Alfalfa mosaic virus is a minor disease but early infection of plants can cause total loss. Symptoms of the disease have been described by Nene *et al.* (1978). Chlorosis of the terminal bud and twisting are the first visible symptoms of the disease followed by necrosis and subsequent proliferation of secondary branches. New secondary branches are stiff and erect with smaller leaflet that shown mild mottle and very few pods are produced. Premature dying is common. Alfalfa mosaic virus (AMV) consists of bacilliform particles of different length. The virus is readily sap- transmissible.

Narrow Leaf

This is a minor disease reported from India, Iran and USA. The disease is caused by bean yellow mosaic virus and symptoms are characterized by yellowing and drying of plants with feathery and deformed leaves. Twisting of the terminal bud occurs 6-7 days after infection followed by the initiation of very narrow leaves from new buds. This finally results in long and terminal branches with very narrow or filiform leaves. The leaves below the proliferated branches turn yellow, show interveinal chlorosis and mosaic depending on genotype tested. The overall height of plant is reduced. Affected plants produced very few distorted flowers that developed into very small pods. The seeds from infected plants are black, small and shrivelled.

Proliferation

This is a minor disease of chickpea reported from Bulgaria, Colombia, India, Iran, Morocco, USA and USSR. It is caused by cucumber mosaic virus. Characteristic symptoms are bushy and stunted plants. Diseased plants produce few flowers and pod, and may die prematurely (Nene *et al.*, 1991).

Chlorotic Dwarf

The chickpea chlorotic dwarf virus (CpCDV, genus *Mastervirus,* family *Geminiviridae*) is commonly found in Pakistan, Iran and Sudan (Horn *et al.*, 1995;

Makkouk *et al.*, 1995, 2001). It has also been documented from India, Egypt, Iraq, Syria, and Yemen (Kumari *et al.*, 2006). CpCDV can cause stunting, internode shortening, phloem browning in the collar region and leaf reddening in desi-type while yellowing in kabuli-type chickpea varieties (Nene and Reddy 1987; Nene *et al.*, 1991; Horn *et al.*, 1993). CpCDV was found to be transmitted by a leafhopper *Orosius orientalis* (Matsumura) in India (Horn *et al.*, 1993) and by *O. albicinctus* (Distant) (Cicadellidae: Hemiptera) in Syria (Kumari *et al.*, 2004) and in Pakistan (Akhtar *et al.*, 2011). It nearly caused 100 per cent yield loss of individual plants when infection occurred before flowering and 75–90 per cent losses when infection occurred during flowering (Horn *et al.*, 1995).

Vigna spp. (*viz.* Mungbean (*Vigna radiata* (L.) Wilczek) and Blackgram (*Vigna mungo* (L.) Hepper) and Cowpea (*Vigna unguiculata* (L.) Walp.)

Yellow Mosaic

Yellow mosaic caused by mungbean yellow mosaic virus (MYMV) is the most serious limiting factor in mungbean and urdbean, soybean and cowpea cultivation in Indian subcontinent. This viral disease is found on several alternate and collateral hosts which act as a primary source of inoculums. Yield losses due to this disease vary from 5 to 100 percent depending upon disease severity, susceptibility of cultivars and population of whitefly (Nene 1972; Singh 1980; Rathi 2002). The infection not only drastically reduces yield but also severely impaired the grain size and quality. The affected plants showed typical symptoms of mild scattered yellow spots on young leaves. The next trifoliate leaf emerging from the growing apex showed irregular alternating yellow and green patches. The leaves showed slight puckering with reduction in size. The size of yellowing areas increased further resulting with complete yellow of apical leaves. Yellowing leads to less flowering and pod development. Early infection often leads to death of plants. Virus is belonging to the family *Geminiviridae* and genus *Begomovirus*. The paired particles of the causal virus measure 30 x 15 nm having ssDNA (Honda *et al.*, 1981). Four species *viz.*, *Mungbean yellow mosaic virus* (MYMV), *Mungbean yellow mosaic India virus* (MYMIV), *Dolichos yellow mosaic virus* (DYMV) and *Horsegram yellow mosaic virus* (HYMV) are known to cause yellow mosaic disease in different leguminous species. All these viruses are bipartite begomoviruses, have geminate (twin) particles, 18-20 nm in diameter, 30 nm long, apparently consisting of two incomplete icosahedra joined together in a structure with 22 pentameric capsomeres and 110 identical protein subunits (Qazi *et al.*, 2007). Estimation of actual losses due to yellow mosaic disease in farmers' field is difficult as these losses vary from year to year and from variety to variety. However, based on the incidence of yellow mosaic disease in mungbean, urdbean and soybean, an annual loss of over US $ 300 million is estimated in these crops (Varma *et al.*, 1992). Yellow mosaic disease occurs in a number of leguminous plants such as mungbean, urdbean, cowpea (Nariani, 1960; Nene, 1973), soybean (Suteri, 1974), horsegram (Muniyappa *et al.*, 1975), lablab bean (Capoor and Varma, 1948) and French bean (Singh, 1979).

MYMV disease is reported to be transmitted by an insect vector, *Bemisia tabaci* Genn and not by seed, soil and mechanical inoculation (Nair and Nene, 1973).

Transmission of the virus is discussed in detail by Singh *et al.* (1998). The effect of disease varies with cultivar to cultivar and is subjected to the genetic makeup of the cultivar.

Cultivation of resistant varieties, manipulation in sowing dates, inter/mixed cropping of mungbean and urdbean with non-host crops like sorghum, pearl millet and maize and application of systemic insecticides such as aldicarb, disyston and foliar application of metasystox has been found effective in controlling the disease by reducing vector control (Vishwa Dhar *et al.*, 2004). Diseased plant should be rouged out to prevent further spread of the disease. Resistant sources in mungbean are Narendra Mung1, Pant Mung 3, PDM 139 (Samrat), PDM-11 (Spring Season), ML 131, ML 267, ML 337, Pusa 105, MUM 2; and urdbean are Narendra Urd1, IPU 94-1 (Uttara), PS 1, Pant U 19, Pant U 30, UG218, WBU 108, KU 92-1 (Spring season), KU 300 (Spring Season)]. In order to prevent whitefly (Bemisia spp.) infestation spray with triazophos 40 EC @ 2.0 ml/l or malathion 50 EC @ 2.0 ml/l or oxydemeton methyl 25 EC @ 2.0 ml/l at 10-15 days intervals if required (Sharma *et al.*, 2011)

Leaf Crinkle

Leaf crinkle disease caused by urdbean leaf crinkle virus is a very serious disease of urdbean. The disease was first reported from India (Williams *et al.*, 1968). The disease causes stunting of plants and crinkling of leaves (Kolte and Nene, 1973). The crinkling is observed on some branches while others remain apparently healthy (Brar and Ratual, 1986). Leaf crinkle disease of urdbean is caused by urdbean leaf crinkle virus (ULCV) belonging to Tospovirus is a common, wide spread, destructive and economically important disease causing systemic infection in blackgram (*Vigna mungo* (L.) Hepper). Disease symptoms include crinkling, curling, and puckering of leaves often coupled with stunting and malformation of floral organs. Enlargement in size followed by crinkle surface of laminae are the characteristics symptoms on affected trifoliate leaves. Pollen production, fertility and subsequent pod formation is severely reduced with affect on seed weight and size of seeds in infected plants leading to decrease in yield (Kadian, 1980; Rishi, 1990; Kolte and Nene, 1973). The disease affects both the vegetative growth and yield components of urdbean plants (Beniwal and Chaubey, 1979; Kadian, 1982 and Kolte and Nene, 1973). The ULCV has been reported to be transmitted by several species of aphids such as *Aphis crassivora, Aphis myzus persicae* insects like *Acyrthosiphon pisum* and *Henosepilachna dodecastigma* Wied (Bardwaj and Dubey, 1986, Dhingra, 1976; Brar and Rataul, 1987; Dhingra and Chenula, 1981) and also by whitefly (*Bemisia tabaci*) (Narayanasamy and Jaganathan, 1973). Research on epidemiological aspects also explains that ULCV disease incidence depends upon the host genotypes, growing seasons and suitable environmental conditions (Ashfaq *et al.*, 2008).

Healthy seed should be used and treat the seeds with imidacloprid 70 WS@ 5ml/kg or give solar Seed treatment by soaking seed in water for 3-4 hours and then exposure to solar heat from 12 to 4 p.m. in May and June. Rogue out the infected plants to avoid contact between healthy and diseased plants during intercultural operations. Application of one foliar spray of insecticide (dimethoate 30 EC @ 1.7ml/ha) on 30 days after sowing and use resistant sources (in mungbean *viz.* D-3-9, K 12, ML 26, RI 59, T44 RII; and in urdbean: HUP 27, 102, 164, 315).

Mosaic Mottle

This disease is caused by Bean Common Mosaic Virus which belongs to potyvirus group. The mosaic mottle of urdbean and mungbean is common in India as well as south East Asian countries (Tsuchizaki *et al.*, 1986). It can be transmitted by sap, mechanically and by seed (Shahare and Raychaudhary, 1963; Nene, 1972). Singh and Nene (1978) also reported its transmission by aphids, *Aphis craccivora* and *A. gossypii*. Host range of the virus is confined to the family *Leguminosae* (Srivastava *et al.*, 1969). However, Urdbean is more susceptible than mungbean. The disease is characterized by a mosaic pattern of irregular broad patches of light and dark green areas and blistering and puckering of leaf blade. The size of the leaf gets reduced and margins show upward rolling. The leaves become rough and brittle. Affected plants show reduction in overall growth and often display excessive branching. In cases of severe infection the whole inflorescence is changed into leaf like structures, thereby causing 100 percent loss in seed yield (Nene, 1972).

Cowpea Aphid Borne Mosaic Disease (CABMD) and Cowpea Golden Mosaic Disease (CPGMD) of Cowpea

The disease is caused by cowpea aphid borne mosaic virus (CABMV) is cosmopolitan, economically significant seed-borne virus of cowpea (Bashir *et al.*, 2002). Virus belonging to the family *Potyviridae* and genus *Potyviru* is a distinctive virus with flexuous filamentous particles of 750 nm long. The virus-infected seed provides the initial inoculum and aphids are responsible for the secondary spread of the disease under field conditions. It can cause a yield loss of 13 - 87 per cent under field conditions depending upon crop susceptibility, virus strain and the environmental conditions (Bashir *et el.*, 1996). The virus symptoms vary with the cowpea genotype and virus strain. Diseased cowpea plants show variable amounts of dark green vein banding or interveinal chlorosis, leaf distortion, blistering and stunting. CPGMV belongs to genus of Begomovirus and is transmitted by whiteflies (*Bemicia spp.*) and it produces intense yellow leaves which after sometime become distorted and blistered.

Field Pea (*Pisum sativum*) and Lentil (*Lens cularis*)

Stunt Disease

Red clover vein mosaic virus (RCVMV) incites a stunt disease of pea (*Pisum sativum* L.) in India (Singh and Khan, 1993) and elsewhere (Smith, 1972). It causes yield reductions ranging from 87.8 to 100 per cent in moderately to severely diseased crops (Khan and Singh, 1997). RCVMV is a member of the carlovirus group which is transmitted in a stylet-borne manner by aphids, primarily the pea aphid (*Acyrthosiphon pisum* L). The virus overwintered mainly in perennial red clover fields and is spread when aphids migrate in the spring and early summer. When peas are harvested, aphids migrate back to clover fields completing the disease cycle (Hampton and Weber, 1983b; Hampton, 1984). Susceptible cultivars are often killed before they bloom if infected by Red clover vein mosaic virus early. Stunting and terminal resetting are symptoms of later infections. Vein chlorosis or necrosis may or may not occur. Low temperatures often have a masking effect on symptoms of pea stunt (Hagedorn and Walker, 1949; Hampton, 1984). The use of tolerant varieties is the best means of controlling this viral disease. Pea stem fly is also a serious pest in the area and has been controlled to certain extent using phosphamidon and dirnethoate (Singh *et al.*, 1988; Singh and Khan, 1997).

Pea Seed-Borne Mosaic

Pea seed-borne mosaic virus (PSbMV) is an economically damaging viral pathogen of field peas and lentils that can cause significant losses in seed yield and quality, especially when infections occur before or during bloom. Symptoms produced in peas include stunting, reduced internod lengths and malformation, and often results in the formation of malformed terminal rosettes. The virus can delay plant maturity, leading to uneven crop maturation. Infected leaves can exhibit clearing and swelling of veins, slight downward curling of leaf margins, chlorosis and / or a mottled or mosaic discoloration. Pods often are deformed, and seeds produced from infected plants can exhibit pronounced discoloration, splitting of seed coats, shrivelling and reduced size. Some varieties may be infected and show no visible symptoms (Stevenson and Hagedorn, 1970; Hampton, 1984). PSbMV is a member of the potyvirus group that is seed-borne and transmissible in a stylet-borne manner by aphids (Gonzales and Hagedorn, 1971). Infected seed has been the primary source of dissemination of this virus worldwide. In the field, the pea aphid (*Acyrthosiphonpisum* L.) is the main vector of the virus spread from infected to healthy plants, although other aphids are known to transmit PSbMV as well. The virus is not known to persist in infected plant hosts but survives from year to year in infected seed of pea, lentil and broad bean (Hampton 1984). Resistant varieties and use of virus free seed are the main control strategies for this virus. Control of aphid populations has not proven a reliable control method (Zimmer and Lamb, 1993).

The evaluation of seed lots for PSbMV is best done with nucleic-acid based or serological techniques. Bioassays in which seeds are planted in the greenhouse and levels of seed-borne PSbMV are determined by assessing plants for visual symptoms of the virus can result in false-negative results. Plants infected with PSbMV sometimes do not develop symptoms or may develop only mild, transient symptoms that are difficult to identify. However, when maximum accuracy is needed, a combination of both approaches should be used, with all plants in a bioassay tested for PSbMV using laboratory-based techniques. PSbMV is seed-borne and seed-transmitted, and it is introduced into new regions through the movement of virus-infected seed. Seed transmission rates as high as 100 percent have been reported for PSbMV in field peas, with transmission rates as high as 30 per cent commonly observed. Seed transmission rates as high as 44 percent have been reported for PSbMV in lentils, although the efficiency of seed transmission of PSbMV in lentils may differ by the strain of the virus. Secondary spread of PSbMV is facilitated by aphids. Aphids transmit the virus from diseased to healthy plants within fields, and they can spread the virus to neighboring fields. When aphid populations are high, even low levels of infected seed can result in severe PSbMV epidemics. More than 20 aphid species, including the pea aphid (*Acyrothosiphon pisum*), are known to transmit PSbMV. Aphids transmit the virus in a non-persistent manner, obtaining and transmitting the virus in short feeding periods.

The use of virus-free seed is the most effective strategy for managing PSbMV, and it is the best method for controlling PSbMV when resistant cultivars are not available. Only seed testing negative for PSbMV in laboratory tests should be used. PSbMV-infected plants do not always express symptoms of PSbMV or may express the symptoms only transiently, which means that diagnosing the disease accurately on the basis of visual symptoms can be difficult. Where PSbMV is detected, seed should not be saved. Infected seed is often symptomless, and infected seed cannot be fully eliminated from seed lots on the basis of appearance or size. Resistance to PSbMV has been identified in lentils and field peas, and resistance has been incorporated into some commercial cultivars of peas developed in other regions. Insecticides can help reduce the secondary spread of PSbMV by reducing populations of aphids that transmit the virus, but insecticides provide only partial control. Insecticides may provide unsatisfactory results, especially when PSbMV-infected seed is planted. Non-persistently transmitted viruses such as PSbMV are carried transiently by aphids, and killing infected aphids with insecticides may result in only a modest reduction in transmission events. Systemic insecticides, which primarily are transported in the sap-conducting (phloem) cells of plants, often provide inconsistent control of non-persistently transmitted viruses such as PSbMV, which are acquired from and transmitted to epidermal cells on the leaf surface, not phloem cells. Insecticides are most likely to be successful when applied to prevent economic aphid populations from being established, especially where seed-borne transmission of PSbMV has occurred in one field and suppression of secondary spread to neighboring fields is desired. In field peas, an insecticide application would be justified if aphids reach the economic threshold level of more than an average of one to two aphids per plant or 10 aphids per sweep. In lentils, insecticide treatment for pea aphid control should be considered (1) when the economic threshold of an average of 30 to 40 aphids per sweep is observed, (2) when few natural enemies are present and (3) when aphid numbers do not decline during a two-day period. If the economic threshold is exceeded, a single application of insecticide usually will be sufficient. Control at the early pod stage provides protection through the pod formation and elongation stages, which are very sensitive to aphid damage (Wunsch *et al.*, 2014).

Pea Streak

Sudden death of pea plants results from an early infection by *Pea Streak Virus* (PSV). Brown or purple stem and leaf spots or streaks are characteristic of a late infection. Only pod symptoms may occur if infection occurred during the early bloom stage. Pods develop necrotic spots or sunken pock marks (Hampton, 1984). PSV is a member of the carlavirus group that is transmitted in a stylet-borne rnanner by aphids, mainly the pea aphid (*Acyrthosiphon pimm* L.). Overwintering occurs primarily in alfalfa, and PSV is transmitted when pea aphids migrate to pea crops in the spring and summer. Persistent of this virus in pea aphids is quite short and is often dissipated after only four or five feedings. This disease is most severe when pea fields are planted next to alfalfa fields (Hampton and Weber 1983a; Hampton 1984). Use of tolerant varieties will help to keep this disease under control. Control of aphids in pea fields and nearby alfalfa fields will reduce the incidence of this virus but not completely eliminate it. Planting pea fields away from alfalfa fields will also help to reduce the incidence of PSV.

Pea Enation Mosaic

If infection from pea enation mosaic virus (PEMV) occurs early, plant distortion and death may occur. Symptoms of later infections include chlorotic flecks, stunting, upward rolling of leaves and pod distortion. Foliage may also develop elongate, chlorotic, translucent lesions. Seed size and quality are often reduced. Tissue proliferation may occur on leaf veins and on pods especially on some of the older susceptible varieties (McWhorter and Cook 1958; Hampton, 1984). PEMV overwinters in leguminous crops or weeds. Alfalfa (*Medicago sativa*) was until recently considered to be the primary reservoir of PEMV, but now has been shown to be a non-host of this virus (Larsen *et al.*, 1996). Aphids, primarily the pea aphid (*Acythosiphon pisum* L.), transmit the virus in a persistent manner (feeding on infected plants, incubation for 8- 12 hours and transmission on subsequent feedings) to host crops. Transmissibility by the aphids is usuaily Iost after several feeding episodes. The virus can also be transmitted by mechanical means (Hagedorn *et al.*, 1964). *Cicer, Lathyrus, Lens, Lupinus. Melilotus, Phaseolus, Pisum, Trifolium* and *Vicia* are known leguminous hosts of PEMV (Hampton, 1984). The primary method of control of PEMV is the use of tolerant varieties.

References

Akhtar, K.P., Ahmad, M., Shah, T.M., Atta, B.M. 2011. Transmission of *Chickpea chlorotic dwarf virus* in chickpea by the leafhopper *Orosius albicinctus* (Distant) in Pakistan. *Plant Protect. Sci.*, **47**: 1-4.

Ali, M. and Gupta, S. 2012. Carrying capacity of Indian agriculture: pulse crops. *Current Science* **102 (6)**: 874-881.

Ashfaq, M., Khan, M. A. and Javed. N. 2008. Characterization of environmental factors conducive for Urdbean leaf crinkle (ULCV) disease development. *Pakistan Journal of Botany* **40(6)**: 2645-2653.

Bashir, M. and Hampton, R.O. 1996. Detection and identification of seedborne viruses from cowpea (*Vigna unquiculata* (L.) walp) germplasm. *Plant Pathol.*, **45**: 54-58.

Bashir, M., Mughal, S. M. and Malik, B. A. 1991. Assessment of yield losses due to leaf crinkle virus in urdbean (*Vigna mungo* (L) Hepper). *Pakistan Journal of Botany* **23**: 140-142.

Beniwal, S. P. S. and Chaubey, S. N. 1979. Urdbean leaf crinkle virus: Effect on yield contributing factors, total yield and seed characters of urdbean (*Vigna mungo*). *Seed Res.*, **7**: 125-181.

Bhardwaj, S.V. and Dubey, G. S. (1986). Studies on the relationship of urdbean leaf crinkle virus and its vectors *Aphis crassivora* and *Acerthosiphon pisum. J. Phytopathol.*, **15**: 83-88.

Bos, L., Hampton, R. O. and Makkouk, K. M. 1988. Viruses and virus diseases of pea, lentil, faba bean and chickpea. p. 591-615. In: *World Crops: Cool Season Food Legumes* (eds. Summerfield, R.J.). Kluwer Academic Publishers, Dodrecht, the Netherlands.

Brar, J.S. and Rataul, H.S. 1986. Some field characteristics of leaf crinkle virus of urdbean, *Vigna mungo* (L) Happer. *Indian J. Virology* **2**: 49-56.

Brunt, A., Crabtree, K. and Gibbs, A. 1990. In: *Viruses of Tropical Plants*. CAB International, Wallingford. UK, p. 707.

Capoor, S.P. and Varma, P.M. 1948. A new virus disease of *Dolichos lablab*. *Current Science* **19**: 248-249.

Dhingra, K.L. 1976. Transmission of urdbean leaf crinkle virus by two aphid species. *Indian Phytopath*. **28(1)**: 80-82.

Dhingra, K.L. and Chenula, V.V. 1981. Studies on the transmission of urdbean leaf crinkle and chickpea leaf reduction viruses by *Aphis crassivora* Koch. *Indian Phytopath*., **34**: 38-42.

Gonzaies, L.C. and Hagedorn, D.J. 1971. The transmission of pea seed-borne mosaic virus by pea seed-borne mosaic virus. *Phytopathology*, **60**: 148-149.

Hagedorn, D.J., Layne, R.E.C. and Ruppel, E.G. 1964. Host range of pea enation mosaic virus and use of *Chenopodium alba* as a local lesion host. *Phytopathology* **54**: 843-848.

Hampton, R.O. and Weber, K.A. 1983. Pea streak virus transmission from alfalfa to peas: virus aphid and virus-host relationships. *Plant Disease* **67**: 305-307.

Hampton, R.O. 1984. Diseases caused by viruses. In: Hagedom, D.J. (ed.), Compendium of Pea Diseases. *The American Phytopaihological Society*, p. 31-37.

Horn, N.M., Reddy, S.V. and Reddy, D.V.R. 1995. Assessment of yield losses caused by chickpea chlorotic dwarf geminivirus in chickpea (*Cicer arietinum*) in India. *European Journal of Plant Pathology* **101**: 221-224.

Horn, N.M., Reddy, S.V., Roberts, I.M. and Reddy, D.V.R. 1993. Chickpea chlorotic dwarf virus: a new leafhopper-transmitted Gemini virus of chickpea in India. *Annals of Applied Biology* **122**: 467-479.

Horn, N.M., Reddy, S.V., Van den Heuvel, J.F.J.M. and Reddy, D.V.R. 1996. Survey of chickpea (*Cicer arietinum* L.) for chickpea stunt disease and associated viruses in India and Pakistan. *Pl. Disease* **80**: 411-414.

Hull, R. 1981. Pea enation mosaic virus. In: E. Kurstak, editor, Handbook of plant virus infections and comparative diagnosis. Elsevier/North-Holland Biomedical Press, Amsterdam. p. 239–256.

Kadian, O.P. 1980. *Studies on leaf crinkle disease of urdbean* (*Vigna mungo* (L.) Hepper), mungbean (*V. radiata* (L.) Wilczek) and its control. Ph.D. Thesis, Dept. Plant Pathology, Haryana Agric. Univ., Hisar. India. p. 177.

Kadian, O.P. 1982. Yield loss in mungobean and urdbean due to leaf crinkle disease. *Indian Phytopath*., **35**: 642-644.

Kaiser, W.J. and Danesh, D. 1971. Biology of four viruses affecting *Cicer arietinum* in Iran. *Phytopathology* **61**: 372-375.

Kannaiyan, J., Nene, Y.L., Reddy, M. ., Ryan, J.G. and Raju, T.N. 1984. Prevalence of pigeonpea diseases and associated crop losses in Asia, Africa and the Americas. *Trop. Pest Manag.*, **30**: 62-71.

Khan, A.T. and Singh, R.N. 1997. Effect of the pea stunt disease on flowering podding, grain stetting and yield in field pea. *Indian Phytopath.* **50**: 112-114.

Kolte, S.J. and Nene, Y.L. 1973. Studies on the symptoms and mode of transmission of leaf crinkle virus of urdbean (*Phaseolus mungo*). *Indian Phytopath.*, **25**: 401-404.

Kraft, J.M. and Hampton, R.O. 1980. Crop losses from pea seed-borne mosaic virus in six processing pea cultivars. *Plant Dis.*, **64**: 922-924.

Kulkarni, N.K., Kumar, P.L., Muniyappa, V., Jones, A.T. and Reddy, D.V.R. 2002. Transmission of Pigeon pea sterility mosaic virus by the eriophyid mite, *Aceria cajani* (Acari: Arthropoda). *Plant Dis.*, **86**: 1297-1302.

Kumari, S.G., Makkouk, K.M. and Attar, N. 2006. An improved antiserum for sensitive serologic detection of chickpea chlorotic dwarf virus. *Journal of Phytopathology* **154**: 129-133.

Kumari, S.G., Makkouk, K.M., Attar, N., Ghulam, W. and Lesemann, D.E. 2004. First report of *Chickpea chlorotic dwarf virus* infecting spring chickpea in Syria.

Larsen, R.C., Kaiser, J.W. and Klein, R.E. 1996. Alfalfa, a non-host of pea enation mosaic virus in Washington State. *Canadian Journal of Plant Science* **76**: 521-524.

Makkouk, K.M., Bashir, M., Jones, R.A.C., Kumari, S.G. 2001. Survey of viruses in lentil and chickpea crops in Pakistan. *Journal of Plant Disease and Protection* **108**: 258–268.

Makkouk, K.M., Dafalla, G., Hussein, M. and Kumari S.G. 1995. The natural occurrence of chickpea chlorotic dwarf geminivirus in chickpea and faba bean in the Sudan. *Journal of Phytopathology* **143**: 465-466.

Mali, V.R. 1996. Important virus diseases of cowpea in India. In: Disease scenario on crop plants (Eds.) Agnihotri, V. P., Prakash, O., Kishun, R. and Mishra, A. K. Internationla Books and Periodicals supply services, Pitampura, Delhi, India p. 95-112.

McWhorter, F.P. and Cook, W.C. 1958. The hosts and strains of pea enation mosaic virus. *Plant Disease Reporter* 42: 51-60.

Mitra, M. 1931. Report of the Imperial Mycologist. *Scientific Rep. Agri. Res. Inst. Pusa,* **19**: 58-71.

Muniyappa,V., Reddy, H.R. and Shivshankar, C. 1970. Yellow mosaic disease of *Dolichos biflorus* Linn. *Current Science* **4**: 176.

Nair, N.G. and Nene, Y.L. 1973. Studies on the yellow mosaic of urdbean (*Phaseolus mungo* L.) caused by mungbean yellow mosaic virus. 2. virus-vector relationships. *Indian J. Farm Science* **1**: 62-70.

Narayanasamy, P. and Jaganathan, T. 1973. Vector transmission of black gram leaf crinkle virus. *Madras Agri. J.*, **60**: 651-652.

Nariani, T.K., 1960. Yellow mosaic of mung (*Phaseolus aureus* L.). *Indian Phytopathology* 13: 24-29.

Nene, Y. L. and Reddy, M.V. 1987. Chickpea disease and their control. In: Saxena,M.C. and Singh, K.B. (eds), The Chickpea. III Viral Diseases, p. 233-270. CAB International, Wallingford.

Nene, Y.L. 1973. Viral diseases of warm weather pulse crops in India. *Plant Dis. Rep.* **57**: 463-467.

Nene, Y.L. 1972. A survey of viral diseases of pulse crops in Uttar Pradesh. *Final Tech. Report. Res. Bull.* No. 4, U.P. Agricultural University, Pantnagar, India, p. 1-91.

Nene, Y.L., Haware, M.P. and Reddy, M.V. 1978. Chickpea diseases: resistance screening techniques. *ICRISAT Information Bulletin*, **3**: 44.

Nene, Y.L., Reddy, M.V., Haware, M.P., Ghanekar, A.M. and Amin, K.S. 1991. Field diagnosis of chickpea diseases and their control. *ICRISAT Information Bulletin*, **28**: 44.

Pande, S., Sharma, M., Kumari, S., Gaur, P.M., Chen, W., Kaur, L., MacLeod, W., Basandrai, A., Basandrai, D., Bakr, A., Sandhu, J.S., Tripathi, H.S. and Gowda, C.L.L. 1999. Integrated foliar diseases management of legumes. In: *International Conference on Grain Legumes: Quality Improvement, Value Addition and Trade*, February 14-16, 2009, Indian Society of Pulses Research and Development, Indian Institute of Pulses Research, Kanpur, India.

Qazi, J., Ilyas, M., Mansoor, S. and Briddon, R.W. 2007. Legume yellow mosaic viruses: genetically isolated begomoviruses. *Mol. Plant Pathology*, **8**: 343-348.

Rathi, Y.P.S. 2002. Epidemiology, yield losses and management of major diseases of *Kharif* pulses in India. In: *PlantPathology and Asian Congress of Mycology and Plant Pathology*, Oct.-1-4, 2002. University of Mysore, Mysore, India.

Reddy, M.V., Kannaiyan, J. and Nene, Y.L. 1984. Increased susceptibility of sterility mosaic infected pigeon pea to powdery mildew. *Int. J. Trop. Plant Dis.* **2**: 35-40.

Rishi, N. 1990. Seed and crop improvement of northern Indian pulses (Pisum and Vigna) throughcontrol of seed- borne mosaic viruses. Final Technical Report, (US-India Fund) Dept. Plant Pathology, CCS Haryana Agric. Univ. Hisar, India, p. 122.

Saxsena, D.R., Moly, Das, S.B., Kataria, V.P. and Bhalla, P.L. 1991. Occurance of chickpea stunt in west Nimar Valley, Madhya Pradesh, India. *Int. Chickpea Newsletter* **24**: 36.

Sharma, O.P., Bambawale, O.M., Gopali, J.B., Bhagat, S., Yelshetty, S., Singh, S.K., Anand, R. and Singh, O.P. 2011. Field guide mungbean and irdbean. NCIPM, LBS Building, IARI Campus, New Delhi, India.

Sharma, O.P., Gopali, J.B., Yelshetty, S., Bambawale, O.M., Garg, D.K. and Bhosle, B.B. 2010. Pests of Pigeonpea and their Management, NCIPM, LBS Building, IARI Campus, New Delhi, India.

Singh, R.N. 1979. Natural infection of bean (*Phaseolus vulgaris*) by mungbean yellow mosaic virus. *Ind J Mycology Plant Pathology* **9**: 124-126.

Singh, H.M., Rizvi, S.M.A. and Singh, R. 1988. Insecticidal control of pea stem fly (*Ophiomyia phaseoli* Tyron). *Indian* J. *Plant Proto* **16**: 41-43.

Singh, J.P. 1980. Effect of virus diseases on growth component and yield of mungbean and Urdbean. *Indian Phytopathology* **8**: 405-08.

Singh, R.A., Gurha, S.N. and Ghosh A. 1998. Diseases of mungbean and urdbean and their management, p. 179-204. In: *Diseases of Field Crops and their Management*. (ed. Thind T.S.), National Agricultural Technology Information Centre, Ludhiana, India.

Singh, R.N. and Khan, A.T. 1993. Association of red clove in mosaic virus wilt a stunt disease of peas (*pisum sativum* L.) in India. *Indian J. Plant Pathol.* **11**: 16-19.

Singh, R.N. and Khan, A.T. 1997. Genotypic and insecticidal management of the stunt disease of field pea. *Indian Phytopath.* **50(3)**: 412-415.

Stevenson, W.R. and Hagedorn, D.J. 1970. Effect of seed size and condition on transmission of three aphid species. *Phytopathology* **61**: 825-828.

Suteri, B.D. 1974. Occurrence of soybean yellow mosaic virus in Uttar Pradesh. *Curr Sci.* **43**: 689-690.

William, F.J., Grewal, J.S. and Amin, K.S. 1968. Serious and new diseases of pulses crop in India in 1966. *Plant Dis. Rept.* **52**: 300-304.

Wounsch, M., Pasche, J., Knodel, J., McPhee, K., Markell, S., Chapara, V., and Pederson, S. 2014. Pea Seed-borne Mosaic Virus (PSbMV) in Field Peas and Lentils, p. 4. In: Plant Disease Management, NDSU Extension services, p. 1704.

Zimmer, R.C. and Lamb, R.J. 1993. Amplification and spread of pea seed-borne mosaic virus in field-grown peas. *Canadian Journat of PIant Pathology* **15**: 17-22.

2016, Diseases of Pulse Crops and their Sustainable Management 429–442
Editors: **Samir Kumar Biswas, Santosh Kumar and Gireesh Chand**
Published by: **BIOTECH BOOKS, NEW DELHI**

Chapter 23

Current Scenario of Web Blight of Urdbean and Mungbean and their Management

***Santosh Kumar*[1]**, *Gireesh Chand*[1]*, *J.N. Shrivastava*[1]*, *Mehi Lal*[2] *and Vivek Singh*[3]**

[1]*Department of Plant Pathology, Bihar Agricultural University, Sabour, Bhagalpur – 813 210, Bihar*
[2]*Plant Protection Section, Central Potato Research Institute Campus, Modipuram, Meerut – 250 110, U.P.*
[3]*Maniyver Kashi Ram University of Agriculture and Technology, Banda, U.P.*

Introduction

Urdbean (*Vigna mungo* L.) Hopper and mungbean (*Vigna radiata* L.) Wilczek are valuable pulse crops of India. Urdbean grain contains about 24 per cent protein, 60 per cent carbohydrate, 1-3 per cent fat and is the richest one in phosphoric acid contents among all pulses while, mungbean contains about 25 per cent protein, 60 per cent carbohydrate, 1-3 per cent fat. Both crops are used as medicine externally and internally particularly in paralysis, rheumatism and infection of nervous system (Singh, 1982). Besides these, Urdbean and mungbean crops are also fixes around 38 kg/ha atmospheric nitrogen in the soil. These crops can be grown through out the

* Corresponding Author: E-mail: santosh35433@gmail.com

year, however, maximum area is falls in *Kharif* season where these crops are intercropped with sorghum, pearl-millet, maize, cotton, rice, caster and pigeonpea etc.

Despite being important pulse crop its productivity is quite low due to various biotic and abiotic factors. Web blight belonging to biotic factor is considered as one of the most important, causes huge amount of losses in both in production as well as productivity in India (Dubey and Patel, 2001). The disease is causes 20-30 per cent losses in India (Shailbala and Tripathi, 2007). Web blight disease of urdbean and mungbean is caused by *Rhizoctonia solani* Kuhn, which perfect stage is *Thanatephorus cucumeris* (Frank) Donk. It causes considerable damage by reducing seed quality and yield. The disease has been reported from Punjab, Haryana, Bihar, Rajasthan, Uttarakhand, Uttar Pradesh, West Bengal, Himanchal Pradesh and Jammu and Kashmir of India (Saksena and Dwivedi, 1973).

Historical Background/Geographical Distribution

The genus *Rhizoctonia* was described by De Candolle (1815) as non-sporulating root pathogen *R. crocorum* D.C. ex Fr. O. Julius Kuhn (1885) observed the pathogen on diseased potato tubers and named it as *Rhizoctonia solani* Kuhn. Duggar (1915) reviewed the earlier literature on *R. solani* and described in detail the morphology and growth habit of the fungus and stated that young hyphal branches are inclined in the direction of growth are invariably somewhat constricted at the point of union with main hyphae. Taxonomy and nomenclature of perfect stage, *Thanatephorus cucumeris* (Frank) Donk of *R. solani* Kuhn is described by Talboto (1965). The genus *Thanatephorus* was proposed by Donk (1956) with *Hypochunus solani* Prills and Delaor *T. cucumeris* (Frank) Donk as type species. The first report on occurrence of *Thanatephorus cucumeris* (Frank) Donk, (*R. solani*) Kuhn on urdbean was reported by Saksena and Dwivedi (1973) and mungbean (Dwivedi and Saksena, 1975) from India. The disease has been reported on common bean (*Phaseolus vulgaris*) from Trinidad, West Indies, Puerto Rico, Jamaica, North America, South America, Argentina, Brazil, Columbia, Central America, Mexico, while in Asia, it is reported from Japan, Philippines, Myanmar, Sri Lanka, India and Pakistan (Webber, 1939; Seksena and Dwivedi, 1973; Alam *et al.*, 1995).The disease has been known to occur in India on other leguminous crop like pigeonpea (Dwivedi and Saksena, 1975) and groundnut (Dwivedi and Dubey, 1986).

Pathogen Biology

Genus *Rhizoctonia* has gained the reputation of being a wide spread, destructive and versatile plant pathogen. R*hizoctonia* belongs to a group of fungi called the "Mycelia Sterilia." These fungi do not produce asexual spores, but grow by producing thin, vegetative strands called hyphae; large masses of hyphae are referred to as mycelium. The hyphae of *Rhizoctonia solani* have the following characteristics: a) Each cell is multinucleate (many nuclei) rather than binucleate; b) Branching at right and acute angles near the distal septum of cells in young vegetative hyphae; c) Formation of a septum in the branch near the point of origin; d) A special type of cross wall within the hyphae, called a dolipore septum.

In general, the growth rate of *R. solani* is very rapid and a typical isolate can covers a 90 mm petri plate in three days. The fungi also produces specialized hyphae composed of compact cells called monilioid cell. The monilioid cells fuse together to produce hard structures called sclerotia, which are black to brown and 3 to 5 mm long and help the fungus to survive under adverse conditions. The hyphae of *R. solani* have many nuclei (commonly 4 to 8) per cell which differentiate from similar group to fungi that have only 2 nuclei per cell. Those fungi with hyphal characteristics similar to *R. solani,* but with only 2 nuclei per cell are called binucleate types and are generally non-pathogenic.

Invasion of host by *R. solani,* sexual spores are formed on specialized structures called basidia. Four spores are produced on each basidium. Basidia are formed when the environment is moist and sufficient growth of the fungus has occurred. Basidiospores are wind-dispersed and germinate in the presence of moisture. Each basidiospore has a single nucleus. The hyphae produced by germinating spores will fuse (anastamose), forming new hyphae with a mixture of different types of nuclei.

Recently anastomosis groups has helped to elucidate the complexity of this large fungal genus without asexual spores with few morphological characteristics to define "*R. solani* isolates", many fungi were placed into this species. However, it is now recognized that *R. solani* has multinucleate cells and other related fungi (2 nuclei per cell), the binucleates, do not belong to that species. Further, when isolates are paired, only related mycelium will fuse and thus anastomosis groups have been defined by a set of tester isolates.

Sclerotia

The sclerotia of *R. solani* are brown to black composed of clusters of melanin encrusted, thick walled cells and rich in nutrients, irregular-shaped structures in soil and on plant tissue formed by repeated branching from short, thick, lateral hyphae. Sclerotia are 4-5 mm in diameter, spherical but flattened when pressed between leaf sheath and culm. Sclerotia and / or mycelium present in soil and / or on plant tissue germinate to produce vegetative hyphae of the fungus that can infect a wide range of food and fiber crops.

Variant of the Pathogen

Ogoshi (1978) divided *R. solani* into 13 ISGs that differ in anastomosis behaviour, cultural appearance and pathogenicity. They are AG-1 1A, AG-1 1B, AG-1 1C, AG-2-1, AG-2-2 IIIB, AG-2-2 IV, AG-3, AG-4, AG-5, AG-6, AG-7, AG-8 and AG-BI. The AG system now considered as the most useful grouping system for *R. solani,* however, the behavior of the fungus cannot be completely understood solely in terms of AGs. For example isolates of AG-1 cause not only aerial / web blight but also sheath blight of rice or leaf blight on many hosts Kuninaga (1986).

Identification

R. solani does not produce spores; hence it is identified only from mycelial characteristics. Its hyphal cells are multinucleate. Also, it produces white to deep brown mycelium when grown on artificial medium. The hyphae are 4-15 μm wide and tend to branch at right angles. A septum near each hyphal branch and a slight constriction at the branch are diagnostic. *R. solani* is sub-divided into anastomosis groups (AG) based on hyphal fusion between compatible strains (Wiese, 1987). The teleomorph of *R. solani* is *Thanatephorus cucumeris*. It forms club-shaped basidia with four apical sterigmata on which oval, hyaline basidiospores are borne. Since, microscopic technique is able to identified *Rhizoctonia* up to the genus level only and it takes at least one week for isolation of fungus. After introduction of PCR machine, a revolution has been occurred in the era of plant diagnostics. The most common region used for these purposes has been the internal transcribed spacer (ITS) region of ribosomal RNA genes. The rDNA region consists of multiple copies (up to 200 copies per haploid genome) arranged in tandem repeats comprising the 18S small subunit, the 5.8S,and the 28S large subunit genes separated by internal transcribed spacer regions (ITS1 and ITS2) (Liew *et al.*, 1998).

Symptoms

The symptoms of web blight on urdbean and mungbean occur on roots, stems, petioles and pods but the disease is most destructive on foliage. During early growth stage, pathogen causes seedling mortality. In severe infection, reddish brown lesion occurs and cause collar rot. The first symptoms appear as small circular brown spot on the leaves. These spot enlarge often show concentric banding and become surrounded by irregular shaped water soaked areas. The lesions expend and coalease and white fungal growth is observed on lower surface of the infected leaves including petioles and young branches of infected plants. The spot become greenish to reddish brown and later turn brown to black. The mycelium on infected leaves appear as spider web thus suggested the name web blight disease (Figures 23.1 and 23.2). If infection occurs on collar region, reddish brown lesion occurs and causes collar rot. It soon girdles the basal portion of stem and this point seedling wilt and collapses and lies flat on the ground.

The pods are also vulnerable to infection at all stages of their development. In initial stage, light irregular specks appeared on the young pods, the spots were dark brown, more or less circular, sunken with darker margin in mature pods. Gradually the spots increased in number and size to cover the entire pods. The grains in the infected pods become small, shrivled and light in colour (Figures 23.3 and 23.4).

Epidemiology of Disease

The pathogen is known to prefer warm wet weather and outbreaks typically occur in the early summer months. A combination of environmental factors has been linked to the prevalence of the pathogen such as: presence of host plant, frequent rainfall/irrigation and increased temperatures in spring and summer. In addition, reductions of drainage of the soil due to various techniques such as soil compaction are also known to create favourable environments for the pathogen. The pathogen is dispersed as sclerotia and these sclerotia can travel by means of wind, water or soil movement between host plants. Dubey (1997) reported that 26-28° C temperature and 90-100 per cent relative humidity (RH) favoured maximum disease development.

Figure 23.1: Typical Symptoms of Web Blight on Leave of Urdbean .

Figure 23.2: Web Blight Symptoms on Mungbean Leaves.

Figure 23.3: infected Pods of Urdbean.

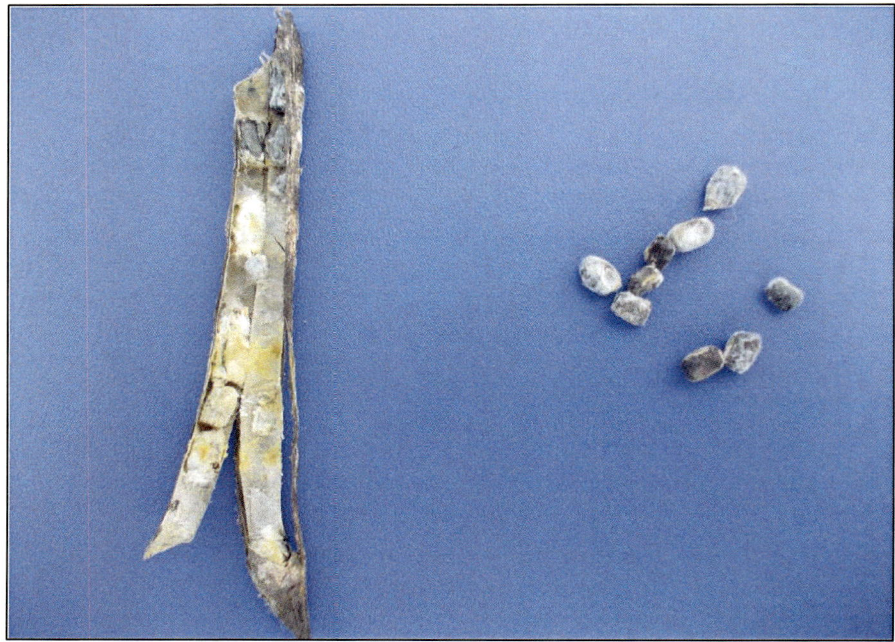

Figure 23.4: Grains of Infected Pods.

Survival

R. solani Kuhn is a soil- borne pathogen. The pathogen is mainly present in/on soil and survives for a long period in the absence of a host either as sclerotia or thick walled brown hyphae in plants debris (Boosalis and Scnaren, 1959). Most of sclerotia occurred on the tap root of diseased plants and in the soil adjacent to the diseased root. Over 80 per cent sclerotia occurred in the top 10 cm of the soil and within 10 cm from the diseased roots. Mycoparasitic and other antagonistic effect of soil microorganism on sclerotial germination and survival of *R. solani* in the soil have been reported by Naiki and Ui (1981). *R. solani* can survive in soil as sclerotia in association with crop residue by pathogenic growth on hosts or by saprophytic growth on fallen dry leaves and organic matter. Primary inoculum sources consist of sclerotia and hyphae or when the teleomorph *T. cucumeris* is present (Baker and Martison, 1970). The pathogen can be disseminated by irrigation water, infected/ infested or contaminated seeds, transplanted material, air currents and rain splash, by soil through farm equipments. In the tropics, where the teleomorph of *R. solani* develops regularly, the pathogen spreads rapidly by the production and dissemination of basidiospores.

Disease Cycle

R. solani is primarily soil-borne and initial inoculums come mainly from soil splashes on to the leaves during heavy rains. The pathogen is capable of extensive growth through soil and survives on crop debris. It can persist in the different soils for many years. The highest inoculums potential is noted in the top 10 cm. The fungus can grow in the soil very fast to a longer distance and can survive there in mycelial or sclerotial form. Once the floral parts are infected, the disease became seed-borne which can lead to spread of the pathogen to uninfected areas (Figure 23.5). The collateral hosts play an important role in initiation and early spread of the disease to the main host. Because of their proximity to the soil carrying primary inoculums and their special micro climate, the weed hosts are first to take the infection and facilitate the production of basidiospores of the pathogen. Thereafter, the pathogen becomes air-borne and cause leaf, stem or pod infection and produces fresh cycle of basidiospores on the main host. The spore development is favoured by night temperature below 24°C and relative humidity above 95 per cent. The basidiospores germinate and penetrate in leaf through the intact surface by formation of infection cushion or by direct penetration through stomata opening. They lose their viability rapid in the storage and after expose to direct sunlight.

Host Range

Scanty information is available in respect of the host range of *R. solani* infecting urdbean and mungbean crops. *R. solani* can infect about 32 plant families. The strain of *R. solani* from weed host has been reported to be pathogenic to several crop plants (Bains and Dhaliwal, 1993). Sharma and Tripathi (2001) have also reported host range of urdbean isolates of *R. solani* and found the wide host range belongs to family *Leguminoceae, Solanaceae, Brassicaceae, Malvaceae and Cucurbitaceae*. In mungbean field, several weeds infected with *R. solani* transmitted the disease in the mungbean (Bain and Dhaliwal, 1993).

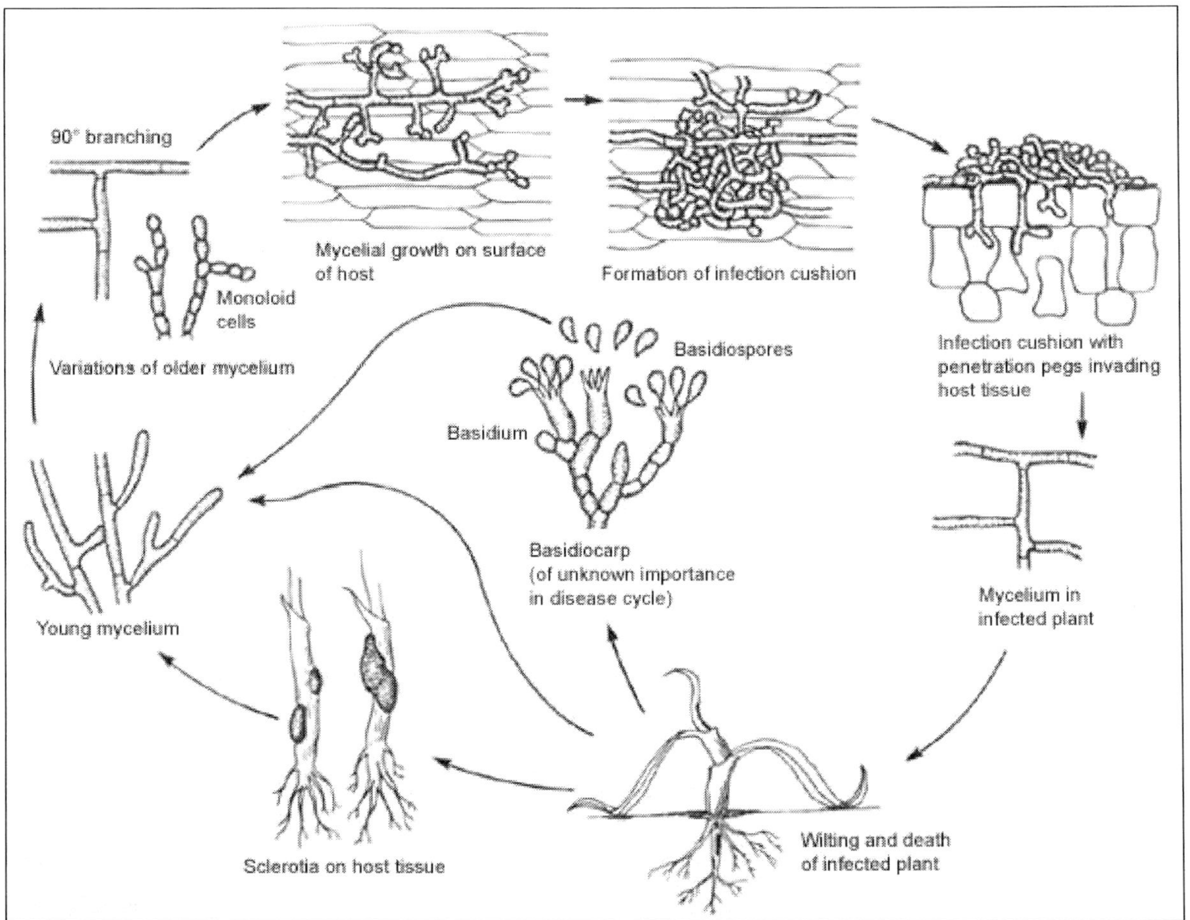

Figure 23.5: Disease Cycle of *Rhizoctonia solani.*

Disease Rating Scale

The web blight disease severity is recorded using 1-9 rating scale (Stonehouse, 1994). The detail of rating scale is as follows:

Grade Scale	*Description*
1	1 per cent No lesion/spot on leaves (highly resistance)
3	1-25 per cent area covered by lesion/spot (moderately resistance)
5	25.1-50 per cent area covered by lesion/spot (moderately susceptible)
7	50.1-75 per cent area covered by lesion/spot (susceptible)
9	75.1-100 per cent area covered by lesion/spot (highly susceptible)

Management

Cultural Practices

Effect of cultural practices in minimizing the web blight severity in urdbean and mungbean is lacking. However, information is available on other crops. Cultural practices mostly affected the environmental condition favourable for the growth; build up of inoculums and their spread. Early sowing, high seed rates, closer row spacing, and traditional use of bushy genotypes result in dense canopies, which favour development of web blight. Thus, techniques which minimize canopy density may alleviate disease intensity.

Manipulation of sowing date may be used to overcome the losses due to web blight, but its efficacy would differ from place to place. A wider row spacing 50 cm or more resulted in decreased aerial blight infection and increased yield of soybean (Joye, 1990). Certain practices *viz.*, proper sanitation, avoid planting at the time of rainy season during the susceptible crop stage, avoid the thick canopy of the crop by using proper seed rate and adopted proper rotation help in controlling the disease to a greater extent. Burying the infected leaves immediately after harvest will reduce the primary inoculums.

Host Resistance

Development of web blight resistant varieties, there is need of screening of existing lines / germplasms / cultivars against web blight of urdbean and mungbean. Although, varietals screening has been reported in the other beans (Schroth and Cook, 1964). Two genotypes HPBU51and P38 were resistant against web blight while HPBU38, HPBU153, LBG626 and UG367 were resistant to mungbean yellow mosaic virus and web blight disease. Genotype P44 was resistant against anthracnose and web blight. Pusa-3 and UPU91-7 resistant to MYMV, web blight and anthracnose (Basandrai *et al.* (1999).

Biological Control

There is growing concern in the entire world, to reduce the amount of chemicals being released in the environment and their persistence and accumulation over period of time in the environment, especially in soil and aquatic ecosystems and to reduce their inherent hazards to plants and animal life. To increase world food production and feed the ever-increasing population, agriculture produce can be augmented with biological control instead of pesticides.

Dubey and Patel (2001) evaluated six isolates of *Trichoderma viride* (Tv), *T. harzianum* (Th) and *T. virens* against *Thanatephorus cucumeris* (Tc) =*R. solani* causing web blight of urdbean. Local isolates of *T. viride* and *T. virens* were superior over others. The antagonistic activity of nine monosporic isolates of *Trichoderma* spp. was evaluated in dual culture and all the isolates significantly reduced the mycelial growth of *R. solani* (Cudom *et al.*, 2003). Shailbala and Tripathi (2010) screened *T. virens* for its antagonistic potential against *R. solani* causing web blight of urdbean by dual culture technique. *T. virens* grew fast and showed better potential after 6 days incubation and attained 78.2 mm growth showing 59.1 per cent inhibition of the radial growth of pathogen. Kumar and Tripathi (2011) evaluated six strain of *T. harzianum* (Th) against *Thanatephorus cucumeris* (Tc) =*R. solani* causing web blight of urdbean and reported that all the strains of *Trichoderma* significantly suppressed the growth of the pathogen. Maximum inhibition 88.8 mm after 72 h of incubation was recorded in strain Th 5 followed by strain Th 1, where inhibition was 85.0 mm. Ray *et al.* (2007) tested efficacy of two fungal bioagents, *T. harzianum*, *T. viride* and one bacterial bio-agent, *Pseudomonas fluorescens* were tested under *in vitro* conditions against *R. solani*. Among the bio-agents, *T. harzianum* inhibited mycelial growth of *R. solani* after 96 h of incubation followed by *T. viride* and *Ps. fluorescens*. After 96 h of incubation 82.43 and 80.36 mm growth of *T. harzianum* and *T. viride* was observed, respectively over no growth of test fungus. However, after 120 h of dual culture with *R. solani* both the bio-agents completely covered the inoculated plates. The bacterial bio-agent *Ps. fluorescens* produced clear inhibition zone and no fungal growth was observed at the site of inhibition zone. After 120 hr of incubation *Ps. fluorescens* overlapped the test pathogen. There is a big opportunities regarding field application of bio-agents for managing the web blight of urdbean and mungbean. If the bio-agents applied as seed treatment revealed better response than the foliar application. Therefore, it is essential to develop a complete bio-control package for managing the disease

Chemical Control

Fungicides play a very important and significant role in disease management because of fact that they are easy to apply, economical, quick acting and effective. The effect of bavistin or benlate @ 2.5 g/kg was found effective in eliminating the seed-borne infection of *T. cucumeris* causing web blight of rice bean (Tiwari, 1993) and dolichos bean (Dwivedi and Dwivedi, 1999). Carbendazim and thiophanate methyl were found best for controlling seedling mortality in mungbean caused by *R. solani* (Singh *et al.*, 1995). Shailbala and Tripathi (2010) screened eight fungicides namely propiconazole, hexaconazole, carbendazim, tebuconazole, epoxiconazole, mancozeb, propineb, thiophanate-methyl were tested at different concentration against web blight of urdbean. Epoxiconazole was found the most effective followed by propiconazole and carbendazim. Epoxiconazole completely inhibited radial mycelia growth at 2.5 μg/ml concentration while propiconazole at 5 μg/ml and hexaconazole and carbendazim at 10 μg/ml concentrations. Kumar and Tripathi (2011) evaluated six systemic and non- systemic fungicides at different concentration in *in vitro*. Among systemic fungicides, bavistin was found the most effective followed by tilt and contaf. Among non- systemic fungicides, captaf was found most effective followed by mancozeb and sulphur. Foliar spray of bavistin 0.05 per cent along with seed treatment with celest (0.2 per cent), raxil (0.2 per cent) or bavistin (0.2 per cent) was highly effective in reducing web blight severity of french bean followed by topsin-M and baycor, significant increase in yield was also observed (Mathew and Gupta, 1996). Five fungicide namely carbendazim (0.2 per cent), propiconazole (0.1 per cent), hexaconazole (0.1 per cent), propineb (0.1 per cent), mancozeb + thiophanate-methyl (0.25 per cent) as seed treatment and foliar sprays were tested at different concentration against web blight of urdbean (*Vigna mungo* cv. PU-19). Propiconazole (0.1 per cent) at a 15 days interval resulted highest reduction in disease severity (30-32 per cent), grain yield (950-1012 kg/ha) and 1000-grain weight (35 g) (Sharma and Tripathi, 2000). Dubey and Dwivedi (1988) conducted experiment under glasshouse conditions during 1984-85; bavistin at 0.2 per cent was the most effective of 6 spray treatments against *Thanatephorus cucumeris* on *Vigna mungo* followed by dithane Z-78 at 0.25 per cent and benlate (benomyle) at 0.2 per cent. Carbandazim and kitazin completely inhibited the mycelial growth of *R. solani* f. sp. *sasakii* at 1ppm concentration (Nene and Thapliyal, 1993; Kumar and Dubey, 2002).

Management through Botanical

The inappropriate and indiscriminate use of pesticides in plant disease management has led to serious environment threat to human life. The raw materials we consume are loaded with substantial quantities of pesticides that are used during their cultivation. There is an all-around compulsion to go for bio-rational alternatives. Plant products are the best alternatives available today.

The presence of antifungal compounds in higher plants has long been recognized as an important factor to disease control (Mahadevan, 1982). Such compounds being biodegradable and selective in their toxicity are considered valuable for controlling some plant diseases (Singh and Dwivedi, 1987). Ansari (1995) reported fungistatic activity of extracts of *Ocimum sanctum*, *Mentha arvensis* and *Eucalyptus* spp. against *R. solani*. Chalavaria (2005) tested nine botanicals namely calotropis, eucalyptus, neem, *Vinca rosea*, mentha, onion, garlic, datura and tulsi against *R. solani* and showed that the garlic extract was found effective in inhibiting cent-percent of mycelial growth. Onion was found next order of effective in inhibiting (50 per cent) mycelial growth of the pathogen which was followed by neem (48.21 per cent), datura and eucalyptus (41.07 per cent). Jhamaria and Sharma (2002) reported that leaf pep (an organic liquid based on natural clove oil) was found effective for controlling web blight of mungbean caused by *R. solani*. Kumar and Tripathi (2012) tested six botanicals namely onion, garlic, neem, bhang, ginger, and datura against *Rhizoctonia solani* causing web blight of urdbean and showed that the onion extract was found effective in inhibiting mycelia *in vitro* and reduction in disease severity, increase in grain yield kg/ha and 1000-grains weight.

Integrated Disease Management

The concept of integrated disease management is the main objective for maintaining and increasing production trend to ensure food security with future dimension while preserving the underlying resources base. The changed agriculture scenario in India has to focus our attention to mange plant diseases, reduce crop losses, avoid wide fluctuations in production and sustain the high levels of productivity. Only chemical is not successful for the control of diseases, biological control has been found as most effective for the management of the wilt. Use of resistant cultivar and judicious use of chemicals has also become very important to save the environment. An integrated control that denotes the rational use of all available control measures should be considered especially with a crop, which is attacked, simultaneously by numerous type and kinds of pathogens (Elad *et al.*, 1980). The three aims of Integrated Disease Management (IDM) are:

1. **Reduction of inoculums:** The disease causing structures *e.g.* spores, cells, sclerotes through careful paddock selection, control of volunteer and alternative hosts and stubble management.
2. **Exclusion of the pathogen:** It can be done through the use of clean seed, farm and paddock hygiene.
3. **Protection of the host:** through the use of varietal resistance and fungicides.

Integration of soil application of *P. glabra* cake, seed treatment with *Trichoderma* plus carboxin plus *Rhizobium* and foliar spray of *P. glabra* leaf extract suppressed web blight disease severity up to 92.7 per cent significantly (Dubey, 2007).

References

Alam, S.S., Qureshi SH, Bashir M. 1995. A report on web blight of mungbean in Pakistan. *Pakistan J. Bot.*, 17(1): 165-167.

Ansari, M.M. 1995. Control of sheath blight of rice by plant extracts. *Indian Phytopath.*, 48: 268-270.

Bains, S.S. and Dhaliwal, H.S. 1993. Development of foci of blighted mungbean plants due to *Rhizoctonia solani* transmitted from weeds. *Indian Botanical Reporter.* 11(1-2): 11-14.

Baker, R, Martison, C.A. 1970. In: *Rhizoctonia solani*, biology and pathology. (Ed. J.R. Parmeter, Jr), University of California Press, Berkeley, p, 172-188.

Boosalis, M.G. and Scharen, A.L. 1959. *Phytopathology*, 49: 192 –198.

Cudom, M.A., Mazza, S.M. and Gutierrez, S.A. 2003. Selection of *Trichoderma* spp. isolates against *Rhizoctonia solani*. *Spanish J. Agric. Res.*, 1(4): 79-82.

Dubey, S.C. 1997. Influence of age of plants, temperature and humidity on web blight development in groundnut. *Indian Phytopath.*, **50**: 119-120.

Dubey, S.C. 2007. Integrating bioagents with botanical and fungicides in different mode of application for better management of web blight and mung bean yield. *Indian J. Agric. Sci.*, **77**(3): 163-163.

Dubey, S.C. and Dwivedi, R.P. 1988. Chemical control of web blight of urdbean. *Farm Science Journal*. **3**: 182-183.

Dubey, S.C. and Patel, B. 2001. Evaluation of antagonists against *Thanatephorus cucumeris* causing web blight of urd and mung bean. *Indian Phytopath.*, 54: 206-209.

Dwivedi, R.P. and Dubey, S.C. 1986. Web blight of groundnut caused by *Thanatephorus cucumeris. Farm Sci.*, J. 1: 33-37.

Dwivedi, V.B. and Dwivedi, R.P. 1999. 51st Annual Meeting and National Symposium on Seed Health Care and Phytosanitation for Sustainable Agriculture. I.I.S.R., Lucknow, Feb. 17-19.

Dwivedi VB, Saksena H.K. 1975. Web blight disease of Arhar (*Cajanus cajan* L.) (Millsp.) caused by *Thanatephorus cucumeris. Indian J. Farm Sci.*, **3**: 114-115.

Elad, Y. Chet, I. Katan, J. 1980. *Trichoderam harziamu*, a biocontrol agent effective agaisnt *Sclerotium rolfsi* and *Rhizoctonia solani*. *Phytopathol.*, 70: 119-121.

Jhamaria, S.L. and Sharma, O.P. 2002. Management of web blight of mungbean through chemicals and plant products. *Indian Phytopath.*, 55: 526.

Joye, G.F. 1990. Management of Rhizoctonia aerial blight of soybean and biology of sclerotia of *Rhizoctonia solani* Kuhn. Ph.D. Thesis, Louisiana State Univ., Baton Rouge, p9.

Kumar, M. and Dubey, S.C. 2002. Relationship of disease intensity with weather and management of web blight of winged bean. *Indian Phytopath.*, 55: 152-157.

Kumar, S. and Tripathi, H.S. 2011. Biological Control Strategies for Management of Web Blight Disease of Urdbean (*Vigna mungo*). *J. Mycol. Pl. Pathol.*, 41: 282-28.

Kumar, S. and Tripathi, H.S. 2012. Evaluation of plant extracts against *Rhizoctonia solani* Kuhn, the incitant of web blight of Urdbean. *Plant Disease Research*, 27 (2): 190-193.

Liew, E.C.Y., MacLean, D.J. and Irwin, J.A.G. 1998. Specific PCR based detection of *Phytophthora medicaginis* using the intergenic spacer region of the ribosomal DNA. *Mycological Research*, 102: 73-80.

Mahadevan, A. 1982. Biochemical aspects of plant disease resistance part I. Preferred inhibitory substance prohibition. Today and Tomorrow Printer and Publishers, New Delhi.

Naiki, J. and Ui, J. 1981. *Rhizoctonia solani* Kuhn root rot of bean, soybean and adzuki bean seedlings. Mem. Fac. Agric. Hokkaido Univ. **12**: 262 – 269.

Nene, Y.L. and Thapliyal, P.N. 1993. Evaluation of fungicides. *In*: Fungicides in Plant Disease Control (3rd ed.). Oxford and IBH Publishing Co. Pvt. Ltd., New Delhi, p 531.

Ogoshi, A. (1978). Studies on the grouping of *R. solani* Kuhn with hyphal anastomosis and on the perfect stage of the groups. *Bull. Nat. Inst. Agric. Sci. Ser. C. Pl. Pathol. Entomol.*, 30: 1– 63.

Ray, A., Kumar, P. and Tripathi, H.S. 2007. Evaluation of bioagents against *Rhizoctonia solani* Kuhn the cause of aerial blight of soybean. *Indian Phytopath.* 60: 532-534.

Saksena, H.K. and Dwivedi, R.P. 1973. Web blight of blackgram caused by *Thanatephorus cucumeris*. *Indian J. Farm Sci.* 1: 58-61.

Schroth, M.N. and Cook, R.K. 1964. *Phytopath.* 54: 670-673.

Shailbala and Tripathi, H.S. 2010. Biological and chemical management of web blight disease of urdbean caused by *Rhizoctonia solani* Kuhn. *J. Pl. Dis. Sci.*, 54: 121-125.

Sharma, J. and Tripathi, H.S. 2000. Biological and chemical control of web blight of urdbean. *Indian Phytopath.* 54: 267-269.

Sharma, J. Tripathi, H.S. 2001. Host range of *Rhizoctonia solani* Kuhn, the causal agent of web blight of urdbean *Vigna mungo* (L.) Hepper. *J. Mycol. Pl. Patho.*, 31: 81-82.

Singh, R. *et al.*, 1995. *Plant Dis. Res.*, 10: 41-43.

Singh, R.K. and Dwivedi, R.S. 1987. Effects of oils on *Sclerotium rolfsii* causing root rot of barley. *Indian Phytopath.*, 40: 531-533.

Stonehouse, J. 1994. Assessment of urdbean web blight diseases using visual keys. *Plant Pathol.*, 43: 519-527.

Tiwari, T.N. 1993. Studies on web blight of rice bean caused by *Thanatephorus cucumeris* (Fronk) Donk Ph.D. thesis, C.S.A. Univ. of Agri. and Tech., Kanpur.

Weber, C.F. 1939. Web blight: A disease of beans caused by *Corticium microsclerotia*. *Phytopathol.*, 29: 559-575.

Wiese, M.V. 1987. *Compendium of wheat diseases*. American Phytopathological Society. p124.

Chalavaria, R. 2005. Estimation of yield losses and botanical management of sheath blight of rice caused by *Rhizoctonia solani* Kuhn. Ph.D. Thesis, G.B. Pant University of Agriculture and Technology, Pantnagar, p53.

Basandrai, A.K., Gartan, S.L., Basandrai, D. and Kalia, V. 1999. Blackgram (*Vigna mungo*) germplasm evaluation against different diseases. *Indian J. Plant Sci.*, 69(7): 236-239.

Duggar, B.M. 1915. *Rhizoctonia crocorum* (Pers.) D.C. *Rhizoctonia solani* Kuhn (*Corticium vagum* B and C) with notes on other species. *Ann. Missouri Botan. Garden*, 2: 403-458.

Shailbala and Tripathi, H.S. 2007. Current status of research on web blight disease of urdbean: a review. *Agri. Rev.* 28 (1): 1-9.

Singh, D.P. 1982. Genetics and breeding of blackgram and greengram. *Tech. Bulletin.* G.B.P.U.A. and T., Pantnagar, p68.

2016, Diseases of Pulse Crops and their Sustainable Management *443–462*
Editors: **Samir Kumar Biswas, Santosh Kumar and Gireesh Chand**
Published by: **BIOTECH BOOKS, NEW DELHI**

Chapter 24

Biological Control of Soil Borne Pathogens of Pulses

A.K. Ratankumar Singh*, S.K. Dutta, S.B. Singh, T. Boopathi and L. Sanajaoba Singh

Crop Protection Division,
ICAR Research Complex for North Eastern Hill Region Mizoram Centre,
Kolasib – 796 081, Mizoram

Introduction

Pulse crops or grain legumes are the major source of protein in the predominantly vegetarian diet of the people of India. In India, probably, grow a wide range of variety of pulses than any other country. These pulses include: chickpea (*Cicer orietinum* L.) pigenopea (*Cajanus cajan* (L.) Millsp.), black gram (*Vigna mungo* (L.) Hepper), green gram (*Vigna radiata* (L.) Wilczek), lentil (*Lens culinaris* Medik), pea (*Pisum sativum* L.), moth bean (*Vigna aconitefolia* (Jacq.) Marechal), horse gram (*Macrotyloma uniflorum* (Lam.) Verdc.) Khesari (grass pea) (*Lathyrus satiuus* L.), dry bean or rajmash (*Phaseoulus vulgaris*), Lablab bean (*Lablab purpureus*), and cowpea (*Vigna unguiculata* (L.) Walp.) (Nene, 1986). Pulses contain 22-24 per cent protein, which is almost twice the protein in wheat and thrice than of rice. Pulses provide significant nutritional and health benefits, and are known to reduce several non-communicable diseases such as colon cancer and cardiovascular diseases (Yude 1993, Jukanti *et al.*, 2012). Among various pulse crops, chickpea dominates with over 40 per cent share of total pulse production

* Corresponding Author: E-mail: ratanplantpatho@gmail.com

followed by pigeonpea (18-20 per cent), mungbean (11 per cent), urdbean (10-12 per cent), lentil (8-9 per cent) and other legumes (20 per cent) (Vision 2030 IIPR).

The present level of pulse production in India during 2010-2011 was 18.1 m ton from 26.3 m ha areas with a productivity of 690 kg/ha against the potential of 2000 kh/ha. The main results for reduction in yield of pulses are due to cultivation by small and marginal farmers in low fertility, problematic soils associated with unpredictable environmental conditions. More than 87 per cent of the total area lies under pulses is grown in rainfed conditions.

With recent changes in the global temperatures the grain yield is likely to be drastically affected by both biotic and abiotic extremities. Among biotic stress, soil and seed borne diseases causing (Table 24.1) drastic impact on most of the economically important pulses of India. In modern agriculture, different approaches may be adopted to prevent, mitigate, control or manage plant soil borne diseases by using ready to use inputs like chemical fertilizers and pesticides. Such hazardous inputs have contributed significantly to the spectacular improvements in crop productivity and quality (Frank *et al.*, 2006). However, the environmental catastrophic caused by excessive use and misuse of agrochemicals, has led to considerable changes in grower's attitudes towards the use of ill-effect of pesticides in agriculture. In consequence, some pest and diseases management researchers have started to focus their efforts on developing alternative eco-friendly inputs to synthetic chemicals for controlling pests and diseases. Therefore, researcher should made the efforts to develop and deliver a cost effective management strategies which is the urgent need of hour and use of bio control agents is certainly an important and encouraging approached toward sustainable pulses production.

MECHANISM OF BIOCONTROL

Plant diseases are the result of interactions among the four components of disease *i.e.* host, pathogen, environment and man. Biological control agents are the organisms that interfering the successful interaction with the disease components to manage the disease. Understanding how the bio control agents work can facilitate optimization of control as well as help to screen for more efficient strains of the agent. Understanding the mechanisms of biological control of plant diseases through the interactions between biocontrol agent and pathogen may allow us to manipulate the soil environment to create conditions conducive for successful biocontrol or to improve biocontrol strategies (Chet, 1987.). Mechanism of some bio-control agents are now understood in detail. Understanding the mechanism of action of a bio control agent may improve the consistency of control either by improving the mechanism or by using the bio- control agents under conditions where it is predicted to be more successful (Zhang *et al.*, 2002). Bio control agents involve a bewildering array of mechanisms in achieving disease control. However, the conclusive evidences for the involvement of a particular factor in biological control are determined by the strict correlation between the appearance of factor and the biological control (Handelsman *et al.*, 1989). Various strategies and mechanisms employed for disease suppression by the biocontrol agents in the soil borne diseases management broadly based on interaction with soil pathogens, host, microbial community. The fungal antagonist,

Table 24.2: List of Antibiotics Effective against the Soil Borne Pathogens of Major Pulse Crops

Sl.No.	*Antibiotics*	*Source*	*Target pathogen*	*Disease*	*References*
1.	2,4 - Diacetyl-phologlucinol (DAPG)	*Pseudomonas fluorescence* F113	*Pythium*	Damping off	Shanahan *et al.* (1992) and Schouten *et al.*, 2004
2.	Phenazine	*P. fluorescens* 2-79	*Fusarium oxysporium*	Wilt	Mazurier *et al.* (2009)
3.	Bacillomycin D	*Bacillus amylo liquefaciens* strain FZB42	*Fusarium oxysporium*	Wilt	Koumoutsi *et al.* (2004)
4.	Xanthobacin A	*Lycobacter* sp. Strain K88	*Aphanomyces cochlioides*	Damping off	Islam *et al.* (2005)
5.	Gliotoxin	*Trichoderma virens*	*Rhizoctonia solani*	Root rot	Wilhite *et al.* (2001)
6.	Zwitermycin AKanosamine	*Bacillus cereus* UW85	*Pythium aphanidermatum*	Damping off	Smith *et al.* (1993) and Pal and Gardener (2006)
7.	Anthranilate	*P. aeruginosa*	*Fusarium oxysporum* f. *sp.ciceris*	Wilt	Anjaiah *et al.* (1998)
8.	Mycostubilin	*Bascillus* BBG100	*Pythium aphanidermatum*	Damping off	Leclere *et al.* (2005)
9	Iturin	*Bascillus subtillus* QST713	*Botrytis, Rhizoctonia solani*	Damping off	Paulitz and Blanger, (2001)
10.	Geldanamycin	*Streptomyces hygroscopicus* var. *geldanus*	*R. solani*	Pea root rot	Rothrock and Gottlieb (1984)

produce multiple antibiotics which can suppress one or more pathogens. For example, *Bacillus cereus* strain UW85 is known to produce both zwittermycin and kanosamine (Pal and Gardener, 2006). The ability to produce multiple antibiotics probably helps to suppress diverse microbial competitors, some of which are likely to be plant pathogens. For example, phenazine antibiotics were shown to play a key role in the biocontrol activity of several *Pseudomonas* species against *F. oxysporum* on diverse crops (Anjaiah *et al.*, 1998; Chin-A-Woeng *et al.*, 2001a, 2001b and Sylvie *et al.*, 2009.). Also the antibiotic, 2,4-diacetylphloroglucinol (DAPG) has substantial activity against pathogenic *F. oxysporum* (Schouten *et al.*, 2004), and populations of DAPG-producing pseudomonads were highly enriched in a soil naturally suppressive to Fusarium wilt of peas (Landa *et al.*, 2002). *Streptomyces hygroscopicus* var. *geldanus* produced the antibiotic geldanamycin, which effectively suppress root rot of pea caused by *Rhizoctonia solani* in the field (79).

Iron Competition

This process is considered to be an indirect interaction whereby pathogens are excluded by depletion of a food base or by physical occupation of site. Generally, nutrient competition has been believed to have an important role in disease suppression, although it is extremely difficult to obtain conclusive evidence. Biocontrol involve suppression of the pathogen by depriving it of nutrients. The typical system involves a siderophore, which is an iron-binding ligand, and an uptake protein, which transports the siderophore into the cell. The fluorescent pseudomonads produce a class of siderophores known as the pseudobactins, which are structurally complex iron-binding molecules. Analyses of mutants lacking the ability to produce siderophores suggest that they contribute to suppression of certain fungal and oomycete diseases. Competition for micro nutrients exists because bio control agents have more efficient utilizing uptake system for the substances than the pathogens (Nelson, 1990). For example, iron competition in alkaline soils may be a limiting factor for microbial growth in such soils. A few studies have demonstrated that siderophore biosynthesis in *P. fluorescens* plays a role in pathogen suppression. Some plant pathogens depend on growth substances or stimulants like fatty acids or peroxidation products of fatty acids, to overcome their dormancy before they can cause infection and bio control agents are known to exert competition for these stimulants there by reducing their disease causing ability (Harman *et al.*, 1994), volatile compounds such as ethanol and acetyldehyde (Paulitz, 1991). Moreover, the co-inoculation of *Pseudomonas* sp. with *Bradyrhizobium/Mesorhizobium* sp. have been found to be caused significant increase in nodules number and weight, plant dried weight of green gram and chickpea as it supported more growth of the siderophore producing Pseudomonads strains (Sindhu *et al.*, 1999, 2002) *Pseudomonas* spp have been found to be produce different types of siderophores such as pyoverdine, pseudobactins, colourless nocardamine, pyochelin, salicylic acid and cephactin (Meyer and Adallalah, 1980, Verma and Chincholkar, 2007). Kseisniak and Kobus (1993) microbial species including bacteria, actinomyces and fungi for siderophore production and highest siderophore producing bacteria belong to genera *Pseudomonas and Erwinia*. Similarly, siderophore production was observed in 74.2 per cent of *Rhizobium/Mesorhizobium* strains infecting pigeonpea, out 31 strains tested (Duhan *et al.*, 1998). Certain fluorescent pseudomonads have also been shown to displace resident fungi and bacteria, in some cases reducing populations of deleterious microorganisms (Kloepper and Schroth, 1981; Yuen and Schroth, 1986). Treatments that enhance plant health can also increase the frequency of manganese-reducing bacteria in the rhizosphere community, thereby increasing manganese availability to the plant, which may in turn enhance resistance to disease (Huber and Wilhelm, 1988; Elmer, 1995). Introduction of the biocontrol agent *B. cereus* strain UW85 can induce dramatic changes in the composition of culturable bacterial communities on soybean roots in the field (Gilbert *et al.*, 1993). So far as the competition for nutrients is concerned, bio control agents compete for the rare but essential micronutrients, such as iron and manganese especially in highly oxidized and aerated soils. In these soils iron is present in ferric form, which is insoluble in water and where the concentration may be as low as 10-8 M, too low to support the microbial growth (Lindsay, 1979). Screening of 563 bacteria obtained from the roots of peas, lentil and chickpea showed that 76 per cent isolates produced siderophores, 5 per cent isolates showed ACC deaminase activity and and per cent isolates were capabale of Indole productions (Hynes *et al.*, 2008).

Table 24.3: *In vitro* Activity of Antagonists against *Fusarium oxysporum* f. sp. *ciceris* and Production of Antimicrobial Metabolites

Antagonisms	*DC (%)*	*EM (%)*	*VC (%)*	*PT*	*CH*	*ST*	*IAA*
P. putida P9	38.7^{d}	38.7^{d}	23.2^{ab}	+	+	+	+
P. putida P10	40.3^{cd}	37.1^{d}	22.5^{ab}	+	++	+	+
P. aueroginosa P11	38.7^{d}	37.1^{d}	22.5^{ab}	–	+	+	+
P. aueroginosa P12	44.9^{abcd}	43.4^{d}	26.3^{a}	–	+	+	+
P. aueroginosa P66	41.1^{bcd}	24.0e	24.8^{ab}	+	++++	–	+
P. aueroginosa P112	41.1^{bcd}	19.4^{e}	7.0^{d}	+	+++	+	+
B. subtilis B1	47.3^{abc}	52.8^{c}	20.1^{bc}	+	–	–	+
B. subtilis B6	50.4a	67.4^{b}	$17^{.8c}$	+	–	–	+
B. subtilis B28	51.2a	78.3^{a}	22.5^{ab}	+	–	–	+
B. subtilis B40	49.7a	69.8^{b}	6.2^{d}	+	–	–	+
B. subtilis B99	48.8^{ab}	75.9^{a}	5.4^{d}	+	–	–	+
B. subtilis B108	44.2^{abcd}	13.2^{f}	3.1^{e}	+	–	–	+
Non infected Control	0^{e}	0^{g}	0^{f}	–	–	–	–

Data are means of four replicates. Means in the column followed by different letters indicate significant differences among treatments at $P \leq 0.05$ according to DMRT.

DC: Percent growth inhibition in dual culture method; EM: Percent growth inhibition in extracellular metabolites method; VC: Percent growth inhibition in volatile compound method; PT: Protease test; CH: Cyanide hydrogen; ST: Siderophore test; IAA: indole acetic acid.

Source: Karimi *et al.* (2012).

The bacterial antagonist in production of both cyanide hydrogen and siderophore, against *Fusarium* wilt of chickpea and their effect on plant growth parameter were studied, in which various strain of *P. aeuroginosa* and *B. subtilis* provided better control than untreated control in seed treatment and soil-inoculation, respectively.

Bio-control by Cell-wall Degrading Enzymes

Extracellular hydrolytic enzymes produced by microbes may also play a role in suppression of plant pathogenic fungi. Chitin and β-1,3-glucans are major constituents of many fungal cell walls. Several studies have demonstrated *in vitro* lysis of fungal cell walls either by *chitinase* or β-*1,3-glucanase* alone or in combination. Recently, genetic evidence for the role of these enzymes in biocontrol has been obtained. A *chitinase* (*ChiA*) deficient mutant of *Serratia marcescens* was shown to have reduced inhibition of fungal germ tube elongation and reduced biocontrol of *Fusarium* wilt of pea seedling in a greenhouse assay (Lam and Gaff n e y, 1993). Furthermore, when ChiA from *S. marcescens* was inserted into the non-biocontrol agent *Escherichia coli*, the transgenic bacterium reduced disease incidence of Southern blight of bean caused by *Sclerotium rolfsii* (Shapira, *et al.*, 1989). Similarly, *Trichoderma harzianum* was transformed with *ChiA* from *S.marcescens* (Haran, *et al.*, 1993.). The transformed strains were more capable of overgrowing on *Sclerotium rolfsii in vitro* than the original strain from which it was derived. Srivastava *et al.* (2001) reported that treatment of *P. fluorescens* decreased the disease severity of *M. phaseolina* (33 per cent) in chickpea by increasing activities of *chitinases and glucanases*. Rajesh *et al.* (2013) reported that Bacillomycin (*bmyB*) and β-*glucanase* produced by *Bacillus* proved to be effective in reducing chickpea wilt incidence (*Fusarium oxysporum* f. sp. *ciceri* race 1, *F. solani* and *Macrophomina phaseolina*) besides significant enhancement in growth (root and shoot length) and dry matter of chickpea plants. *Streptomyces* sp. S160 enhanced the growth and helped in inducing resistance against charcoal rot disease of chickpea caused by *M. phaseolina* by increasing activity of defense-related enzymes in chickpea plants, leading to the synthesis of defense chemicals in plants (Yadav *et al.*, 2013).

Suppresion Mechanism Interaction with Host/Colonization on Host

It seems logical that a biocontrol agent should grow and persist, or "colonized," on the surface of the root to protect and it considered the colonization is widely believed to be essential for biocontrol. Rhizosphere organisms provide an initial barrier against pathogens attacking the root (Weller 1988) and microorganisms that can grow in the rhizosphere are ideal for use as biocontrol agents. For example, in the suppression of damping-off of peas by *Burkholderia cepacia*), there is a significant relationship between population size of the biocontrol agent and the degree of disease suppression (Parke *et al.*, 1986 and Parke, 1990). It has been proposed that the growth of both the AM fungi and root pathogens depends on host photosynthates and that they compete for the carbon compounds reaching the root (Sharma *et al.*, 1992; Linderman 1994). Cordier *et al.* (1996) showed that *Phytophthora* development is reduced in AM fungal-colonized and adjacent uncolonized regions of AM root systems, and that in the former the pathogen does not penetrate arbuscule-containing cells. Plant growth promoting rhizobacteria (PGPR), such as *Pseudomonas* and *Bacillus* strains, are the major root colonizers (Manikanda *et al.*, 2010; Joseph *et al.*, 2007), and can elicit plant defenses (Kloepper *et al.*, 2004). Combined inoculation with *G. intraradices*, *P. putida* and *P. polymyxa* caused the greatest reduction in galling, nematode multiplication and root-rot disease complex (*Meloidogyne incognita* and *Macrophomina phaseolina*) of chickpea had adverse effects on root colonisation by *G. intraradices*, while root colonisation by arbuscular mycorrhizal fungus was increased in the presence of *P. putida* and *P. polymyxa* (Akhtar and Siddiqui 2007) and *G. intraradices* reduced in root-rot of pea caused by *Aphanomyces etueiches* (Bodker *et al.*, 1998.)

Bio control by Induced Resistance

Biocontrol agents induce a sustained change in the host plant, increasing its tolerance to infection by a pathogen, a phenomenon known as induced resistance. It is clear that induced resistance imparted by biocontrol agents involves the same suite of genes and gene products shown in the well documented plant response known as systemic acquired resistance (SAR). SAR is typically a response to a localized infection or an attenuated pathogen, which is manifested in subsequent resistance to a broad range of other pathogens (Uknes *et al.*, 1992; Ryals *et al.*, 1996). These defense responses may include the physical thickening of cell walls by lignification, deposition of callose, accumulation of antimicrobial low-molecular-weight substances (*e.g.*, phytoalexins), and synthesis of various proteins (*e.g.*,chitinases, glucanases, peroxidases, and other pathogenesis related (PR) proteins). The elicitors responsible for inducing resistance are not known in detail. *Trichoderma* species produce a 22 kDa xylanase that, when injected in plant tissues, will induce plant defence responses including K^+, H^+ and Ca^{2+} channelling, PR protein synthesis, ethylene biosynthesis, and glycosylation and fatty acylation of phytosterols (Bailey and Lumsden, 1998). Pectic oligogalacturonides released after hydrolysis by a nonpathogenic binucleate *Rhizoctonia* isolate may act as elicitors of defence responses in French bean (Jabaji *et al.*, 1999). Seed bacterized with *B. subtilis* significantly reduced the incidence of pigeonpea wilt and an increase in phenylalanine ammonia-lyase (PAL) and peroxidase activities in host plant (Podile and Laxmi, 1998 and Harish *et al.*, 1998.). Recently, many strains of PGPR have been shown to be effective in controlling plant diseases by inducing plant systemic resistance as Paromarto and co worker (1988) also implied that induced resistant is the mechanism of biocontrol of *Rhizoctonia solani* on soybean by bi-nucleated *Rhizoctonia solani*. Plants colonized by these strains are more resistant to foliar diseases, even though the PGPR is present only on roots. The level of chitinases and glucanases increased by 6.6- to 7-fold at post-inoculation; thereafter, little decrease in the activity of PR proteins was observed when root-colonizing populations of *P. fluorescens* were treated at different inoculum concentrations. Inoculation of root tips of chickpea by *P. fluorescens*, 2,6-dichloroisonicotinic acid, and *o*-acetylsalicylic acid induced systemic resistance against charcoal (Srivastava *et al.*, 2001). *Trichoderma* biocontrol revealed that mycoparasitism-associated coiling and chitinase production as well as secondary metabolism are affected by the internal cAMP level; in addition, a cross talk between regulation of light responses and the cAMP signalling pathway was found in *Trichoderma atroviride*. Although induced systemic disease resistance has been studied mainly in laboratories and greenhouses, some recent reports have indicated that microbial-induced SAR can protect crops from pathogen infection under field conditions by treatment of these beneficial microorganisms (Wei *et al.*, 1996). Two antagonistic fungi *viz.*, *Trichoderma harzianum* and *T. viride* were found to be effective against natural incidence of wilt and wet root rot of chickpea when it apply in soil one week before sowing than seed treatment in reducing the disease (Prasad *et al.*, 2002). Seeds coated with *T. viride* increased fresh and dry weight of shoot, root and nodules of broad beans (Woo *et al.*, 2006). In another study by Kumar *et al.* (2007), *P. fluorescens* inhibited the mycelial growth of the *M. phaseolina* resulted reducing the disease severity and significantly increased the biomass of the chickpea plants, shoot length, root length and protein content of the chickpea seeds. A number of chemical elicitors of SAR and ISR may be produced by the PGPR strains upon inoculation, including salicylic acid, siderophore, lipopolysaccharides, and 2,3-butanediol, and other volatile substances (Ongena *et al.*, 2004 and Ryu *et al.*, 2004).

Formulations and Commercialized Products of Biocontrol Agents

Biocontrol formulation is blending of active ingredients (a.i.) such as fungal and bacterial propagules with the inert material such as diluents and surfactants in order to alter the physical characteristics to a more desirable form. A final formulation must be easy to handle in application and storage, retain stability during storage and transportation, good shelf-life application, affordable and marketability. Various inert carriers used in *Trichoderma* and *Pseudomonas* based formulation are celaton, vermiculite, talcum powder, diatomaceous earth, molasses, gypsum, lignite, kaolin, peat wheat bran, clay perlite, calcium chloride, rice flour, sugarcance pressmud,polymer, alginate etc. (Khan *et al.*, 2001, Zaid and Singh, 2004 and Singh and Joshi, 2007 and Juniad, 2013). Different methods of application *viz.*, soil, seed treatment, seedling dip and foliar spray of the bio control agent were adopted against major diseases of pulses (Manikanda *et al.*, 2010). One of the draw back in the commercialization of bio agents is the loss of viability of the propagules over the time. This is because low storage temperature and moisture free environment slows the metabolic growth of viable propagules and prevents the accumulation of metabolic toxins and depletion of nutrients. Although the number of bio control products in plant disease management is increasing, these products still represent only 1 per cent of the agricultural control measures while fungicides account for 15 per cent of total chemicals used in agriculture (Friavel *et al.*, 2005). In recent years many small and large entrepreneurs have entered into the commercial production of bio control agents resulting into the entry of various bio- control products into the world market. Unlike the bio control of insects, bio control of plant diseases is relatively new. Historically, the first bio control agent of bacterium called *Agrobacterium radiobacter* strain K 84 was registered with the United States Environmental Protection agency (EPA) for the control of crown gall in 1979. Ten years later the first fungus *Trichoderma harzianum* ATCC 20476 was registered with the EPA for the control against pea and soybean diseases. Currently, a total of 14 bacteria and 12 fungi have been registered with the EPA for the control of plant diseases (Fravel, 2005). Most of these are sold commercially as one or more products. The maximum bio control products were registered and commercialized from Europe and America whereas; the technology of

Table 24.4: List of Commercialized Bio-control Products and their Target Organisms in Global and India

Sl.No.	*Bio-control Agent*	*Product Bame*	*Target Pathogen*	*Crop*	*Manufacturer Country*
			Global		
1.	*Bacillus pumillus GB34*	YieldShield	Soil borne and root rot diseases	Soybean	Gustafson Inc. USA
2.	*Streptomyces lyticus*	Actinovate	Soil borne diseases	All pulses	Natural Industries Inc.,USA
3.	*Bacillus subtulis MB1600*	Subtilex	*Fusarium* spp., *Rhizoctonia* spp., and *Pythium* spp.	All pulses	Becker Underwood, USA
5.	*B. subtilis* strainFZB24	Rhizo Plus	*R. solani, Fusarium* spp., *Sclerotinia* and *Verticillium*	All pulses crop	KFZB Biotechnik Berlin, Germany
6.	*Bacillus subtilis* GB03	System-3	Seedling pathogens	Beans, Pea, Soybean	Helena ChemicalCo.,Memphis USA
7.	*Bascillus subtillus* strain GB34	GB-34	*Rhizoctonia, Fussarium*	Soybean	Gustafon, USA
8.	*Bascillus subtillus* strain GB03	Kodiac	*Rhizoctonia, Aspergillus*	Soybean, peas	Growth Products, USA
10.	*Pseudomonas aureofaciens* strain TX-1	Bio–Jet	*Pythium, Rhizoctonia solani*	Pulses crops	EcoSoil System, CA
11.	*Streptomycine griseoviridis*	Mycostop	Soil borne pathogens	Pulses	Kemira Oy, Finland
12.	*Trichoderma harzianum T-22*	Root Shield	Soil borne pathogens	Pulses seedling	Bio works, USA
13.	*Gliocladium catenulatum* strain JI446	Prima Stop	Soil borne pathogens	All Pulses	Kemira Agro Oy, Finland
			India		
14.	*Trichoderma* spp.	Basderma	Soil borne pathogen	All Pulses	Basaras Biocon, India Pvt. Ltd, Chennai
15.	*T. viride*	Sun derma	*R. solani, Macrophomina phaseolina*	All Pulses	Sun Agro Biosystem Pvt. Ltd, Chennai
15.	*T. viride*	Ecofit	*R. solani Macrophomina phaseolina*	All Pulses	Hoechst Scherig Agr Evo. Ltd. Mumbai
16.	*T. viride*	Trieco	*Root rot, Wilt*	All Pulses	Ecosense Lab.(I)Pvt. Ltd. Mumbai
17.	*T. viride*	Antagon	Soil borne pathogens	All Pulses	Pralshar Biotech.(P) Ltd. Goa
18.	*T. harzianum*	Bioguard	*Rhizoctonia solani, Sclerotiarum, Fusarium*		PKVK, Puducheery

production and commercialization is still in its infant phase in India. Some of the commercially available bio control products available in the world and India markets are shown in Table 24.4. In India, more than 40 stakeholders from different provinces have registered for mass production of PGPR with Central Insecticide Board, Faridabad, Haryana through collaboration with Tamil Nadu Agricultural University, Coimbatore, India for the technical support and information (Ramakrishnan *et al.*, 2001).

Conclusion and Future Prospects

India needs around 32 million tons of pulses by 2030, to feed the estimated population of about 1.68 billion (IIPR, Vision 2030). Hence, India needs to produce the required quantity, besides remain competitive to protect indigenous pulses production. In the present crop production scenario, the bio control is of utmost importance, but its potential is yet to be exploited fully mainly because the research in this area is still confined to the laboratory and very little attention has been paid to produce the commercial formulations of bio agent. With people turning more health conscious biological control seem to the best alternative to disease suppression and its bring the disease suppression with no environmental hazards. Moreover, whatever has been commercially produced has not been used efficiently by the farmers owing to the lack of information regarding its use. So to popularize the concept of biological control extension at various levels direction needs to be improved. Most of the bio agents perform well in the laboratory conditions but fail to perform to their fullest once applied either in seed or to soil. Probably, it may be attributed to the micro physio-chemical and ecological constraints that lessen the efficacy of bio agents. To overcome this problem, genetic engineering and other molecular tools offer a new possibility for improving the selection and further characterization of bio control agents (Joshi *et al.*, 2006). Various methods that can contribute to increase the efficacy of bio agent include mutation or protoplasm fusion utilizing poly ethylene glycol. Combinations of beneficial fungal and bacterial strains that interact synergistically are currently being devised and numerous recent studies show a promising trend in the field of inoculation technology. There is also an urgent need to mass produce the bio agents, understand their mechanism of action and to evaluate the environmental factors that favour the rapid growth of bio control agents.

References

Agrawal, S.C. and Prasad, K.V.V., 1997. *Diseases of Lentil*. Oxford and IBH Publishing Co. Pvt. Ltd.,New Delhi, India, p. 155.

Akhtar, M.S. and Siddiqui, Z.A., 2007. Biocontrol of a chickpea root-rot disease complex with *Glomus intraradices, Pseudomonas putida* and *Paenibacillus polymyxa*. *Australasian Plant Pathology*, **36**: 175–180.

Ali Siddiqui, I., Ehetshamul-Haque, S., and Shahid Shauka, S., 2001. Use of rhizobacteria in the control of root rot–root knot disease complex of Mungbean. *Journal of Phytopathology*, **6**: 337–346.

Allen, D.J., Brachara, R.A. and Smithson, J.B., 1998. Diseases of common bean In: *The Pathology of Food and Pasture Legumes* (Eds.) DJ Allen and J M Lenne. CAB International, Wallingford, UK and ICRISAT, Pantencheru, India, p. 179-266

Altomare, C., Norvell, W.A., Bjorkman, T. and Harman, G.E., 1999. Solubilization of phosphate and micro nutrients by the plant growth promoting fungus *Trichoderma harzianum* Riafi. *Applied Environmental Microbiology*, **65**: 2926-2933.

Anjaiah, V., Koedam, N., Nowak-Thompson, B., Loper, J.E., Höfte, M., Tambong, J.T. and Cornelis, P., 1998. Involvement of phenazines and anthranilate in the antagonism of *Pseudomonas aeruginosa* PNA1 and Tn*5* derivatives toward *Fusarium* spp. and *Pythium* spp. *Molecular Plant-Microbe Interactions*, **11**: 847–854.

Bailey, B.A. and Lumsden, R.D. 1998. Direct effects of *Trichoderma* and *Gliocladium* on plant growth and resistance to pathogens. In: *Enzymes, biological control and commercial applications Trichoderma* and *Gliocladium* (Eds.) Harman, G.E. and Kubicek, C.P. Taylor and Francis. London UK. p. 185–204

Bender, C.L., Rangaswami, V. and Loper, J., 1999. Polyketide production by plant associated Pseudomonads. *Annual Review of Phytopathology*, **37**: 175–196.

Bodker, L., Kjoller, R. and Rosendahl, S., 1998. Effect of phosphate and arbuscular mycorrhizal fungus *Glomus intraradices* on disease severity of root rot of peas (*Pisum sativum*) caused by *Aphanomyces euteiches*. *Mycorrhiza*, **8**: 169–174.

Chaube, H.S., Mishra, D.S., Varshney and Singh, U.S., 2003. Bio control of plant pathogens by fungal antagonists: a historical background, present status and future prospects. *Annual Review of Plant Pathology*, **2**: 1-42.

Chaudhary, R.C., Neetu Shukla and Prajapati, R.K. Biological control of soil borne diseases: An update in ppulse crops In: *Ecofriendly Management of Plant Diseases* (Eds.) Shahid A. and Udit Narain, Daya Publishing House, p. 178-200.

Chaudhary, R.G. and Naimuddin, 2000. Pea diseases in Indian perspective and their economic management. In: *Advance in Plant Diseases Management* (Eds.) Udit Naraain *et al.*, Advance Publishing Concept, New Delhi, pp. 47-60.

Chaudhary, R.G., 2004. Biological Control of soil borne diseases of pulses crops. In: *Pulses in New Perspective.* (Eds.) Ali *et al.*, Indian Society of Pulses Research and Development, IIPR, Kanpur, India, p. 345-361.

Chet, I., Harman, G.E. and Baker, R., 1981. *Trichoderma hamatum*: Its hyphal interaction with *Rhizoctonia solani* and *Pythium* spp. *Microbial Biology*, **7**: 29-38.

Chet, I., 1987. Trichoderma application, mode of action, and potential as biocontrol agent of soil-borne pathogenic fungi. In: *Innovative Approaches to Plant Disease Control.* (ed.), John Wiley, New York, p. 137-160.

Chin, A., Woeng, T.F.C., Thomas Oates, J.E., Lugtenberg, B.J.J. and Bloemberg, G.V., 2001a. Introduction of the *phzH* gene of *Pseudomonas chlororaphis* PCL1391 extends the range of biocontrol ability of phenazine-1-carboxylic acid-producing *Pseudomonas* spp. strains. *Molecular Plant Microbe Interactions*, **14**: 1006-1015.

Chin-A-Woeng, T.F.C., Van Den Broek, D., De Voer, G., Van Der Drift, K.M.G.M., Tuinman, S., Thomas-Oates, J.E. and Lugtenberg, B.J.J., 2001b. Phenazine-1-carboxamide production in the biocontrol strain *Pseudomonas chlororaphis* PCL1391 is regulated by multiple factors secreted into the growth medium. *Molecular Plant Microbe Interactions*, **14**: 969–979.

Cordier, C., Gianinazzi, S. and Gianinazzi Pearson, V., 1996. Colonisation patterns of root tissues by *Phytophthora nicotianae* var. *parasitica* related to reduced disease in mycorrhizal tomato. *Plant and Soil*, **185**: 223-232.

Dwivedi, D. and Johri, B.N., 2003. Antifungals from fluorescent pseudomonads: Biosynthesis and regulation. *Current Science*, **85**: 1693-1703.

Erwin, D.C., 1957. *Fusarium* and *Verticillium* wilt of *Cicer arientinum*. *Phytopathology*. **47**: 10

Franks, A., Ryan, P.R., Abbas, A., Mark, G.L. and O'Gara, F., 2006. Molecular tools for studying plant growth-promoting rhizobacteria (PGPR): Molecular techniques for soil and rhizosphere microorganisms. CABI Publishing, Wallingford

Fravel, D.R., 2005. Commercialization and implementation of bio control. *Annual Review of Phytopathology*, **43**: 337-359.

Fridlender, M., Inbar, J. and Chet, I., 1993. Biological control of soil borne plant pathogens by a b-1,3-glucanase producing *Pseudomonas cepacia*. *Soil Biology and Biochemistry*, **25**: 1211–1221.

Goldman, G.H., Hayes, C.K. and Harman, G.E., 1994. Molecular and cellular biology of biocontrol by *Trichoderma* sp. *Trends Biotechnology*, **12**: 478-482.

Gurha, S.N., Singh, Gurdip and Sharma, Y.R., 2003. Diseases of chickpea and their management. In: *Chickpea Research in India* (Eds.) Masood Ali *et al.*, Indian Institute of Pulses Research, Kanpur, India, p. 195-227.

Gurha, S.N., 2004. Chana mein rog prabandhan. In: *Dalhan* (Eds.) Masood Ali *et al.*, Bharatiya Dalhan Anusandhan Sansthan, Kanpur, India, p. 201-212.

Handelsman, J. and Parke, J.L., 1989. Mechanism in bio control of soil borne plant pathogens In: *Plant Microbe Interactions, Molecular and Genetic Perspective*. Vol. 3rd. (Eds.) T, Kosuge and E. W. Nester, McGraw Hill, New York, p. 27-61.

Haran, S., Schickler, H., Peer, S., Logeman, S., Oppenheim, A. and Chet, I., 1993. Increased constitutive chitinase activity in transformed *Trichoderma harzianum*. *Biological Control*, **3**: 101-108.

Harish, S., Manjula, K. and Podile, A.R., 1998. *Fusarium udum* is resistant to the mycolytic activity of a biocontrol strain of *Bacillus subtilis* AF 1. *FEMS Microbiology Ecology*, **25(4)**: 385–390.

Harman, G.E. and Nelson, E.B., 1994. Mechanisms of protection of seed and seedlings by biological control treatments: Implications for practical disease control. In: *Seed Treatment: Progress and Prospects.* T. Martin, (ed.,) BCPC, Farnham, UK. p. 283-292.

Haware, M.P., 1998. Diseases of Chickpea In: *The Pathology of Food and Pasture Legumes* (Eds.) Allen DJ and Lenne JM. CAB International, Wallingford, UK and ICRISAT, Pantencheru, India, p. 473-516.

Hayes, C.K., Klemsdal, S., Lorito, M., Di Pietro, A., Peterbauer, C., Nakas, J.P., Tronsmo, A. and Harman, G.E., 1994. Isolation and sequence of an endochitinase endochitinase encoding gene from a cDNA library of *Trichoderma harzianum*. *Gene,* **138**: 143-148.

Hebbar, K.P., Davey, A.G., Merrin, J., McLoughlins, T.J. and Dart, P.J., 1992. *Pseudomonas cepacia* a potential suppressor of maize soil borne disease: Seed inoculation and maize root colonization. *Soil Biology and Biochemistry,* **24**: 999–1007.

Hynes, R.K., Lang, G.C., Hirkala, D.L. and Nelson, L.M., 2008. Isolation, selection and characterization of beneficial rhizobacteria from pea, lentil and chickpea grown in western Canada. *Canada Journal of Microbiology,* **54**: 248-258.

Islam, T.M., Hashidoko, Y., Deora, A., Ito and Tahara, S., 2005. Suppression of damping-off disease in host plants by the rhizoplane bacterium *Lysobacter* sp. strain SB-K88 is linked to plant colonization and antibiosis against soilborne peronosporomycetes. *Applied Environmental Microbiology,* **71**: 3786-3796.

Jabaji, H.S., Chamberland, H. and Charest, P.M., 1999. Cell wall alterations in hypocotyls of bean seedlings protected from *Rhizoctonia* stem canker by a binucleate Rhizoctonia isolate. *Mycological Research,* **103**: 1035-1043.

Joseph, B., Patra, R.R. and Lawrence, R., 2007. Characterization of plant growth promoting rhizobacteria associated with chickpea (*Cicer arietinum* L.). *International Journal of Plant Production,* **1(2)**: 141-151.

Joshi, R., Mc Spadden and Gardener, B.B., 2006. Identification and characterization of novel genetic markers associated with biological control activities in *Bacillus subtilis*. *Biological Control,* **96**: 145–154

Jukanti, A.K., Gaur, P.M., Gowda, C.L.L. and Chibbar, R.N., 2012. Nutritional quality and health benefits of chickpea (*Cicer arietinum* L.): a review. *British Journal of Nutrition,* **108**: 11-26.

Junaid, J.M., Dar, N.A., Bhat, T.A., Bhat, A.H. and Bhat, M.A, 2013. Commercial biocontrol agents and their mechanism of action in the management of plant pathogens. *International Journal Modern Plant and Animal Science,* **1(2)**: 39-57.

Karimi, K., Amini, J., Harighi, B. and Bahramnejad, B., 2012. Evaluation of biocontrol potential of *Pseudomonas* and *Bacillus* spp. against Fusarium wilt of chickpea. *Australian Journal of Crop Science,* **6(4)**: 695-703.

Kloepper, J.W., Leong, J., Teintze, M. and Schroth, M.N., 1980. *Pseudomonas* siderophores: A mechanism explaining disease suppression in soils. *Current Microbiology,* **4**: 317-320.

Kloepper, J.W., Ryu, C.M. and Zhang, S., 2004. Induce systemic resistance and promotionplant growth by *Bacillus* spp. *Phytopathology,* **94**: 1259-1266.

Koumoutsi, A., Chen, X.H., Henne, A., Liesegang, H., Gabriele, H., Franke, P., Vater, J. and Borris, R., 2004. Structural and functional characterization of gene clusters directing nonribosomal synthesis of bioactive lipopeptides in *Bacillus amyloliquefaciens* strain FZB42. *Journal of Bacteriology*, **86**: 1084-1096.

Kraft, J.M., Larssen, R.C. and Inglis, D.A., 1998. Diseases of peas In: *The Pathology of Food and Pasture Legumes* (Eds.) Allen, D.J. CAB International, Wallingford, UK and ICRISAT, Pantencheru, India, p. 473-516.

Kumar, V., Kumar, A. and Kharwar, R.N., 2007. Antagonistic potential of fluorescent pseudomonads and control of charcoal rot of Chickpea caused by *Macrophomina phaseolina*. *Journal of Environmental Biology*, **28(1)**: 15-20.

Lam, S.T., and Gaffney, T.D., 1993. Biological activities of bacteria used in plant pathogen control. In: *Biotechnology in Plant Disease Control*. I. Chet, (ed.) John Wiley, New York, p. 291-320.

Landa, B.B., Mavrodi, O.V., Raaijmakers, J.M., McSpadden- Gardener, B.B., Thomashow, L.S. and Weller, D.M., 2002. Differential ability of genotypes of 2,4-diacetylphloroglucinol- producing *Pseudomonas fluorescens* to colonize the roots of pea. *Applied Environmental Microbiology*, **68**: 3226–3237.

Leclere, V., Bechet, M., Adam, A., Guez, J.S., Wathelet, B., Ongena, M., Thonart, P., Gancel, F., CholletImbert, M. and Jacques, P., 2005. Mycosubtilin over-production by *Bacillus subtitlis subtilis* BBG100 enhances the organism's antagonistic and biocontrol activities. *Applied Environmental Microbiology*, **71**: 4577-4584.

Linderman, R.G., 1992. Vesicular-arbuscular mycorrhizae and soil microbial interactions. In: *Mycorrhizae in Sustainable Agriculture*. (Eds) Bethlenfalvay, G.J. and Linderman, R.G. ASA Special Publication No. 54, Madison, p. 45–70.

Liu, L., Kloepper, J.W. and Tuzun, S., 1995. Induction of systemic resistance in cucumber against *Fusarium* wilt by plant growth promoting rhizobacteria. *Phytopathology*, **85**: 695–698.

Lorito, M., Hayes, C.K., Zonia, A.S.F., Del, S.G., Woo, S.L. and Harman, G.E., 1994. Potential of genes and gene products from *Trichoderma* sp.and *Gliocladium* sp. for the development of biological pesticides. *Molecular Biotechnology*, **2**: 209- 217.

Manikanda, R., Saravanakumar, D., Rajendran, L., Raguchander, T. and Samiyappan, R., 2010. Standardization of liquid formulation of *Pseudomonas fluorescens* Pf1 for its efficacy against *Fusarium* wilt of tomato. *Biological Control*, **54**: 83-89.

Mazurier, S., Corberand, T., Lemanceau, P. and Raaijmaker, J.M., 2009. Phenazine antibiotic produced by fluorescent pseudomonads contribute to natural suppressiveness to *Fusarium* wilt. *ISME J.*, **3**: 977-991.

Medvecky, B.A., Ketterings, Q.M. and Nelson, E.B., 2007. Relationship amongst soil borne seedling diseases, *Lablab purpureus* L. and maize stover residue management. Bean insects pests and soil characterization in Trans Nzoia district, Kenya. *Applied Soil Ecology*, **35**: 107-119.

Mukherjee, S. and Tripathi, H.S., 2000. Biological and chemical control of wilt complex of French bean. *Journal of Mycology and Plant Pathology*, **30(3)**: 380-385.

Mukhopadhyay, A.S.N. and Mukherjee, P.K., 1998. Biological control of plant diseases: status in India in Biological suppression of Plant Diseases: *Phytopathogens, Nematodes and weeds*. Eds. Singh, S.P. and Husain, S.S., **7**: 1-20.

Nelson, E.B., 1990. Exudate molecules initiating fungal responses to seed seeds and roots. *Plant and Soil*, **129**: 61-73.

Nene, Y.L. and Reddy, M.V., 1987. Chickpea diseases and their control. In: *The Chickpea* (Eds.) M.C. Saxena and K.B. Singh. CAB, International, Wallingford, UK, p. 233-270.

Nene, Y.L., 1986. Opportunistic for research on disease of pulses crops. *Indian Phytopathology*, **39(3)**: 333-342.

Ongena, M., Duby, F., Rossignol, F., Fouconnier, M.L., Dommes, J. and Thonart, P., 2004. Stimulation of the lipoxygenase pathway is associated with systemic resistance induced in bean by a nonpathogenic *Pseudomonas* strain. *Mol. Plant-Microbe Interact.*, **17**: 1009-1018.

Ordentlich, A., Elad, Y. and Chet, I., 1988. The role of Chitinase of *Serratia marcescens* in the bio control of *Sclerotium rolfsii*. *Phytopathology*, **78**: 84-92.

Pal, K.K. and Gardener, B.M., 2006. Biological Control of Plant Pathogens. *The Plant Health Instructor*, p. 1-25.

Pal, K.K., Tilak, K.V.B.R., Saxena, A.K., Dey, R. and Singh, C.S., 2000. Antifungal characteristics of a fluorescent Pseudomonas strain involved in the biological control of *Rhizoctonia solani*. *Microbiology Research*, **155**: 233- 242.

Parke, J.L., Moen, R., Rovlra, A.D. and Bowen, G.D., 1986. Soil water flow affects the rhizosphere distribution of a seed-borne biological control agent, *Pseudomonas fluorescens*. *Soil Biology and Biochemistry*, **18**: 583-588.

Parke, J.L., 1990. Population dynamics of *Pseudomonas cepacia* in the pea spermosphere in relation to biocontrol of Pythium. *Phytopathology*, **80**: 1307-1311.

Patel, S.I. and Patel, R.L., 2012. Organic amendments in the management of pigeonpea wilt caused by *Fusarium udum*. *Journal Environment and Ecology*, **30(3)**: 549-551.

Paulitz, T.C. and Belanger, R.R., 2001. Biological control in greenhouse systems. *Annual Review Phytopathology*, **39**: 103-133.

Paulitz, T.C., 1991. Effect of *Pseudomonas putida* on the stimulation of *Pythium ultimum* by seed volatiles of pea and soybean. *Phytopathology*, **81**: 1282-1287.

Podile, A.R. and Laxmi, V.D.V., 1998. Seed Bacterization with *Bacillus subtilis* AF 1 Increases Phenylalanine Ammonia-lyase and Reduces the Incidence of Fusarial Wilt in Pigeonpea. *Journal of Phytopathology*, **146 (5-6)**: 255 –259.

Poromarto, S.H., Nelson, B.D. and Freeman, T.P., 1988. Association of binucleate *Rhizoctonia* with soybean and mechanism of biocontrol of *Rhizoctonia solani*. *Phytopathology*, **88**: 1056-1067.

Prasad, R.D., Rangeshwaran, R., Anuroop, C.P. and Rashmi, H.J., 2002. Biological control of wilt and root rot of chickpea under field conditions. *Annals of Plant Protection Sciences*, **10**: 72-75.

Ramakrishnan G, Nakkeeran S, Chandrasekar G, and Doraiswamy S.2001. Biocontrol agents- novel tool to combat plant diseases. In: *the III Asia Pacific Crop Protection Conference*, New Delhi, India. p. 20-39.

Reddy, M.V., Sharma, S.B. and Nene, Y.L., 1998. Pigeonpea diseases management In: *The Pathology of Food and Pasture Legumes* (Eds.) Allen D J and Lenne J M. CAB International, Wallingford, UK and ICRISAT, Pantencheru, India, p. 517.

Rothrock, C.S. and Gottlieb, D., 1984. Role of antibiosis in antagonism of *Streptomyces hygroscopicus* var. *geldanus* to *Rhizoctonia solani* in soil. *Canada Journal of Microbiology*, **30**: 1440-1447.

Ryals, J.A., Neuenschwander, U.H., Willits, M.G., Mollna, A., Steiner, H.X. and Hunt, M.D., 1996. Systemic acquired resistance. *Plant Cell*, **8**: 1809-1819.

Saxena, D.R., Saxena, Moly and Khare, M.N., 1998. Disease of lentil and their management. In: *Disease of Field Crop and their Management* (Ed.) T S Thind, National Agricultural Technology Information Centre, Ludhiana, India, p. 271-283.

Schouten, A, van den Berg, G., Edel-Hermann, V., Steinberg, C. and Gautheron, N. *et al.*, 2004. Defense responses of *Fusarium oxysporum* to 2,4-DAPG, a broad spectrum antibiotic produced by Pseudomonas fluorescens. *Molecular Plant Microbe Interaction*, **17**: 1201-1211.

Schroth, M.N. and Hancock, J.G., 1981. Disease suppressive soil and root colonizing bacteria. *Science*, **216**: 1376-1381.

Shanahan, P., O'Sullivan, D.J., Simpson, P., Glennon, J.D. and O'Gara, F., 1992. Isolation of 2,4-Diacetylphloroglucinol from a fluorescent pseudomonad and investigation of physiological parameters influencing its production. *Applied Environmental Microbiology*, **58**: 353-358.

Shapira, R., Ordentlich, A., Chet, I., and Oppenheim, A.B., 1989. Control of plant diseases by chitinase expressed from cloned DNA in *Escherichia coli*. *Phytopathology*, **79**: 1246-1249.

Sharma, A.K., Johri, B.N. and Gianinazzi, S., 1992. Vesicular-arbuscular mycorrhizae in relation to plant disease. *World Journal of Microbiology Biotechnology*, **8**: 550–563.

Siddique, Z.A. and Mahmood, I., 1996. Biological control of *Heterodera cajani* and *Fusarium udum* by *Bacillus subtuilis*, *Bradyrhizobium japonicum* and *Glomus fasciculatum* Pigeonpea. *Fundamental Application of Nematology*,**18(6)**: 559-566.

Siddiqui, Z.A., 2006. PGPR: Prospective biocontrol agents of plant pathogens. In: *PGPR: Biocontrol and Biofertilization* (Ed.) Z A Siddiqui), Springer, The Netherlands, p. 111-142.

Sindhu, S.S., Gupta, S.K. and Dadarwal, K.R., 1999. Antagonistic effect of *Pseudomonas* spp. on pathogenic fungi and enhancement of growth of green gram (*Vigna radiata*). *Biol. Fertil. Soils.*, **29**: 62-68.

Sindhu, S.S., Suneja, S., Goel, A.K., Paramar, N. and Dardarwal, 2002. Plant growth promoting effects of *Pseudomonas* sp. on co incoluation with *Mesorhizobium* sp. *Cicer* strain under sterile and wilt sick soil conditions. *Applied Soil Ecology*, **19**: 57-64.

Singh, Rajesh Kumar, Kumar, D. Praveen, Singh, Pratiksha, Solanki, Manoj Kumar, Srivastava, Supriya, Kashyap, Prem Lal, Kumar, Sudheer, Srivastava, Alok K., Singhal, Pradeep K. and Arora, Dilip K., 2013. Multifarious plant growth promoting characteristics of chickpea rhizosphere associated *Bacilli* help to suppress soil-borne pathogens. DOI: DOI 10.1007 / s10725-013-9870-z

Singh, V. and Joshi, B.B., 2007. Mass multiplication of *Trichoderma harzianum* on sugarcane press mud. *Indian Phytopathalogy*, **60**: 530-53.

Smith, K.P., Havey, M.J. and Handelsman, J., 1993. Suppression of cottony leak of cucumber with *Bacillus cereus* strain UW85. *Plant Disease*, **77**: 139-142.

Smith, S.N., Armstrong, R.A., Baker, M., Bird, R.A. Chohan, R.A., Hartel, N.A. and Whipps, J.M., 1999. Determination of *Coniothyrium minitans* conidial and germling lectin affinity by flow of cytometry and digital microscopy. *Mycological Research*, **103**: 1533-1539.

Smith, S.N., Chohan, R.A., Armstrong, R.A. and Whipps, J.M., 1998. Hydrophobicity and surface charge of conidia of the mycoparasite *Coniothyrium minitans*. *Mycological Research*, **102**: 243-299.

Srivastava, A.K., Singh, T., Jana, T.K. and Arora, D.K., 2001. Induced resistance and control of charcoal rot in *Cicer arietinum* (chickpea) by *Pseudomonas fluorescens*. *Canadian Journal of Botany*, **79**: 787–795.

Suleman, P., L-Musallam, A. and Menezes, C.A., 2002.The effect of biofungicide mycostop on *Ceratocystis radicicola*, the causal agent of black scorch on date palm. *Biocontrol*, **47**: 207–216.

Sylvie, M., Therese, C., Philippe, L. and Jos, M.R., 2009. Phenazine antibiotics produced by fluorescent pseudomonads contribute to natural soil suppressiveness to Fusarium wilt. *The ISME Journal*, **3**: 977–991.

Tiwari, A.K., 1996. Biological control of chick pea wilt complex using different formulations of *Gliocladium virens* through seed treatment. *Ph.D. Thesis*, submitted to G. B. Pant University of Agriculture and Technology, Pantnagar, UK, India, p. 167.

Uknes, S., Mauch-Manl, E., Moyer, M., Potter, S., Williams, S., Dincher, S., Chsndler, D., Slusarenko, A., Ward, E. and Ryals, J., 1992. Acquired resistance in Arabidopsis. *Plant Cell*, **4**: 645-656.

Validov, S., Mavrodi, O., Fuente, Ldl, Boronin, A., Weller, D., Thomashow, K. and Mavrodi, D., 2005. Antagonistic activity among 2,4-diacetylphloroglucinol producing fluorescent *Pseudomonads* sp. *FEMS Microbiology Letters*, p. 242-249.

Van Loon, L.C., Bakker, P.A.H.M., and Pieterse, C.M.J., 1998. Systemic resistance induced by rhizosphere bacteria. *Annual Review of Phytopathology*, **36**: 453-483.

Vishwa Dhar and Chaudhary, R.G., 1998. Disease of pigeonpea and field pea and their management. In: *Disease of Field Crop and their Management* (Ed.) TS Thind, National Agricultural Technology Information Centre, Ludhiana, India, p. 217-238.

Vision-2030, 2013. Indian Institute of Pulses Research, Indian Council Agricultural Research, Kanpur, UP.

Vock, N.T., Langton, P.W., and Pegg K.G., 1980. Root rot of Chickpea caused by *Phytophthora mnegaspernia var. soajae* in Queensland. *Australian Plant Pathology*, **9**: 117

Wei, G., Kloepper, J.W. and Tuzum, S., 1996. Induced systemic resistance to cucumber diseases and increased plant growth by plant growth-promoting rhizobacteria under field conditions. *Phytopathology*, **86**: 221-224.

Weller, D.M., 1988. Biological control of soil borne plant pathogens in the rhizosphere with bacteria. *Annual Review of Phytopathology*, **26**: 379-407.

Wilhite, S.E., Lunsden, R.D. and Strancy, D.C., 2001. Peptide synthetase gene *in Trichoderma virens*. *Applied Environmental Microbiology*, **67**: 5055-5062.

Woo, S.L., Scala, F., Ruocco, M. and Lorito, M., 2006. The molecular biology of the interaction between *Trichoderma* spp., phytopathogenic fungi, and plants. *Phytopathology*, **6**: 181–85

Xue, A.G.A.D., 2000. Effect of seed borne *Mycosphaerella pinodes* seed treatment on emergence foot rot severity and yield of field pea. *Canadian Journal of Plant Pathology*, **22(3)**: 248-253.

Yadav, A.K. and Yandigeri, M.S. and Vardhan, S., 2013. *Streptomyces* sp. S160: a potential antagonist against chickpea charcoal root rot caused by *Macrophomina phaseolina* (Tassi) Goid. *Annals of Microbiology*, p. 1-10.

Yude, C., Kaiwei, H., Fuji, L. and Jie, Y., 1993. The potential and utilization prospects of kinds of wood fodder resources in Yunnan. *Forestry Research*, **6**: 346-350.

Zaid, W.N. and Singh, U.S., 2004.Development of improved technology for the mass multiplication and delivery of Fungal (*Trichoderma*) and Bacterial (*Pseudomonas*) bio agents. *Journal of Mycology and Plant Pathology*, **34**: 732-745.

Zaidi, N.W. and Singh, U.S., 2004. Use of farm yard manure for mass multiplication and delivery of biocontrol agents, *Trichoderma harzianum* and *Pseudomonas fluorescens*. *Asian Agric. Hist.*, **52**: 165-172.

2016, Diseases of Pulse Crops and their Sustainable Management 463–483
Editors: Samir Kumar Biswas, Santosh Kumar and Gireesh Chand
Published by: BIOTECH BOOKS, NEW DELHI

Chapter 25

Microbial Approach to Control Fusarium Wilt for Sustainable Pigeon Pea Production

Deepak Kumar Verma[1]*, Balaram Mohapatra[2], Diganggana Talukdar[3], Vipul Kumar[4], Shikha Srivastava[5], Mamta Sahu[6], Mukesh Mohan[6] and Bavita Asthir[7]

[1]Department of Agricultural and Food Engineering,
[2]Environmental Microbiology Lab, Department of Biotechnology,
Indian Institute of Technology, Kharagpur – 721 302, West Bengal
[3]Department of Plant Pathology,
Assam Agricultural University, Jorhat – 785 013, Assam
[4]Department of Plant Pathology, [6]Department of Agricultural Biochemistry,
C.S. Azad University of Agriculture and Technology, Kanpur, Uttar Pradesh
[5]Department of Botany,
Deen DyalUpadhyay Gorakhpur University, Gorakhpur – 273 009, U.P.
[7]Department of Biochemistry, PAU, Ludhiana – 141 004, Punjab

Introduction

Increasing demand of food by the overwhelming population of the world has led to serious environmental impact. For balanced food supply and fulfilling food

* Corresponding Author: E-mail: deepak.verma@agfe.iitkgp.ernet.in, rajadkv@rediffmail.com

demand, crop yield plays a crucial role and it depends on plant health and its management. One of the major concern for plant health is plant diseases; a prime biotic constraint, leading to significant crop losses and hence productivity worldwide. Plant diseases caused by variety of causal agents (bacteria, fungi, viruses and nematodes) reduce crop yields and is dependent on climatic conditions. It is estimated and supported by the Consultative Group on International Agricultural Research (CGIAR; http://www.cgiar.org/) that 10-15 per cent of the yield loss occurs in developing countries due to disease attack, and loss is severely affected if post-harvest damage is considered. In 2004, Food and Agricultural Organization (FAO) estimated that the loss effect is more serious in developing countries due to high food demand and around 800 million people do not have enough food, with 1.3 billion live on less than one dollar a day. Waddington *et al.* (2010)studied regarding production constraints due to diseases for major crops (sorghum, chickpea, pigeon pea and cowpea) in 13 Asian and African farming systems and reported that losses caused by diseases ranged from 3-14 per cent, whereas yield loss due to all biotic stresses ranged from 16- 37 per cent. The disease causing agents are diverse and affect plant growth at different stages *viz.* seedling, flowering and post-harvesting. The diseases also induce diverse symptoms at various growth stages critically affecting plant immunity. Along with it, introduction of new manmade crops or cultivars and crop intensification has also introduced serious epidemics around the world (Buddenhagen, 1977). So, plant diseases need to be controlled and more emphasis should be given to plant health management to maintain the quality and abundance of food to mitigate the food demand. Different approaches may be used to control plant diseases like good agronomic practices (use of clean and high-quality seed, optimally fertilized soil, irrigation and management practices), use of micro-biocides. These breakthroughs to agriculture have contributed spectacularly in crop productivity over the few decades. But in the recent trend, due to use of these chemicals, environmental pollution has become one of the world's major concerns, due to its effects of bio-accumulation, bio-magnification and bio-diversity loss. Detection and replacement of these toxic chemical compounds in the environment, particularly in water and soil and their biological effects on organisms has therefore become increasingly important (Mohapatra *et al.*, 2013). So a sustainable and eco-friendly approach of crop production is the only way out for plant health management. The most recent, effective and holistic approach is Integrated disease management (IDM) which combines biological, physical and chemical control strategies which proved to be more effective. Among these, researches have been focusing more on microbial control of disease management as a swift alternative to chemical fungicides due to its high specificity to plant pathogens, easy degradability after usage, and low cost of production. Recently, microbial approaches as a useful technique of organic and eco-friendly agriculture have been initiated by using antagonistic microorganisms to combat the various diseases in most of the crops as an emerging tools for pigeon pea diseases which is devoid of chemical substances and made possible to control the targeted organism without being harmful to humans or any beneficial organisms in natural eco-systems. Under increased disease incidence use of promising microorganisms or their cellular products has been attracted attention recently and more emphasis has been put on the aspect of microbial disease control (MDC). As most of the pathogens are soil born and belong to microbial groups (mainly bacteria and fungi), controlling them with the use of target specific microorganisms are promising and attempted intensively (Shan-da and Baker, 1980; Baker, 1988). Among the bacterial control agents; *Pseudomonas* (de-Freitas and Germida, 1991), *Bacillus* (Utkhede, 1984; Silo-suh *et al.*, 1994), *Agrobacterium* (Kerr, 1980), *Alcaligenes* (Martinetti and Loper, 1992), *Burkholderia* (Homma *et al.*, 1989), *Pantoea*(Sandra *et al.*, 2001) and *Streptomyces* (Cook *et al.*, 1996) are promising groups. Along with bacterial groups, many soil fungal genera like: *Trichoderma* spp., *Gliocladium* spp., *Taloromyces* spp., *Ampelomyces* spp., and *Aspergillus* spp., are used against phytopathogenic fungi and nematodes for reducing disease incidence in most of the pulse crops and among all, pigeon pea is found to be attacked by potential pathogens and having severe disease consequences (Brimner and Boland, 2003).

Pigeon Pea at a Glance

The pigeon pea [*Cajanus cajan*, (L.) Millspaugh] is a diploid (2n = 22, 44, or 66 chromosomes), belonging to the family *Fabaceae* and also known as red gram, Congo pea, gungo pea, no eye pea, dhal, gandul, gandure, frijol de árbol, and poiscajan, occurs in several varieties (Long and Lakela 1976). It is an important grain legume crop of rain-fed agriculture in the semi-arid tropics (Mallikarjuna *et al.*, 2011), which have probably originated in India, but may have come from Africa and both are centers of diversity for the genus *Cajanus* (van der Maesen 1990). Today, pigeon pea is the 5^{th}prominent pulse crop in the world and 2nd most important pulse crop after chickpea in India (Patel and Patel, 2012). This pulse crop is widely cultivated in all tropical and semi-tropical regions of both the old and the new world including Florida, Puerto Rico, and the U.S. Virgin Islands (Long and Lakela 1976; Mallikarjuna *et al.*, 2011). World production of pigeon pea is estimated as 4.3 million tons[2] and about 82 per cent of this is grown in India. Along with cereal food, it is used as both proteinaceous food crops as nutritional alternative for human consumption, animal feed and also grown as forage / cover crop and it symbiotically fixes 90 kg nitrogen per hectare (Adu-Gyamfi *et al.*, 1997). As a protein food supplement, it contains high levels of protein and mostly some important essential amino acid like, methionine, lysine and tryptophan with phenyl alanine+tyrosine found to be of higher in content (110.4 mg / g of protein).

Pigeon Pea Diseases at a Glance

Many pigeon pea natural enemies (*viz. Alternaria* spp, *Colletotrichum* spp., *Cercospora indica, Sclerotium rolfsii, Rhizoctonia* spp, *Fusarium* spp, *Phytophthora* spp, *Xanthomonas* spp, *Pseudomonas* spp) reported by various researchers (Kannaiyan *et al.*, 1984; Hillocks *et al.*, 2000; Joshi *et al.*, 2001; Maisuria *et al.*, 2008) and some of them are summarized in Table 25.1. As a pulse (protein) crop, Pigeon pea suffers from over 210 pathogens, among them 83 fungi, 4 bacteria, 19 viruses and mycoplasma and 104 nematodes are spreading in 58 countries (Reddy *et al.*, 1990; Nene *et al.*, 1996). The microbial agents used against the pigeon pea pathogens belong to the genera *Bacillus, Pseudomonas* and *Rhizobium* as bacteria and fungi such as non-pathogenic and non-host *Fusarium* species (Chérif *et al.*, 2007). Both *in vitro* and field study showed significant reduction of disease incidence (Chérif *et al.*, 2007). The mechanism for microbial management of pigeon pea (Chérif and Benhamou, 1990; Chérif *et al.*, 2002;

Table 25.1: Major Diseases of Pigeon Pea, Pathogenic Agents, and their Distribution

Disease	*Pathogenic Agent*	*Distribution*
	Fungal Diseases	
Alternaria blight	*Alternaria* spp., *Alternaria alternata* (Fries) Keissler, *Alternaria tenuissima* (Kunze ex Persoon) Wiltshire	India, Kenya, and Puerto Rico
Anthracnose	*Colletotrichumcajani* Rangel, *Colletotrichum truncatum*, *Colletotrichum graminicola* (Ces.) Wilson	Brazil, India, Puerto Rico and USA (Hawaii)
Botrytis gray mold	*Botrytis cinerea* Persoon ex Fries	Bangladesh, India, Nepal, and Sri Lanka
Cercospora leaf spot	*Mycovellosiella cajani* (*Cercospora cajani*), *Cercospora indica*, *Cercospora instabilis*, *Cercospora thirumalacharii*	Bangladesh, Brazil, Colombia, Dominican Republic, Guatemala, India, Jamaica, Kenya, Malawi, Mauritius, Nepal, Nigeria, Puerto Rico, Sri Lanka, Tanzania, Trinidad and Tobago, Uganda, Venezuela, Zambia, and Zimbabwe
Collar rot	*Sclerotium rolfsii* Saccardo, *Athelia rolfsii* [teleomorph] = *Corticium rolfsii*	India, Pakistan, Puerto Rico, Sri Lanka,Trinidad and Tobago, USA, and Venezuela.
Dry root rot	*Macrophomina phaseolina* (Tassi), Goidanich *Rhizoctonia bataticola* (Taub.) Butler	India, Jamaica, Myanmar (Burma), Nepal, Sri Lanka, and Trinidad and Tobago
Fusarium leaf blight	*Fusarium pallidoroseum* (*Fusarium semitectum*)	India and Puerto Rico
Fusarium wilt	*Fusarium udum* Butler, *Gibberella indica* [teleomorph]	Bangladesh, Ghana, Grenada, India, Indonesia, Kenya, Malawi, Mauritius, Myanmar (Burma), Nepal, Nevis, Tanzania, Thailand, Trinidad and Tobago, Uganda, and Venezuela
Phoma stem canker	*Phoma cajani* (Rangel)	Brazil and India
Phyllosticta leaf spot	*Phyllosticta cajani* Sydow	Brazil, India, Jamaica, Puerto Rico, Sri Lanka, and Trinidad and Tobago
Phytophthora blight	*Phytophthora drechsleri* Tucker *f. sp. cajani*	Dominican Republic, India, Kenya, Panama, and Puerto Rico
Powdery mildew	*Leveillula taurica* [Teleomorph], *Oidiopsis taurica* [Anamorph], *Ovulariopsis ellipsospora*	Ethiopia, India, Kenya, Malawi, Sri Lanka, Tanzania, Uganda, and Zambia

Contd...

Table 25.1–*Contd...*

Disease	*Pathogenic Agent*	*Distribution*
Rust	*Uredo cajani* Sydow	Bermuda, Colombia, Guatemala, India, Jamaica, Kenya, Nigeria, Puerto Rico, Sierra Leone, Sri Lanka, Tanzania, Trinidad and Tobago, Uganda, and Venezuela
	Bacterial Diseases	
Bacterial leaf spot and stem canker	*Xanthomonas campestris* pv. *cajani*	Australia, India, Myanmar (Burma), Panama, Puerto Rico, and Sudan
Halo blight	*Pseudomonas amygdali* pv. *phaseolicola*	Australia, Ethiopia, and Zambia
	Viral Diseases	
Phyllody	Mycoplasma-like organismVector: Not known	India, Myanmar (Burma), and Thailand
Sterility mosaic	Vector:Eriophyid mite *Aceria cajan i*Channabasavanna	Bangladesh, India, Myanmar (Burma), Nepal, and Sri Lanka
Witches' broom	Mycoplasma-like organismVector: Leaf hopper *Empoasca* spp.	Australia, Bangladesh, Costa Rica, Dominican Republic, El Salvador, Haiti, Jamaica, New Guinea, Panama, Puerto Rico, Taiwan, Trinidad and Tobago, and USA
Yellow mosaic	Mung bean yellow mosaic virus (MBYMV) Vector	India, Jamaica, Nepal, Puerto Rico, and Sri Lanka
	Nematode Diseases	
Dirty root (Reniform nematode)	*Rotylenchulu sreniformis* Linford and Oliveira	Fiji, India, Jamaica, Puerto Rico, and Trinidad and Tobago
Pearly root (Cyst nematode)	*Heterodera cajani* Koshy	Egypt and India
Root-knot (Root-knot nematode)	*Meloidogyn eacronea*Coetzee, *Meloidogyne arenaria* (Neal) Chitwood, *Meloidogyne incognita* (Kofoid and White) Chitwood, *Meloidogyne javanica*(Treub) Chitwood	

Source: Reddy *et al.* (1993b).

2003; Fuchs *et al.*, 1997) may include as following; 1) Antibiosis, 2) Antifungal enzyme production, 3) Competition for infection sites, 4) Direct / Hyper parasitism, 5) Induced systemic resistance or Enhanced host resistance, and 6) Saprophytic competition for nutrients. The said mechanisms for microbial management of pigeon pea have different modes of action which are not necessarily exclusive of one another, and many of these mechanisms may be synergistically active and used by the same microbial agent (Chérif *et al.*, 2002; Mandeel and Baker, 1991) which should lead to control of the major diseases of pigeon pea. The major diseases that assume significant importance include wilt (*Fusarium udum* Butler), Powdery mildew (Reddy *et al.*, 1993a), sterility mosaic (Pigeon pea sterility mosaic virus) and phytophthora blight (*Phytophthora drechsleri*) (Kannaiyan *et al.*, 1984). The major disease, fusarium wilt (FW) and its microbial control has been describedin details.

Fusarium Wilt

Fusarium udum (Butler) is the causal soil borne pathogenic fungal agent of *Fusarium* wilt (FW) severely affecting seed yield, production, economy and demand in India (Kannaiyan *et al.*, 1984; Ajay *et al.*, 2013). The loss is dependent on the stage at which crop is attacked, stage at which the plants wilt and environmental parameters, where the loss can be 100 per cent, if disease develops at pre-pod stage, about 67 per cent and 30 per cent when wilt occurs at crop maturity and pre-harvest stage, respectively (Kannaiyan and Nene, 1981; Sheldrake *et al.*, 1984) and the total loss is estimated to be approximately 97,000 ton per year in India (Saxena *et al.*, 2002). In Indian scenario, the disease incidence in percentage varies from 0.1 to 22.6 per cent (Kannaiyan *et al.*, 1984; Upadhyay and Rai, 1992) reported minimum in Rajasthan and maximum in Maharashtra, respectively. Fusarium wilt disease incidence is believed to have increased significantly over the time (Gwata *et al.*, 2006) with an average of 10-15 per cent incidence and 16–47 per cent of crop loss (Prasad *et al.*, 2003). Reddy *et al.* (1993a) reported the annual loss due to this disease is estimated at US $71 million with an economic loss of 470, 000 ton of grain in India and 30,000 ton of grain in Africa (Joshi *et al.*, 2001). Worldwide crop loss due to wilt reported by Kannaiyanet al. (1984) and it was found to be 15.9 per cent (0-90 per cent), 36.6 per cent (0-90 per cent) and 20.4 per cent (0-60 per cent) in Kenya, Malawi and Tanzania respectively with annual loss estimated at US $ 5 million in each of the countries with 96 per cent of disease incidence in Tanzania (Mbwaga, 1995).

Pathogenic Agent

Fusarium wilt disease of pigeon pea was firstly reported in India by Butler in 1906 (Butler, 1906). The causal organism was described as *F. udum* by him in 1990 (Butler, 1910)and was subsequently described as multiple names *F. butleri*, *F. uncinatum*, *F. lateritium*var. *uncinatum*, *F. oxysporum* f. sp. *udum*, *F. lateritium* f. sp. *cajani*and *F. udum*f. sp. *Cajani* (Dhar *et al.*, 2005). Due to having well distinguished prominent hook shaped macro-conidia the name *F. udum*was accepted as an imperfect state (Booth, 1971). *F.udum*is a host specific (pigeon pea) (Padwick, 1940; Subramanian, 1963; Booth, 1971) with consistent pathogenic variability and both cultural and morphological differences. Shit and Gupta (1980) collected seven isolates of *F. udum*from pigeon pea from various part of India and confirmed the varied cultural

Figure 25.1: *Fusarium* Wilt Symptoms in the Pigeon Pea Field Appear during Flowering and Podding of Early Developmental Stages. Pictures adopted from Reddy *et al.* (1993b).

characters such as aerial mycelium, texture and sporulation ability. Pawar and Mayee (1983) published a research paper that revealed five category of *F. udum* on the basis of virulence differences. Patil (1984) reported 9.4-12.0 x 3.1-3.3 μm size of conidia, 19.2 x 3.5-5.0 μm of macro conidia and it was mostly found to be whitish in the basal medium. Madhukeshwara and Sheshadri (2001) described six isolates of *F. udum* with different colony characteristics, pigmentation and sporulation. 195 isolates of *F. udum* has been isolated (IIPR, 2007-08) and revealed that 135 were highly pathogenic (>50 per cent wilt), 33 moderately pathogenic (30-50 per cent wilt) and 32 were weak pathogenic (<30 per cent wilt) agent.

Distribution

Currently, Fusarium wilt diseases is considered as highly destructive (Nene *et al.*, 1989) and distribution occurs in severalcountries *viz.* Bangladesh, Ghana, Grenada, Grenada, India, Indonesia, Kenya, Malawi, Mauritius, Myanmar (Burma), Nepal, Nevis, Tanzania, Thailand, Tobago, Trinidad, Uganda, Venezuela and Zambia where field losses of over 50 per cent are more prevalent and common in India, east Africa and Malawi (Kannaiyan *et al.*, 1984; Kimani, 1991; Reddy *et al.*, 1993a; Marley and Hillocks, 1996; Ajay *et al.*, 2013).

Disease Symptoms

The first indication of Fusarium wilt disease symptoms usually appear in the field during early developmental stages (Figure 25.1) and when the crop is flowering and podding stages (Prasad *et al.*, 2003) but never visible until later in crop developmental stages of pigeon pea (Reddy *et al.*, 1990; Hillocks *et al.*, 2000). The pathogen *F. udum* enters to the host vascular system through wounds leading to

Figure 25.2: A) Development of dark purple bands on the stem surface extending upwards from the base, B) Prominent internal browning and blackening on 1-2 months old plants die from wilt, C) Visible black streaks in xylem strands on the main stem or primary branches when it split open, and D) Die-back symptoms with a purple band on branches extending from tip of the plant to downwards and starts drying. Pictures adopted from Reddy *et al.* (1993b).

progressive chlorosis of leaves and branches, wilting and collapse of the root system (Jain and Reddy, 1995; Butler, 1906). Loss of turgidity in leaves and interveinal clearing appear as initial visible symptoms of the pathogen in which slight chlorosis shown by leaves and before wilting, sometimes look like bright yellow (Reddy *et al.*, 1990). The disease is distinguished from termite damage, drought, and phytophthora blight due to partial wilting of the plant as if there is water shortage even though the soil may have adequate moisture associated with lateral root infection, while total wilt is due to tap root infection (Nene, 1980; Reddy *et al.*, 1993).

Systemic symptoms appears on xylem strands as black streaks and this results in development of dark purple bands on the stem surface of partially wilted plants extending upwards from the base (Figure 25.2A) and can be visible when the main stem or primary branches are split open (Figure 25.2C) (Reddy *et al.*, 1990; Reddy *et al.*, 1993). The intensity of browning or blackening decreases from the base to the tip of the plant and sometimes lower branches starts drying, even if there is no band on the main stem. These branches have die-back symptoms with a purple band extending from tip downwards, and intensive internal xylem blackening depicted in Figure 25.2D (Reddy *et al.*, 1993). When young plants (1-2 months old) die from wilt, they may not show the purple band symptom (Figure 25.2B), but shows prominent internal browning and blackening as a serious disease consequence.

Microbial Control

The adverse effect of most fungicides and pesticides to the farmland and more reliability on sustainable agriculture, microbe mediated control has been a promising and attractive alternative for management of plant health and soil borne pathogens. For an eco-friendly and sustainable management of Fusarium wilt disease, there are many novel microorganism *viz. Aspergillus* spp., *Bacillus* spp., *Trichoderma* spp., *Pseudomonas* spp. and *Pantoea* spp., were evaluated against *F. udum* (Upadhyay and Rai, 1981; Bhatnagar, 1996; Somasekhara *et al.*, 1996; 1998; Gundappagaland Bidari, 1997; Biswas and Das, 1999; Prasad *et al.*, 2002; Khan and Khan, 2002;Anjaiah *et al.*, 2003; Sawant *et al.*, 2003; Roy and Sitansu, 2005; Dhar *et al.*, 2006; Maisuria *et al.*, 2008; Ram and Pandey, 2011). Many beneficial rhizobacteria as bio-inoculants have been reported by Pusey (1989), Upadhyayand Rai, (1992), Bapat and Shar (2000), Siddiqui *et al.* (2005), Siddiqui, (2006), Siddiqui and Shakeel (2007). It has been shown that fungal or bacterial antagonists of pathogen inoculated to soil reduces Fusarium wilt and its pathogenesis (Bapat and Shar, 2000; Singh *et al.*, 2002; Anjaiah *et al.*, 2003; Mandhare and Suryawanshi, 2005; Maisuria *et al.*, 2008). Inoculation with *T. harzianum* reported a disease control of 22-61.5 per cent by Prasad *et al.* (2002).Upadhyay and Rai, (1981) reported many species of fungi mostly *A.niger*, *A. flavus*, and *A terreus*which could be used for suppression of the population of *F. udum*. In a different study of Khan and Khan (2002) observed that rhizospheric application of *B. subtilis*, *P. fluorescens*, *A. awamori*, *A. niger*and *Penicillium digitatum* resulted in significant decline of *F. oxysporum* in the vicinity of root. Anjaiah *et al.* (2003) reported that disease incidence of wilt was drastically reduced after inoculation of *P. aeruginosa* (PNA1) to booth chick-pea and pigeon pea in naturally infested soil. Soil antagonistic bacteria also have to be known for their function through suppressing the wilt through induction of resistance (Upadhyay and Rai, 1981; 1992). Mahesh *et al.* (2010) recommended integrated management in a combined way (systemic fungicide, bio-control agent and FYM) as the most effectuate treatment of *F. udum* to control its infestation in a large scale management process. In the context of variable pathogenesis and multiple action of bio-inoculants, Bhatnagar (1996) studied the antifungal activity of three *Trichoderma* spp. against wilt pathogen at different pH, temperatures and C/N ratios and found that all of them were equally efficient and showed maximum antagonistic properties at 35 ± 2 °C temperature and pH of about 6.5.

Apparently, Somasekhara *et al.* (1996) worked on two delivery systems (seed treatment and foliar application) by using six isolates of *Trichoderma* spp. and studied their efficacy which was found to be extreme on the 35th day of inoculation. As the plant is resistance to dry period, Gundappagaland Bidari (1997) used *T. viride* for seed treatment to resistant cultivar and became effective in integrated disease management of pigeon pea under dry land cultivation. As *Trichoderma* spp. are known to be producer of extracellular volatile compound, which was found to be most toxic to wilt pathogen (Pandey and Upadhaya 1997). Somasekhara *et al.* (1998) observed that non-volatile antibiotics of *T. viride* was highly toxic followed by *T. harzianum*, and *T. koningii*. *In vitro* condition of reducing pathogenesis was studied by Biswas and Das (1999) and among five *Trichoderma* spp. tested, *T. harzianum* was most effective antagonist followed by *T. hamatum, T. longiconis* and *T. koningii*. They also noted that of pigeon pea treated with *T. harzianum* seed failed to reduce pathogen growth while soil amendment with *T. harzianum* in maize meal: sand applied at 40-60 g/kg soil resulted a significant reduction of wilt up to 90 per cent. Comparison of different biological control agents and their products showed reduced wilt incidence of pigeon pea, and seed treatment with its formulated cell mass at 8 g/kg of seed recorded the lowest wilt incidence (Sawant *et al.*, 2003). Khan and Khan (2002) confirmed the differential behavior of multiple bio-control agents (*Trichoderma, Bacillus, Pseudomonas*) for controlling FW and recorded 17-48 per cent of decrease disease incidence. Many mutational full and recombinant bio-inoculants have been tried in this field to reduce the wilt incidence and found to be successful. Roy and Sitansu (2005) found that among the recombinant, *T. harzianum*, 50Th3II and 125Th4I reduced the wilt disease in non-sterilized soil, while 75Th4IV was in sterilized soil with a percentage of 36.51, 33.86 and 33.33 per cent, respectively. Combined effect of *T. viride, T. harzianum* and *Gliocladium virens* showed upto 35.5-57.3 per cent of reduction in disease incidence in Fusarium wilt of pigeon pea Dhar *et al.* (2006). Ram and Pandey, (2011) suggested the combined use of *T. viride* and *P. fluorescens* for reduction of growth of *F.udum*. Several *Bacillus* spp. have also been proved to be used as bio-control agents for reduction of growth of pathogen and disease incidence around the globe (Siddiqui, 2007). Isolation of an indigenous *Bacillus* spp. from the disease suppressive soil of the same environment may increase the probability of efficiency of the strains and can be used as potential agents for disease suppression (Cook and Baker, 1983; Weller *et al.*, 1985). Several Bacillus isolates (B603, B613, B615) were used as bio-control agents against *Fusarium* for its antifungal activity in both pot and field experiments and found to be effective in terms of reduction in fungal growth and disease incidence (Siddiqui and Shakeel, 2007).

Opportunities and Challenges

Opportunities

Pigeon pea is a source of protein for vegetarian diet and resource poor farmers in the rainfed tropics. It has built in resilience to withstand drought and can yield even under very low input conditions. Significant efforts to broaden the genetic base and introduction of various traits for desirable biotic and abiotic stress are one of the important aspects of "Microbial Approach to Control *Fusarium* Wilt Diseases for

Sustainable Pigeon Pea Production". There is renewed interest to exploit more wild relatives from the secondary gene pool, and such efforts would have a big impact on broadening the genetic base of variation of pigeon pea and introduction of useful biotic, abiotic and agronomic traits. The possibility of exploiting wild relatives from the tertiary gene pool has opened up new vistas for the broadening of the genetic base of variation and for improvement in pigeon pea. Development of genomic resources has gained new impetus with community effort, and the development of genome-wide markers may open avenues for molecular marker-assisted gene introgressions and breeding. It has been learned from the biological research conducted over the long years that microbial agents are the way of eco-friendly and sustainable approach for agriculture. But, bio-control researchers need to be look forward to define new area which will help the bio-control technologies and applications. Currently, fundamental knowledge in computing, molecular biology, biotechnology, statistics and chemistry have led to new research aimed at characterizing the functions of bio-control agents, pathogens, and host plants at sub-cellular and ecological levels. In the present crop production scenario, the bio-control agents are of supreme importance, but its potential is still to be exploited and very little attention has been paid to produce the commercial formulations of it. The commercially produced formulations have not been used efficiently by the farmers due to lack of information regarding its use. There are many research challenges in the area of exploration of microbial agents for controlling plant diseases and discussed in five major points below. Many of the challenges need to be addressed by the scientific community to solve the issue of use of multiple microbial agents, their combined action of killing the plant pathogen.

Challenges

There are several challenges have arisen reported by many researchers which are as follows:

In Field Application

The potential antagonists from *in vitro* tests often fail to effectively work in field. Several factors like, organic matter, pH, nutrient level, and moisture level of the soil influence its efficacy (Lee *et al.*, 1999). Due to the variations in environmental factors, a good bio-control agent fails in field under *in vitro* conditions. To achieve the success, agents must be enriched and isolated from such surroundings. Likewise, the method of application influences the success of field trials. There are three means of applying the antagonistic agent, they are; **1)** seed inoculation,**2)** vegetative part inoculation, and **3)** soil inoculation. Lee *et al.* (1999) showed in a field trial that the control efficacy of *Burkholderia cepacia* strain N9523 against *Phytophthora capsici*, was higher by soil-drenching than by wounded stem inoculation. Dawar *et al.* (2008) showed the bio-control potential of different microbial antagonists, *i.e. R. meliloti*, *A. niger* and *Trichoderma harzianum* and *B. thuringiensis* by coating the seeds with glucose, sucrose and molasses. This method has successfully reduced the infection of root rot fungi, *i.e. Macrophomina phaseolina*, *Rhizoctonia solani* and *Fusarium* spp. The highest suppression capacity was shown by seed treatment with *T. harzianum* using 2 per cent of glucose.

In Mixtures of Multiple Antagonists and their Efficacy

Association of several microorganisms is needed to control most pathogens in field. Most of the microbial agents are specific for a target pathogen. Whipps (1997) reported positive, synergistic, interaction between *Trichoderma* spp. and antagonist *P. syringae* for combined applications in the control of plant pathogens. The mechanism could be explained with the positive interaction of lipopeptides of the bacterial antagonist and cell wall-degrading enzymes of the fungal bio-control agent (Fogliano *et al.*, 2002). Multiple traits antagonizing of the pathogen can be achieved with a higher level of protection by combining the strains of microorganisms. Combination of *P. putida* strain WCS358, (produces pseudobactin siderophore) with *P. putida* strain RE8, systemic resistance against *F. oxysporum* was significantly enhanced (de Boer *et al.*, 2003). A mixture of three different bio-inoculants applied as a seed treatment, showed growth promotion and decrease multiple cucumber diseases (Raupach and Kloepper, 1998). Another innovative approach for controlling pathogenesis could be the cocktails containing strains that cross talk with each other for maximum antibiotic production and disease control (Becker *et al.*, 1997; Davelos *et al.*, 2004).

In Genetic Manipulation

Molecular genetics and genomics are the tools shows new possibilities for improving the characterization, selection and management of microbial control. Development in functional genomics-proteomics can give us the expression of crucial genes of BCA's during mass production, application and mechanism of action. Molecular techniques for modification of strains to improve their ability to control soil borne diseases are new and promising one. Deletion of one or more genes, coming from different antagonists, could be added, expressed in regulatory way can enhance their ability of action. Four methods are usually employed for genetic manipulation of fungi, they are; **1)** conventional mutagenesis by chemicals (Howell and Stipanovic, 1983) or by ultraviolet light (Baker, 1989), **2)** protoplast fusion (Harman *et al.*, 1989), **3)** transposon mutagenesis (Brown and Holden, 1998), and **4)** transformation. The role of siderophore production in the rhizosphere has been studied with molecular methods (Haas, 2003) and a strain of *P. fluorescens* was genetically modified to utilize ferric siderophores (Moenne Loccoz *et al.*, 1996). A new and challenging genetic transformation of bio-control agents could be the insertion of genes for abiotic stresses (genes for increased tolerance or resistance to cold, heat, drought, high salinity, heavy metal). Apart from that, if genetically modified (GM) antagonists could be used efficient control over plant pathogen could be achieved.

In Whole-genome Analysis

The revolutionary high through put DNA sequencing of whole genomes have resulted tremendous success for understanding the mechanism of action of BCAs. 40 microbial genome sequencing has been completed (http://www.genomesonline.org/organisms? Organism.Domain=BACTERIAL) and another 200 are in progress. Among these are several rhizosphere-inhabiting bacteria such as *P. aeruginosa P. putida, P. fluorescens, P. syringae, Sinorhizobium meliloti, Mesorhizobium loti, Rhizobium leguminosarum, B. subtilis* and *Streptomyces coelicolor* (Kaneko *et al.*, 2000;Kunst *et al.*, 1997). Building of a genomic encyclopedia of the rhizobacterial strain *P. fluorescens* SBW25 (PfSBW25) on the basis of short-run noncontiguous sequence (Spierset al.,2001) and by both Illumina and Ion torrent platform. Comparative and functional genomics by genome assembly and annotation facilitate the identification of genes in rhizosphere bacteria and expressed on the seed or in the rhizosphere, which are involved in the regulation and production of secondary metabolites. The construction of artificial chromosome (BAC and YAC) libraries gene expression study and identification of genes of interest is of great value, especially bacteria whose genome has not been sequenced, but acting as BCAs (Rondon *et al.*, 1999).

In Formulation and Methods of Application

In organic farming using microbial control agents (MCAs), the narrowness of the technology is represented by without compromising the yield and lowering the selling price of the product. For biological control, the major difficulty is to reach the market and to compete with the chemical fungicides and its shelf life. All problems can be faced with a better solution of formulation of bio-control agents. Until now formulation are being carried out without a methodology. The most advantage of formulation include greater efficacy, increased safety, lower production costs, increased shelf life, ease of handling and compatibility with agricultural practices.

Future Research Needs

As the world is constantly facing the food problem, development of well adapted and resistant pigeon pea varieties is the most efficient and economical method for control of fusarium wilt. Several germplasm have been identified for the plant. The resistant genotypes with high yield and good agronomic traits can be adopted by the farmers very easily and reliably. Knowledge on mode of inheritance, disease spreading ability, its pathogenesis efficiency of fusarium wilt and other agronomic traits need to be well understood. Mapping of the *Fusarium* genes completely is recommended for future. This will help the ignorance of developing the resistant pigeon pea cultivars through the use of marker-assisted selection which found to be a hazy task. Study of morphovars, serovars of the pathogens will help to identify the presence of different races of *F. udum* for better knowledge of its disease pathogenesis in near future.

Conclusion

Pigeon pea is an important protein rich food of vegetarian diet. It is a favorite crop of small holder farmers as the crop can tolerate and yield high under drought conditions when many other crops fail. Pigeon pea yield has reached a plateau and is susceptible to arrange of diseases caused by virus, fungi and bacteria. Crop improvement programs are looking for increased genetic diversity by tapping wild relatives from different gene pools. But plant health management through eco-friendly approach has been a promising and new area of research these days providing a new insight for better health and increased productivity in plants by reducing the disease incidence. Biological control of plant pathogens dates back to 13th century where it was used for potato scab disease, but current research on microbial management of plant disease is becoming more and more attractive and dynamic by many scientists after the International Symposium held at Berkeley on soil borne plant pathogens: their biological control in 1963. Despite many decades of research in biological control, bio-pesticides represent about 1 per cent of the global pesticide market and only 0.1-

0.5 per cent of the new microbial beneficial agents have been discovered and cultured. Microbial control agents still represent a specialized market and have demand among the farming community, growers and regulators. Microbial agents are generally characterized by a narrow spectrum of activity, with a positive impact on environmental but not from a commercialization perspective, so improvised research techniques in making better strains are becoming restricted. Despite the difficulties encountered by the development of bio-fungicide, fungistatic and bacteriocides there are many reasons for being optimistic. Different strategies are now currently being established to speed up the production and commercialization of bio-control agents and their effectiveness of control. As key elements of IPM programs, use of microorganisms can play an important role in a more complex vision of crop protection. Mixtures of antagonists, genetically modified (GM) antagonistic microbes, enhancement of production, formulation, and application technologies and study of genetic island of microbial agents will be useful tools to accomplish this difficult and exciting task of making it a more farmers friendly, Sustainable, easily acceptable and efficient agents of using it in a organic way of farming. An increased research focus on the bacterial, fungal groups associated with plants will strengthen our understanding of the complex interactions between these organisms and develop useful products for management of pathogens in the agricultural farmlands.

References

Adu-Gyamfi, J.J., Yoneyama, I.O., Devi, T.G. and Katayama, K., 1997.Nitrogen management and biological nitrogen fixation in sorghum/pigeon pea intercropping on alfisols of the semi-arid tropics. *Soil Sci. Plant Nutr.*, **43**: 1061-1066.

Ajay, B.C., Prasad, P.S., Gowda, M.B., Ganapathy, K.N., Gnanesh, B.N., Fiyaz, R.A., Veerakumar, G.N., Babu, H.P., Venkatesha, S.C. and Ramya, K.T., 2013. Inheritance of resistance to Bangalore race of *Fusarium* wilt disease in pigeonpea (*CajanuscajanL.*). *Austr. J. Crop Sci.*, **7(10)**: 1520-1524.

Anjaiah, V., Cornelis, P. and Koedam, N., 2003. Effect of genotype and root colonization in biological control of Fusarium wilts in pigeonpea and chickpea by *Pseudomonas aeruginosa* PNA1. *Can. J. Microbiol.*, **49**: 85–91.

Baker, K.F., 1987. Evolving concepts of biological control of plant pathogens. *Ann. Rev. Phytopathol.*, **25**: 67-85.

Baker, R., 1989. Some perspectives on the application of molecular approaches to bio-control problems. *In*: Biotechnology of Fungi for Improving Plant Growth (eds. Whipps, M. and Lumsden, R.D.). Cambridge University Press, Cambridge, Great Britain, p. 220–223.

Baker, R., 1989. Some perspectives on the application of molecular approaches to biocontrol problems. *In*: Biotechnology of Fungi for Improving Plant Growth (eds. Whipps M. and Lumsden, R.D.). Cambridge University Press, Cambridge, Great Britain, p. 220–223.

Bapat, S. and Shar, A.K., 2000. Biological control of Fusarium wilt of pigeon pea by *Bacillus brevis*. *Can. J. Microbiol.*, **46(2)**: 125-132.

Becker, D.M., Kinkel, L.L. and Schottel, J.L., 1997. Evidence for interspecies communication and its potential role in pathogen suppression in a naturally occurring disease suppressive soil. *Can. J. Microbiol.*, **43**: 985–990.

Bhatnagar, H., 1996. Influence of environmental conditions on antagonistic activity of *Trichoderma* spp. against *Fusariumudum*. *Indian J. Mycol. Plant Pathol.*, **26**: 58-63.

Biswas, K.K. and Das, N.D. 1999. Biological control of pigeon pea wilt caused by *Fusariumudum* with *Trichoderma* spp. *Ann. Pl. Prot. Sci.*, **7**: 46-50.

Booth, C., 1971. The genus *Fusarium*. Common Wealth Mycological Institute, Kew, Surrey, U.K., p. 114.

Brimner, T.A. and Boland, G.J., 2003. A review of the non-target effects of fungi used to biologically control plant diseases. *Agri. Ecosy. Env.*, **100**: 3-16.

Brown, J.S. and Holden, D.W., 1998. Insertional mutagenesis of pathogenic fungi. *Curr. Opin. Microbiol.*, **1**: 390–394.

Buddenhagen, I.W., 1977. Resistance and vulnerability of tropical crops in relation to their evolution and breeding. *Ann. New York Aca. Sci.*, **287**: 309-326.

Butler, E.J., 1906. The wilt disease of pigeonpea and pepper. *Agril. J India.*, **1**: 25–26.

Butler, E.J., 1910. The wilt disease of pigeonpea and parasitism of *Neocosmospora vasinfecta* Smith. Memories of the Department of Agriculture in India, Botanical Series, **2**: 1-64.

Chérif, M. and Benhamou, N., 1990. Cytochemical aspects of chitin breakdown during the parasitic action of a *Trichoderma* sp. on *Fusarium oxysporum* f.sp. *radicisly copersici*. *Phytopathol.*, **80**: 1406-1414.

Chérif, M., Arfaoui, A. and Rhaiem, A., 2007. Phenolic compounds and their role in bio-control and resistance of chickpea to fungal pathogenic attacks. *Tunis. J. Plant Prot.*, **2**: 7-21.

Chérif, M., Sadafi, N., Benhamou, N., Hajlaoui, M.R., Boubaker, A. and Tirilly, Y., 2002. Ultrastructure and cytochemistry of *in vitro* interactions of the antagonistic bacteria *Bacillus cereus* X16 and *B. thuringiensis*55T with *Fusariumroseum*var. *sambucinum*. *J. Plant Pathol.*, **84**: 83-93.

Chérif, M., Sadfi, N. and Ouellette, G.B., 2003. Ultrastructure and cytochemistry of *in vivo* interactions of the antagonistic bacteria *Bacillus cereus* X16 and *B. thuringiensis* 55T with *Fusarium roseum* var. *sambucinum*, the causal agent of potato dry rot. *Phytopathol. Medite.*, **42**: 41-54.

Cook, R.J., Bruckart, W.L., Coulson, J.R., Goettel, M.S. and Humber, R.A., 1996. Safety of microorganisms intended for pest and plant disease control: a framework for scientific evaluation. *Biologi. Control*, **7**: 333-351.

Cook, R.J. and Baker, K.F., 1983. The Nature and Practice of Biological Control of Plant Pathogens. Ame. Phytopat. Soci., St. Paul, Minnesota.

Davelos, A.L., Kinkel, L.L. and Samac, D.A., 2004. Spatial variation in frequency and intensity of antibiotic interactions among *Streptomycetes* from prairie soil. *Appl. Environ. Microbiol.*, **70**: 1051–1058.

Dawar, S., Hayat, S., Anis, M. and Zaki, M.J., 2008. Effect of seed coating material in the efficacy of microbial antagonists for the control of root rot fungi on okra and sunflower. *Pak. J. Bot.*, **40 (3)**: 1268-1278.

de Boer, M., Bom, P., Kindt, F., Keurentjes, J.J.B., van der Sluis, I., van Loon, L.C. and Bajjer, P.A.H.M., 2003. Control of Fusarium wilt of radish by combining *Pseudomonas putida* strains that have different disease-suppressive mechanisms. *Phytopathol.*, **93**: 626–632.

de-Freitas, J.R. and Germida, J.J., 1991. *Pseudomonas cepacia* and *Pseudomonas putida* as winter wheat inoculants for biocontrol of *Rhizoctonia solani*. *Can. J. Microbial.*, 37: 780-784.

Dhar, V., Mishra, S. and Chaudhary, R.G. 2006. Differential efficacy of bioagents against *Fusarium udum* isolates. *Indian Phytopathol.*, **59**: 290-293.

Dhar, V., Reddy, M.V. and Chaudhary, R.G., 2005. Major Diseases of Pigeon pea and their Management. *In*: Advances in Pigeon pea Research (eds. Ali, M. and Kumar, S). Indian Institute of Pulses Research, Kanpur, India, p. 229-261.

Fogliano, V., Ballio, A., Gallo, M., Scala, F. and Lorito, M., 2002. *Pseudomonas lipodepsi* peptides and fungal cell wall-degrading enzymes act synergistically in biological control. *Mol. Plant-Microbe Interact.*, **15**: 323–333.

Fuchs, J.C., Moënne-Loccoz, Y. and Défago, G., 1997. Nonpathogenic *Fusarium oxysporum* strain Fo47 induces resistance to Fusarium wilt in tomato. *Plant Dis.*, **81**: 492-496.

Gundappagal, R.C and Bidari, V.B., 1997. *Trichoderma viride* in the integrated management of pigeon pea wilt under dryland cultivation. *Advances in Agril. Res. India*, 7: 65-69.

Gwata, E.T., Silim, S.N. and Mgonja , M., 2006. Impact of a new source of resistance to *Fusarium* wilt in pigeonpea. *J. Phytopathol.*, **154**: 62–64.

Haas, H., 2003. Molecular genetics of fungal siderophore biosynthesis and uptake: the role of siderophores in iron uptake and storage. *Appl. Micro. Biotechnol.*, **62**: 316–330.

Harman, G.E., Taylor, A.G. and Stasz, T.E., 1989. Combining effective strains of *Trichoderma harzianum* and solid matrix priming to provide improved biological seed treatment systems. *Plant Dis.*, **73**: 631–637.

Hillocks, R. J., Minja, E., Silim, S.N. and Subrahmanyam, P., 2000. Diseases and pests of pigeon pea in eastern Africa. *Inter. J. Pest Man.*, **46**: 7–18.

Homma, Y., Kato, Z., Hirayama, F., Konno, K., Shirahama, H. and Suzui, T., 1989. Production of antibiotics by *Pseudomonas cepacia*as an agent for biological control of soilborne plant pathogens. *Soil Biol. Biochem.*, **21**: 723-728.

IIPR (2007-08) Annual Report 2007-08 of Indian Institute of Pulses Research, IIPR, Kanpur, Uttar Pradesh, India, p. 33.

Jain, K.C. and Reddy, M.V., 1995. Inheritance of resistance to Fusarium wilt in pigeon pea (*Cajanus cajan* L.). *Indian J. Genet. Plant Breed.*, **55**: 434–437.

Joshi, P.K., Parthasarathy Rao, P., Gowda, C.L.L., Jones, R.B., Silim, S.N., Saxena, K.B. and Kumar, J., 2001. The world chickpea and pigeonpea economies: facts, trends, and outlook. Patancheru 502 324, Andhra Pradesh, India: International Crops Research Institute for the Semi-Arid Tropics.

Kaneko, T., Nakamura, Y., Sato, S., Asamizu, E., Kato, T., Sasamoto, S., Watanabe, A., Idesawa, K., Ishikawa, A. and Kawashima, K. *et al.*, 2000. Complete genome structure of the nitrogen-fixing symbiotic bacterium *Mesorhizobium loti*. *DNA Res.*, **7**: 331-338.

Kannaiyan, J. and Nene, Y.L., 1981. Influence of wilt at different growth stages on yield loss in pigeonpea. *Trop. Pest Man.*, **27**: 141.

Kannaiyan, J., Nene, Y.L., Reddy, M.V., Ryan, J.G. and Raju, T.N., 1984. Prevalence of pigeon pea diseases and associated crop losses in Asia, Africa and the Americas. *Trop. Pest Man.*, **30**: 62-71.

Kerr, A., 1980. Biological control of crown gall through production of agrocin 84. *Plant Dis.*, **64**: 25-30.

Khan, M.R. and Khan, S.M., 2002. Effect of root-dip treatment with certain phosphate solubilizing microorganisms on the *Fusarial*wilt of tomato. *Biores. Technol.*, **85**: 213-215.

Kimani, P.M., 1991. Pigeonpea improvement research in Kenya: An overview. *In*: Proceedings of the First Eastern and Southern Africa Regional Legumes (Pigeonpea) workshop (eds. Singh L., Silim S. N., Ariyanagam R. P. and Reddy, M. V.), 25-27 Jun 1990, Nairobi, Kenya. Eastern Africa Regional Cereals and Legumes (EARCAL) Program, International Crops Research Institute for the Semi-Arid Tropics, p. 108-117.

Kunst, F., Ogasawara, N., Moszer, I., Albertini, A.M., Alloni, G., Azevedo, V., Bertero, M.G., Bessieres, P., Bolotin, A. and Borchert, S., 1997. The complete genome sequence of the Gram-positive bacterium *Bacillus subtilis*. *Nature*, **390**: 249-256.

Lee, J.Y., Kim, B.S., Lim, S.W., Lee, B.K., Kim, C.H. and Hwang, B.K., 1999. Field control of *Phytophthora* blight of pepper plants with antagonistic rhizobacteria and DL-α-amino-*n*butyric acid. *Plant Pathol. J.*, **15(4)**: 217-222.

Liogier, H.A., 1988. Descriptive flora of Puerto Rico and adjacent islands, Spermatophyta Vol.-2. Editorial de la Universidad de Puerto Rico, Río Piedras, PR. p. 481.

Long, R.W. and Lakela, O., 1976. A flora of Tropical Florida. Banyan Books, Miami, FL. p. 962.

Madhukeshwara, S.S. and Sheshadri, V.S., 2001. Studies on the variation in the growth phase of different isolates of *Fusariumudum* Butler. *Curr. Res.*, **30**: 157-158.

Mahesh, M., Saifulla, M., Sreenivasa, S. and Shashidhar, R.K., 2010. Integrated management of pigeon pea wilt caused by *Fusarium udum* Butler. *EJBS*, **2(1)**: 1-7.

Maisuria, V.B., Gohel, V., Mehta, A.N., Patel, R.R. and Chhatpar, H.S., 2008. Biological control of *Fusarium* wilt of pigeonpea by *Pantoea dispersa*, a field assessment. *Ann. Micro.*, **58(3)**: 411-419.

Mallikarjuna, N., Saxena, K.B. and Jadhav, D.R., 2011. Cajanus. *In*: *Wild Crop Relatives: Genomic and Breeding Resources, Legume Crops and Forages* (ed. Kole, C.). Springer-Verlag Berlin Heidelberg, pp. 21-23.

Mandeel, Q. and Baker, R., 1991. Mechanisms involved in biological control of Fusarium wilt of cucumber with strains of non-pathogenic *Fusarium oxysporum*. *Phytopathol.*, **81**: 462-469.

Mandhare, V.K. and Suryawanshi, A.V., 2005. Application of *Trichoderma* species against pigeonpea wilt. *JNKVV Res. J.*, **32(2)**: 99-100.

Marley, P.S. and Hillocks, R.J., 1996. Effect of root-knot nematodes (*Meloidogyne* spp.) on fusarium wilt in pigeon pea (*Cajanuscajan*). *Field Crops Res.*, **46**: 15-20.

Martinetti, G. and Loper, J.E., 1992. Mutational analysis of genes determining antagonism of *Alcaligenes* sp. strain MFAI against the phytopathogenic fungus *Fusarium oxysporum*. *Can. J. Micro.*, **38**: 241-247.

Mbwaga, A.M., 1995. Fusarium wilt screening in Tanzania. *In*: Improvement of pigeon pea in Eastern and Southern Africa (eds. Silim S. N., King S. B. and Tuwaje S.) Annual Research Planning Meeting 1994, 21-23 Sep. 1994, Nairobi, Kenya. Patancheru 502 324, Andhra Pradesh, India: International Crops Research Institute for the Semi-Arid Tropics. p. 101-102.

MoenneLoccoz, Y., McHugh, B., Stephens, P.M., McConnel, F.I., Glennon, J.D., Bowling, D.N. and Ogara, F., 1996. Rhizosphere competence of fluorescent *Pseudomonas* sp. B24 genetically modified to utilize additional ferric siderophores. *FEMS Microbiol. Ecol.*, **19**: 215–225.

Mohapatra, B., Verma, D.K., Sen, A., Panda, B.B. and Asthir, B., 2013. Bio-fertilizers: A gateway to sustainable agriculture. *Popular Kheti*, **1(4)**: 97-106.

Nene, Y.L., 1980. Proc. Consultants Group Discussion on Resistance to soil borne diseases in Legumes, *ICRISAT*, India, p. 167.

Nene, Y.L., Kannaiyan, J., Reddy, M.V., Zote, K.K., Mahmood, M., Hiremath, R.V., Shukla, P., Kotasthane, S.R., Sengupta, K., Jha, P.K., Haque, M.F. and Grewal, J.S. 1989. Multi-locational testing of Pigeon pea for broad based resistance to *Fusarium* wilt resistance in India. *Indian Phytopath.*, **42**: 449-453.

Nene, Y.L., Sheeila, V.K. and Sharma, S.B., 1996. A world list of chickpea and pigeon pea pathogens. Fifth Edition, ICRISAT, Patancheru, Andhra Pradesh, India, p. 19-20.

Padwick, G.W., 1940. Genus *Fusarium* 5: *Fusarium udum* Butler, *F. vasinfectum* Atk. And *F. lateritium* var. *uncinatum* Wr. *Indian J. Agric. Sci.*, **10**: 863-878.

Pandey, K.K. and Upadhyay, J.P., 1997. Selection of potential bio-control agents based on production of volatile and non-volatile antibiotics. *Veg. Sci.*, **24**: 144-146.

Patel, S.I. and Patel, B.M., 2012. Pigeonpea wilt and its management: a review. *AGRES – An International e-Journal*, **1(4)**: 400-413.

Patil, P.Y., 1984. Study on pigeon pea wilt. *Ph.D. dissertation*, Mahatma Phule Agricultural University, Rahuri, India.

Pawar, N.B. and Mayee, C.D., 1983. Virulence differences in *Fusarium udum* isolates from Maharashtra State. *ICPN*, **2**: 41-42.

Prasad, P., Eswara Reddy, N.P., Anandam, R.J. and Lakshmikantha, R.G., 2003. Isozymes variability among *Fusarium udum* resistant cultivars of pigeon pea (*Cajanus cajan* (L.) (Millsp). *Acta Physiol. Plant*, **25**: 221–228.

Prasad, R.D., Rangeshwaran, R., Hegde, S.V. and Anuroop, C.P., 2002. Effect of soil and seed application of *Trichoderma haezianum* on pigeon pea wilt caused by *Fusariumudum* under field conditions. *Crop Prot.*, **21**: 293-297.

Pusey, P.L., 1989. Use of *Bacillus subtilis* and related organisms as Biofungicides. *Pesti Chem. Sci.*, **27**: 133-140.

Ram, H. and Pandey, R.N., 2011. Efficacy of bio-control agents and fungicides in the management of wilt of pigeon pea. *Indian Phytopath.*, **64(3)**: 269-271.

Raupach, G.S. and Kloepper, J.W., 1998. Mixtures of plant growth promoting rhizobacteria enhance biological control of multiple cucumber pathogens. *Phytopat.*, **88**: 1158–1164.

Reddy, M.V., Raju, T.N., Sharma, S.B., Nene, Y.L. and McDonald, D., 1993a. Handbook of pigeobpea diseases, (In En. Summaries in En.Fr.). Information Bulletin No. 14. Patancheru, A.P. 502 324, India: International Crops Research Institute for the Semi-Arid Tropics.

Reddy, M.V., Raju, T.N., Sharma, S.B., Nene, Y.L. and McDonald, D., 1993b. Handbook of pigeonpea diseases. (In En. Summaries in En., Fr.) Information Bulletin No. 42. Patancheru, A.P. 502 324, India: International Crops Research Institute for the Semi-Arid Tropics.

Reddy, M.V., Sharma, S.B. and Nene, Y.L., 1990. Pigeon pea: Disease management. *In*: The Pigeon pea(eds. Y. L. Nene, S. D. Hall and V. K. Sheila), CAB International, Wallingford, U.K., p. 303-347.

Rondon, M.R., Raffel, S.J., Goodman, R.M. and Handelsman, J., 1999. Toward functional genomics in bacteria: analysis of gene expression in *Escherichia coli* from a bacterial artificial chromosome library of *Bacillus cereus*. *Proc. Natl. Acad. Sci. USA*, **96**: 6451-6455.

Roy, A. and Sitansu, P., 2005. Biological control potential of some mutants of *Trichoderma harzianum* and *Gliocladium virens* against wilt of pigeon pea. *J. Interacademicia.*, **9**: 494-497.

Sandra, A.I., Wright, C.H., Zumoff, L.S. and Steven, V.B. 2001. *Pantoeaagglomerans* strain EH318 produces two antibiotics that inhibit *Erwinia amylovora in vitro*. *Appl. Environ. Microbiol.*, **67**: 282-292.

Sawant, D.M., Kolase, S.V. and Bachkar, C.B. 2003. Efficacy of bio-control agents against wilt of pigeon pea. *J. Maharastra Agric. Univ.*, **28**: 303-304.

Saxena, K.B., Kumar, R.V. and Rao, P.V., 2002. Pigeon pea nutrition and its improvement. *J. Crop Prod.*, **5**: 227–260.

Shan-da, L. and Baker, R., 1980. Mechanisms of biological control in soil suppressive to *Rhizoctonia solani*. *Phytopat.*, **70**: 404-411.

Sheldrake, R., Narayanan, A. and Kannaiyan, J., 1984. Some effects of the physiological state of pigeonpea on the incidence of wilt disease. *Trop. Grain Legum. Bull.*, **11**: 24-25.

Shit, S.K. and Gupta, P.K.S., 1980. Pathogenic and enzymatic variation in *Fusarium oxysporum* f. sp. *udum*. *Indian J. Microbiol.*, **20**: 46-48.

Siddiqui, S., Siddiqui, Z.A. and Iqbal, A., 2005. Evaluation of fluorescent pseudomonads and *Bacillus* isolates for the biocontrol of wilt disease complex of pigeon pea. *Wor. J. Microbiol. and Biotechnol.*, **21**: 729-732.

Siddiqui, Z.A., 2006. PGPR: Prospective biocontrol agents of plant pathogens. *In*: *Biocontrol and biofertilization* (ed. Siddiqui Z. A.). Springer, Amsterdam, The Netherlands, p. 111-142.

Siddiqui, Z.A. and Shakeel, U., 2007. Screening of *Bacillus* isolates for potential biocontrol of the wilt disease complex of pigeon pea (*Cajanus cajan*) under greenhouse and small-Scale field conditions. *J. Plant Pathol.*, **89(2)**: 179-183.

Silo-suh, L.A., Lethbridge, B.J., Raffel, S.J., He, H., Clardy, J., and Handelsman, J., 1994. Biological activities of two fungistatic antibiotics produced by *Bacillus cereus* UW85. *Appl. Environ. Microbial.*, **60**: 2023-2030.

Singh, R., Singh, B.K., Upadhyaya, S.K., Rai, B. and Lee, Y.S., 2002. Biological control of Fusarium wilt disease of Pigeonpea. *J. Plant Pathol.*, **18(5)**: 279-283.

Somasekhara, Y.M., Anilkumar, T.B. and Siddaramaiah, A.L., 1996. Bio-control of pigeon pea [*Cajanus cajan*(L.) Millsp.] wilt (*Fusarium udum* Butler). *Mysore J. Agric. Sci.*, **30**: 159-163.

Somasekhara, Y.M., Siddaramaiah, A.L. and Anilkumar, T.B., 1998. Evaluation of *Trichoderma*isolates and their antifungal extracts as potential bio-control agents against pigeon pea wilt pathogen, *Fusarium udum* Butler. *Curr. Res.*, **27**: 158-160.

Spiers, A., Field, D., Bailey, M. and Rainey, P.B.. Notes on designing a partial genomic database: the PfSBW25 encyclopedia, a sequence database for *Pseudomonas fluorescens* SBW25. *Microbiolo.*,**147**: 247-253.

Subramanian, S., 1963. *Fusarium* wilt of pigeon pea I. Symptomatology and infection studies. *Proc. Indi. Aca. Sci.*, Section B, **57**: 134-138.

Upadhyay, R.S. and Rai, B., 1981. Effect of cultural practices and soil treatments on incidence of wilt disease of pigeon pea. *Plant and Soil*, **62**: 309-312.

Upadhyay, R.S. and Rai, B., 1992. Wilt of pigeon pea. *In*: *Plant Diseases of International Importance* (eds. Singh A. N., Mukhopadhyay J., Kumar J. and Chaube H. S.), Volume I, Cereals and Pulses (New Jersey, USA: Prentice-Hall Inc.), p. 388-414.

Utkhede, R.S., 1984. Antagonism of isolates of *Bacillus subtilis* to *Phytophthora cactorum*. *Can. Bot.*, **62**: 1032-1035.

Van der Maesen, L.J.G., 1990. Pigeon pea: origin, history, evolution, and taxonomy. *In*: *The pigeon pea* (eds. Nene Y. L., Hill S. H. and Sheila V. K.). CAB International. Wellingford, UK. p. 15-46.

Waddington, S.R., Li, X., Dixon, J., Hyman, G. and de Vicente, C., 2010. Getting the focus right: production constraints for six major food crops in Asian and African farming systems. *Food Sec.*, **2**: 27-48.

Weller, D.M. and Thomashow, L.S., 1993. Use of rhizobacteria for biocontrol. *Curr. Opin. Biotechnol.*, **4**: 306-311.

Whipps, J.M., 1997. Developments in the biological control for soilborne plant pathogens. *Adv. Bot. Res.*, **26**: 1–134.

2016, **Diseases of Pulse Crops and their Sustainable Management** *485–493*
Editors: **Samir Kumar Biswas, Santosh Kumar and Gireesh Chand**
Published by: **BIOTECH BOOKS, NEW DELHI**

Chapter 26

Plant Parasitic Nematodes and its Integrated Management Approaches

Amlan Sushree[1], Mamta Bharti[1], S.K. Biswas[1] and Santosh Kumar[2]

[1]Department of Plant Pathology, CSA University of Agriculture and Technology, Kanpur – 208 002, U.P.
[2]Department of Plant Pathology, Bihar Agricultural University, Sabour – 813 210, Bihar

Introduction

Nematodes belong to the phylum Nematoda of the kingdom Animalia inhabited in very broad range of environment. Members of this phylum have existed on this planet for almost a billion years, making them one of the most ancient types of animals. Over 25,000 species of nematodes have been described so far and are difficult to distinguish them. Out of these more than half are parasitic to both plants and animals (Cobb, 1919).

In the year 1984, a 26 million year old fossil of a drosophila was found to carry a stylet bearing nematode. The people of Egyptian civilisation knew about the roundworms and the guinea worms were parasitic to human. The fact that nematodes could cause plant disease came into the forefront when the first plant parasitic

* Corresponding Author: E-mail: shona.nayak29@gmail.com

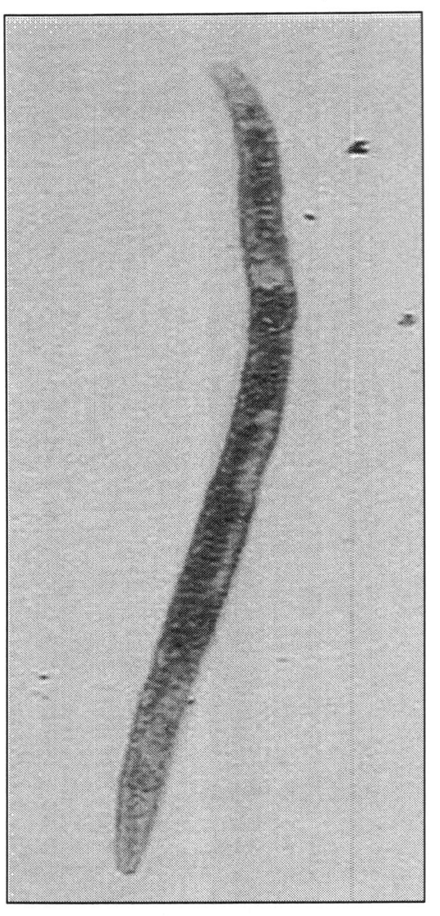

Figure 26.1

nematodes were discovered in wheat seeds by Needham (1743 AD) in England, which is now well known as *Anguina tritici*. "Plant nematology" emerged as an important scientific discipline with the discovery of root-knot nematodes on greenhouse cucumber by Berkeley in the year 1855 in England and cyst nematodes causing "beet-tired" disease on sugar beets by Schacht (1859) in Germany. In the early 1900's, pioneering activities of Nathan A. Cobb, a USDA scientist established agricultural nematology as a new branch of science. He studied the structure of various different nematodes and classified them. He also coined the term "Nematology", developed new techniques that simplified the study of nematodes. Hence, he is rightly called as the "Father of US Nematology". Chemical control of nematodes for the first time was achieved through soil fumigation by Carter 1940's. Atkinson reported nematode-fungus association by showing that the severity of Fusarium wilt of cotton is enhanced in the presence of root knot nematodes as early as 1892. Hunger (1901) showed that bacterial wilt of tomato was facilitated by root knot nematode. Hewitt and his associates (1958) gave a head start to the study of nematodes as a vector for virus transmission by showing that the grapevine fan leaf virus was transmitted by a nematode. Today plant parasitic nematodes have been established as a group of major agricultural pathogens causing production losses throughout the world. About 77 billion US dollars is loss worldwide each year due to damage caused by nematodes (Sasser and Freckman, 1987). Soil that is usually thought as a safe environment is actually a hostile world to a microscopic nematode. The presence of voracious predators and frequent changes in soil temperature and moisture make life difficult in soil. Moreover the death or harvest of favourable host leading to fallowing of land affects deeply the survival of nematode in soil. A nematode population must be able to combat with these factors of nature to thrive in soil environment. The research on evolution has enabled these organisms to evade these biotic and abiotic stresses by employing various survival strategies. They have adapted to these adverse conditions and developed their behaviour, physiology and metabolism so that, they can successfully overcome these obstacles. A thick cuticle around the nematode body provides protection from predation. But this defence structure is very easily breached by pathogens which have evolved specialised mechanism to penetrate it. The plant parasitic nematode evades predation by living inside host plant. Some nematodes might also limit their mobility in the soil environment to escape. By limiting movement and time in the soil, a nematode can reduce its chance of predation or pathogenesis. Plant parasitic nematodes might be an ectoparasite spending most of its time in the soil or endoparasite, spending an eternity of life within the host tissues. The endoparasitic nematodes have some degree of protection from predation, but as their host plant succumbs to disease, the risk of death increases. Vice versa occurs with the nematodes moving from host to host that reduce the risk of perishing with their host, but have a greater chance of encountering a predator or pathogen. Therefore the biology of plant-parasitic nematode is required to formulate an effective management strategy to reduce crop loss.

Characteristics of Typical Plant Parasitic Nematode

1. **Size** : a) length: - 0.2mm to about 12mm
   b) Width: - 0.01mm to 0.5mm
2. **Shape**: eel like, round, lemon, pear or kidney shaped
3. **Segmentation** : body is unsegmented
4. **Appendages** : absent
5. **Visibility** : microscopic in nature and cannot be viewed with naked eyes
6. **Symmetry** : body is bilaterally symmetrical whereas radial symmetry is found in anterior portion.
7. **Physiology** : All systems except respiratory and circulatory systems are present. The function of circulatory system is carried out by the pseudocoelomic fluid and respiration is is carried by the process of diffusion through the body cuticle.
8. **Body cavity** : pseudocoelom.
9. **Reproduction** : sexual or parthenogenetic
10. **Occurance** : greatest population found in top 15 to 30cm of soil, near the region around the host plant.
11. **Habitat** : ectoparasite or endoparasite, sedentary or migratory
12. **Symptoms**: a) above ground - growth reduction, wilting, yellowing, stem or leaf or seed galls, devitalised buds, lesions and rots, twisting and distortion of leaves.

    b) Below ground – root knots, galls, lesions, root proliferation, injured root tips, root rots.

Nematode-Host Interaction

Nematodes interact with plants mainly to find food and evade its natural enemies. The nematodes first probe the plant roots and then they puncture the root at a susceptible site by the help of stylet, a specialized spearlike structure causing direct mechanical injury. These injuries results in only slight d amage to the plant but when the nematode population is beyond the threshold level it becomes significant.

Usually the major part of damage is caused by the release of enzymes such as cellulase, amylase and protease at feeding. The nematode enters the host cells, releases its digestive enzymes and removes the entire cell contents. This leads to various disorders at cellular level like cell death and tissue disintegration, hyperplasia and hypertrophy. Some nematodes may also induce the formation of giant cells and nurse cells. Externally root galls or knots are formed and root proliferation is also seen. The root system of the host is weakened. Absorption of water and minerals is disrupted, impairing the plant from carrying out its normal metabolism. Thus the resultant effect of the nematode feeding is manifestation of symptoms and causation of disease in the plants.

The nematodes predispose its host to attack of other pathogens by causing mechanical injury while probing and inserting their stylet. The punctured holes become the portals of entry for various other plant pathogens. The pathogens previously unable to enter the plants easily get inside the host and cause disease. The nematodes have often been associated with disease complexes. They are involved with bacteria like for *e.g. Anguina tritici* and *Corynebacteriun tritici* causing tundu disease of wheat and with fungi for example *Fusarium* is mostly associated with nematodes causing severe wilts even in resistant varities. It has also been observed that the plant parasitic nematodes also play as vectors for the transmission of plant virus like NEPO, NETU and TOBRA viruses.

Association of Plant Parasitic Nematode with Other Plant Pathogens

Nematodes are soil dwellers spending most parts of its lifecycle in soil. They are surrounded by a number of other organisms like fungi, bacteria and other pathogenic organisms. Nematodes feed on all parts of plants like root, stem, leaf, seed etc. They provide an opportunity for other plant pathogens to invade the root. In certain cases, these nematodes and surrounding pathogenic organism enter into an association. It has been observed that the resistance of economically important cultivars is also affected by some nematode-plant pathogens association.

Interaction between Nematode and Phytopathogenic Bacteria

Bacteria often enters into plant through natural openings like stomata, lenticels and hydathodes. Nematode-bacteria association has developed in 2 ways

1. When nematode feeds on any plant, it leaves behind a series of injury on its surface and bacteria enters through these pores causing severe damage to host. *e.g.* root knot nematode-*Ralstonia sp.* or *Pseudomonas sp.*
2. Nematodes sometimes act as vectors of bacterial diseases spreading the disease from one plant to another *e.g. Aguina tritici-Corynebacterium tritici*

Interaction between Nematode and Virus

Numerous virus – nematode complexes have been identified by Hewit, Raski and Goheen. Hewitt and his co-workers(1958) were the first to discover that *Xiphinema index* was the vector of grapevine fan leaf virus. When a plant parasitic nematode interact with virus usually the viral component uses the nematode as vector for its transmission. Nematode acquires the virus at the time of its feeding on virus infected plant.

Nematode	*Virus Transmitted*
Xiphinema spp.	NEPO virus *e.g.* grapevine fanleaf virus
Paratrichodorus spp.	NETU virus *e.g.* tobacco rattle virus

Whenever a nematode feeds on a virus infected plant it acquires the virus. After the virus have been acquired it is has been seen that it can persists in its body for longer periods than in vitro. There are two types of mechanisms involved in virus transmission

1. Virus and its vector close biological association
2. Mechanical retention of virus in its vector

The virus is retained either in the cuticle lining of oesophagus or stylet of the nematode and the virus particles are released into plant cell with the help of oesophagus when it feeds on a healthy host plant.

Interaction between Nematode and Fungi

Atkinson was the first to report nematode-fungus association by showing that the severity of Fusarium wilt of cotton is enhanced in the presence of root knot nematodes as early as 1892. Since then several studies have been carried out in several other cultivated crops such as soybean, lentil, chickpea, pea, pigeonpea etc by several workers. Examples of some other nematode fungus association are *Meloidogyne javanica and Fusarium* sp. in chickpea(Upadhyay and Dwivedi, 1989), *Rotylenchulus* sp. and *Fusarium* sp. in pea, *Meloidogyne* sp. and *Rhizoctonia* sp. in soybean (Sitaramaiah and Singh, 1989).

Mechanism of Development of Disease in Pulse Crops

Wilt is a severe threat to pulse cultivation throughout the world. With the attack of nematode the risk of wilt increases. In some cases, these associations have been so acute that they have resulted in breaking resistance against wilt pathogen.

Fusarium oxysporum causes wilt disease in most of the pulses and is often associated with a soil sickness problem in india. Among the nematodes, the root-knot nematode, cyst namatode, lesion nematode, reniform nematode commonly attack these crops. Nematodes often interact with other soil-inhabiting plant pathogens to form disease complex. Nematode-fungal relationships have been reported in pulses very frequently. The resultant disease is much more severe than any disease caused by the pathogens individually. Alone a nematode can seldom cause such a massive loss. The synergistic association between these organisms not only suppresses the plant growth but also interferes in the nodulation and nitrogen fixing activities and kill the plants when they are still very young. This multi pathogenic disease also adversely affects the resistance of the crop. The most significant effect of this interaction is breaking of reisistance to the fungal pathogens which has disrupted many breeding programmes. For example breakdown of *Fusarium* wilt resistance in Avrodhi variety of chickpea (Upadhyay and Dwivedi, 1989)

It is highly essential to understand the mechanism behind the disease complexes. The primary cause is nematode attack due to which the plant is weakened. Seeing the devastating effects of this association a lot of steps are now being taken to curb the ill effects of this disease complex by its economic management. A sustainable management approach to restrict the nematode population below the economic threshold has recently gained popularity among the cultivators.

Sustainable Approach for Management of Nematode Diseases

Plant nematodes are not typically controlled using just one method but instead they can manage using a combination of methods in an integrated pest management system so as to reduce the risk of pest resistance, human health and environment (Barfield and Stimac, 1980).

- ☆ **Crop Rotation :** Rotation with resistant or non-host crops for two to three years generally is a very effective method to limit nematode population in soil. It has proven to be an excellent control of root-knot nematodes. The cropping system is so designed that the plants and nematodes can't grow together. *i.e.* the plants are non-hosts like growing non-leguminous crop like Rice. When such plants are grown in alternate years, the nematode is starved and their population decreases dramatically.
- ☆ **Maintenance of field sanitation :** It is important, to keep the crops free of weeds or volunteer plants, affected plants, crop debris so as to check the spread of nematodes.
- ☆ **Fallowing and flooding :** Fallowing refers to keeping the land free from any sort of vegetation for a certain period of time. This disturbs the nematode population by depriving them from feeding. Flooding of the fields decreases oxygen content in soil resulting in the death of nematodes. It was seen by Hollis and Jhonson in 1975 that non sterile flooded soils had low nematode population than sterilised soils. But both these practices are can't be carried out for long time period as its not economically viable.
- ☆ **Use of trap crops :** The crops that are highly susceptible to the attack of nematodes are frequently used as trap crops. These crops are planted alongwith the main crop. As the crops grow they get highly infested by the nematodes and at a particular growth stage the trap crops are destroyed before the nematode completes its lifecycle. *e.g.* Crotolaria.
- ☆ **Use of antagonistic plants :** The phenomenon in which one organism inhibits the growth of another organism through competition or secretions of toxic substances is called as antagonism. The plants which inhibit the growth of nematode population in soil are termed as antagonistic plants. *E.g.* Mustard (Triffit,1930; Ellenby, 1951), Marigold (Oostenbrink and associates, 1957), Asparagus (Rhode and Jenkins, 1958), Sesame (Atwal and Mangar), Neem.

Antagonistic Plant	*Toxic Secretion*
Mustard	Allyl isothiocyanate
Marigold	α terthienyl compounds
Asparagus	Asparaguistic acid
Neem	Nimbidin, thionemone

- **Resistant Varieties :** It is perhaps the best method of controlling any kind of plant pathogen as it nullifies the cost that is involved in protecting the crop against them. However, these varieties are usually resistant to only one or two species. The application of this method has been limited as numerous new intra-specific races and biotypes are evolving day by day. *e.g.* K-851, ML-80 *Meloidogyne* resistant variety of mung.
- **Incorporation of Organic Matter :** Incorporating organic matter into the soil suppress the development of nematodes and reduce their population. Linford (1937) reported that there was a marked reduction in the population of root knot nematodes during the decomposition of organic amendments. It stimulates populations of beneficial microorganisms like bacteria, fungi, and other soil inhabitants that are antagonistic to nematodes (Khan *et al.*, 1974 Badra *et al.*, 1979).
- **Treatment of Propagation Material :** Nematode infected plant parts can be disinfected by hot water treatment. The first time such process used by Marcinowshl in 1909. He treated begonias and ferns @ 50° C for 5 minutes to destroy *Aphlenchoides fragariae.* The temperature and period of exposure varies with the plant being treated. Temperature of 50° to 52° C is most commonly used. The seeds before sowing may be immersed in brine solution so as to separate the galls from healthy ones. This is called the floatation test.
- **Heat Treatment of Soil :** Soil solarisation is the most commonly used method of nematode control. This is because it is both effective and environment friendly option available to the grower. Nematodes are usually killed by exposure to temperatures of 104° to 130°F (40° to 54°C). A deep summer ploughing in the hot month of May exposes the cysts and galls. A light irrigation results in emergence nematodes from inside the cysts. These newly hatched nematodes die due to excessive hot conditions. Mulching the soil and covering it with tarp, usually with a transparent polyethylene cover, to trap solar energy has also been found to be effective for controlling nematodes upto some extent.
- **Use of botanicals :** These are the naturally occurring compounds that possess nematicidal or nematostatic or nematicidal properties and are being frequently used to control the nematodes in soil. Botanicals are alternative means of control that has shown promising results in controlling the nematode population in soil. Neem and neem products are effective against different phytoparasitic nematode in various pulses *e.g.* neem seed kernel

extracts (NSKE) and Econeem. These formulations were not phytotoxic even at the highest concentration. Biocontrol of nematode has been emphasized against chemical as the nematicides have hazardous effect on environment which in some cases further leads to biomagnifications. The application of neem cakes as organic amendment is also found to be effective. Even latex of some plant are known to possess some nematicidal property.

- ☆ **Use natural predators and pathogens of nematodes :** The biological control method also employs the use of natural predators (Cobb,1920) and pathogens of nematodes *e.g. Trichoderma* sp.,*Paeciliomyces lilacinus* (Jatala,1986). The *Trichoderma* sp. releases chitinolytic enzymes that degrades the chitinous membranes of nematodes which makes an effective tool of pathogens through direct parasitism. Nematode-trapping fungi are natural enemies of nematodes possessing sophisticated structures such as nets, knobs, branches or rings that help in trapping nematodes. *e.g. Athrobotrys sp., Glomus sp.* The evidence that there is reduction in the nematode population by the use of nematode trapping fungi was given by Stirling in 1991. Protozoan *Theratromyxa weberi* has been reported to prey on the second stage juvenile of golden cyst nematode and some other nematodes also (Weber *et al.*, 1952). *Hypoaspis aculieifer* a mite is reported to be predaceous on root knot and cyst nematode (Inserra and Davis 1983). Various formulations of *Bacillus thuringiensis* was found to be toxic to eggs and larvae of *Meloidogyne* sp. *in vitro* (Prasad *et al.*, 1972). *Pseudomonas penetrans* has been reported to be an obligate parasite of a large number of nematodes.
- ☆ **Chemical Control :** Control of root-knot nematodes through the use of chemicals is highly effective and practical, particularly on a field basis and where crops of relatively high value are involved. It may be the only alternative where crop rotation cannot be practiced or resistant varieties are unavailable. Both fumigant and non fumigant chemicals are available for nematode control.

Nematodes pose a serious threat to the pulse production in India. These techniques for management alone are not sufficient for mitigating this problem. Thus a holistic approach including all the available resources of management in optimum levels is required for the efficient control of nematode problem in field. Scientists have proven in several occasions that the integrated management techniques not only give better yields but also reduce the risks of environmental misbalances that are often created by application of harmful chemicals.

References

Atwal, A.S. and Mangar, A., 1969. *Indian J. Ento.* **31**: 286.

Cobb, N.A., 1919. The orders and classes of nemas. *Contrib. Sci. Nematol.* **8**: 213-216.

Cobb, N.A., 1920. *Science,* **57**: 640-641.

Hewitt, W.B., Raski, D.J. and Goheen, A.C., 1958. *Phytopathology* **48**: 586-595.

Inserra, R.N. and Davis, D.W., 1983. *J. Nematol.*, **15**: 324-325.

Jatala, P., 1986. *Annual Review Phytopath.* **24**: 453-489.

Khan, A.M., Alam, M.M. and Ahmed, R., 1974. Mechanism of control of plant parasitic nematodes as a result of application of oil cakes to the soil. *Ind. J. Nematol.*, **4**: 93-96.

Mathews, D. J., 1920. Report of the work of the W. B. Randall research assistant. Nursery and Market Garden Industry Development Society, Ltd. Exp. and Research Sta. Cheshunt, Herts. Ann. Rept. **5**: 18-21.

Needham, 1743. *Philos.Tarms R. Soc.*, **42**: 173-174, 643-641.

Oostenbrink, M., 1972. Evaluation and integration of nematode control methods. In *Economic Nematology*, J. M. Webster, editor, p. 497-514.

Prasad, S.S.V., Tilak, K.V.B.R. and Gollakata, K.G., 1972. *J. Invertebr. Pathol.*, **20**: 377.

Rhode, R.A. and Jenkins, W.K., 1958. *Bull Univ. Md. Agric. Exp. Sta A-97*, p. 19.

Sasser, J.N. and Freckman, D.W., 1987. A world perspective on nematology: the role of the society.

Sitaramiaah and Singh, 1989. Upadyay and Dwivedi, K. *Textbook of Plant Nematology*. Host-parasite relationship, **6**: 37.

Triffit, M.J., 1930. *Helminth.* **8**: 19-48.

Upadyay and Dwivedi, K. 1989. *Textbook of Plant Nematology*. Host-parasite relationship, **6**: 37.

Weber, A.Ph, Zwillenberg L.O. and Vander Lann, P.A., 1952. *Nature*, **169**: 834-835.

2016, **Diseases of Pulse Crops and their Sustainable Management** *495–510*
Editors: **Samir Kumar Biswas, Santosh Kumar and Gireesh Chand**
Published by: **BIOTECH BOOKS, NEW DELHI**

Chapter 27

Major Nematode Problems in Pulses and its Management in India

Tamoghna Saha* and S.N. Ray
Department of Entomology,
Bihar Agricultural University, Sabour, Bhagalpur – 813 210, Bihar

Introduction

Pulses constitute a major source of dietary protein of a large section of vegetarian population of the world. They occupy an important position in human diet and are good sources of vegetable proteins (17-42 per cent) in the cereal based diets of the people. They are known to supply large quantity of nutritious green fodder for the livestock (Anon, 1999). There are several constraints for the production of pulses such as the genetic inability for high yields even under favorable conditions, pests and diseases, poor soil factors and also inadequate plant nutrition. Among these constraints, plant parasitic nematodes are one of the major factors affecting the productivity of pulses. Various nematodes can attack pea, including the pea cyst nematode (*Heterodera goettingiana*), root-knot nematode (*Meloidogyne incognita*), and root-lesion nematode (*Pratylenchus penetrans*). The life cycle of different nematode species vary, and may include feeding on the outside of roots or penetration and development within roots. The cyst nematode *Heterodera caiani* (Koshy.) is one of the major limiting factors in the production of pulses, *viz.*, pigeonpea, cowpea, green

* Corresponding Author: E-mail: tamoghnsaha1984@gmail.com

gram, black gram, which causes stunting of plants, reduction in number of nodules on roots, with heavy crop yield losses to the tune of 30.1 to 67.5 per cent (Saxena and Reddy, 1987). Root Lesion Nematode (*P. penetrans*) is estimated to occur at damaging levels in at least 40 percent of Western Australia's cropping paddocks. Yield losses to wheat and barley crops from these nematodes are estimated at 5-15 per cent annually (http://archive.agric.wa.gov.au/). In India, the root-knot nematode is reported to reduce its yield of chickpea from 17 per cent to 60 per cent depending on nematode inoculum density and soil types (Ali, 1995). *Meloidogyne* spp., root knot nematode is one of the most harmful nematode pests in both tropical and subtropical crop production regions and cause extensive economic damage worldwide (Sikora and Fernandez, 2005). Root-knot in chickpea has been reported in various states of India (Jamal, 1976). Surveys were conducted in Uttar Pradesh, Rajasthan and Gujarat in collaboration with ICRISAT to find out nematodes associated with pigeonpea. In Kanpur area, major nematodes encountered in pigeonpea fields are *Meloidogyne javanica* and *Rotylenchulus reniformis*. *Heterodera cajani* was recorded from Fatehpur, Kanpur and Lucknow. Survey conducted in Gujarat revealed that the presence of *H. cajani* in the locations of Bharuch, Kheda, Surat and Vadodera districts in 69 per cent. The population densities of *R. reniformis* were 2-3 times more in Bharuch and Surat districts than Vadodara and Kheda. In mungbean, highest population of *R. reniformis* followed by *M. javanica* was recovered from root samples of mungbean. Keeping in view, nematode not only suppresses the plant growth but also interferes in the nodulation, nitrogen fixation and adversely affects the overall yield. Hence, there is an urgent need for an eco-friendly substitute for nematode management.

The details of theses nematodes and their management strategies on pulse crops are given below

Root Lesion Nematode (RLN)

The lesion nematodes rank third behind root-knot and cyst nematodes as the nematodes of greatest economic impact in crops worldwide. This is not only due to their wide host range (more than 400 crop plant species), but their distribution in almost every temperate and tropical environment (http://mint.ippc.orst.edu/rootnemaid.htm).

Plant-parasitic nematodes in the genus *Pratylenchus* are commonly called either root-lesion nematodes or lesion nematodes. These parasites can be seen only with the aid of a microscope. They are transparent, eel-shaped, and about 1/64 inch (0.5 mm) long (Richard, 2010).

Host of Root Lesion Nematode (RLN)

RLN species have a wide host range, although the nematodes multiply to varying degrees on different crops. For example, field pea and faba bean are poor hosts (*i.e.* resistant) to *P. neglectus*, while chickpea and wheat are good hosts (*i.e.* susceptible). The differences between cultivars within a crop species, as well as differences between nematode species (*e.g.* field pea is more susceptible to *P. penetrans* than it is to *P. neglectus*). Root lesion nematode will multiply at varying rates depending on the susceptibility of the host plant. For example, *P. neglectus* will reach high levels on susceptible hosts like most wheat and chickpea cultivars, and intermediate levels on moderate hosts like barley. Poor hosts (field pea, faba bean, lupin, lentil and triticale) will restrict *P. neglectus* multiplication and reduce nematode densities in the soil. These poor or moderate hosts can be used in rotations to reduce nematode numbers and limit yield losses in subsequent susceptible crops (MacLeod *et al.*, 2005)

Biology

Pratylenchus species have a vermiform (eel- or pencil-shape) body that measures about 0.5 mm (1/64 inch) long and 0.02 mm (1/1000 inch) in diameter. Root-lesion nematodes are classified as migratory endo- and ecto-parasites, meaning that they may become entirely embedded within root tissue (endoparasitic) and migrate from cell to cell within that tissue, and that they also may feed on the root surface without actually moving from the rhizosphere into the internal root tissue (ectoparasitic). Root-lesion nematodes never lose the ability to leave the root to migrate back into soil. Root-lesion nematodes possess a stylet that allows them to puncture, feed upon, and penetrate cells of the root epidermis and the root cortex. These nematodes have the ability to enter mature as well as immature segments of roots (Figure 27.1).

Pratylenchus species feed only on living root tissue but may deposit eggs in soil as well as inside root tissue. Females deposit about one egg per day and first-stage juveniles molt to become second-stage juveniles within the egg. Three additional molts within 35 to 40 days result in the adult stage. All juvenile and adult stages are parasitic. The number of nematodes in root tissue increases greatly during the growing season.

P. neglectus and *P. thornei* are parthenogenic, meaning that females produce fertile eggs without copulation with a male. Populations of these two species are comprised nearly or entirely of females. In contrast, *P. penetrans* is an amphimictic species, meaning that a male and female must mate before fertile eggs are produced. Populations of *P. penetrans* therefore include nearly equal proportions of males and females. Life cycles range from 45 to 65 days depending upon temperature, moisture and other environmental variables. Reproduction of *P. neglectus* and *P. thornei* is greatest at temperatures between 68°F and 77° F but may continue slowly at soil temperatures as low as 45° F. These species are therefore well adapted for multiplication during most of the year, especially at soil depths greater than one foot, where temperatures throughout the year are typically a rather constant 50° F to 55° F. Maximum population densities in fields with deep silt loams have been measured at soil depths as great as the third foot. Population levels of *Pratylenchus* often decline during long fallow periods between crops but high rates of survival have also been reported at soil depths greater than six to ten inches, particularly in fields where weeds or volunteer cereals (*i.e.*, any green growth) are allowed to become established between the times of harvest and planting (Richard, 2010).

Symptoms of Root Lesion Nematode Infection

As most of the pulses are resistant to *P. neglectus*, there are few observations of RLN symptoms for these crops. Chickpea, however, is susceptible and can show distinctive root symptoms. The roots of chickpea exhibit dark brown-orange lesions,

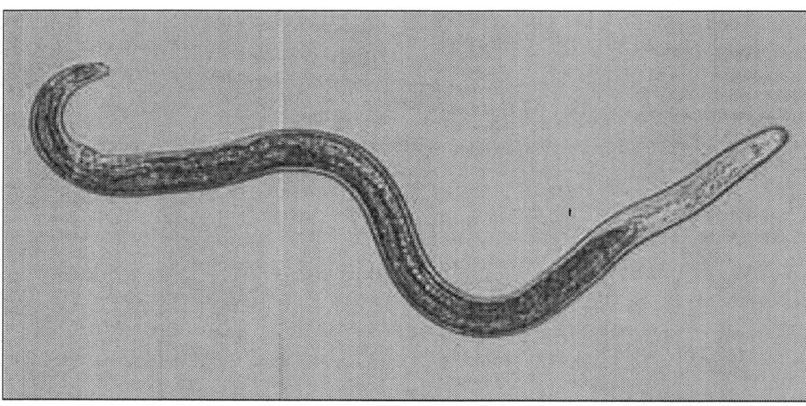

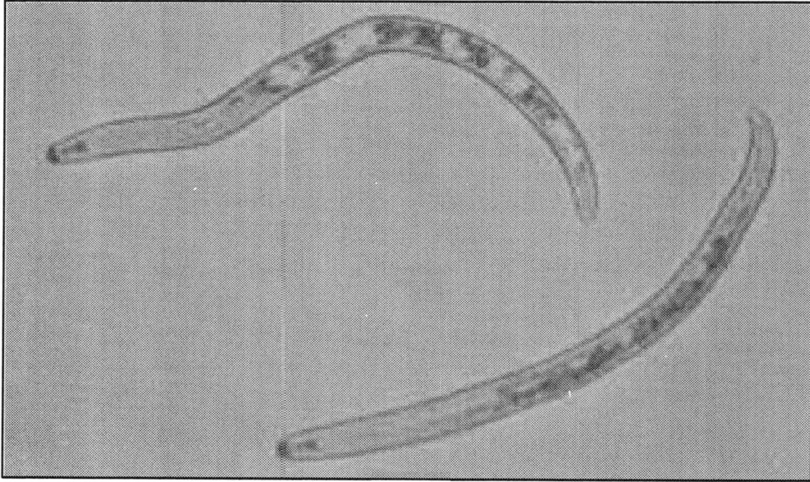

Figure 27.1: Adult Female (Above) and Juveniles (Below) of Root Lesion Nematode (*P. neglectus*).

and the lateral roots will be severely stunted and reduced in number. Above-ground symptoms are indistinguishable. Plants may show a combination of symptoms, all of which can be confused with, or exacerexacbated by, nutrient deficiencies. These include uneven growth and stunting. Affected plants are often more prone to wilting under water stress.

Once root symptoms have been detected, laboratory testing of plants or soil is still necessary for detection of the RLN species and levels. For this requires extraction of nematodes from roots or soil, followed by microscopic examination of the nematodes present. Nematode levels can be determined when soil is moist, and plants can be examined and tested during the growing season (MacLeod *et al.*, 2005).

Root lesion nematode penetrates in the root tissues results in lesions that favor greater colonization by root-rotting fungi and by saprophytic bacteria, fungi, and nonparasitic nematodes. These secondary organisms cause more intense rotting and discoloration than that caused by the root-lesion nematode alone. Cortical degradation and reduced root branching often are not visible until plants are six or more weeks old, and these root symptoms are often confused with those caused by *Pythium* or *Rhizoctonia* root rot. Interactions of root-lesion nematodes, fungal pathogens, other plant-parasitic nematodes, and insect pests have been reported (Richard, 2010).

Management

Maximum numbers of RLN in the soil can significantly reduce the yield potential of intolerant crops. It is therefore important to monitor and manage nematode numbers as effectively as possible. Their populations can rapidly multiply on a good host, but it may take two or more successive years of a poor host for numbers to decline below damaging levels, especially if they were initially high (MacLeod *et al.*, 2005)

No chemical treatments available for nematode control. Eradication of RLN is not possible. RLN can only be managed within the cropping system by use of crop rotation and other cultural practices (*e.g.* good nutrition, weed control, early sowing). The first step is to determine the levels and identity of RLN species:

- ✫ To inspect crops for uneven growth, and symptoms resembling nitrogen deficiency;
- ✫ To examine roots for distinguishing symptoms, and;
- ✫ To take soil and plant samples from unthrifty patches of the crop to submit for laboratory tests to identify

Rotation with poor host (*i.e.* resistant) crops and cultivars, combined with crop resistance, is the key to management of RLN. Some of the management strategies are

- ✫ Exploit resistant crops and cultivars in rotations to reduce nematode levels and yield loss;
- ✫ Avoid susceptible hosts such as chickpea if nematode levels are high;
- ✫ Control susceptible weeds (particularly wild oat, barley grass, brome grass and wild radish) and susceptible volunteers (especially cereals and chickpea) to reduce RLN build-up and carry-over and
- ✫ Make certain adequate nutrition (especially nitrogen, phosphorus and zinc).

Root Knot Nematode

Root knot nematodes (genus: *Meloidogyne,* Greek word means melon, apple or gourd shaped female) are sedentary endoparasites of diverse crops. Root knot nematodes (*Meloidogyne* spp.) are one of the most important polyphagous pests in agriculture. Among the top five plant pathogens affecting world's food production, root knot nematodes are one of the most devastating pathogen of crops. Infestation on crops greatly impact their health, yield and quality. They are adapted to parasitize on large number of plants and over 3000 wild and cultivated plant species are reported to be affected (Hussey and Janssen, 2002). They are distributed worldwide over a wide range of geographical conditions of tropical, subtropical and temperate regions of the world. Several weed species (226 species belonging to 43 families) are known to act as hosts of root knot nematodes worldwide (Rich *et al.*, 2008). The effect of nematode infection on plant root induce typical symptoms, popularly known as 'root knot' or 'root gall' of varying sizes depending on the species of root knot nematode and the host. Considering its economic and social significance, research interest is growing at faster rate on its different aspects of biology, ecology, management etc.

The root –knot nematodes (*Meloidogyne* spp.) are the common important group of plant parasitic nematodes which is posing a threat to pulse crops including mung bean (*Vigna radiate*). In mung bean, avoidable losses under field conditions due to *M. javanica* (Treub) Chitwood ranged from 42.1 to 93.4 per cent (Gupta and Verma, 1990). Similarly, the yield loss of 18 to 65 per cent and 23 to 49 per cent due to *M. incognita* (Kofoid and White) Chitwood and *M. javanica* respectively on mung bean has been reported from Uttar Pradesh (Sharma *et al.*, 2000).

Historical Background

In India, first report of plant parasitic nematode believed to be root knot nematode (*tea eelworm*) infecting tea was from Kerala (Barber, 1901). Fourteen species have been recorded from India. So far out of the 14 species in India, four have been described for the first time. The species list of *Meloidogyne* known so far from India is given below (Khan *et al.*, 2014):

1. *M. incognita* (Kofoid and White) Chitwood, 1949
2. *M. javanica* (Treub) Chitwood, 1949
3. *M. arenaria* (Neal, 1889) Chitwood, 1949
4. *M. hapla* Chitwood, 1949
5. *M. africana* Whitehead, 1968
6. *M. brevicauda* Loos, 1963
7. *M. thamesi* Chitwood *in* Chitwood, Specht and Havis, 1952 (Goodey, 1963)
8. *M. exigua* Goeldi, 1887
9. *M. graminicola* Golden and Birchfield, 1968
10. **M. indica* Whitehead, 1968
11. *M. graminis* (Sledge and Golden) Whitehead, 1968
12. **M. lucknowica* Singh, 1969
13. **M. triticoryzae* Gaur, Saha and Khan, 1993
14. **M. piperi* Sahoo, Ganguly and Eapen, 2000

Host Crop of Root Lesion Nematode

Root knot nematodes are considered one of the most important plant-parasitic nematode primarily due to their wide host range and distribution. More than 3000 plant species are parasitized by the root knot nematodes (Hussey and Janssen, 2002).

The reports on occurrence of root knot nematode (*Meloidogyne* spp.) on various crops in different states of India are tabulated in Table 27.1.

Table 27.1: Distribution of Root Knot Nematodes (*Meloidogyne* spp.) Infesting Crops in India (Khan *et al.*, 2014)

Meloidogyne spp.	*Hosts*	*State/Union References Species Territory*	*References*
M. incognita	Pigeonpea	Andhra Pradesh	Ali and Askary (2001)
M. incognita	Cowpea	Chhattisgarh	Sahu *et al.* (2011)
M. javanica	Pigeonpea	Andhra Pradesh	Ali and Askary (2001)
M. javanica	Cowpea	Chhattisgarh	Sahu *et al.* (2011)
M. javanica	Pulses	Orissa	Das *et al.* (1979)

Biology

The root-knot nematode is parthenogenic, that is a single female can reproduce without males and a new generation can occur every 28 days if conditions are ideal. Inside the gall, the enlarged female appears as a shiny white body, the size of a pinhead. She deposits 300 to 500 eggs in a protective jelly-like material. These glistening white to yellow egg masses are present on the root surfaces. Juveniles emerge from the eggs in the soil and penetrate between and through cells at the center of the root, usually near the growing tip. These larvae actively feed and remain at this same site. The juvenile stage can over-winter even under very unfavorable conditions. New adult nematodes develop from the larvae and start the cycle again.

Symptoms of Root Knot Nematode Infection

Basically root knot nematodes are parasites of roots or underground stem or pods. The diseases caused by them are not of epidemic in nature but rather is akin to slow decline in yields spreading very gradually and steadily year after year. Some areas within a field may be severely affected, whereas plants in other parts may not show any signs of the disease. The above-ground parts of the diseased plants exhibit symptoms typical of mineral deficiency or drought injury due to the slow debility of affected roots to function properly for nutrient, water uptake and translocation even when adequate fertilizers and moisture are present in the fields. The other symptoms may include dieback, yellowing, wilting and premature shedding of the foliage with severe stunting, depending upon the initial nematode population in the field. Chlorosis of the foliage lowers the quality of the crop resulting in severe losses. Poor emergence and death of young seedlings may occur in heavily infested soil but death of grown up plants is rarely seen unless some other co-inhabiting pathogenic fungi or bacteria is involved in the development of disease complex. The below-ground symptoms include galls or knots on the roots or tubers of numerous crops. These galls vary in size from pin-head to large size which in case of heavy infection may coalesce to form large secondary galls. The size of galls also depends upon the host plants and nematode species (Khan *et al.*, 2014)

Management

Cultural control

Among different crop rotations, sorghum - chickpea + mustard sequence recorded highest chickpea equivalent yield of 1918 kg/ha followed by maize-based system. Lowest root knot nematode population was recorded in maize – chickpea sequence whereas maximum population was recorded in mungbean - chickpea/mustard system. As regards the lesion nematodes, the trend was reverse. It thrived well in cereal – chickpea system and populations were low in legume – chickpea + mustard system. In urdbean based cropping systems, it was observed that the urdbean + sorghum – wheat and urdbean – wheat + mustard cropping systems were best in decreasing the nematode population up to 83.3 and 75.6 per cent in a year (Anon, 2009)

Host Plant Resistance

A total of 1860, 510, 350, 526, 410 and 360 accessions of chickpea, pigeonpea, mungbean, urdbean, fieldpea and lentil, respectively were screened against root knot nematode, *Meloidogyne javanica*. The resistant accessions of different pulse crops are given in Table 27.2.

Table 27.2: Genotypes/Accessions found Promising against Nematodes (Anon, 2009)

Crop	*Nematode*	*Resistant Donors*
Chickpea	*Meloidogyne javanica* *M. javanica* and *M. incognita*	BG 369, BGM 481, 483, GL 88341, GMS 815, BGM 481, 483, GL 88341, GMS 815
Pea	*M. javanica*	DDR 16, HFP 4, HFP 9510
Lentil	*M. javanica*	IPL 100, KLS 214, DPL 14
Pigeonpea	*M. javanica*	MTH 9, T 9, UB 218, MA 7, TT 10, H 84-5,
Mungbean	*M. javanica*	GM 85-2, 85-68, ML 323,HUM 7, MH 98-1, PMB 34, LGG 460, Ganga 4, ML 5, Pusa 105
Urdbean	*M. javanica*	TPU 3, TPU 4, WBU 105, PLU 648, AKU 7, UG 218, UG 562, KU 304

Biological Control

Two endoparasitic fungi belonging to *Catenaria* sp. were isolated from the root knot nematode juveniles collected from chickpea fields. Similarly, one endoparasitic fungus was isolated from *H. cajani*. One nematode trapping fungi identified as *Pestolosia* spp. was isolated from the soil. Out of eight fungi isolated from the egg masses of root knot nematode and cysts of *H. cajani*, two fungi *Fusarium solani* and *Aspergillus niger* were found most effective in reducing the number of galls, egg masses and soil population of root knot nematode in chickpea. Number of galls was 10 and 30 in *F. solani* and *A. niger* treated soil as compared to 64 galls per root system in plants without fungus treatment.

Trichoderma viride and *T. harzianum* applied as seed treatment @ 10 g/kg seed or soil application @ 10 g/kg soil were found effective in reducing the galling in chickpea and increasing the plant growth. The egg infection was observed slightly more (65 per cent) with *T. harzianum* than *T. viride* where egg infection was observed 57 per cent (Anon, 2009)

Chemical Control

Chickpea: Seed treatment with Posse 3 per cent and Bavistin 0.2 per cent were effective in reducing *M. javanica* infection in chickpea (Figure 27.2). Similarly, Nimbecidine @ 0.2 per cent was superior to other neem products in managing the nematode population in chickpea. In a trial conducted at farmers fields, seed treatment with neem seed powder @ 5 per cent and latex 1 per cent were found effective for the control of nematode population and higher yield in chickpea.

Pigeonpea : Seed treatment with neem seed powder 5 per cent was found effective in reducing the nematode infection and increasing the biomass and yield in pigeonpea cv. Narendra Arhar 1 followed by Neemarin (2 per cent).

Mungbean : Among various organic amendments (neem cake, castor cake and poultry manure), lowest root knot counts and highest yield were recorded in poultry manure @ 10 q/ha treated soil followed by neem cake.

Urdbean : In a nematode management trial in urdbean, 12.5 NSP + 1 tonne FYM/ha in combination with seed treatment with carbosulfan @ 3 per cent w/w was found best and recorded highest biomass (16.32 q/ha), yield (3.83 q/ha), minimum gall index and soil population of root knot and cyst nematode (Anon, 2009).

Pigeon Pea Cyst Nematode

Pearly root of pigeonpea caused by *Heterodera cajani* has been recorded from major pulse growing states, *viz.*, Andhra Pradesh, Bihar, Delhi, Haryana, Rajasthan, Uttar Pradesh and Tamil Nadu. In Tamil Nadu, H. *cajani* was first observed on pigeonpea plants showing poor growth (Janarthanan, 1972). Subsequently, the survey carried out by All India Co-ordinated Research Project (Nematodes) in pulses growing districts such. as Thiruchirapalli, Madurai, Coimbatore and Periyar revealed the presence of H. *cajani* infestation, in mung, field bean, and Urd (Velayutham, 1988).

Plant Host of Pigeon Pea Cyst Nematode

Cajanus cajan (pigeon pea), *Cyamopsis tetragonoloba* (cluster bean), *Lablab urpureus* (hyachinth bean), *Vigna aconitifolia* (aconite-leaved kidney beans), *Vigna radiata* (bean, mung), *Vigna mungo* (black gram), *Phaseolus vulgaris* (common bean), *Pisum sativum* (pea), *Sesamum indicum* (sesame) and *Vigna unguiculata* (cowpea) (Sharma and Nene, 1985).

Biology and Transmission

The adult females are lemon-shaped and sedentary in habit, and remain attached to roots semi-endoparasitically. These are white to slightly brown, with a neck and posterior cone-like elevation on which the vulva is situated; turns into a cyst of same size and shape. Posterior part of body protruding outside the root usually appears with small rounded egg sac attached to it. Cephalic region has two annules; the second is larger than the first. Stylet is of medium strength, in two equal parts; basal

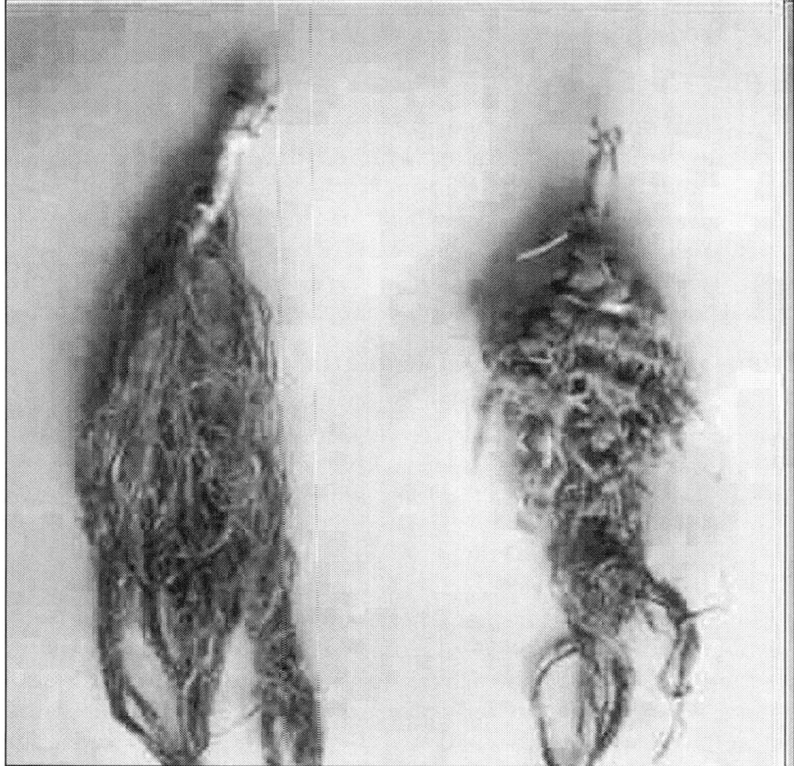

Figure 27.2: Root Knot Nematode Infesting Chickpea.

knobs round to slightly anteriorly flattened. Median esophageal bulb large rounded, with well developed valve plates. The excretory pore is placed posterior behind the median bulb. Oesophageal glands extend over the intestine. Ovaries are paired and convoluted. Uterus with several eggs filling most of the body. Vulva is a large transverse slit on a cone-shaped elevation of the body. Anus is close to vulva. An egg sac is present. The size of egg sacs varies between 0.5-2 times the sizes of cyst. Egg sacs are yellow, occasionally purple. Few to 200 eggs (average 54) are found in the egg sacs (Koshy and Swarup 1971). Eggs may be retained inside the female body but many are laid in a gelatinous matrix forming egg sacs. Eggs are oval, 95-115 μm long, and 37-48 μm wide. Eggshell is hyaline, without surface markings. Morphological characteristics of 2nd-, 3rd- and 4th-stage juveniles of *H. cajani, H. avenae* and *H. mothi* are described and compared by Taya and Bajaj (1986). Males are vermiform.

Symptoms

Heterodera cajani is primarily a root-parasite. The most important characteristic symptom is the presence of cysts on the root surface. Identification of "pearly root" caused by the presence of white females is a useful symptom of *H. cajani* infestation in pigeon pea at the vegetative stage (Sharma, 1993). The symptoms of nematode injury include stunting, reduced leaf lamina size and yellowing on cotyledonary leaves (Gaur and Singh, 1977). Flowers and pods are reduced in size and number and the root system may also be poorly developed. Foliage symptoms are generally not apparent even in heavily infested soils, but a reduction in height and vigor of the infected plants can be discerned by careful comparison with healthy plants.

List of Symptoms/Signs (http://www.plantwise.org/KnowledgeBank/Datasheet)

Fruit - reduced size

Leaves - abnormal colours

Roots - cysts on root surface

Roots - reduced root system

Stems - stunting or rosetting

Whole plant - dwarfing

Management

Cultural Control

H. cajani has a very restricted host range and is mostly confined to the Fabaceae. Growing the host crop in rotations every 4-5 years with non-hosts can help to maintain populations below threshold levels. The significance of double crop (intercrop and sequential crop), single crop (rainy season crop fallow) and rotations on the densities of *H. cajani* and other nematodes was studied by Sharma *et al.* (1996). Mean population densities of *H. cajani* were about eight times lower in single crop systems than in double crop systems, with pigeon pea as a component intercrop. Plots planted to sorghum, safflower and chickpea in the preceding year contained fewer *H. cajani* eggs and juveniles than did plots previously planted to pigeon pea, cowpea or *Vigna radiata*. Continuous cropping of sorghum in the rainy season and safflower in the post-rainy season markedly reduced the population density of *H. cajani*. Intercropping sorghum with a tolerant pigeon pea cultivar could be effective in increasing the productivity of traditional production systems in *H. cajani* infested regions (Sharma *et al.*, 1996).

Resistant Varieties

The use of resistant varieties is the most economical method of controlling plant parasitic nematodes. A greenhouse technique to screen pigeon pea for resistance to *H. cajani* is described by Sharma *et al.* (1991).

Biological Control

Plant parasitic nematodes have many natural predators and parasites in the soil which provide opportunities for biological control. Combinations of biocontrol agents and vesicular arbuscular mycorrhizal fungi have been used to manage the wilt disease complex of pigeon pea caused by *H. cajani* and *Fusarium udum* [*Gibberella indica*]: *Bacillus subtilis, Bradyrhizobium japonicum* and *Glomus fasciculatum* (Siddiqui and Mahmood, 1995); *Paecilomyces lilacinus, Verticillium chlamydosporium* and *Gigaspora margarita* (Siddiqui and Mahmood, 1995b); and *Trichoderma harzianum, V. chlamydosporium* and *Glomus mosseae* (Siddiqui and Mahmood, 1996).

Chemical Control

Cyst nematodes are highly resistant to chemical control because the female, after death, transforms into a tough brown sac protecting the eggs which remain within.

Fensulfothion, when evaluated as a seed treatment for pigeon pea, reduced nematode populations, with high concentrations and long soaking period being the most effective (Zaki and Bhatti, 1986). Seed soak treatments with dimethoate, acephate and quinalphos did not give effective control (Velayutham, 1988). Certain chemicals stimulate juvenile emergence and can be used with proper knowledge of the biology of the parasites to control them. Experimental studies of various synthetic chemical treatments showing some control of *H. cajani* include the following:

T. viride gave better control than carbendazim, mancozeb and ziram on *Vigna radiata* in pot experiments (Mishra and Gupta, 1991).

Reniform Nematode

Reniform nematode, *Rotylenchus reniformis* Linford and Oliveira, 1940 was first discovered from roots of cowpea (*Vigna Sinensis*) grown in a pineapple field in the Island of Ohu, Hawaii. In India this nematode has been reported from almost all vegetable growing tracts of the country and crop losses caused to cowpea were estimated to be of 10 per cent (Anonymous, 2004). Cowpea genotypes have been found to be good hosts for *R. reniformis* in India (Rao and Ganguly, 1996) but there is morphological variation in reniform nematode populations from different parts of the country (Rao and Ganguly, 1998). Most likely, reniform nematode originated in a tropical area. It has the unique ability to survive for long periods in very dry soil, much like what is encountered in many tropical areas.

Host Plant of Reniform Nematode

Pea, Pigeon pea and Black gram, Soybean, French bean and Lentil (Vadhera *et al.*, 1995)

Biology

Eggs hatch one to two weeks after being laid. The first-stage juvenile molts within the egg, producing the second-stage juvenile (J2) that emerges from the egg. The infective stage is reached one to two weeks after hatching. Once root penetration occurs, one or two more weeks are required for females to reach maturity. The male, which remains outside of the root, can inseminate the female before female gonad maturation. Sperms are stored in the spermatheca. Soon after female gonad maturation, the eggs are fertilized with sperm, and about 60 to 200 eggs are deposited into a gelatinous matrix. Typically, there is an equal number of females and males in a population. Some populations of reniform nematodes reproduce parthenogenetically (egg production without fertilization).The life cycle of this nematode is usually shorter than three weeks, but depends on soil temperature. However, it can survive at least two years in the absence of a host in dry soil through anhydrobiosis, a survival mechanism that allows the nematode to enter an ametabolic state and live without water for extended periods of time (Radewald and Takeshita, 1964).

Symptoms

The reniform nematode causes root necrosis resulting in severe root pruning and subsequent dwarfing of plants. Fibrous or feeder roots are mostly attacked which may reduce the absorption ability and other physiological functions of the plant. This may lead to stunting, yellowing and wilting. In some cases, the nematode may produce lesions, distortions or cracks on storage roots which reduce marketability (http://keys.lucidcentral.org/keys).

Management

Cultural Control

- ☆ Crop rotation.Non-host crops or resistant crops can be planted when nematode population is high.
- ☆ Use of organic amendments.
- ☆ Use of trap and antagonistic crops. Planting Tagetes erecta and Crotolaria spectabilis in nematode infested soil has been found effective against the nematode.

Biological Control

Paecilomyces lilacinus, a fungal egg parasite was found effective against the reniform nematode (Gapacin and Valdez, 1976).

Chemical control

Several nematicides have been reported to be effective against the reniform nematode. Examples are Nemagon, Mocap, Dasanit, Nemacur, Furadan, Temik, Vydate (Birchfield and Martin, 1968)

Conclusion

Nematodes are found throughout the pulse growing region. Have soil tested in a laboratory to determine the population densities of root-lesion nematodes or root knot nematodes and to find out which species are present in your each of your paddocks. If root-lesion nematodes are detected, there are two strategies to put in place: 1) for long-term management, include resistant crops in your rotations to keep populations of root-lesion nematodes at low levels, for example, sorghum when *P. thornei* is present. More than one resistant crop may be needed in sequence to reduce high populations of root-lesion nematodes. 2) Grow a variety with tolerance to the species of nematode identified to ensure that yields are maximized.

References

Ali, S.S. 1995. *Nematode problems in chickpea,* Indian Institute of Pulse Research, Kanpur, p. 184.

Ali, S.S. and Askary, T.H. 2001. Taxonomic status of phytonematodes associated with pulse crops. *Curr. Nematol.* **12**: 75–84.

Anonymous, 1999. Survey of Indian Agriculture. *The Hindu,* p. 61-65.

Anonymous, 2004. Third Annual Report (2003-2004), ICAR Network Project on Reniform Nematodes, *Rotylenchulus reniformis,* Division of Nematology, Indian Agricultural Research Institute, New Delhi, India, p. 67-68.

Anonymous, 2009. 25 Years of Pulses Research at IIPR. Indian Institute of Pulses Research. Printed at: Army Printing Press, 33 Nehru Road, Sadar Cantt., Lucknow. p. 146-152.

Birchfield and Martin, 1968. Evaluation of nematicides for controlling nematodes of sweetpotatoes. *Plt. Dis. Rept.,* **52**: 127-131.

Das, S.N., Mishra, C.D. and Mohanty, K.C. 1979. Occurrence and host preference of some plant parasitic nematodes on pulse crops. *Indian J. Nematol.* **9**: 74.

Gapasin, R.M. and Valdez, R.B. 1979. Pathogenicity of *Meloidogyne* spp. and *Rotylenchulus* reniformis on sweetpotato. *Annals of Tropical Research* **1**: 20-26.

Gaur, H.S. and Singh, I. 1977. Pigeon-pea cyst nematode, *Heterodera cajani,* associated with the moong crop in the Punjab State. *Journal of Research, Punjab Agricultural University* **14**: 509.

Gupta, D.C. and Verma, K.K. 1990. Studies on avoidable losses in mung bean (*Vigna radiata*) due to root-knot nematode, *Meloidogyne javanica* and its control under field conditions. *Indian J. Nematol.* **20**: 148-145.

Hussey, R.S. and Janssen, J.G.W. 2002. In: Plant Resistance to Parasitic Nematodes (J.L. Starr, R. Cook and J. Bridge, eds.) CAB International, Wallingford Oxon, UK.

Jamal, A. 1976. Studies on the relationship between *Meloidogyne incognita* and galling behavior of *Cicer arietinum* roots. *Current Science* **45**: 2 30-231.

Janartanan R. 1972. Occurrence of pigeonpea cyst nematode in Tamil Nadu, Indian. *Journal of Nematod.* **2**: 215.

Khan, M.R., Jain, R.K., Pal, S. and Ghule, T. 2014. Root Knot Nematodes in India- A Comprehensive Monograph. AICRP Plant Parasitic Nematodes with Integrated Approach. *Indian Agricultural Research Institute,* New Delhi, p. 78.

Koshy, P.K. and Swarup, G. 1971. Investigations on the life history of the pigeon-pea cyst nematode, *Heterodera cajani. Indian Journal of Nematology,* **1**: 44-51.

MacLeod, B., Vivien, V., Roger, J. and Jean, G. 2005. Diseases of pulses. *Producing Pulses in the Northern Agricultural Region.* **4656**: 98-103.

Mishra, S.M. and Gupta, P. 1991. The effect of various pesticidal chemicals on *Heterodera cajani* associated with mung bean. *Current Nematology,* **2(1)**: 63-66.

Radewald, J.D. and Takeshita, G. 1964. Desiccation studies on five species of plant-parasitic nematodes of Hawaii. *Phytopathology,* **54**: 903-904.

Rao, G.M.V.P. and Ganguly, S. 1996. Host preference of six geographical isolates of reniform nematode, *Rotylenchulus reniformis. Indian Journal of Nematology,* **20**: 19-22.

Rao, G.M.V.P. and Ganguly, S. 1998. Geographical variations in morphobiometrics of reniform nematode, *Rotylenchulus reniformis. Indian Journal of Nematology,* **28**: 56-71.

Rich, J.R., Brito, J.A., Kaur, R. and Ferrell, J.A., 2008. Weed species as hosts of Meloidogyne: A review. *Nematropica,* **39**: 157–185.

Richard, W.S. 2010. Root-lesion nematodes: Biology and management in Pacific Northwest wheat cropping systems. Basin Agricultural Research Center, Oregon State University, A Pacific Northwest Extension Publication, p. 1-9.

Sahu, R., Chandra, P. and Poddar, A.N. 2011. Community analysis of plant parasitic nematodes prevalent in vegetable crops in district Durg of Chhattisgarh, *India. Res. J. Parasitol.,* **6**: 83–89.

Saxena, R. and Reddy, D.R. 1987. Crop losses in pegeonpea and mung bean by pigeonpea cyst nematode, Heterodera cajai. *Indian Journal of Nematology,* **17**: 19-94.

Sharma, S.B. 1993. Pearly root of pigeonpeas caused by *Heterodera cajani. Indian Journal of Nematology* **21**: 169.

Sharma, S.B. and Nene, Y.L. 1985. Additions to host range of pigeonpea cyst nematode, Heterodera cajani. *International Pigeonpea Newsletter* **4**: 42.

Sharma, S.B., Ashokkumar, P. and McDonald, D. 1991. A greenhouse technique to screen pigeonpea for resistance to *Heterodera cajani. Annals of Applied Biology,* **118(2)**: 351-356.

Sharma, S.B., Rego, T.J., Mohiuddin, M. and Rao, V.N. 1996. Regulation of population densities of *Heterodera cajani* and other plant-parasitic nematodes by crop rotations on vertisols in semi-arid tropical production systems in India. *Journal of Nematology,* **28(2)**: 244-251.

Sharma, S.B., Sharma, H.K., and Pankaj 2000. Nematodes problem in india In: *Crop pest and Disease management- challenges for he millennium* (Ed.D.Prasad And S.N. Puri), Joyti Publishers, New Delhi, p. 267-275.

Siddiqui, Z.A. and Mahmood, I. 1995. Biological control of *Heterodera cajani* and *Fusarium udum* by *Bacillus subtilis, Bradyrhizobium japonicum* and *Glomus fasciculatum* on pigeonpea. *Fundamental and Applied Nematology,* **18(6)**: 559-566.

Siddiqui, Z.A.and Mahmood, I. 1996. Biological control of *Heterodera cajani* and *Fusarium udum* on pigeonpea by *Glomus mosseae, Trichoderma harzianum,* and *Verticillium chlamydosporium. Israel Journal of Plant Sciences,* **44(1)**: 49-56.

Sikora, R.A. and Fernandez, E. 2005. Nematode Parasites of Vegetables. In: Plant Parasitic Nematodes in Subtropical and Tropical Agriculture, ed. Luc. M., Sikora, R.A., Bridge, J. 2nd Edition CABI Publishing Wallingford, UK, p. 319-392.

Taya, A.S. and Bajaj, H.K. 1986. Larval stages of *Heterodera avenae* Woll, *H. cajani* Koshy and *H. mothi* Khan and Husain. *Indian Journal of Nematology,* **16**: 160-162.

Vadhear, I., Shukla, B.N. and Bhatt, J. 1995. Infectivity of *Rotylenchulus reniformis* to French bean and some plant species. *Ind J. Mycol. Plant Path.,* **25**: 318-320.

Velayutham, B., 1988. Efficacy of nematicidal seed treatment in the control of the pigeon pea cyst nematode, *Heterodera cajani* Koshy, 1967 affecting pigeonpea, Cajanus cajan L. Indian Journal of Nematology, **18(2)**: 365-366.

Zaki, F.A. and Bhatti, D.S. 1986. Control of pigeon-pea cyst nematode, *Heterodera cajani* Koshy, 1967 by chemical seed treatment. *Indian Journal of Nematology,* **16(1)**: 106-108.

Internet Website

http: / / archive.agric.wa.gov.au /

http: / / mint.ippc.orst.edu / rootnemaid.htm

http: / / www.missouribotanicalgarden.org

http: / / www.plantwise.org / KnowledgeBank / Datasheet

http: / / keys.lucidcentral.org / keys

2016, Diseases of Pulse Crops and their Sustainable Management 511–518
Editors: Samir Kumar Biswas, Santosh Kumar and Gireesh Chand
Published by: BIOTECH BOOKS, NEW DELHI

Chapter 28

Host Plant Resistance in Pulse Crop Diseases Management

Vinod Kumar Singh[1]*, Santosh Kumar[2], Lalan Sharma[3] and Dipak T. Nagrale[3]

[1]*Jute Research Station, Katihar, Bihar Agricultural University, Sabour – 854 105, Bihar*
[2]*Department of Plant Pathology, Bihar Agricultural University, Sabour – 813 210, Bihar*
[3]*National Bureau of Agriculturally Important Microorganisms, Kushmaur, Maunath Bhanjan, Mau – 275 101, U.P.*

Introduction

The major challenge is food security to fulfill requirement of foregoing world population. In that condition, the agricultural productivity has to be increased without any environmental problems. In food production, pulse crops have immense significance in food production as well as nutritional values. They are wonder crops because *Firstly* pulses have relatively higher protein content (containing 20 to 30 percent) in the seed, which makes the human diet more balanced in its nutritive value and are rich in lysine but deficient in sulpher containing amino acids, methionine and cystein, while the cereals are rich in methionine and cystein but deficient in lysine. Therefore, pulses and cereals are complementary to each other nutritionally. Another unique attribute of pulses is their ability to develop a symbiotic relationship

* Corresponding Author: E-mail: vinod546@gmail.com

in their root nodules with the nitrogen-fixing bacteria like *Rhizobium* species enriching the soil fertility. This way they also maintain soil health and crop productivity. Again pulses and cereals are complementary to each other for judicious utilization and maintenance of soil fertility. Further, pulses have high capacity to extract and utilize minerals by virtue of their inherent deep root system. These can be grown under diverse agro-ecological conditions, and play vital role in stress conditions, crop rotation, mixed cropping and intercropping.

Of the major and most important food legumes grown world over like chickpea (*Cicer arietinum* L.), lentil (*Lens culinaris*) and field pea (*Pisum sativum* L.) and are grown in cool (rabi) season. Production and productivity of pulses are being hampered by several kinds of biotic (diseases, insect-pests, nematodes and weeds) and abiotic stresses such as draught, frost, flood, water logging etc. Significant diseases of plants in agriculture are caused by diverse group of plant pathogens, including viruses, bacteria, fungi, nematodes, and others. Among biotic stresses, crop diseases are the serious threat. According to estimates made in India nearly 10-15 per cent of food legumes production is lost due to crop diseases alone. Although, effective chemical control measures are known, and are using but these measures are quite costly and hazardous to human and environmental health. Plant diseases can also be managed by cultivation practices such as crop rotation, tillage, planting density, purchase of disease-free seeds and cleaning of equipment, but plant varieties with inherent (genetically determined) disease resistance are generally the first choice for disease control.

Although, resistant varieties can be the simplest, practical, effective and economical method of plant disease control. The use of resistant varieties not only ensure protection against diseases but also save the time, energy and money spent on other measures of control in reference of resource poor farmers. Resistance has been recognized as the most desirable and efficient strategy for the management of rust disease (Bond *et al.*, 1994); breeding for disease resistance is continuous process since plants were first domesticated, but it requires continual effort to develop a resistance crop plant. This is because pathogen populations are often under natural selection for increased virulence, new pathogens can be introduced to an area, cultivation methods can favor increased disease incidence over time, changes in cultivation practice can favour new diseases, and plant breeding for other traits can disrupt the disease resistance that was present in older plant varieties. A plant line with acceptable disease resistance against one pathogen may still lack resistance against other pathogens.

Steps of Breeding for Disease Resistance

- ☆ Identify and characterize potential of plant pathogen that may be riskful to continuously growing high yielding varieties.
- ☆ Introduction of resistant varieties of crop plants.
- ☆ Screening of resistant breeding sources (plants that may be less desirable in other ways, but which carry a useful disease resistance trait). Ancient plant varieties and wild relatives are very important to preserve because they are the most common sources of enhanced plant disease resistance.

- ☆ Crossing of a desirable but disease-susceptible plant variety to another variety that is a source of resistance, to generate plant populations that mix and segregate for the traits of the parents.
- ☆ Growth of the breeding populations in a disease-conducive setting. This may require artificial inoculation of pathogen onto the plant population. Careful attention must be paid to the types of pathogen isolates that are present, as there can be significant variation the effectiveness of resistance against different isolates of the same pathogen species.
- ☆ Selection of disease-resistant individuals. Breeders are trying to sustain or improve numerous other plant traits related to plant yield and quality, including other disease resistance traits, while they are breeding for improved resistance to any particular pathogen.
- ☆ Development of multilines of crop plants.

Each of the above steps can be difficult to successfully accomplish, and many highly refined methods in plant breeding and plant pathology are used to increase the effectiveness and reduce the cost of resistance breeding. These are-

Crop Rotation

Usually, pathogens attack specific plant species or crop plant or a family of plants. For example, common disease problems on pulses do not affect other crops such as *Botrytys cinerea* on chickpea. By increasing the diversity of crops grown within a rotation, the pathogens in the soil or on residue from the previous crop usually will not infect the subsequent crop grown. In this way, crop rotation can be one of the most effective biological control options available to producers. Although, crop rotation does not provide effective control/management for broad range having plant pathogens like *Rhizoctonia, Sclerotium, Fusarium, Pythium* and so on.

Host Plant Resistance

Resistance means that as we know to check or restrict or limit the activity of plant pathogens. So what we can say, a crop plant that has potential (morphological, physiological, biochemical, and genetic configuration) to limit plant pathogen activity. It is recorded and observed that a plant may be susceptible in early growth stage and may not be in later stage. It means a definite period is needed to crop plant to synthesize a bio-molecules and environmental makes conducive for them. Since, there are some examples that can be used in host resistance to particular pathogen. Early or late maturity of the crop may prevent physical contact of the pathogen with the host. Mechanical and anatomical barriers such as thick cuticle, waxy bloom on leaves and stem, stomatal regulation prevent penetration of spores. Resistance that is specific to certain races or strains of a pathogen species is often controlled by single R genes and can be less durable but completely effective is called species specific resistance. The *broad-spectrum resistance* against an entire pathogen species is often quantitative and only incompletely effective, but more durable, and is often controlled by many genes that segregate in breeding populations. However, there are numerous exceptions to the above generalized trends, which were given the names vertical resistance and horizontal resistance, respectively by J.E. Vanderplank (1996). It has been determined that resistance of a crop to a physiological race of the pathogen depends not only on the genotype of the host for resistance, but also upon the genotype of the pathogen for virulence or aggressiveness. In this concern, Flor (1946) proposed the gene-for-gene hypothesis, according to which, for every gene for resistance in the host, there is a corresponding gene for pathogenicity in the pathogen.The plant resistance genes can be broadly divided into eight groups based on their amino acid motif organization and their membrane spanning domains (Mayank *et al.*, 2012, Table 28.1).

Table 28.1: Classes of Resistance Genes

Sl.No.	*Resistance Gene Classes*	*Examples*
1.	NBS – LRR - TIR	*N, L6, RPP5*
2.	NBS – LRR - CC	*I2, RPS2, RPM1*
3.	LRR - TrD	*Cf-2, Cf-4, Cf-9*
4.	LRR – TrD, Kinase	*Xa21*
5.	TrD - CC	*RPW8*
6.	NBS – LRR – TIR – NLS – WRKY	*RRS1R*
7.	LRR – TrD – PEST - ECS	*Ve1, Ve2*
8.	Enzymatic R-genes	*Pta, Rpg1, Hm1*

Major classes of plant resistance genes – LRR: Leucine rich repeats; NBS: Nucleotide-binding site; TIR: Toll/Interleukin-1-receptors; CC: Coiled coil; TrD: Trans-membranedomain; PEST: Amino acid domain; ECS: Endocytosis cell signaling domain; NLS: Nuclear localization signal; WRK: Amino acid domain; HCtoxinreductase: Helminthosporium carbonum toxin reductase enzyme.

Non Host Resistance

There are thousands of species of plant pathogenic microorganisms, but only a small minority of these pathogens has the capacity to infect a broad range of plant species. Most pathogens instead exhibit a high degree of host-specificity. Non-host plant species are often said to express *non-host resistance*. The term *host resistance* is used when a pathogen species can be pathogenic on the host species but certain strains of that plant species resist certain strains of the pathogen species. There can be overlap in the causes of host resistance and non-host resistance. Pathogen host range can change quite suddenly if, for example, the capacity to synthesize a host-specific toxin or effector is gained by gene shuffling / mutation or by horizontal gene transfer from a related or relatively unrelated organism. For example, blast disease of rice is caused by a fungal pathogen (*Pyricularia grisea*) and it is not capable to infect to maize or soybean (same season crop). Here, we can say that these resistant plants may have definite morphological, physiological and metabolism that reduces chances of infection.

Epidemics and Population Biology

Plants in native populations are often characterized by substantial genotype diversity and dispersed populations that is grow in a mixture with many other plant species. They also have undergone millions of years of plant-pathogen co-evolution. Hence as long as novel pathogens are not introduced from other parts of the globe, natural plant populations generally exhibit only a low incidence of aggressive plant pathogen. In agricultural systems, humans often cultivate single plant species at high density, with numerous fields of that species in a region, and with significantly reduced genetic diversity both within fields and between fields. In addition, rapid travel of people and cargo across large distances increases the risk of introducing new pathogens against which the plant has not been selected for resistance. Climate change can alter the viable geographic range of native or alien pathogen species and cause significant crop losses to become a problem in areas where the disease was previously less important. These factors make modern agriculture particularly prone to disease epidemics. Common solutions to this problem include constant breeding for disease resistance, use of pesticides to suppress recurrent potential epidemics, use of border inspections and plant import restrictions, maintenance of significant genetic diversity within the crop gene pool, and constant surveillance for disease problems to facilitate early initiation of appropriate responses. Some pathogen species are known to have a much greater capacity to overcome plant disease resistance than others, often because of their ability to evolve rapidly and to disperse broadly.

Marker-Assisted Selection and Trait Pyramiding for Horizontal Resistance

Marker assisted selection (MAS) is the ability to select for and breed for a desirable trait with a marker, or suite of markers, from within a plant genotype without the need to express the associated phenotype. Therefore, MAS offers great opportunity for improved efficiency and effectiveness in the selection of plant genotypes with a desired combination of traits. This approach relies upon the establishment of a tight linkage between a molecular marker and the chromosomal location of the gene(s) governing the trait to be selected in a particular environment. Once this has been achieved, selection can be conducted in the laboratory and does not require the expression of the associated phenotype. For example, using MAS, disease resistance can be evaluated in the absence of the disease and in early stages of plant development. Sequence tagged sites (STS) are ideal markers for MAS. STS markers are mapped loci for which all or part of the corresponding DNA sequence has been determined. The sequence information is used to design PCR primers for amplification of all or part of the original sequence. They are more robust and reproducible than the arbitrary sequences they are designed from, such as RAPD markers, as they are developed from the known sequences and produce an amplicons from longer primers. Differences in the lengths of amplified fragments serve as genetic markers for the locus. If no length polymorphism is detected, the amplified fragments can be cleaved with restriction enzymes to observe subsequent length differences. This technique is often referred to as cleaved amplified polymorphic sequences or CAPS (Jarvis *et al.*, 1994). The use of converted locus-specific PCR markers is also referred to as a specific polymorphic locus amplification test (SPLAT), as well as sequence characterized amplified region (SCAR) markers and allele specific associated primer (ASAP) markers. SPLAT markers are designed from sequencing the insert of a polymorphic

RFLP marker (Gale and Witcombe, 1992), whereas SCAR and ASAP markers are developed from sequencing specific RAPD markers (Gu *et al.*, 1995; Ford *et al.*, 1999; Paran and Michelmore, 1993). The conversion of more technically-demanding RFLP markers into PCR based markers (*e.g.* SPLAT) may provide a more rapid, cost-effective and efficient tool in breeding for disease resistance against pulse crop diseases.

Quantitative Trait Loci Mapping

When resistance is governed by multiple genes and or co-dominantly inherited genes, a more holistic genome mapping approach may be undertaken to identify genomic locations, interaction and subsequent molecular markers for accurate trait selection.

Association Mapping

Association or linkage disequilibrium (LD) mapping was used successfully to discover genetic determinants to traits initially in humans is now being used in plants (Flint-Garcia *et al.*, 2003; Thornsbury *et al.*, 2001). Using association mapping, entire genomes can be scanned for markers associated with qualitative and quantitative traits. The association mapping approach may allow plant breeders to break out of restrictive F1-derived mapping populations and employ any plant population including those from breeding programs or germplasm collections to conduct marker-trait association studies (Flint-Garcia *et al.*, 2003). Gebhardt *et al.* (2004) clearly summarized the four potential benefits of the association mapping approach: (1) it allows assessment of the genetic potential of specific genotypes before phenotypic evaluation; (2) it allows the identification of superior trait alleles in germplasm; (3) it can assist in high resolution QTL mapping; and (4) it can be used to validate candidate genes responsible for individual traits (Gebhardt *et al.*, 2004). The important issues to consider in designing and implementing any association mapping studies in plants are: (1) determination of the population structure (Pritchard *et al.*, 2000); (2) estimation of nucleotide diversity (Zhu *et al.*, 2003); (3) estimates of haplotype frequencies and LD (nonrandom association of alleles at different loci) (Flint-Garcia *et al.*, 2003); and (4) precise evaluation of phenotypes (Neale and Savolainen, 2004).

Trait Mapping

Many simply inherited traits have been placed on lentil genetic maps. By knowing the map position of a gene, the presence of the gene can be diagnosed using flanking DNA markers without waiting for the gene effect to be present in the phenotype (Paterson *et al.*, 1991). Bulked segregant analysis (BSA), first described by Michelmore *et al.* (1991) is a method used to identify molecular markers linked to phenotypic traits controlled by single major genes. This method relies on the availability of bulked DNA samples collected from individuals that segregate for the two extreme divergent phenotypes within a single population. One bulk contains the DNA of the trait being targeted, while the other contains DNA from individuals lacking the trait. DNA polymorphisms between the bulks are therefore, likely to be linked to genes that govern the trait. In lentil, this method has been used to identify markers that are tightly linked to genes for resistance to Fusarium vascular wilt and Ascochyta blight (Chowdhury *et al.*, 2001; Eujayl *et al.*, 1998b; Ford *et al.*, 1999).

Future Needs - Host Plant Resistance

Key needs in the area of genetic resistance and plant breeding include cooperation with, and support for, classical plant breeding programs, and research support in the following areas:

1. Natural and novel gene source identification and development;
2. Classical and novel gene transfer techniques;
3. Genetic studies on HPR, tolerance traits and virulence factors;
4. Bioengineering studies to facilitate molecular approaches to gene study;
5. Transfer and expression of foreign genes in plants; and
6. Certain other potential areas such as microtoxin use *in vitro*, and immunization systems.

References

Bond, D. A., Gellis, G. S., Rowland, G. G. Le Guen. J., Robertson, L. D., Kholid, S. A. and Le-Juan, L. 1994. Present status and future strategy in breeding Faba bean (*Vicia fabae* L) for resistance to biotic stress. *Euphytica* **73**: 151-166.

Chowdhury, M.A., Andrahennadi, C.P., Slinkard, A.E. and Vandenberg, A. 2001. RAPD and SCAR markers for resistance to ascochyta blight in lentil. *Euphytica,* **118**: 331–337.

Eujayl, I., Baum, M., Erskine, W., Pehu, E. and Muehlbauer, F.J. 1997. The use of RAPD markers for lentil genetic mapping and the evaluation of distorted F2 segregation. *Euphytica,* **96**: 405–412.

Flint-Garcia, S.A., Thornsberry, J.M. and Buckler, E.S. 2003. Structure of linkage disequilibrium in plants. *Annu Rev Plant Biol.,* **54**: 357–374.

Ford, R., Pang, E.C.K. and Taylor, P.W.J. 1999. Genetics of resistance to ascochyta blight (*A. lentis*) of lentil and the identification of closely linked RAPD markers. *Theor Appl Genet.,* **98**: 93–98.

Gale, M.D. and Witcombe, J.R. 1992. DNA markers and marker mediated applications in plant breeding, with particular reference to pearl millet breeding. In: J.P. Moss (Ed.), Biotechnology and Crop Improvement in Asia, p. 323–332. ICRISAT, Patancheru.

Gebhardt, C., Ballvora, A., Walkemeir, B., Oberhagemann, P. and Schuler, K. 2004. Assessing genetic potential in germplasm collections of crop plants by marker-trait association: A case study for potatoes with quantitative variation of resistance to late blight and maturity type. *Mol. Breed,* **13**: 93–102.

Gu, W.K., Weeden, N.F., Yu, J. and Wallace, D.H. 1995. Large-scale, cost-effective screening of PCR products in marker-assisted selection applications. *Theor. Appl. Genet.,* **91**: 465–470.

Haware, M.P. and Nene, Y.L. 1982. *Plant Dis.* **66**: 250-251.

Jarvis, P., Lister, C., Szab´o, V. and Dean, C. 1994. Integration of CAPS markers into the RFLP map generated using recombinant inbred lines of *Arabidopsis thaliana*. *Plant Mol Biol*, **24**: 685–687.

Kannaiyan, J. and Nene, Y.L. 1976. Reaction of lentil germplasm and cultivars against three root pathogens *Indian Juornal of Agricultural Sciences*, **46(4)**: 165-7.

Mayank, A.G., Jelli, V. Chandrama, P.U., Akula, N. Shashank, K.P. and Se, W.P. 2012. Plant disease resistance genes: Current status and future directions. *Physiological and Molecular Plant Pathology* **78:** 51–65.

Michelmore, R.W., Paran, I. and Kesseli, R.B. 1991. Identification of markers linked to disease-resistance genes by bulked segregant analysis: A rapid method to detect markers in specific genomic regions by using segregating populations. *PNAS USA* **88**: 9828–9832.

Neale, D.B. and Savolainen, O. 2004. Association genetics of complex traits in conifers. *Trends Plant Sci*. p. 325–330.

Nene, Y.L. 1988. Multiple disease resistance in grain legumes. Annual Review of *Phytopathology*, **26**: 203-17.

Paran, I. and Michelmore, R.W. 1993. Development of reliable PCR based markers linked to downy mildew resistance genes in lettuce. *Theor Appl Genet* **85**: 985–993.

Paterson, A.H., Tanksley, S.D. and Sorrells, M.E. 1991. DNA markers in plant improvement. *Adv Agron*, **46**: 39–90.

Pritchard, J.K., Stephens M. and P. Donnelly, 2000. Association mapping in structured populations. *Am J Hum Genet*., **67**: 170–181.

Thornsbury, J.M., Goodman, M.M., Doebley, J., Kresovich, S., Nielsen, D.and Buckler, E.S. 2001. Dwarf polymorphisms associate with variation in flowering time. *Nat Genet*., **28**: 286–289.

Zhu, Y.L., Song, Q.J., Hyten, D.L., Van Tassell, C.P., Matukumalli, L.K., Grimm, D.R., Hyatt, S.M., Fickus, E.W., Young, N.D. and Cregan, P.B. 2003. Single-nucleotide polymorphisms in soybean. *Genetics*, **163**: 1123–1134.

2016, Diseases of Pulse Crops and their Sustainable Management 519–542
Editors: Samir Kumar Biswas, Santosh Kumar and Gireesh Chand
Published by: BIOTECH BOOKS, NEW DELHI

Chapter 29

Diagnosis of Pulse Diseases and Biotechnological Approach for their Management

***Ravi Ranjan Kumar*[1], *Mahesh Kumar*[1], *M.S. Nimmy*[2], *Vinod Kumar*[1], *Sweta Sinha*[1], *Md. Shamim*[1] *and Dharamsheela*[1]**

[1]*Department of Molecular Biology and Genetic Engineering, Bihar Agricultural University, Sabour, Bhagalpur – 813 210, Bihar*
[2]*National Research Centre for Plant Biotechnology, PUSA, New Delhi – 110 012*

Introduction

Pulses being rich source of protein form an integral part of vegetarian diet in rainfed arid and semi arid climate of Indian sub continent. They maintain soil fertility through biological nitrogen fixation, improve soil organic matter content by leaf fall at the time of maturity and occupy prominent place in various cropping systems and crop mixtures (Kumar *et al.*, 2011). Thus pulses play a vital role in providing protein rich food to human beings and in sustaining both soil health and crop production on long-term basis. India has the distinction of being the largest producer of pulses in the world, accounting for 37 per cent of the area and 27 per cent of the world's production (FAOSTAT, 2012). Their growth and productivity is under constant threat from environmental challenges in the form of various biotic and abiotic stress factors. Pluses are frequently exposed to a plethora of several biotic constraints including bacteria, fungi, viruses, nematodes as well as from herbivores because plants lack a circulatory system and antibodies, they have evolved a defense mechanism that is distinct from the vertebrate immune system. All these stress factors pose a challenge for plants, preventing them from reaching their full genetic potential and limit crop productivity worldwide (Cramer *et al.*, 2011).

Farmers often faced challenges of multiple disease attack in the field and use to apply chemical pesticides / fungicides which results in generation of more resistant pathogen attacking same crop. The traditional method of identifying plant pathogens through visual examination is possible only after major damage done to the crop, so treatments become major limiting factor. Pathogens capable of causing systemic infections on their host plants are usually transmitted through vegetative propagation from infected mother plants to next progeny. When seedlings get infected, it may cause severe loss of the crop after germination or when transported to new areas, resulting in the invasion and establishment of new diseases and contamination of that soil and environment. In areas where the pathogens already exist, infected seedlings or planting materials serve as inoculums sources of pathogens that can be transmitted by insects and other vectors, causing outbreaks of disease. Pathogens, located inside the cells, cause systemically spread throughout the plant, the diseases they cause are often very difficult to control by conventional measures such as pesticide sprays.

The early detection and identification of plant borne pathogens is an integral part of successful disease management and this is especially important in relation to the importation of foreign plant material. The rapid identification of a plant pathogen, allows for the appropriate control measures to be applied prior to the further spread of the disease or its introduction (Table 29.1). Molecular and biotechnological techniques are providing the potential tools to characterize pathogen at all stages, from the earliest recogni-tion events to the changes that occur during resistant and susceptible responses. However over the last 50 years, several techniques have been developed which have found application in plant pathogen diagnosis; these include the use of monoclonal antibodies and enzyme linked immunosorbant assay (ELISA), which drastically increased the speed in which pathogen antigens could be detected *in vivo*, and DNA-based technologies, such as the polymerase chain reaction (PCR) which enable regions of the pathogen's genome to be amplified several million fold, thus increasing the sensitivity of pathogen detection. Despite such advances, cultural diagnosis still predominates, largely due to the technical experience and costs associated with the more recent techniques. Using Real Time PCR, it is possible not only to detect the presence or absence of the target pathogen, but also possible to quantify the number of the pathogen in the sample. The DNA Microarray technology, originally designed to study gene expression and generate single nucleotide polymorphism (SNP) profiles, is currently a new and emerging pathogen diagnostic technology, which in theory, offers a platform for unlimited multiplexing capability.

Breeders of grain legumes, like other breeders, seek to exploit multiple plant breeding approaches to combine both traditional and modern techniques. More traits are still selected by conventional means in field sites where most important diseases can be manipulated for the purpose of the selection. Marker assisted selection is an

Table 29.1: Visualization of Causal Organisms in Pulse Diseases (Chandrashekara *et al.*, 2012)

Causal Organism	*Examples*	*Description*	*Reproduction Mmethod*	*Morphological Visualization Method*
Bacteria	Bacterial blight, bacterial wilt, leaf spot	Single cell organism, small size	Asexual	Microscope (100-1000x magnification)
Fungi	Fusarium wilt, root rot, Aschochyta blight, Rust, Powdery mildew, Alterneria blight	Thread like filament, tiny size	Spore, cell division	Microscope (20-250x magnification)
Virus	SMV, YVMV, Necrosis virus, stunt virus, Alfalfa mosaic	RNA/DNA as genetic material, very tiny size	Host as a machinery system	Electron microscope (20,000-1,00,000x magnification)
Nematode	Root knot, dirty root, root lesion	Tiny Roundworms	Eggs	Microscope (1-60x magnification)
Insect Pests	Aphids, termites, grass hopper, thrips	Segmented bodies, outer covering mostly made up of chitin	Sexual	Naked eye, simple microscope

excellent tool for breeding, wherein genetic marker(s) tightly linked with the desired trait/gene(s) are utilized for indirect selection for that trait in segregating/non-segregating generations. In this section, we would discuss about the identification of pathogens through molecular methods and the development of advance genomics and breeding approaches in pulses which could assist in development of resistant pulse crops against various diseases. In this chapter, the molecular method for identification of causal organisms and advancement of resistance against various pests and diseases is being discussed.

1. Molecular Diagnosis of Pulse Diseases

Food production faces severe challenges because of various factors either directly or indirectly affecting the farming system of crops. Global warming and urbanization of agricultural lands are the major factors, which decrease the land availability for cultivation. Further problems, which are likely to be worsened by climate change, result from the action of pests and diseases that can cause severe crop losses. Environmental change and globalization of trade promotes the emergence of new diseases. About 42 per cent of the world's total agricultural crop is destroyed yearly by diseases and pests. However, these losses can be reduced, if plant pathogens are correctly diagnosed and identified in the early stage of the disease.

There are several methods to identify plant pathogens but each methods have their own limitation. In traditional method: identification of plant pathogens includes visual expression of symptoms and these symptoms may be small necrotic or chlorotic spots called local lesions develop at the site of infection. Typical leaf symptoms of disease include mosaic patterns, chlorotic or necrotic lesions, vein clearing, yellowing, vein banding, stripes or streaks, leaf rolling and curling. This is often possible only after major damage has already been done to the crop, so treatments will be of limited or no use.

After the introduction of polymerase chain reaction (PCR) and advancement of molecular biology, plant pathology, virology and biotechnology have made the development kits that can detect the plant diseases early, either by identifying the presence of the pathogen molecules (RNA, DNA and protein) during infection. These techniques require minimal processing time and are more accurate in identifying pathogens. And while some require laboratory equipment and training, other procedures can be performed on site by a person with no special training. Using Real Time PCR, it is possible not only to detect the presence or absence of the target pathogen, but also possible to quantify the number of the pathogen in the sample. Enumerating the pathogen upon detection is crucial to estimate the potential risks with respect to diseases development and provides a useful basis for diseases management decisions. Crops can be attacked by many pathogens which, in addition, often occur in complexes. In that situation diagnosis of many diseases require simultaneous detection and quantification of several targets.

Molecular based detection of the plant pathogen become common laboratory practices. There are few techniques evolved with the advances in biosystematics and molecular biology, and besides conventional PCR other technologically advanced methodologies such as the second generation PCR known as the real time PCR and microarrays. This techniques allows unlimited multiplexing capability have the potential to bring pathogen detection to a new and improved level of efficiency and reliability (Mumford *et al.*, 2006). However, while the specificity and sensitivity of detection of pathogens are greatly improved and pathogen detection is becoming simpler and faster, there are still major challenges, technical and economic nature. In this chapter we highlighted mainly major techniques for the detection of the common pathogen with reference to pulse crops.

2. Polymerase Chain Reaction (PCR)

The polymerase chain reaction (PCR) is one of the most important molecular biology tools for the detection of plant pathogens. It allows the amplification of millions of copies of specific DNA sequences by repeated cycles of denaturation, polymerisation and elongation at different temperatures using specific oligonucleotides (primers), deoxyribonucleotide triphosphates (dNTPs) and a thermostable *Taq* DNA polymerase in the adequate buffer (Mullis and Faloona, 1987). The presence of a desired amplicon size of the DNA indicates the presence of the target pathogenin the sample. Advances in PCR-based methods, such as real-time PCR, allow fast, accurate detection and quantification of plant pathogens in an automated reaction. Main advantages of PCR techniques include high sensitivity, specificity and reliability. Moreover, it reduces the diagnosis time from weeks to hours by escaping the pathogen isolation step from the infected material. This characteristic particularly used in the symptomless plants for identification of pathogen.

PCR-Based Methods

Conventional PCR

Conventional PCR may be used for the identification of fungal pathogens at different taxonomic levels (genus, species or strain) depending on the specificity of the primers. Recently, there are several reports available for identification of fungus based on the specific sequences of the ITS region, for example: *Sclerotium rolfsii* (Jeeva *et al.*, 2010) and *Colletotrichum capsici* (Torres-Calzada *et al.*, 2011). With the advancement of the PCR increases the sensitivity, specificity and throughput and allowing the quantification of fungi in infected plants or environment.

Nested-PCR

Nested PCR approach mainly used to increases the sensitivity and / or specificity of pathogen detection. In this method; PCR is performed in two consecutive rounds of amplification. Primarily two external primers are used to amplify a large amplicon and that is then used as a target for a second round of amplification using two internal primers (Porter-Jordan *et al.*, 1990). This method has been extensively used for molecular characterization and / or detection of the plant pathogen in the field condition and it has been reported from the numerous fungi (Mercado-Blanco *et al.*, 2001; Grote *et al.*, 2002; Ippolito *et al.*, 2002; Aroca and Raposo, 2007; Langrell *et al.*, 2008; Hong *et al.*, 2010; Meng and Wang 2010; Qin *et al.*, 2011; Wu *et al.*, 2011).

Multiplex PCR

Generally, multiplex PCR is used for detection of more than two pathogens. It is based on the use of several PCR primers in the same reaction and allowing the sensitive detection of different DNA targets and reducing time and cost. Based on the different ampliconssize of PCR product in the agarose gel; enables the different pathogen. Multiplex PCR technique has been used for the simultaneous detection and differentiation of *Podosphaera xanthii* and *Golovinomyces cichoracearum*in sunflower (Chen *et al.*, 2008); and for distinguishing among eleven taxons of wood decay fungi infecting hardwood trees (Guglielmo *et al.*, 2007).

Magnetic Capture Hybridisation-Polymerase Chain Reaction (MCH-PCR)

Most of the time quality of the DNA is not good or very less due to the phenolic compound present in the pulse crop and fruits. This will be hindered the PCR amplification, sensitivity toward the target sequence and its specificity. In that situation Magnetic Capture Hybridization–PCR will be used as an alternative approach. In this technique, magnetic beads are coated with a biotinilated oligonucleotide that is specific to a DNA region of the pathogen of interest. Pathogen DNA and magnetic beads-oligomer hybridised takes place and conjugate is separated from inhibitorycompounds. After the magnetic capture-hybridisation, PCR amplification was carried outusing species-specific primers. Langrell and Barbara (2001) used this method to detect *Nectria galligena*in apple and pear trees.

Polymerase Chain Reaction-Enzyme Link Imminosorbant Aassay (PCR-ELISA)

This serological-based PCR method uses forward and reverse primers carrying at their 5′ end biotin and an antigenic group (*e.g.* fluorescein), respectively (Landgraf *et al.*, 1991). PCR amplified DNA can be immobilized on avidin or streptavidin-coated microtiter plates via the biotin moiety of the forward primer and then can be quantified by an ELISA specific for the antigenic group of the reverse primer (*e.g.* anti-fluorescein antibody detected by colorimetric reactions). PCR-ELISA method is as sensitive as nested PCR. In addition, it does not require electrophoretic separation and/or hybridisation, and can be easily automated. All reactions can be performed in 96-well microtiter plates for mass screening of PCR products making them very suitable for routine diagnostic purposes. This procedure has been used for detectionand differentiation of *Didymella bryoniae* from related *Phoma*species in cucurbits (Somai *et al.*, 2002) and for detection of several species of *Phytophtora* and *Pythium* (Bailey *et al.*, 2002).

Reverse Transcription-Polymerase Chain Reaction (RT-PCR)

RT-PCR is used in the case of the RNA virus which is not detected by the normal PCR. This method is also used in the differentiation of living or dead pathogen from the field sample that are not normally detected by the normal methods. Since mRNA is degraded rapidly indead cells, the detection of mRNA by RT–PCR is considered an accurate indicator of cell viability (Sheridan *et al.*, 1998).

***In situ* PCR**

In this technique PCR and*in situ* hybridisation (ISH) used to amplify the specific gene sequences within intact cells or tissues (Long, 1993; Nuovo, 1992).The improved sensitivity of this technique allows the localization of one target copy per cell (Haase *et al.*, 1990; Nuovo *et al.*, 1991). Major limitation of this technique is background noisethat is very high becauseof nonspecific DNA synthesis during *in situ* PCR on tissue sections (Nuovo *et al.*, 1994). Apart from that, it is a time-consuming technique and technically demanding procedures such as light microscopy. *In situ* PCR technique was used for detection of the *Blumeria graminis* spores and mycelia on barley leaves (Bindslev *et al.*, 2002).

Polymerase Chain Reaction-Denaturing Gradient Gel Electrophoresis (PCR-DGGE)

In this method target DNAs were isolated from the plant pathogen used for the amplification and then subjected to denaturing electrophoresis. Sequence variants of particular fragments migrate at different positions in the denaturing gradient gel, allowing a very sensitive detection of polymorphisms in DNA sequences. This method is mainly applied for the analysis of the genetic diversity of microbial communities without the need of any prior knowledge of the species (Portillo *et al.*, 2011)

Real-time PCR

Currently, Real-time PCR is most important tool for detection of plant pathogens. In this technique amplification process has been monitored during reaction by the measuring the fluorescent intensity. The fluorescent signal will increases proportionally to the number of amplicons generated and to the number of targets present in the sample (Wittwer *et al.*, 1997).

Many are the advantages of real time PCR over conventional PCR that are mentioned in following

i) It does not require the use of post PCR processing (electrophoresis, colorimetric reaction or hybridisation), ii) avoiding the risk of cross contamination, iii) Reduction of the assay labour and material costs and iv) Increase the sensitivity and specificity and allows the accurate quantification of the target pathogen. In addition, real-time PCR is a high throughput method for the analysis of a large number of samples. Another advantage of real-time PCR is the capability to perform multiplex detection of two or more pathogens in the same reaction. The fluorescent dye used for the detection of pulse pathogen that is mentioned in the Table 29.2.

Table 29.2: Detection of Pulse Pathogens using Real Time PCR

Pathogen	*Real-time Chemistry*	*Crops*	*References*
Fusarium oxysporum	SYBR Green	Chickpea, pea	Jiménez-Fernández *et al.*, 2010
*Fusarium oxysporum*f. sp. *ciceris*	SYBR Green	Chickpea	Jiménez-Fernández *et al.*, 2011
Macrophominaphaseolina	SYBR Green	Chickpea Soybean Pigeon pea	Babu *et al.*, 2011
Macrophomina phaseolina	TaqMan MGB	Chickpea Soybean Pigeon pea	Babu *et al.*, 2011
Phialophora gregata	TaqMan	Soybean	Malvick and Impullitti, 2007

Development of portable real-time PCR machines help in diagnosis of large number of pathogen infected sample in the field conditions providing a realistic option to perform molecular tests at the same place of the collection of samples (*e.g.* portable Cepheid Smart Cycler).

Fingerprinting

Fingerprinting approaches allow the screening of random regions of the fungal genome for identifying species-specific sequences when conserved genes have not enough variation to successfully identify species (McCartney *et al.*, 2003). Fingerprinting analyses are generally used to study the phylogenetic structure of fungal populations. However, these techniques have been also useful for identifying specific sequences used for the detection of fungi at very low taxonomic level, and even for differentiate strains of the same species with different host range, virulence, compatibility group or mating type.

DNA Arrays

A DNA array is a collection of species-specific cDNAs or oligonucleotides (known asprobes) immobilized on a solid support that is subjected to hybridisation with a labeled target DNA. For preparation of the microarrays chips uses silicon orglass, or microscopic beads in where thousands of sample spots (less than 200 ìm in diameter) are immobilised via robotisation. DNA micro- and macro-arrays are generally used for gene expression profiling but are also powerful tools for identification and differentiation of plant pathogens (Anderson *et al.*, 2006; Lievens and Thomma, 2005).Currently, it is one of the most suitable techniques to detect and quantify multiple pathogens present in a sample (plant, soil, or water) in a single assay (Lievens *et al.*, 2005a).The specificity of the DNA array technology allows an accurate SNP detection. This characteristic is crucial for diagnostic application since closely related pathogens may differ in only a single base pair polymorphism for a target gene. This technology has been applied for detecting oomycete plant pathogens by using specific oligonucleotides designed on the ITS region (Anderson *et al.*, 2006; Izzo and Mazzola, 2009).

Sequencing

For the identification of the new pathogens (bacteria, fungus and viruses) species have been elucidated by the sequencing approaches. However, the use of sequence databases to identify organisms based on DNA similarity may have some drawbacks including flawed and incomplete sequences, sequences associated with misidentified organisms from the data base. An effort for generating and archiving high quality data by the researchers community should be the remedy of this drawback (Kang *et al.*, 2010). Initially, sequencing was carried out using Sanger sequencing method. Now, Sanger sequencing method has been partially replaced by several "next-generation"sequencing (NGS) technologies. The NGS technology is able to produce a high number of short sequences from multiple organisms in short time. The next-generation technologies commercially available today include the 454 GS20 pyrosequencing-based instruments (Roche Applied Science), the Solexa 1G analyzer (Illumina, Inc.), and the SOLiD instrument (Applied Biosystems).

DNA Barcoding

DNA barcoding is a taxonomic method that uses a short genetic marker in the organism DNA to identify an organism as belonging to a particular species. The updated fungal species identification platforms are available: *Fusarium*-ID was created as a simple, web-accessible BLAST server that consisted of sequences of the *TEF 1á*gene from representative species *Fusarium* (Geiser *et al.*, 2004). Almost all the *Fusarium* species multiple marker loci sequences have been progressively included for the identification of the strain and analyses of phylogenetic. Apart from the above data base, two additional platforms have been constructed: First, *Fusarium* Comparative Genomics Platform (FCGP), which includes five genomes from four species. This supports genome browsing and data analysis, and shows computed characteristics of multiple gene families and functional groups. Second platform is Fusarium Community Platform (FCP). It is an online research and education forum. All together, these platforms form the Cyber infrastructure for *Fusarium* (CiF; http://www.fusariumdb.org/) (Park, *et al.*, 2011). Apart from these data base, some of important data base have been shown in Table 29.3.

3. Development of Genomics for Genetic Improvement of Pulses against Biotic Stresses

Plants have a well developed and sophisticated system to combat against the attack of pathogenic microorganisms. To suppress this defense, pathogens deliver effectors into host cells. However, plants can also detect these effectors and suppress their effects (Jones and Dangl, 2006). Resistance can be conferred by single, race-specific resistance genes (R genes), usually conferring complete resistance, or by a number of minor genes resulting in a broad-spectrum incomplete resistance. The identification of genes underlying resistance is challenging and a detailed understanding of the interaction between plants and their pathogens at the genetic, histological and molecular level could be helpful to reach this objective. Knowledge of the mechanism of resistance conferred by a gene can help in the identification of the gene (Rubiales *et al.*, 2015). Parallel investigations of the molecular determinants of pathogenicity and virulence are critical to our understanding of resistance gene function. Studies on pathogen variation an epidemiology are integral to characterizing and manipulating the evolution of different types of resistance genes (Michelmore, 1995).

The first step in defense is the recognition of the pathogen by the plant, which activates signal transduction cascades that subsequently trigger transcription of plant defense genes (Park *et al.*, 2008). Gene expression studies provide information about the genes and metabolic pathways differentially regulated during plant-pathogen interactions and contribute to the identification of candidate resistance genes involved in each of these steps of the defense response (Rispail *et al.*, 2013; Barilli *et al.*, 2014). Knowledge of the genes involved in defense can be useful for marker assisted selection (MAS) and also to select genes whose altered expression through transformation can result in an increased resistance (Rubiales *et al.*, 2015).

The development of new and less expensive sequencing techniques is facilitating *de novo* sequencing of genomes and transcriptomes in pulses at a relative low cost.

Table 29.3: Database of Fungus for Identification and Comparative Analysis of Isolated Strain of the Fungus

Database	*Web Address*	*Identification of Fungi Group*	*Based on the*
Phytophthora Database	(http://www.phytophthoradb.org/)	*Phytophthora*	ITS region,rRNA genes, β-tubulin, enolase, HSP protein 90, TEF1β; (Park, J. *et al.*, 2008)
Phytophthora-ID	(http://phytophthora-id.org/)	Phytophthora	ITS and the *cox* I and *cox* II spacer regions (Grunwald *et al.*, 2011)
UNITE	http://unite.ut.ee/index.php	Ectomycorrhizal fungi	ITS database (Koljalg *et al.*, 2005)
TrichOKey	http://www.isth.info/tools/molkey/index.php	*Hypocrea* and *Trichoderma* species	ITS (Druzhinina *et al.*, 2005)
BOLD	http://www.boldsystems.org	Oomycetes	ITS and *cox* I (Robideau *et al.*, 2011)

Next generation sequencing techniques (NGS) can be used with reduced complexity transcriptome techniques such as SSH, cDNA-AFLP, SuperSAGE to increase tremendously the amount of transcripts identified compared to cloning and Sanger sequencing approaches (Almeida *et al.*, 2013).

The combination of Super SAGE with NGS has already been applied to study the molecular aspects of resistance to ascochyta blight in several legumes (Madrid *et al.*, 2013). The combination of SSH with NGS has also started to be applied in soybean to study the genes involved during the symbiotic interaction (Barros de Carvalho *et al.*, 2013)

In addition, RNAseq technique that involves the sequencing of all transcripts expressed in a given situation is not only a powerful tool to develop *de novo* transcriptomes but also to compare the whole transcriptome expression profiling in two different situations. Moreover, in the case of plant-pathogen interactions it offers the added advantage of simultaneously studying changes in gene expression in both the plant host and the infecting pathogen/pest. RNA seq has been extensively used to study plant-pathogen interactions in other crops or in legumes for other traits (Gao *et al.*, 2013; Zhu *et al.*, 2013) but rarely to study plant-pathogen interaction in legumes (Kim *et al.*, 2011).

4. Development of Resistance through Biotechnological Approaches

Despite the systematic and continuous breeding efforts through conventional methods, substantial genetic gain in pulses production and productivity could not be achieved. The major yield constraint in pulses is high genotype x environment (GxE) interactions on the expression of important quantitative traits leading to slow gain in genetic improvement and yield stability of pulses (Kumar *et al.*, 2011), besides severe losses caused by susceptibility of pulses to biotic and abiotic stresses. These issues require an immediate attention, and overall, a paradigm shift is needed in the breeding strategies to strengthen our traditional crop improvement programmes. One way is to utilize genomics tools in conventional breeding programmes involving molecular marker technology in selection of desirable genotypes or growing of transgenic crops (Ranalli, 2011). The use of transgenic crops is especially required for those traits that are not easy to improve genetically through conventional approaches because of the lack of satisfactory sources of desirable gene(s) in crossable gene pools. However, the ongoing debate on biosafety and ethical issues involving use of transgenic crops for commercial cultivation suggests that molecular marker–aided conventional methods of breeding may be the main short-term option for increasing productivity. There are the following biotechnological approaches through which remarkable landmarks can be achieved to develop disease resistance in pulses

Strategies for Development of Disease Resistance Pulses

Marker Assisted Selection (MAS)

MAS is most useful for traits that are difficult to select *e.g.* disease resistance, salt tolerance, drought tolerance, heat tolerance, quality traits (aroma of *basmati* rice, flavour of vegetables). The approach involves selecting plants at early generation with a fixed, favourable genetic background at specific loci, conducting a single large scale marker assisted selection while maintaining as much as possible the allelic segregation in the population and the screening of large populations to achieve the objectives of the scheme. No selection is applied outside the target genomic regions, to maintain as much as possible the Mendelian allelic segregation among the selected genotypes. After selection with DNA markers, the genetic diversity at un-selected loci may allow breeders to generate new varieties and hybrids through conventional breeding in response to targets set in breeding programme (Datta *et al.*, 2011).

Material Required for MAS

Molecular markers, a set of authentic lines carrying trait of interest and a population to validate themarkers to be used *e.g.*, F2 or BC_1F2 for each of the individual traits / genes (Datta *et al.*, 2011). Following are the basic pre requisites for MAS:

- ☆ Evaluating molecular markers that are linked to the trait of interest
- ☆ Validation of markers in parents and breeding population
- ☆ Designing and validation of new markers in case of non availability of the markers
- ☆ Designing of selection scheme and breeding strategy
- ☆ Fix the minimum population to be assayed to capture all beneficial alleles
- ☆ Progeny testing for fixation of traits.

Limitations of MAS

- ☆ Cost factor
- ☆ Requirement of technical expertise
- ☆ Automated techniques for maximum benefit
- ☆ *Per se,* DNA markers are not affected by environment but traits may be affected by the environment and show G x E interactions. Therefore, while developing markers, phenotyping should be carried out in multiple environments and implications of G x E should be understood and markers should be used judiciously.
- ☆ DNA marker has to be validated for each of the breeding population.

Identification of Disease Resistance Genes through Molecular Markers in Pulses

DNA based markers have shown great promises in expediting plant breeding methods. The identification of molecular markers closely linked with resistance genes would facilitate expeditious pyramiding of major genes into elite background, making it more cost effective. Once the resistance genes are tagged with molecular marker the selection of resistant plant in the segregating generations becomes easy (Datta *et al.*, 2011). Major success of identification of disease resistance genes is observed in chickpea and pigeonpea due to availability of genomic resources but still minor pulses like lentil, mungbean and urd bean needs an attention. A chickpea linkage map was established with help of 354 molecular surveyed among 130 recombinant inbred lines derived from a *C. arietinum* × *C. reticulatum* (Winter *et al.*, 2000). DNA markers associated with two closely linked genes for resistance to fusarium wilt race 4 and 5 in chickpea were also identified from a population of 131 recombinant inbred lines derived from a wide cross between *Cicer arietinum* and *Cicer reticulatum* (Benko-Iseppon *et al.*, 2003). These markers will pave the way for MAS and searching other useful genes. Gowda *et al.* (2009) identified flanking markers for chickpea fusarium wilt resistance genes in a RIL population. Reddy *et al.* (2009) performed bulk segregant analysis on a segregating population of ICPL 7035 x ICPL 8863 for identification of RAPD markers associated with pigeonpea sterility mosaic disease resistance. Dhanasekhar *et al.* (2010) identified two RAPD markers OPF04700 and OPA091375 were linked with the open and tall plant type gene in pigeonpea F2 population of the cross between TT44- 4 and TDI2004-1through bulk segregant analyses. Kotresh *et al.* (2006) used bulk segregant analysis with 39 RAPD primers which led to identification of two markers (OPM03 704 and OPAC11 500) that were associated with Fusarium wilt susceptibility allele in a pigeonpea F2 population derived from GS1 x ICPL87119. Saxena (2010) assessed the DNA polymorphism in a set of 32 pigeonpea lines screened with 30 SSR markers. Based on polymorphism of marker alleles, higher genetic dissimilarity coefficient and phenotypic diversity for *Fusarium* wilt and sterility mosaic disease resistance data, five parental combinations were identified for developing genetically diverse mapping populations suitable for the development tightly linked markers for *Fusarium* wilt and sterility mosaic disease resistance. Tullu *et al.* (2003) tagged anthracnose resistance gene LCt-2 of lentil cultivar PI 320937 with RAPD and AFLP markers. Taran *et al.* (2003) identified two molecular markers associated with Ascochyta blight resistance in lentil *viz.*, UBC 2271290 linked with ral1 gene and RB18680 linked with AbR1 and a marker (OPO61250) linked with Anthracnose resistance gene were utilized for identifying lines that possessed pyramided genes in a population of 156 RILs developed from a cross between 'CDC Robin' and a breeding line '964a-46'. These markers can be converted into more robust SCAR markers for routine use in marker assisted selection. Basak *et al.* (2004) developed molecular marker linked to yellow mosaic virus (YMV) resistance gene in *Vigna* sp. from a population segregating for YMV disease resistance. Maiti *et al.* (2010) identified molecular markers CYR1and YR4 in a F2 population for screening of MYMIV resistance genes. CYR1 co-segregated with MYMV resistance gene in F2 plants and F3 progenies. These two markers can be used simultaneously with the help of a multiplex PCR reaction. Nguyen *et al.* (2001) converted a RAPD marker into a SCAR (SCARW19) for selecting ascochyta blight resistance gene of lentil accession ILL5588. Rubeena *et al.* (2003) identified QTLs for ascochyta blight resistance in lentil. Further validation is required to use these markers for MAS. Hamwieh *et al.* (2005) mapped microsatellite markers identified from a genomic library of lentil. The linkage spanning about 751cM, consisting of 283 marker loci was derived from 86 recombinant inbred lines derived from the cross ILL 5588 × L 692-16-1(s) using 41 microsatellite and 45 amplified fragment length polymorphism markers. The average marker distance was 2.6 cM. Two flanking markers (SSR marker SSR59-2B at 8.0 cM and AFLP marker p17m30710 at 3.5 cM) were linked with fusarium resistance (Table 29.4).

Table 29.4: Trait Mapping in Pulses for Biotic Stress

Trait	*Gene/QTL*	*Source*
Chickpea		
Fusarium wilt	*foc-0, foc-1,foc-2, foc-3,foc-4, foc-5, QTL*	Varshney *et al.*, 2015
Ascochyta blight	QTL	Varshney *et al.*, 2015
Rust	QTL	Varshney *et al.*, 2015
Botrytis gray mold	QTL	Varshney *et al.*, 2015
Pigeonpea		
Fusarium wilt	BSA	Varshney *et al.*, 2015
SMD resistance	QTL, BSA	Varshney *et al.*, 2015
Lentil		
Fusarium wilt	QTL	Hamweih *et al.*, 2005
Aschochyta Blight	QTL	Rubeena *et al.*, 2006

Marker Assisted Backcross Breeding

A backcross breeding programme is aimed at gene introgression from a "donor" line into the genomic background of a "recipient" line. The potential utilization of molecular markers in such programmes has received considerable attention in the recent past. Markers can be used to assess the presence of the introgressed gene (foreground selection) when direct phenotypic evaluation is not possible, or too expensive, or only possible late in the development. Markers can also be used to accelerate the return to the recipient parent genotype at other loci ("background selection"). It is assumed that the introgressed gene can be detected without ambiguity, and the theoretical study was restricted to background selection only. The use of molecular markers for background selection in backcross programmes has been tested experimentally and proved to be very efficient. Introgressing the favourable allele of QTL by recurrent backcrossing can be a powerful mean to improve the economic value of a line, provided the expression of the gene is not reduced in the recipient genomic background. Yet, recent results show that for many traits of economic importance QTLs have rather small effects. In this case, the economic improvement resulting from the introgression of the favourable allele at a single QTL may not be competitive when compared with the improvement resulting from conventional breeding methods over the same duration. Marker assisted introgression of superior QTL alleles can then compete with classical phenotypic selection only if several QTLs could be manipulated (Datta *et al.*, 2011). Efforts are being made to introgress resistance to different races independently as well as pyramiding of resistance to two races for Fusarium Wilt in some elite varieties in India. ICRISAT (India) is pyramiding resistances for *Foc1* and *Foc3* from WR 315 and 2 QTLs for Ascochyta blight (AB) resistance from ILC 3279 line into C 214 (Varshney *et al.*, 2013).

Genomics Assisted Breeding Approach

The advent of markers based on simple sequence repeats (SSRs) and single nucleotide polymorphisms (SNPs) and the availability of high-throughput (HTP) genotyping platforms have further accelerated the generation of dense genetic linkage maps and the routine use of the markers for marker-assisted breeding in several crops (Collard and Mackill, 2008). However, despite the routine use of markers for genome-wide profiling and trait-specific marker-assisted selection (MAS), breeding of crops with many traits of interest such as yield, improved nutritive value and resistance to several biotic and abiotic stresses is still a challenge due to complex inheritance of these traits. Plant genomics has enormous potential to revolutionize crop improvement by providing extensive knowledge from the analysis of genomes which in turn can be used for rapid and efficient plant breeding towards crop improvement (Kumptala *et al.*, 2012).

The advent of NGS technologies has changed the dynamics and the pace of genomic research in pulses against biotic stresses because of their rapid, inexpensive and highly accurate sequencing capabilities. Unlike Sanger sequencing method which depends upon capillary electrophoresis, these NGS technologies are highly dependent on massive parallel sequencing, high resolution imaging, and complex algorithms to deconvolute the signal data to generate sequence data. NGS technologies offer a wide variety of applications such as whole genome de novo and re-sequencing, transcriptome sequencing (RNA-seq), microRNA sequencing, amplicon sequencing, targeted sequencing, chromatin immunoprecipitated DNA sequencing (ChIP-seq), methylome sequencing etc (Varshney *et al.*, 2015).

To facilitate crop improvement, NGS and other accessory technologies can be used for whole genome sequencing, transcriptome sequencing, genome wide and candidate gene marker development, targeted enrichment and sequencing and other applications. These NGS technologies even hold promise for a methodological leap towards genotyping by sequencing (GBS) and genetic mapping applications. Analysis of NGS data from genome wide association studies, transcriptomics and epigenomics in combination with data from proteomics, metabolomics and other 'omics' can provide an integrative systems biology approach to understand the regulation of complex traits (Table 29.5).

Table 29.5: Availability of Molecular Markers in Pulses (Varshney *et al.*, 2013)

	Chickpea	*Pigeonpea*	*Lentil*	*Mung bean*	*Urd bean*
Species name	*Cicer arietinum*	*Cajanus cajan*	*Lens culinaris*	*Vigna radiata*	*Vigna mungo*
Ploidy level	2n=2×=16	2n=2×=22	2n=2×=20	2n=2×=22	2n=2×=22
Genome size	740 Mbp	833.07 Mbp	4063Mbp	579 Mbp	574Mbp
SSR markers	~2000	~4000	~200	~300	~100
SNP markers	9000	10000	—	—	—

Transgenics Approaches

Challenges of food security has directed scientific communnity towards gene revolution which involves direct modification of traits in an organism by transferring desired genes using transgenic approach/ genetic transformation'. In contrast to classical breeding, genetic engineering offers an excellent tool for incorporating gene(s) of unrelated organisms into plant cells. These processes take less time thus accelerating the process of genetic improvement of crop plants. In addition, this exciting technology also allows access to an unlimited gene pool without the constraint of sexual compatibility (Ortiz, 1998). Pulse crops engineered to suit the environment better through incorporation of genes for tolerance to biotic and abiotic stresses have been suggested to represent an improvement in crop production. Over the past few decades, breeding possibilities have been broadened by genetic engineering and gene transfer technologies, as well as by gene mapping and identification of the genome sequences of model plants and crops which resulted in efficient transformation and generation of transgenic lines in a number of crop species (Sanghera *et al.*, 2011; Gosal *et al.*, 2009).

The presence of very rigid and thick cell walls coupled with efficient DNA repair systems has made the process of genetic transformation of some legume crops quite challenging. Despite these difficulties, it is possible to transform many pulse crops, such as chickpea and pigeonpea, provided that an efficient protocol of *in-vitro* regeneration is available. Legume transformation has been mainly performed by *Agrobacterium tumefaciens* but in some cases, especially in common bean, micro-particle bombardment has been used to deliver exogenous DNA to embryogenic or organogenic tissues (Chandra and Pental, 2003). All these efforts toward improving *in vitro* regeneration and transformation procedures allow the creation of transgenic lines with improved resistance to pests and diseases mainly in chickpea and pigeonpea.

One of the main transformation strategies was the introduction of derivatives of *cry1* genes from *Bacillus thuringiensis* conferring resistance to many pod borer insects. To improve the efficiency of resistance, the strategy has moved from the transfer of a single *cry1* gene to constructs containing *cry* genes with different modes of action. This strategy has worked well for corn and cotton to increase the durability of the resistance and is especially important for insect resistance management (Acharjee and Sarmah, 2011).

The first successful genetic transformation of nuclear genome of chickpea was reported in 1997 using the *cry1Ac* gene (Kar *et al.*, 1997). Subsequently, various research groups within India initiated genetic transformation of chickpea using *Cry1Ac* gene and reported generation of transgenic chickpeas (Sanyal *et al.*, 2005, Indurker *et al.*, 2007; Mehrotra *et al.*, 2011). A second gene, *Cry2Aa*, was also introduced in chickpea to facilitate pyramiding with existing *Cry1Ac* lines (Acharjee *et al.*, 2010). Mehrotra *et al.* (2011) generated pyramided *Cry1Ac* and *Cry1Ab* gene chickpea; however, pyramiding two or more genes with different mode of action is preferred. Transgenic pigeon pea consisting chimeric cry1AcF (encoding cry1Ac and cry1F domains) gene and its resistance towards *Helicoverpa armigera* were also observed by Ramu *et al.* (2011).

Conclusions

With advancement in molecular biology, researchers and farmers alike will be able to improve plant disease diagnosis. Efforts are already underway to produce better diagnostic kits to detect pathogens in crops important to developing countries. The Whole genome sequencing of both model and crop legumes are expected to offer new perspectives in legume species to develop resistance against various biotic as well as abiotic stress. Until 2005, there was a shortage of genomic resources in legume crops and therefore these crops were often referred to as orphan legume crops. Nevertheless, the collaborative and coordinated efforts of the legume community made during the last decade, contributed to development of large-scale genomic resources in these crops. As a result, pulses have become 'genomic resource rich' crops which can be used to understand the genetics of traits of several traits and as a result, approaches like MABC and MARS are being used in these crops. Genomic Selection seems to be a potential approach to be used very soon in pulse crops. While genome sequence has become available in pigeonpea and chickpea, molecular breeding approaches will have major milestones to combat against diseases and pests. Although conventional breeding based exclusively on phenotypic selection remains the mainstay for most breeding programs in pulse crops, adoption of molecular methods is increasing and in some cases is superseding conventional approaches.

References

Almeida, N.F., Leitao, S.T., Rotter, B., Winter, P., Rubiales, D. and Vaz Patto, M.C. 2013. Transcriptional profiling of grass pea genes differentially regulated in response to infection with *Ascochyta pisi*, In: *First Legume Society Conference*, Novi Sad, Serbia p140.

Anderson, N., Szemes, M., O'Brien, P., De Weerdt, M.; Schoen, C., Boender, P. and Bonants, P., 2006. Use of hybridization melting kinetics for detecting *Phytophthora* species using three-dimensional microarrays: demonstration of a novel concept for the differentiation of detection targets. *Mycological Research* 110: 664-671.

Aroca, A. and Raposo, R. 2007. PCR-based strategy to detect and identify species of *Phaeoacremonium* causing grapevine diseases. *Applied and Environmental Microbiology* 73(9): 2911–2918.

Aroca, A. Rapos, R. and Lunello, P. 2008. A biomarker for the identification of four *Phaeoacremonium* species using the beta-tubulin gene as the target sequence. *Applied Microbiology and Biotechnology* 80(6): 1131-1140.

Babu, B.K., Mesapogu, S., Sharma, A., Somasani, S.R., and Arora, D.K. 2011. Quantitative real-time PCR assay for rapid detection of plant and human pathogenic *Macrophomina phaseolina* from field and environmental samples. *Mycologia* 103(3):466-473.

Baile,y A.M., Mitchell, D.J., Manjunath, K.L., Nolasco, G. and Niblett, C.L. 2002. Identification to the species level of the plant pathogens *Phytophthora* and *Pythium* by using unique sequences of the ITS1 region of ribosomal DNA as capture probes for PCR Elisa. *Fems Microbiology Letters* 207(2):153-158.

Barilli, E., Rubiales, D. and Castillejo, M.A. 2012. Comparative proteomic analysis of BTH and BABA-induced resistance in pea (*Pisum sativum*) toward infection with pea rust (*Uromyces pisi*). *J. Proteomics* 75: 5189–5205.

Barros de Carvalho, G.A., Silva Batista, J.S., Marcelino-Guimaraes, F.C., Costa do Nascimento, L., and Hungria, M. 2013. Transcriptional analysis of genes involved in nodulation in soybean roots inoculated with *Bradyrhizobium japonicum* strain CPAC 15, *BMC Genomics* 14: 153

Basak, J., Kundagrami, S., Ghoose, T. K. and Pal, A. 2004. Development of yellow mosaic virus (YMV) resistance linked DNA marker in *Vigna mungo* from populations segregating for YMV-reaction. *Mol. Breed.* 14: 375-382.

Benko-Iseppon, A., Winter, M., Huettel, P., Staginnus, B., Muehlbauer, C. and Kahl, F.J.G. 2003. Molecular markers closely linked to fusarium resistance genes in chickpea show significant alignments to pathogenesisrelated genes located on Arabidopsis chromosomes 1 and 5, *Theor. Appl. Genet.* 107:379-386.

Bindslev, L., Oliver, R.P. and Johansen, B. 2002. *In situ* PCR for detection and identification of fungal species. *Mycological Research* 106 (3):277–279.

Chen, R.S., Chu, C., Cheng, C.W., Chen, W.Y. and Tsay, J.G. 2008. Differentiation of two powdery mildews of sunflower Helianthus annuus by a PCR-mediated method based on ITS sequences. *European Journal of Plant Pathology* 121 (1):1-8.

Chen, W., Seifert, K.A. and Lévesque, C.A. 2009. A high density COX1 barcode oligonucleotide array for identification and detection of species of *Penicillium subgenus Penicillium*. *Molecular Ecology Resources* 9 (1): 114-129.

Collard, B.C. and Mackill, D. J. 2008 Marker-assisted selection: an approach for precision plant breeding in the twenty-first century, *Philos Trans R Soc Lond B Biol Sci.* 363 (1): 557-72.

Cramer, G.R, Urano, K., Delrot, S., Pezzotti, M. and Shinozaki, K. 2011. Effects of abiotic stress on plants: a systems biology perspective, *BMC Plant Biol.* 11:1-14.

Datta, D., Gupta, Sanjeev, Chaturvedi, S.K. and Nadarajan, N., 2011, Molecular Markers in Crop Improvement. Indian Institute of Pulses Research, Kanpur, 1-54

Dhanasekar, P., Dhumal, K.N. and Reddy K.S. 2010. Identification of RAPD markers linked plant type gene in pigeonpea. *Indian J. Biotech.* 9: 58-63.

Druzhinina, I. S., Kopchinskiya, A.G., Komoja, M., Bissettb, J., Szakacs, G. and Kubiceka, C.P. 2005. An oligonucleotide barcode for species identification in *Trichoderma* and *Hypocrea*. *Fungal Genetics and Biology* 42(10): 813-828.

Gao, L., Tu, Z.J., Millett, B.P., and Bradeen, J.M. 2013. Insights into organspecific pathogen defense responses in plants: RNA-seq analysis of potato tuber-Phytophthora infestans interactions, *BMC genomics* 14:340.

Geiser, D.M., Jiménez-Gasco, M., Kang, S., Makalowska, I., Veeraraghavan, N., Ward, T.J., Zhang, N., Kuldau, G.A. and O'Donnell, K., 2004. *Fusarium*-ID v.1.0: A DNA sequence database for identifying *Fusarium*. *European Journal of Plant Pathology* 110: 5-6.

Gosal, S.S., Wani, S.H., Kang, M.S. 2009. Biotechnology and drought tolerance. *J. Crop Improv.* **23**:19–54.

Gowda S.J.M., Radhika P., Kadoo N.Y., Mhase L.B., and Gupta V.S., 2009, Molecular mapping of wilt resistance genes in chickpea, *Mol. Breed.*, **24**, 177-183.

Grote, D., Olmos, A., Kofoet, J.J., Tuset, E., Bertolini, E. and Cambra, M. 2002. Specific and sensitive detection of *Phytophthora nicotianae* by simple and nested-PCR. *European Journal of Plant Pathology* 108(3): 197-207.

Grünwald, N.J., Martin, F.N., Larsen, M.M., Sullivan, C.M., Press, C.M., Coffey, M.D., Hansen E.M. and Parke J.L., 2011, *Phytophthora*-ID.org: A sequence-based *Phytophthora* identification tool. *Plant Disease* 95 (3): 337-342.

Guglielmo, F., Bergemann, S.E., Gonthier, P., Nicolotti, G. and Garbelotto, M. 2007. A multiplex PCR-based method for the detection and early identification of wood rotting fungi in standing trees. *Journal of Applied Microbiology* 103: 1490- 1507.

Haase, A.T., Retzel, E.F. and Staskus, K.A. 1990. Amplification and detection of lentiviral DNA inside cells. *Proceedings of the National Academy of Sciences of the United States of America* 87(13): 4971-4975.

Hamwieh, A., Udupa, S.M., Choumane, W., Sarker, A., Dreyer, F., Jung, C., and Baum, M. 2005 A genetic linkage map of Lens sp. based on microsatellite and AFLP markers and the localization of fusarium vascular wilt resistance, *Theor. Appl. Genet.* 110: 669-677.

Hong, S.Y., Kang, M.R., Cho, E.J., Kim, H.K. and Yun, S.H., 2010, Specific PCR detection of four quarantine *Fusarium Species* in Korea. *Plant Pathology Journal* 26 (4): 409-416.

Indurker, S., Misra, H.S., Eapen, S. 2007. Genetic transformation of chickpea (*Cicer arietinum* L.) with insecticidal crys tal protein gene using particle gun bombardment, *Plant Cell Rep.* 26:755-63.

Ippolito, A., Schena, L. and Nigro, F. 2002 Detection of *Phytophthora nicotianae* and *P. citrophthora* in citrus roots and soils by nested PCR, *European Journal of Plant Pathology* 108(9): 855-868.

Jeeva, M.L., Mishra, A.K., Vidyadharan, P., Misra, R.S. and Hegde, V., 2010, A species-specific polymerase chain reaction assay for rapid and sensitive detection of *Sclerotium rolfsii*. *Australasian Plant Pathology* 39(6): 517-523.

Jiménez-Fernández, D., Montes-Borrego, M., Jimenez-Diaz, R.M., Navas-Cortes, J.A. and Landa, B.B. 2011. *In planta* and soil quantification of *Fusarium oxysporum* f. sp. *ciceris* and evaluation of Fusarium wilt resistance in chickpea with a newly developed quantitative polymerase chain reaction assay. *Phytopathology* 101(2): 250-262.

Jiménez-Fernández, D., Montes-Borrego, M., Navas-Cortés, J.A., Jiménez-Díaz, R.M. and Landa, B.B. 2010. Identification and quantification of *Fusarium oxysporum* in planta and soil by means of an improved specific and quantitative PCR assay. *Applied Soil Ecology* 46 (3): 372–382.

Jones, J. D. G. and Dangl, J. L. 2006. The plant immune system. *Nature* 444: 323–329

Kang, S., Mansfield, M.A., Park, B., Geiser, D.M., Ivors, K.L., Coffey, M.D., Grünwald, N.J., Martin, F.N., Lévesque, C.A. and Blair, J.E., 2010, The promise and pitfalls of sequence-based identification of plant-pathogenic fungi and oomycetes. *Phytopathology* 100(8):732-737.

Kar, S., Basu, D., Das, S., Ramakrishnan, N.A., Mukherjee, P., Nayak, P., Sen, S.K., 1997, Expression of *CryIA*(C) gene of *Bacillus thurigenesis* in transgenic chickpea plants inhibits development o f pod borer (*Heliothis armigera*) larvae, *Transgenic Res.* 6:177-185.

Koljalg, U., Larsson, K.H., Abarenkov, K., Nilsson, R.H., Alexander, I.J., Eberhardt, U., Erland, S., Hoiland, K., Kjoller, R., Larsson, E., Pennanen, T., Sen, R., Taylor, A.F.S., Tedersoo, L., Vralstad, T. and Ursing, B.M. 2005. UNITE: A database providing web-based methods for the molecular identification of ectomycorrhizal fungi. *New Phytologist* 166 (3):1063-1068.

Kotresh, H., Fakrudin, B., Punnuri, S.M., Rajkumar, B.K., Thudi, M., Paramesh, H., Lohithaswa, H. and Kuruvinashetti, M.S. 2006 Identification of two RAPD markers genetically linked to a recessive allele of a Fusarium wilt resistance gene in pigeon pea (*Cajanus cajan* L. Millsp.), *Euphytica* 149: 113-120.

Kumar, J., Choudhary, A.K., Solanki, R.K. and Pratap, A. 2011. Towards marker-assisted selection in pulses: a review, *Plant Breeding* 130, 297—313.

Kumpatla, S.P., Buyyarapu, R., Abdurakhmonov, I.Y. and Mammadov, J.A., 2012, Genomics-Assisted Plant Breeding in the 21st Century: Technological Advances and Progress, *Plant Breed*, Dr. Ibrokhim Abdurakhmonov (Ed.), ISBN: 978-953-307-932-5.

Landgraf, A., Reckmann, B. and Pingoud, A. 1991. Direct analysis of polymerase chainreaction products using enzyme-linked-immunosorbent-assay techniques. *Analytical Biochemistry* 198: 86-91.

Langrell, S.R.H. and Barbara, D.J. 2001. Magnetic capture hybridization for improved PCR detection of *Nectria galligena* from lignified apple extracts. *Plant Mol. Biol. Reporter* 19(1): 5–11.

Lievens, B., Grauwet, T.J.M.A., Cammue, B.P.A. and Thomma, B.P.H.J. 2005. Recent developments in diagnostics of plant pathogens: a review. *Recent Research Developments in Microbiology* 9: 57-79.

Lievens, B. and Thomma B.P.H.J., 2005. Recent developments in pathogen detection arrays:

Long, A.A. 1998. In-situ polymerase chain reaction: foundation of the technology and today's options. *European Journal of Histochemistry* 42: 101–109.

Madrid, E., Palomino, C., Plotmer, A., Horres, R., Rotter, B., Winter, P., Krezdorn, N., and Torres, A.M. 2013. Deep SuperSage analysis of the *Vicia faba* transcriptome in response to *Ascochyta fabae* infection., *Phytopathol. Mediterr.* **52**: 166–182

Malvick, D.K. and Impullitti, A.E. 2007. Detection and quantification of *Phialophora gregata* in soybean and soil samples with a quantitative, real-time PCR assay. *Plant Disease* 91(6): 736-742.

McCartney, H.A., Foster, S.J., Fraaije, B.A. and Ward, E. 2003. Molecular diagnostics for fungal plant pathogens. *Pest Management Science* 59(2): 129–142.

Mehrotra, M., Singh, A.K., Sanyal, I., Altosaar, I., Amla, D.V. 2011. Pyramiding of modified cry1Ab and cry1Ac genes of *Bacillus thuringiensis* in transgenic chickpea (*Cicer arietinum* L.) for improved resistance to pod borer insect. *Helicoverpa armigera, Euphytica* 182:87-102

Meng, J. and Wang, Y. 2010, Rapid detection of *Phytophthora nicotianae* in infected tobacco tissue and soil samples based on its *Ypt1* gene, *Journal of Phytopathology* 158 (1):1–7.

Mercado-Blanco, J., Rodríguez-Jurado, D., Pérez-Artés, E. and Jiménez-Díaz, R.M. 2001. Detection of the non defoliating pathotype of *Verticillium dahliae* in infected olive plants by nested PCR, *Plant Pathology* 50 (5): 609-619.

Michelmore, R. 1995. Molecular approaches to manipulation of disease resistance genes, *Annu. Rev. Phytopathol.* 15: 393-427.

Mullis, K.B. and Faloona, F.A. 1987, Specific synthesis of DNA in vitro via a polymerasecatalyzed chain-reaction. *Methods in Enzymology* 155: 335-350.

Mumford, R.A., Boonham, N., Tomlinson, J. and Barker, I. 2006 Advances in molecular phytodiagnostics – new solutions for old problems. *Eur. J. Plant Pathol.* 116:1–19.

Nguyen, T.T., Taylor, P.W.J., Brouwe, J.B.R., Pang, E.C.K. and Ford, R. 2001 A novel source of resistance in lentil (*Lens culinaris ssp. culinaris*) to ascochyta blight caused by Ascochyta lentis, *Australasian Plant Pathol.* **30**: 211–215.

Nuovo, G.J. 1992. *PCR in situ hybridization: Protocols and applications.* Raven Press, ISBN 0881679402, New York.

Nuovo, G.J., MacConnell, P. and Gallery, F. 1994. Analysis of nonspecific DNA synthesis during in situ PCR and solution-phase PCR, *Genome Research* 4:89-96.

Nuovo, G.J., MacConnell, P., Forde, A. and Delvenne, P. 1991. Detection of human papillomavirus DNA in formalin fixed tissues by in situ hybridization after amplification by the polymerase chain reaction, *American Journal of Pathology,* 139(4): 847-854.

Ortiz, R. 1998, Critical role of plant biotechnology for the genetic improvement of food crops: perspectives for the next millennium, *Electronic J Biotech.* 1-12.

Park, B., Park, J., Cheong, K.C., Choi, J., Jung, K., Kim, D., Lee, Y.H., Ward, T.J., O'Donnell, K., Geiser, D.M. and Kang, S. 2011. Cyber infrastructure for Fusarium: three integrated platforms supporting strain identification, phylogenetics, comparative genomics and knowledge sharing. *Nucleic Acids Research,* 39(1): 640-646.

Park, C., Peng, Y., Chen, X., Dardick, C., Ruan, D., Bart, R., Canlas, P.E. and Ronald, P.C. 2008. Rice XB15, a protein phosphatase 2C, negatively regulates cell death and XA21-mediated innate immunity. *PLoS Biol.* 6: e231.

Park, J., Park, B., Veeraraghavan, N., Blair, J.E. 2008. *Phytophthora* database: forensic database supporting the identification and monitoring of *Phytophthora*. *Plant Disease* 92 (6): 966-972.

Porterjordan, K., Rosenberg, E.I., Keiser, J.F., Gross, J.D., Ross, A.M., Nasim, S. and Garrett, C.T. 1990, Nested polymerase chain reaction assay for the detection of cytomegalovirus overcomes false positives caused by contamination with fragmented DNA. *Journal of Medical Virology* 30(4): 85-91.

Portillo, M.C., Villahermosa, D., Corzo, A. and González, J.M., 2011, Microbial community fingerprinting by differential display-denaturing gradient gel electrophoresis. *Applied and Environmental Microbiology* 77(1): 351–354.

Qin, L., Fu, Y., Xie, J., Cheng, J., Jiang, D., Li, G. and Huang, J. 2011. A nested-PCR method for rapid detection of *Sclerotinia sclerotiorum* on petals of oilseed rape *Brassica napus*. *Plant Pathology* 60(2): 271-277.

Ramu, S.V., Rohini, S., Keshavareddy, G., Gowri Neelima, M., Shanmugam, N.B., Kumar, A.R.V., Sarangi, S.K., Ananda Kumar, P. and Udayakumar, M., 2011, Expression of a synthetic cry1AcF gene in transgenic Pigeon pea confers resistance to *Helicoverpa armigera*, *J. Appl. Entomol*. 1-13.

Ranalli, P. 2011 Breeding Methodologies for the Improvement of Grain Legumes, In: *PK. Jaiwal and R.P Singh* (*eds*.), *Improvement Strategies/or Leguminosae Biotechnology* 3-21.

Reddy, L.P., Reddy, B.V., Rekha Rani, K., Sivaprasad, Y., Rajeswari, T. and Reddy, K. R. 2009 RAPD and SCAR marker linked to the sterility mosaic disease resistance gene in pigeonpea (*Cajanus cajan* L. Millsp.), *The Asian and Australasian J. Plant Sci. and Biotech.* 3:16-20.

Rispail, N. and Rubiales, D. 2014. Identification of sources of quantitative resistance to *Fusarium oxysporum* f. sp. medicaginis in *Medicago truncatula*, *Plant Disease* 98: 667–673.

Robideau, G.P., de Cock, A.W.A.M., Coffey, M.D., Voglmayr, H. 2011. DNA barcoding of oomycetes with cytochrome c oxidase subunit I and internal transcribed spacer. *Molecular and Ecological Resources* 11:1002–1011.

Rubeena, P., Taylor, W.J. and Ades, P.K. 2003 QTL mapping of resistance in lentil (*Lens culinaris* ssp. *culinaris*) to ascochyta blight (*Ascochyta lentis*), *Plant Breed*. 125: 506–512.

Rubeena, Taylor, P.W.J., Ades, P.K. and Ford, R. 2006. QTL mapping of resistance in lentil (Lens culinaris ssp. culinaris) to aschochyta blight (*Aschochyta lentis*), *Plant Breed*. 125(5):506-512.

Rubiales, D., Fondevilla, S., Chen, W., Gentzbittel, L., Higgins, T. J. V., Castillejo, M.A., Singh, K. B. and Nicolas, R., 2015, Achievements and Challenges in Legume Breeding for Pest and Disease Resistance, *Critical Reviews in Plant Sciences* 34:1-3, 195-236, DOI: 10.1080/07352689.2014.898445.

Sanghera, G.S., Wani, S.H., Singh, G., Kashyap, P. L. and Singh, N. B. 2011. Designing Crop Plants for Biotic Stresses using Transgenic Approach, *Vegetos* 24 (1): 1-25.

Sanyal, I., Singh, A.K., Kaushik, M., Amla, D.V. 2005. *Agrobacterium* mediated transformation of Chickpea (*Cicer arietinum* L.) with *Bacillus thuringiensis cry*1Ac gene for resistance against podborer insect, *Helicoverpa armigera, Plant Sci.* 168:1135-1146.

Saxena, R.K., Saxena, K.B., Kumar, R.V., Hoisington, D.A., and Varshney, R.K., 2010, SSR-based diversity in elite pigeonpea genotypes for developing mapping populations to map resistance to *Fusarium* wilt and sterility mosaic disease, *Plant Breed* 129: 135–141.

Sheridan, G.E.C., Masters, C.I., Shallcross, J.A. and Mackey, B.M. 1998 Detection of mRNA by reverse transcription-PCR as an indicator of viability in *Escherichia coli* cells. *Applied Environmental Microbiology* 64(4): 1313–1318.

Somai, B.M., Keinath, A.P. and Dean, R.A., 2002, Development of PCR-ELISA detection and differentiation of *Didymella bryoniae* from related *Phoma* species. *Plant Disease* 86(7): 710–716.

Taran, B., Buchwald, L., Tullu, A., Banniza, S., Warkentin, T.D. and Vandenberg, A. 2003 Using molecular markers to pyramid genes for resistance to ascochyta blight and anthracnose in lentil (*Lens culinaris* Medik), *Euphytica* 134: 223-230.

Torres-Calzada, C., Tapia-Tussell, R., Quijano-Ramayo, A., Martin-Mex, R., Rojas-Herrera, R., Higuera-Ciapara, I., Perez-Brito, D. 2011. A species-specific polymerase chain reaction assay for rapid and sensitive detection of *Colletotrichum capsici*. *Molecular Biotechnology* 49 (1): 48-55.

Tullu, A., Buchwaldt, L., Warkentin, T., Taran, B., and Vandenberg, A. 2003. Genetics of resistance to anthracnose and identification of AFLP and RAPD markers linked to the resistance gene in PI320937 germplasm of lentil (*Lens culinaris* Medikus), *Theor. Appl. Genet,*. **106**: 428-434.

Varshney, R.K., Kudapa, H., Pazhamala, L., Chitikineni, A., Thudi, M., Bohra, A., Gaur, P.M., Janila, P., Fikre, A., Kimurto, P. and Ellis, N. 2015. Translational Genomics in Agriculture: Some Examples in Grain Legumes, *Critical Reviews in Plant Sciences* 34:1-3, 169-194, DOI:10.1080/07352689.2014.897909.

Varshney, R.K., Mohan, S.M., Gaur, P.M., Chamarthi, S.K., Singh, V.K., Srinivasan, S., Swapna, N., Sharma, M., Singh, S., Kaur, L. and Pande, S., 2013, Marker-assisted backcrossing to introgress resistance to Fusarium wilt (FW) *race 1* and Ascochyta blight (AB) in C 214, an elite cultivar of chickpea, *Plant Gen.* DOI: 10.3835/plantgenome 2013.10.0035.

Winter, P., Benko-Iseppon, A.M., Huttel, B., Ratnaparkhe, M., Tullu, A., Sonnante, G., Pfaff, T., Tekeoglu, M., Santra, D., Sant, V.J., Rajesh, P.N., Kahl, G., and Muehlbauer, F. J., 2000. A linkage map of the chickpea (*Cicer arietinum* L.) genome based on recombinant inbred lines from *C. arietinum* × *C. reticulatum* cross: localization of resistance genes for fusarium wilt race 4 and 5, *Theor. Appl. Genet.* 101: 115-1163.

Wittwer, C.T., Herrmann, M.G., Moss, A.A. and Rasmussen, R.P. 1997 Continuous fluorescence monitoring of rapid cycle DNA amplification. *Biotechnique,* 22(1): 130–138.

Wu, C.P., Chen, G.Y., Li, B., Su, H., An, Y.L., Zhen, S.Z. and Ye, J.R. 2011 Rapid and accurate detection of *Ceratocystis fagacearum* from stained wood and soil by nested and realtime PCR. *Forest Pathology* 41(1): 15-21.

Zhu, Q.H., Stephen, S., Kazan, K., Jin, G., Fan, L., Taylor, J., Dennis, E. S., Helliwell, C.A. and Wang M.B. 2013 Characterization of the defense transcriptome responsive to *Fusarium oxysporum*-infection in Arabidopsis using RNA-seq, *Gene* **512**: 259–266.

Index

D

E

Q

R